Perspectives on Peripatetic Physics

Philosophia Antiqua

A SERIES OF STUDIES ON ANCIENT PHILOSOPHY

VOLUME 176

The titles published in this series are listed at *brill.com/pha*

Perspectives on Peripatetic Physics

From Theophrastus to Alexander

Edited by

Gweltaz Guyomarc'h
Alessio Santoro

BRILL

LEIDEN | BOSTON

Open access of this volume has been funded by the ANR, project "Peripatos", code: ANR-20-CE27-0002.

The Library of Congress Cataloging-in-Publication Data is available online at https://catalog.loc.gov

Typeface for the Latin, Greek, and Cyrillic scripts: "Brill". See and download: brill.com/brill-typeface.

ISSN 0079-1687
ISBN 978-90-04-73930-7 (hardback)
ISBN 978-90-04-73931-4 (e-book)
DOI 10.1163/9789004739314

This book is printed on acid-free paper and produced in a sustainable manner.

Contents

Acknowledgements

This volume is the result of two conferences held in Lyon as part of the ANR "Peripatos" project (ANR-20-CE27-0002): one in 2022, dedicated specifically to Theophrastus' physics, and the other in 2024, focusing on Peripatetic physics after Theophrastus, up to Alexander. We extend our gratitude to all the speakers at these conferences, as well as to the institutions that generously funded them: Agence Nationale de la Recherche, Université Jean Moulin Lyon 3, Institut de Recherches Philosophiques de Lyon, Centre Léon Robin (Paris), CAPHI (Rennes), UMR STL (Lille), and UMR Sphere (Paris). We express our heartfelt thanks to all the authors of the contributions gathered below for their collegial cooperation. We are especially grateful to the members of the "Peripatos seminar", who contributed to the lively discussions during the two conferences: Annick Jaulin, Fabienne Baghdassarian, Claire Louguet, Charlotte Murgier and David Lefebvre. Our warm thanks also go to Chad Jorgenson for his attentive and generous work on the language, which greatly enhanced the quality of the volume.

GG & AS

Abbreviations

Plato

Phlb.	*Philebus*
Resp.	*Respublica*
Theaet.	*Theaetetus*
Tim.	*Timaeus*

Aristotle

APo.	*Analytica posteriora*
APr.	*Analytica Priora*
DA	*De anima*
DC	*De caelo*
EE	*Ethica Eudemia*
EN	*Ethica Nicomachea*
GA	*De generatione animalium*
GC	*De generatione et corruptione*
HA	*Historia animalium*
IA	*De incessu animalium*
Insomn.	*De insomniis*
MA	*De motu animalium*
Metaph.	*Metaphysica*
Meteor.	*Meteorologica*
Meteor.	*Meteorologica*
PA	*De partibus animalium*
Phys.	*Physica*
Pol.	*Politica*
Rhet.	*Rhetorica*
Sens.	*De sensu*
Top.	*Topica*

Pseudo-Aristotle

Aud.	*De audibilibus*
Col.	*De coloribus*

Mund.	*De mundo*
Probl.	*Problemata*

Theophrastus

CP	*De causis plantarum*
FHS&G	Fortenbaugh William, Pamela M. Huby, Robert W. Sharples, and Dimitri Gutas. 1992. *Theophrastus of Eresus, Sources for his Life, Writings, Thought, and Influence*, Leiden–Boston: Brill.
HP	*Historia plantarum*
Metaph.	*Metaphysica*
Od.	*De odoribus*
Sens.	*De sensibus*

Eudemus

W	Wehrli, Fritz. 1955. Revised ed. 1969. *Die Schule des Aristoteles, Vol. VIII: Eudemos von Rhodos*, Basel: Schwabe.

Strato

Sharples	Sharples, Robert W. "Strato of Lampsacus: The Sources, Texts and Translations". In: Marie-Laurence Desclos and William W. Fortenbaugh. (eds.). 2011. *Strato of Lampsacus*. Text, Translation, and Discussion, New Brunswick: Transaction Publishers. Revised ed.: London–New York: Routledge, 2017, 5–229.
W	Wehrli, Fritz. 1950. Revised ed. 1969. *Die Schule des Aristoteles, Vol. V: Straton von Lampsakos*, Basel: Schwabe.

Alexander

DA	*De anima*
Fat.	*De fato*
In Metaph.	*In Metaphysica commentaria*
In Meteor.	*In Meteorologicorum libros commentaria*
In Sens.	*In librum De sensu commentarium*

In Top.	*In Topicorum libros commentaria*
Mant.	*Mantissa*
Mixt.	*De mixtione*
Quaest.	*Quaestiones et solutiones*
Prov.	*De providentia*
Temp.	*De tempore*

Other Commentators

CAG	*Commentaria in Aristotelem Graeca*
In Cat.	*In Aristotelis Categorias commentaria*
In DA	*In libros Aristotelis De anima commentaria*
In DC	*In Aristotelis De caelo commentaria*
In EN	*In Ethica Nicomachea quae supersunt commentaria*
In Metaph.	*In Aristotelis Metaphysica commentaria*
In Meteor.	*In Aristotelis Meteorologicorum libros commentaria*
In Phys.	*In Aristotelis Physicorum libros commentaria*
Metaphr.	*Metaphrasis in Theophrastum*

Others

De aet. mundi	Philo of Alexandria, *De aeternitate mundi*
Deipn.	Athenaeus, *Deipnosophistae*
DK	Diels, Hermann and Walther Kranz. 1959–1960. *Die Fragmente der Vorsokratiker*, Hildesheim: Weidmann.
DL	Diogenes Laertius, *Vitae philosophorum*
Exp.	Theon of Smyrna, *Expositio rerum mathematicarum ad legendum Platonem utilium*
Falcon	Falcon, Andrea. 2012. *Aristotelianism in the First Century* BCE*: Xenarchus of Seleucia*, Cambridge: Cambridge University Press.
LS	Long, Anthony A. and David N. Sedley. 1987. *The Hellenistic Philosophers*, Cambridge: Cambridge University Press.
ND	Cicero, *De Natura deorum*
NQ	Seneca, *Naturales Quaestiones*
SMT	Galen, *De simplicium medicamentorum temperamentis ac facultatibus*
SVF	Von Arnim, Hans. 1903–1924. *Stoicorum Veterum Fragmenta*, Leipzig: Teubner.
W	Wehrli, Fritz. 1944–1978. *Die Schule des Aristoteles*, Basel: Schwabe.

Contributors

Han Baltussen
The University of Adelaide.

Rafaëlle Beauvallet
Université Paris I Panthéon-Sorbonne / UMR Sphere.

Mai-Lan Boureau
KU Leuven.

István Bodnár
Eötvös University, Budapest / Central European University, Vienna.

Frans A.J. de Haas
Universiteit Leiden.

Andrea Falcon
Università degli Studi di Milano.

Ludovica M. Fazio
Università degli Studi di Torino.

Pavel Gregorić
Institute of Philosophy, Zagreb.

Gweltaz Guyomarc'h
Université de Caen Normandie / UR "Identité et Subjectivité".

Orna Harari
Tel Aviv University.

Myrto Hatzimichali
Homerton College, Cambridge.

Katerina Ierodiakonou
Université de Genève / National and Kapodistrian University of Athens.

Inna Kupreeva
University of Edinburgh.

David Lefebvre
Sorbonne Université / Centre Léon Robin.

Robert Mayhew
Seton Hall University.

Federico M. Petrucci
Università degli Studi di Torino.

Robert Roreitner
Charles University, Prague.

Alessio Santoro
Institut de Recherches Philosophiques de Lyon.

Richard Schorlemmer
Universität Siegen.

Introduction

Gweltaz Guyomarc'h and Alessio Santoro

The study of the physical world occupies a prominent position within Aristotle's scientific enterprise. Over half of the works that make up the *Corpus Aristotelicum* deal with some aspect or other of nature, ranging from general investigations into the nature of motion, time and place to giving scientific accounts of winds and earthquakes, from the study of sense perception, sleep and breathing to an enquiry into the different kinds of animal reproduction. The term "physics" in the title of this volume should be understood in this broad sense. Its proper domain is demarcated by the notion of *kinēsis* (change or motion), which is intrinsic to Aristotle's conception of nature (*physis*) as the inner principle of change.[1] Specifically, physics is concerned with all those beings, and only those beings, which are subject to motion or other kinds of change, be they eternal (like celestial bodies) or perishable (like animals and plants). That is, it is concerned with beings that are possessed of matter and perceptible through our senses.[2] That the physical realm should be regarded as a single area of investigation is clear from the famous opening chapter of the *Meteorology*, where Aristotle situates the enquiry he is embarking upon within his overall scientific agenda.[3] Meteorology follows up on the general enquiry into the first causes of nature and the kinds of natural motion in the *Physics*, the investigation of the movements of the stars in *De caelo* I–II and the study of the transformations of the four elements, as well as of generation and corruption, in Aristotle's work of the same name and in *De caelo* III–IV. Meteorology, in turn, is to be followed by an enquiry into common and specific traits of animals and plants, which we find in Aristotle's zoological works, in the short treatises grouped under the title *Parva naturalia* and in his lost work *On Plants*—or perhaps in Theophrastus' botanical treatises.[4] Once this enquiry has also been completed, Aristotle concludes, "we may say that the whole of our original undertaking will have been carried out".[5] Thus, as this brief over-

1 For a thorough analysis of this conception of nature, see Waterlow (1982a), chap. 1.

2 Cf. Aristotle, *Metaphysics* XII 1, 1069a30–b2, and *Physics* II 7, 198a27–31.

3 On this "large-scale map of Aristotle's natural philosophy", see Burnyeat (2004), 7–24.

4 See Falcon (2005), chap. 1, for a detailed analysis of the boundaries and internal unity of Aristotle's study of nature. James Lennox's suggestion that Aristotle may be referring to a colleague such as Theophrastus (5, n. 10) is explored in greater depth in Falcon (2024).

5 Aristotle, *Meteorology* I 1, 339a8–9, transl. Webster.

view makes clear, these investigations are different parts of one and the same scientific project, characterised by a comprehensive and systematic analysis of the natural world.

Aristotle's physics exerted such a strong influence on the development of science in the West that it constitutes the pre-eminent instance of a scientific paradigm. In the summer of 1947, the young Thomas Kuhn was preparing a set of lectures on "the origins of seventeenth-century mechanics"[6] and, in doing so, turned his attention to Aristotle's *Physics*. It was then that Kuhn experienced what he himself termed his "enlightenment",[7] an episode now commonly referred to as his "Aristotle experience" or even his "epiphany".[8] Challenging a cumulative view of scientific knowledge, Kuhn describes how he became aware not only of the radical difference between Aristotelian and modern physics, but also of the former's internal "integrity".[9] Of course, Kuhn was not the first to highlight Aristotle's role in the development of Western physics—one need only think of Koyré,[10] to whom Kuhn refers in many footnotes—but this moment of insight subsequently led to the development of the concept of a paradigm and, with the addition of a tradition-bound practice, to the notion of normal science, as laid out in *The Structure of Scientific Revolutions*.[11] The consolidation of this scientific tradition in the Middle Ages is obvious, with the tenth-century Greek encyclopaedic dictionary known as the *Suda* famously characterising Aristotle as the "scribe of nature".[12] In the Arabic-speaking world, Aristotle's physical works had been circulating for more than two centuries when the Andalusian philosophers Avempace (Ibn Bājja, d. 1139) and Averroes (Ibn Rušd, d. 1198) produced their influential commentaries on the *Physics*, *De generatione et corruptione*, *De caelo*, *Meteorology* and a collection of Aristotle's zoological treatises entitled *Book of Animals*.[13] As for the Latin West, the impact of Aristotle's science is epitomised by Johannes de Sacrobosco's *De sphaera mundi*, an astronomy textbook combining Aristotle's *De caelo* with Ptolemy's *Almagest* which was in use in European universities from the twelfth

6 Kuhn (1977), xi.
7 Ibid.
8 See, e.g., Wray (2021), chap. 1.
9 Kuhn (1977), xi and xiv.
10 Koyré (1957), where the expression "Closed World" primarily refers to the Aristotelian view.
11 See Kuhn (1977), xvii–xviii.
12 The expression is already found in Eusebius, *Praeparatio Evangelica* XV, 9.9.2: ὁ τῆς φύσεως, ὥς φασι, γραμματεύς.
13 On the different stages of the Syriac and Arabic translations of the *Corpus Aristotelicum*, see D'Ancona (2022).

to the seventeenth centuries—that is, before Copernican astronomy replaced the Aristotelian-Ptolemaic paradigm.[14]

Nonetheless, this vast medieval and early modern legacy must be set against the relative lack of success of Aristotelian physics in antiquity[15]—"relative" because Aristotelian physics was indeed known beyond the confines of the Peripatetic school: traces of its influence can be found in other philosophical traditions (e.g. in Posidonius[16]), as well as in scientific corpora, such as those of Galen or Ptolemy.[17] On the whole, however, there is little evidence, in Hellenistic and post-Hellenistic times, of an explicit discussion of Aristotle's physics in philosophical schools other than the Peripatos. Indeed, a renewal of interest did not occur until late antiquity. Even then, within the Neoplatonic philosophical curriculum, the study of Aristotle's thought was viewed as complementary, but ancillary, to that of Plato, and was thus confined to a selection of works from which most parts of physics—notably biology—were excluded.[18] This harmonising approach to Plato and Aristotle is also a feature of Themistius,[19] who paraphrased the *Physics* and the *De caelo*, as well as of Simplicius,[20] whose lengthy commentaries on these works are a gold mine of sources on the Peripatetics—as Han Baltussen shows in his opening contribution on Theophrastus. One of the central hypotheses of the project from which this volume originated is that the Aristotelian paradigm only attained the prominence we have described because philosophers and scientists within the Peripatetic school cultivated, renewed and criticised Aristotle's thought, thereby ensuring its transmission. It is thus by taking into account the work of the Peripatetics that it becomes possible to resolve the apparent contradiction between the limited success of Aristotelian physics in the Hellenistic and

14 See Taub (2024) for a reassessment of the historical accuracy of this label.

15 See Falcon (2018), 195, who was probably inspired by Pierre Duhem's remark, which is worth quoting: "Qu'un tel système ait exercé sur les esprits la séduction puissante qu'éprouveront, au Moyen-Âge, la plupart des philosophes arabes ou chrétiens, on le comprend aisément. Plus volontiers, on serait surpris en constatant que les successeurs immédiats d'Aristote se sont montrés, en général, rebelles à cette influence [...]" (Duhem (1913), 242).

16 Posidonius' extensive knowledge of the Peripatetic tradition is also evidenced by the fact that he was familiar with Theophrastus' *On the Coming to Be of the Elements* (Simplicius, *In DC* 700.3–8).

17 See Gottschalk (1987), 1164–1171.

18 See Falcon (2014).

19 The absence of Themistius from the present volume is not accidental, as we share the view that he is more closely aligned with the Neoplatonists and do not consider him an authentic member of the Peripatetic tradition.

20 Menn (2022), 4–9.

post-Hellenistic periods and its flourishing in the Middle Ages. In order to fully explain this success, we must take into account the Peripatetic development of Aristotelian physics, rather than merely its initial formulation.

Indeed, the comprehensive and systematic investigation of nature remained a central concern for Aristotle's first heirs. Theophrastus of Eresus, his immediate successor at the head of the school, explored an impressive variety of topics, ranging from the boundaries of teleological explanations (in his *Metaphysics*)[21] to the nature of fire—which is examined in Mai-Lan Boureau's contribution to this volume—and from physiological phenomena such as sweat, dizziness, fatigue,[22] paralysis or fainting[23] to somewhat peculiar groups of animals (e.g. *On Creatures that Retreat into Holes*, *On Creatures that Appear in Swarms*, *On Creatures that Are Said to Be Grudging*).[24] He is famous above all for his botanical treatises, the *History of Plants* and the *Causes of Plants*, which are explicitly modelled on Aristotle's zoological works.[25] His contemporary, Eudemus of Rhodes, was such a careful and perceptive interpreter of Aristotle's *Physics* that Simplicius, who benefited greatly from his elucidations, considered him "the most genuine of Aristotle's students".[26] The chapters by István Bodnár and Robert Mayhew will deal, respectively, with Eudemus' theory of cosmic motion and his discussion of infinity. The third head of the school, Strato of Lampsacus, placed even more emphasis on the study of nature, earning him the nickname "the physicist".[27] According to Cicero, Strato refuses to invoke divine agency in his explanations of the universe, but "investigating the individual parts of the world, teaches that whatever is or comes to be is brought about or has been brought about by natural weights and movements".[28] Frans De Haas' contribution will shed light on a particularly interesting part of Strato's worldview, namely his conception of time. After Strato, evidence on physics becomes scanty, while ethics seems to take centre stage between his first- and last-known Hellenistic successors, Lyco of Troas and Diodorus of Tyre. However, the bulk of the collection known as *Problemata physica* likely dates to the early Peripatos and may reflect the direction that Strato imparted to the school.[29] This natur-

21 In particular, Theophrastus, *Metaphysics* 10a22–11b23.
22 Fortenbaugh, Sharples and Sollenberger (2003).
23 Frs. 346 and 345 FHS&G, respectively.
24 Frs. 366–370, 359, 362 FHS&G.
25 Falcon (2024).
26 Simplicius, *In Phys.* 411.15–16 (Fr. 59 W). On this description, see Menn (2022), 20.
27 DL V 58; Cicero, *De natura deorum* I 35 (Fr. 19A Sharples).
28 Cicero, *Academic Books* (*Lucullus*) II 121 (Fr. 18 Sharples), transl. Sharples.
29 Although it is hard to date it precisely, the *Problemata* most probably antedates Alexander. Cf. Bodnár (2015), 5.

alistic direction may also be echoed in the short treatise *De coloribus*, which Richard Schorlemmer's chapter will present in a new light. Finally, as Andrea Falcon's contribution will show, a concern to defend the eternity of the world was still a crucial part of Critolaus' philosophical agenda as a Peripatetic.

It is customary to point out a distinct development within the Peripatetic school in the first century BCE. This transformation occurred within the context of what David Sedley has described as "the most significant change to occur in the entire history of ancient philosophy"[30]—namely, its "decentralisation" from Athens, which involved a shift towards Rome in particular. The symbol of this shift is the sack of Athens by Sulla in 86 BCE, which is said to have led to the destruction of the Academy and the Lyceum.[31] The creation of a diaspora severed the direct connection between disciples and the heirs of the schools' founders, resulting in textual commentary becoming an essential part of philosophical activity. Various accounts affirm that, in the case of the Peripatos, this event led to the transfer of Aristotle's library to Rome and to the relocation of the school to Alexandria, where Andronicus of Rhodes is held to have worked on his edition of Aristotle before moving to Rome.[32] A sign of this rupture is the fact that he and Boethus of Sidon are regarded by Neoplatonic commentators as the last scholarchs of the school in a direct line "from Aristotle onwards".[33] Thus, this general narrative has its roots in antiquity: even within the Peripatos, ancient sources distinguish between "early" and "later" Peripatetics.[34]

Among these "later" Peripatetics, physics continued to enjoy great prominence. In opposition to his teacher Andronicus—whose views are reconstructed in Myrto Hatzimichali's chapter—Boethus placed it at the beginning of the

---

30 Sedley (2003), 31.

31 Plutarch, *Sulla* 12.4; Lynch (1972), 192–207. The questions of the destruction and the disappearance of the schools (which are two distinct issues) remain open. For the Academy, see the excellent overview in Bonazzi (2020).

32 See Strabo, *Geog.* 13, 1, 54; Plutarch, *Sulla* 26.3; Cicero, *Ep. ad Atticum* 4.10.1 (22 April, 55 BCE) famously refers to "Faustus' library", but he does not mention any Peripatetic work (see Barnes (1997), 49). For the hypothesis regarding Alexandria, see Athenaeus, *Deipn.* I, 3a–b; Al-Fārābī, quoted by Ibn Abī Uṣaybiʿa, *ʿUyūn al-anbāʾ fī ṭabaqāt al-aṭibbāʾ*, and Perkams (2019); Rashed (2021), esp. ccciii–ccclxv; Verde (2022), xxiii, n. 53. For Andronicus' work, see the excellent synthesis by Hatzimichali (2018).

33 Elias, *In Cat.* 113.19–20 and 117.22; Ammonius, *In DI* 5.28–29; *In A.Pr.* 31.12–13. Beyond the issue of counting the "eleventh scholarch" (Andronicus or Boethus? On this, see Moraux (1973), 53; Chiaradonna and Rashed (2020), 4) the testimony is probably unreliable (Lynch (1972), 202–204). We refer to it only to illustrate the awareness of a rupture within the school already in antiquity.

34 Aspasius, *In EN* 44.20–45.16. See Strabo, *Geog.* 13, 1, 54 (τοῖς μὲν πάλαι [...] τοῖς δ' ὕστερον); Plutarch, *Sulla* 26.3.

philosophical curriculum,[35] while, in his *Compendium*, Nicolaus of Damascus devoted considerable attention to it.[36] This period also marks the beginning of commentaries on the *Physics*, some of which were perhaps already composed by Andronicus and Boethus,[37] although there is more solid evidence for later commentaries by Adrastus and Alexander of Aphrodisias.[38] With few exceptions, such as Nicolaus, the biological corpus is largely absent. However, the *De anima*—and particularly the question of sensation—continued to attract attention. The *De caelo* also attracted careful study: it was meticulously discussed and commented on by Xenarchus of Seleucia, Alexander of Aegae, Aspasius, Herminus and Alexander of Aphrodisias. Cosmological questions remained very much alive, as evidenced by the contributions to this volume by Andrea Falcon on eternalism and by Ludovica Maria Fazio and Federico Maria Petrucci on Adrastus.

Nonetheless, it would be a mistake to imagine that the early Peripatetics pursued Aristotle's project with inventiveness, while the later ones were mere exegetes, preoccupied only with finding meaning in texts whose language they no longer fully understood.[39] This view is questionable, first, because the early Peripatetics were already engaged in working with Aristotle's texts—as evidenced, for instance, by Eudemus' so-called "paraphrase" of the *Physics* (a term that should be used with caution, as demonstrated below by István Bodnár)—and, second, because post-Hellenistic Peripatetics continued to practice a living physics: Xenarchus of Seleucia is known for his heterodox theses in cosmology.[40] Moreover, contrary to his reputation as a marginal figure, Xenarchus was not alone: Sosigenes and Adrastus also sought to update Aristotle's celestial model, which they regarded as obsolete.[41] The same holds true for Alexander of Aphrodisias, to whom this volume devotes special attention. Long considered a mere armchair philosopher, solely engaged in commenting on Aristotle's texts

35 Philoponus, *In Cat.* 5.15–10; Elias, *In Cat.* 117.15–25, 118.9–13 (Boethus, Frs. 2a–b Chiaradonna and Rashed). On Elias' arguments, cf. Moraux (1973), 145. On the reliability of this testimony, cf. Chiaradonna in Chiaradonna and Rashed (2020), 169.

36 See Frs. 2–3 Drossaart-Lulofs. For the dating of Nicolaus, see the doubts raised by Fazzo (2008b) and the clear response by Falcon (2016), 112–115. On the particular emphasis Nicolaus places on Aristotelian biology, see Gottschalk (1987), 1123.

37 For Andronicus, see Moraux (1973), 114. On the question of whether Boethus wrote a commentary on the *Physics*, see Kupreeva (2018), 272, and Chiaradonna and Rashed (2020), 7.

38 For Adrastus, cf. Simplicius, *In Phys.* 122.33 ff. For Alexander, cf. Rashed (2011).

39 The question of language is not merely a jest. On this, see Sedley (1997), 113–116.

40 Falcon (2011).

41 Sharples (2010), 191–195; Kupreeva (2016), 146–150.

and indifferent to the scientific discoveries of his time,[42] he is, as the contributions by Orna Harari, Inna Kupreeva and Rafaëlle Beauvallet demonstrate, quite the opposite: his exegetical activity did not hinder a dynamic engagement with philosophical problems—one that was sometimes critical of Aristotle and attentive to empirical phenomena.

From this perspective, it is important to add some nuance to an excessively dualistic representation of the history of the school as divided between Hellenistic and post-Hellenistic Peripatetics, from Theophrastus to Alexander.[43] The question, then, is what constitutes the unity of the Peripatos in physics—and, indeed, whether it is even possible to identify a unifying criterion. A philosophical tradition can exist only if, on the one hand, it does not merely rehash the thought of its founder, while, on the other hand, it does not fragment into entirely heterogeneous positions. Broadly speaking, three main causes led the Peripatetics to reinvigorate Aristotle's thought. The first is that Aristotelian philosophy itself encourages the pursuit of scientific enquiry.[44] As the chapters by Robert Roreitner and Katerina Ierodiakonou clearly illustrate, Theophrastus was driven by the project of extending, for instance, Aristotle's theory of perception—both in terms of its physiological explanation and at a more general theoretical level. This was already recognised in antiquity. As Boethius states: "On all matters about which [Theophrastus] argues after his master, he touches lightly on those which he knows have been said by Aristotle earlier, but follows up more diligently other things not dealt with by Aristotle." (Fr. 72 FHS&G). To be an Aristotelian is to affirm that the actualisation of human nature requires the fulfilment of our "desire to know". The second cause is scientific progress itself, as new astronomical or physiological discoveries demanded responses from Aristotelians. Sosigenes and Adrastus, who were already mentioned in connection with astronomy, exemplify this engagement. The last cause is the polemical context and the need to respond to other schools, including with regard to issues that Aristotle himself had not addressed.[45] This is particularly evident in the *De mundo*, as Pavel Gregorić's chapter below illustrates, showing how the treatise engages specifically with Stoic cosmology. Likewise, as David Lefebvre argues, it was in response to the intellectual challenges of his time that Aspasius developed a theory of providence.

---

42 Moraux (2001), 401–402 and 420; Towey (2019), 215 and 217.

43 See Lefebvre (2016), 15: "Over the course of these 200 years, there was no official *aggiornamento* within the Peripatos as there was in the Academy. Thus there is no such thing as an 'old,' a 'middle,' and a 'new' Peripatos."

44 On the idea of scientific progress in Aristotle, see Angioni (2010).

45 Verde (2022) is entirely devoted to exploring this line of enquiry.

These three causes contributed to making the Peripatos relatively tolerant of a certain form of heterodoxy. Strato has been known since antiquity for his independent thinking. It is significant that Cicero describes him as emerging "out of left field" or "unexpectedly".[46] Simplicius echoes this when he refers to the "new path" (καινοτέραν [...] ὁδόν) taken by Strato.[47] Similarly, Critolaus is criticised (once again by Cicero) for not adhering to the principles of his ancestors.[48] These external attacks are not solely motivated by malice, as when Galen chastises the "philosophers of today who call themselves Aristotelians and Peripatetics", despite their great ignorance of Aristotle.[49] They also testify to the existence within the Peripatos of significant divergences from Aristotle, which, however, did not lead to any expulsions (so to speak) from the school.[50] Even Alexander, who is often regarded (incorrectly) as being more concerned with exegetical and systematic issues than with living philosophical inspiration, develops an Aristotelian theory of fate and providence, topics which are notably absent from the Aristotelian corpus.

Beyond these divergences, it might be tempting to try to identify what Minio-Paluello called an "Aristotelian core"[51]—as if there existed a certain number of theses that form an unvarying doctrinal body and serve as an identifying characteristic. In fact, one would likely end up with a very limited core,[52] not to mention that this core would vary depending on which authors are being compared. It would be even smaller if it were reduced to concepts such as actuality and potentiality, the categories, or the soul as the form of the body.[53] Moreover, it would be easy to find genuine Platonists who borrow these concepts.[54] An alternative approach would be to seek the unity of the Peripatos not in the physical theses themselves, but in the scientific practice of the Peripatetics, their methods and the epistemic norms governing their enquiries. Yet this procedural approach, as we might call it, suffers from the opposite flaw of

46 "Ecce tibi e transverso Lampsacenus Strato" (*Acad.* 2.121 = Fr. 18 Sharples).

47 Simplicius, *In Phys.* 788.35–789.2.

48 Cic. *Fin.* 5.14 = Fr. 1B in Sharples (2010) = Fr. 11 W.

49 Galen, *De semine* 516.5–9 K.

50 On Xenarchus, see Falcon (2021), 201: "his departures, no matter how significant they are, do not necessarily make him a rebellious follower of Aristotle." On these divergences, see more broadly Baltussen (2016b), chap. 5.

51 Minio-Paluello (1972), 420.

52 Think of the comparison between Boethus and Alexander, on which see Rashed (2017), 22–26, and Chiaradonna and Rashed (2020).

53 See Baltussen (2016b), 110.

54 The same problem also arises for specialists in the Platonic tradition. See, e.g., Boys-Stones (2018), 4, 8, 16 and 83.

the "substantial" approach. As the discussions during the two Lyon conferences that gave rise to this volume have shown, it proves too deflationary to account for the endurance of the Peripatetic movement.

A third solution, then, would be to adopt a relational approach, one that treats interactions between corpora as the source of identity. In other words, to use contemporary metaphysical vocabulary, belonging to the Peripatos would need to be treated as an entirely relational property. Seen in this way, "Aristotelianism" becomes the name for the mereological sum of these relations, as they perdure in part throughout the tradition. Such an approach offers the heuristic advantage of freeing research from the need to judge in advance what does or does not merit the label "Aristotelian", since it would instead treat the relations between corpora as primitive. These relations include those of the Peripatetics with Aristotle, as an epistemic authority who supports divergence, as well as those of the Peripatetics with each other. Indeed, all the contributions gathered in this volume vividly illustrate this network of relations that weaves together the unity of the Peripatos. Accordingly, this book does not claim to provide exhaustive coverage of Peripatetic physics. Rather, each chapter reflects the complex unity of the Peripatos and the intricate relationships that connect all of Aristotle's heirs. In this sense, the volume offers various perspectives on Peripatetic physics.

CHAPTER 1

Simplicius as a Source for Theophrastus' *Physics*: Evidence, Challenges, Prospects

Han Baltussen

1 Introduction

Extensive scholarship has gone into demonstrating the importance of sixth century Platonist Simplicius of Cilicia for our knowledge of Greek philosophy in general and part of his writings—his vast commentaries on Aristotle's treatises—includes valuable evidence relevant for Theophrastus.[1] Simplicius had great respect for both scholarchs, calling Theophrastus "the best of Aristotle's students" (*In Cat.* 14, 15b10–12 = 153A FHS&G). Since we have some 3,000 pages of his work, his contribution to the history of Greek philosophy is enormous, even though his reports should not be taken at face value. For Simplicius, authors like Aristotle and Theophrastus gave access to evaluative reports on the early Greek thinkers and food for thought on philosophical problems of a fundamental nature. Both parts require careful exegesis. An important aspect of his comments on Theophrastus concerns the question how much the successor to Aristotle informs us about, and contributes to, the study of early Greek philosophers. This aspect was certainly a primary concern of Diels when he created the Simplicius edition for the CAG series (starting with the *Physics* commentary, vols. 9–10, 1882–1885), his pioneering *Doxographi Graeci* (1879), studies which led to the innovative edition of the Presocratics (*Vorsokratiker* 1923). The renewed interest in Simplicius in recent decades by those studying late Platonism and its background has also led to drawing attention to his frequent citing and quoting of the earliest Greek thinkers.[2] The extent to which he had access to any of these works is not easy to determine, but several statements of his strongly suggest he did have some available.[3] Since Hermann Diels' impressive reconstruction of the transmission of philosophical views (*Doxographi Graeci*

1 See, e.g., Wiesner (1987), Mansfeld (1994), Luna (2001), Golitsis (2008), Baltussen (2002a–c), (2008), (2015b), (2018), (2022), Hadot (2020), Menn (2022a).

2 Hadot (2022), Golitsis (2008), Baltussen (2008).

3 E.g. Parmenides, Empedocles, and Anaxagoras, for which see below and Baltussen (2008), ch. 2.

© HAN BALTUSSEN, 2025 | DOI:10.1163/9789004739314_003

1879), his role in the so-called doxographical tradition has gained greater delineation. This is the reason why the relationship between the work *ta physika* and *physikai doxai* is a "subordinate but important issue" in the scholarship.[4]

But the commentator's value for Theophrastus' own work on nature still awaits a fuller treatment. In his huge corpus Simplicius cites and quotes authors from the whole history of Greek philosophy, offering valuable materials regarding the first Greek thinkers, formerly referred to as the "Presocratics". Perry, commenting on the role of Simplicius for Parmenides in his doctoral dissertation (1983), could still point out that Simplicius had for too long been used as a "quarry for Greek philosophers".[5] His direct access to early philosophical views is indeed valuable, as it allows us to compare that material to his access to these thinkers *via* the first two scholarchs of the Peripatos.

This chapter aims to expand the analysis of Simplicius' value for Theophrastus. It offers a synthesis of the scholarship on Simplicius' use of his sources by reviewing his handling of Theophrastus in the commentary on Aristotle's *Physics*. Furthermore, it advances the analysis by examining significant aspects of his treatment in detail and by asking how this can assist in improving our understanding of Theophrastus' own *Physics*. This work stands in a special relation to Theophrastus' *Physikai doxai*, a (lost) collection of views on physics which also discussed the earlier natural philosophers: these are the texts in FHS&G on "physical doxography" as the section title indicates, which are placed among the materials for physics. Simplicius represents nearly half of the evidence considered relevant to that *doxai*-collection.[6] My analysis aims to test this claim, suggesting Simplicius' role is important, but more so regarding Theophrastus' *Physics*. My analysis will emphasise how Simplicius in his sprawling commentaries on Aristotle offers valuable clues regarding the *structure* and *content* of Theophrastus' *Physika* (the *treatise* rather than the broad area of study mentioned above). The specific questions I will be asking are: 1. what evidence do we have for his *Physika*? And more importantly, 2. what does Simplicius contribute to this body of evidence? For both lines of enquiry,

4 Partial quotation from Gottschalk (1967), 19 who raised it regarding Steinmetz' hypothesis on the *Physics*. Steinmetz saw the *physikōn doxai* as a "collective title of monographs" though not "consistently" so (Gottschalk, ibid. 19 n. 2).

5 Perry was supervised by John McDiarmid, himself a former student of Harold Cherniss, whose famous monographs *Aristotle's Criticism of Presocratic Philosophy* (1935) and *Aristotle's Criticism of Plato and the Academy* (1944) had been instrumental to re-evaluate the sources for the Early Greek Philosophers ('Presocratics').

6 I am preparing a short commentary on this material (texts 224–245 FHS&G for Brill Commentaries).

the discussion will be selective and limited to the most interesting and significant materials, so that we may learn how useful the testimony of Simplicius is. Among the useful evidence a few other authors play a role, like Plutarch, Alexander of Aphrodisias and Porphyry. I will end with articulating further challenges for the fuller evaluation of this commentator's contributions to early and Hellenistic Greek philosophy.

2 Sorting Out the Sources

Theophrastus' research on the natural world is varied and difficult to unify (the claim that he was mostly focusing on ἄψυχα is inaccurate, given his works on plants and human psychology). The only complete works extant are his extensive studies on botany, several short treatises on the inanimate world (for instance, on winds, stones, and weather signs), or a few long fragments on physiological topics such as sweat, dizziness and fatigue.[7] The remaining fruits of his penmanship are represented by ca. 224 titles, listed by the third century CE biographer Diogenes Laertius, supplemented by a wide range of later sources in the form of passing references, citations, and quotations. While Diogenes' lists provide us with titles and contextual information, the composite nature of the lists reveals some duplicates and irregularities which at times create interpretive problems.[8]

Given the limited knowledge we have about Theophrastus' *Physika*, attempting a reconstruction of his study of the natural world ("physics") faces several challenges which relate to the scope as well as the nature of the evidence: the many titles of his works (DL, V 42–48) suggest a wide-ranging interest in natural philosophy but only a limited number of extant works (*opuscula*). There is much that we do not know. The loss of so many works also reinforces the secondary role he has played in transmission of Aristotle's writings. Three issues demand our attention. The *first* issue is the discrepancy between the number of titles preserved and the number of works we can match up with these: Diogenes Laertius lists two entries for a work entitled *Physika*, one in 16 books (DL, V 46), another in 18 books (DL, V 48). *Secondly*, evidence for the titles and works we do have (mostly in DL and Simplicius) poses all kinds of other problems: some may refer to a whole treatise, others to a part (book), some may be doublets

7 On the short works (or *opuscula*) see Burnikel (1974). On the physiological works, see Fortenbaugh, Sharples and Sollenberger (2003) with new editions, translations and commentary. Several shorter texts still await new editions and translations.

8 See esp. the commentary on DL V by Sollenberger (1985).

(with variations) etc. *Third and last*, the meaning and place of many titles has not always been fully established.

As to the later references, the range of authors is so wide that interpreting their mention of Theophrastus' works and ideas also sets challenges. The materials have been collected in two volumes edited by Fortenbaugh, Sharples, Huby and Gutas and published in two volumes (1992), which goes by the acronym FHS&G. But as we will see this collection was intended as a "workbook" and if we want to use it for creating a synthetic picture of Simplicius' engagement with Theophrastus, the material needs to be rearranged. In this landscape of sources for Theophrastus' *Physika*, Simplicius is unique in that he offers us actual quotations from, and comments on, a work he refers to as *ta physika*. So, one would assume that his offerings are the closest we can get to anything genuine on Theophrastus' *Physika*. Yet it will be wise to withhold judgement and establish whether Simplicius' mention of *ta physika* refers to a single work, several works, or simply the subject matter of "things to do with nature".

The list of relevant sources for the *doxographica* has undergone some changes since Diels' overview, compared to FHS&G (by author in chronological order):

Authors	Diels *Phys. Op.* (25)	*FHS&G* (30)
Cicero	1: 18	240
Plutarch	1: 22	243, 245
Philo	1: 12	–
Galen	2: 5a, 14	231, 239
Taurus (*ap*. Philop)	2: 11, 21	241A–B
Alexander of Aphrod.	2: 6, 23	*227C*
Diogenes Laertius	2: 6a, 17	*227B*, *227D*, 227E, 233, 237
Stobaeus [DG: Aëtius]	2: 16, 19	232, 236
Proclus	2: 10, 20	242
Simplicius	10: 1, 2, 3, 4, 5, 7, 8, 9, 13, 15	224, 225, 226A–B, 227A, 228A, *228B*, 229, 230, 234, 235, 238
–	–	Paulus (Kitāb) 244, Šahrastānī 241C

The chronological range here from Cicero (first c. BCE) to Simplicius (sixth c. CE) shows a certain gap in the transmission from Theophrastus (fourth c. BCE) but Epicurean material via Lucretius (ca. 100 BCE) would partially fill the gap for some topics.[9] I have highlighted certain numbers in italics which (as we shall see) should be considered "misplaced" in DG and FHS&G, since they are not evidence for *Phys. Op.* (the work listing opinions of natural philosophers), but rather cite a *Physics*. This may lead to a reduction of the number of texts in the last column by four (total: 26).[10]

The recent completion of the English translation of his *Commentary* on Aristotle's *Physics* in the London Translation Project is a major boon to anyone wanting to engage with his work. In the following pages a brief section will deal with the debate around the confusion that has existed concerning the work *ta physika* and the *physikai doxai* (see Mansfeld & Runia 2020). In this debate, the two works sometimes interconnect but the main reason for discussing them here is to clarify how we can distinguish between them more.

The reconstruction offered by Steinmetz' monograph (1964) which used the title "Die Physik des Theophrastos" (study of nature or the treatise *Physics*?) was impressive but it has met with criticism for stretching the evidence. Steinmetz included all the extant writings on nature in his analysis and aimed to build a cohesive single work incorporating all surviving titles. As such it provides a good survey of the material, but it is not generally accepted in all its details and implications. By involving all surviving smaller works in constructing a larger compound work Steinmetz thought he could represent the "Physik" of Theophrastus. He discussed the specific problem of the titles and treatises transmitted, including the relation to the doxographical work (1964, 334 ff.). As one reviewer pointed out, his reconstruction fails to convince precisely on this issue of trying to achieve a perfect fit between titles and works, linked to a claim about the larger *Physics*.[11] A better question to ask is what the evidence tells us about a more narrowly construed *physika*. In short, we should establish his physics empirically from the firmest evidence instead of working with a preformed notion of the work.

9 On Lucretius channelling Epicurus' doxographical material, see Sedley (1998a).

10 Some potential additions may also be considered, as I will discuss in my FHS&G commentary (Forthcoming). Mansfeld (1992, 64 n. 7) already noted that "the greater part of these, however, should be attributed to Theophrastus' *Physics*" (with reference to Steinmetz [1964], 334 ff.).

11 See Gottschalk (1967), 25. Pertinent criticisms of Steinmetz's reconstruction in Gottschalk (1967), 18–25.

3 Recovering Titles and Works Related to Physics

In broad terms there is agreement that Theophrastus followed the main plan of Aristotle's description of the physical world.[12] The work seems to adhere to the same objectives of coherence and comprehensiveness in ideas, and he also must have dealt with the subject on a similar scale as Aristotle had done, at least thematically. His dealings with Aristotle's *Physika* were characterised by Wehrli as follows: "Ferner versuchte auch Theophrast, die Physik des Aristoteles systematisierend auszubauen".[13] This seems a rather generous reading: "auszubauen" suggests extending, but indications are that he was not only elaborating to extend, but also to clarify. It is fair to say with Wehrli that he did this "systematisierend", in that he wanted to ensure the works offered a coherent doctrine. Such an ambition included corrections, additions, and extensions. Aristotle's main plan for the study of nature is reflected in several *pragmateiai* which cover the whole of nature: *Lectures on physics*, *De caelo*, *De generatione et corruptione* and *Meteorology*.[14] As Aristotle explains himself in the introduction to his *Meteorology*, the student of nature should study the natural phenomena in a certain order.[15]

There are, however, also indications that Theophrastus could no longer be certain of a comprehensive system encompassing the whole of nature.[16] His metaphysical view on the presence of (some degree of) disorder in our world (*Metaph.* 7a19–b9) bears witness to his conviction that teleology has its limits.[17] In fact, we see in his botanical works that he had a particular eye for the anomalies in the phenomena he studied, which he tries to accommodate within the Aristotelian framework whenever possible.[18] They were probably intended as a kind of curriculum suitable for studying physics, starting with the first causes of nature and natural motion (*Phys.*), next treating the move-

---

12 An assumption found in many studies on the remains of his doxographical works: Kahn (1960), 25. See also the well-known statement by Boethius on his method (72A FHS&G), quoted at the end of the paper (n. 63).

13 Wehrli (1958), 89, in his commentary to Eudemus.

14 Simplicius, *In Phys.* 2.23–24 τὰ βιβλία τῆς ἐπιγραφομένης φυσικῆς ἀκροάσεως.

15 See Aristotle, *Meteor.* I 1 for the general framework Aristotle had in mind and compare the remarks of Alexander in his commentary on *Meteor.* 1.1–4.11 Hayduck (CAG III.2) and comments by Sharples (1995). Cf. also text below and n. 17.

16 Cf. Regenbogen (1940), col. 1554: "Es bereitet sich mit T. im Philosophen ein Gelehrtentypus vor, der gerade mit dem wachsenden Erkenntnisumfang eine leichte Resignation bezüglich der Möglichkeit letzter synthetischer Erkenntnisse verbindet".

17 *Metaph.* esp. b5 μέχρι πόσου τὸ τεταγμένον, 10a21 ff. See also Repici (1990), 182 ff.

18 See also my remarks in Baltussen (1993), 107.

ments of the stars (DC), the elements and their mutual transformations (DC and GC), growth and decay (GC), and the meteorological phenomena occurring in the sublunar sphere.[19]

There is some ancient evidence regarding Andronicus' idea of the presumed "*Corpus Aristotelicum*" to support this.[20] It shows the existence of a discussion about the curriculum and the sequence of topics almost immediately after Aristotle, and it proved to be a recurrent issue until very much later.[21] This sequence can assist us, if we want to assume that Theophrastus and Eudemus by and large followed their teacher in this respect.[22] An additional point worth making is that the sequence also illustrates that physics comprised more than a work *ta physika*, so that the references to φυσικά in later sources should be approached with some caution.

The evidence for the books of Theophrastus' *Physics* and *Physical Opinions* can be used to establish a basis for the number, content and title(s) of Theophrastus' work(s) on physics (in the wider sense), although it is by no means complete. It comprises the following texts and passages (a full list in FHS&G no. 137):

- the catalogue in DL, V 44–48 lists many works on physics including
- self-references
- later references, from Galen (2nd c.) to Simplicius (6th c.) to Šahrastānī (Arabic; n.d.)

Steinmetz' book regarding the reconstruction of the *Physics* is an important starting point, and while I will be taking issue with some points in his argument, I acknowledge that his book contains much that is valuable. I mainly disagree with certain claims about the evidence (Steinmetz 1964, 11–13) which strike me as inaccurate or lacking persuasion. For instance, he claims "aus den ersten fünf Büchern sind Zitate *mit genauer Angabe* der Fundstelle erhalten" (p. 11; italics mine), after which he goes on to present the relevant passages. I will give the "Zitate" below to show, among other things, that Steinmetz' remark applies only to books I, III and V.[23]

---

19 Algra (1995), 145 (with n. 55) referring to *Meteor.* 338a20–26.

20 Cf. Algra (1995), 145 on the late Arab versions of the list of Ptolemaeus (references to earlier literature in n. 56).

21 On the renewed evaluation of the order of the Aristotelian works by, e.g., Adrastus (approx. 200 years later) see Moraux (1984), 314–317 and below. See also Ross (1936), 1–2 who listed various meanings of phrases that refer to a *Physics*.

22 For Eudemus see Baltussen (2006).

23 He also could have been clearer about the fact that we have two references to the περὶ κινήσεως "on motion"—on which see below pp. 22–23.

But first some comments on the evidence for the doxographical work, to clarify the issues in distinguishing the *Physika* from the *Physikai doxai*. The titles on the so-called doxographical works are not presenting us with a clear-cut case. The main issue arises from presuppositions about what kind of writing the *physikai doxai* might be: some say a history, others a more doctrinal review.[24] If these are wrong or unfounded, any conclusions building on them will need to be re-examined.

If we consider the main evidence for the titles that might be referring to the work in which Theophrastus collected views on physics (*doxai*, hence 'doxographical'), the short references in much later sources do not produce a coherent picture. Apart from titles that contain reference to the study of nature (*ta physika*), there is a second confusion about the doxographical work as to whether its most common title was describing "physical opinions" (*physikai doxai*) of "opinions of natural philosophers" (*physikōn doxai*). The problem is one of emphasis, and may be resolved by considering the grammatical aspect, as Mansfeld has pointed out: the problem may turn on the presence or absence of a definite article in a genitive plural: (a) περὶ φυσικῶν δόξων "On physical opinions'" (241A FHS&G = Taurus in Philop. *Against Proclus* 6.8; cf. DL, V 48); (b) περὶ τῶν φυσικῶν δόξων "On opinions of natural philosophers" (cf. DL, V 45). In favour of the first (and *contra* the second) Mansfeld has also pointed out that the second title would require an extra τῶν for the Greek, περὶ τῶν ⟨τῶν⟩ φυσικῶν δόξων, to be read as "on *the* views of *the* natural philosophers".[25] Clearly these details should not be overlooked (though in the second case, haplography may at least explain its omission).

3.1 *Evidence for* Phys. Dox. *and* Phys.

Simplicius mentions Theophrastus as his source as part of his extended survey when commenting on Aristotle's *Phys.* A.1–2 (*In Phys.* 20.19; cf. *In DC* 1.8). Near the end of the discussion, Simplicius remarks (*In Phys.* 28.30–32): "This brief review of investigations into first principles has been arranged, not in chronological order, but by way of 'kinship' (*kata syngeneian*)". This intriguing comment already suggests a method of ordering information that has its precedent in Aristotle and Theophrastus but also shows that Simplicius is aware of it. He was also aware of certain *doxai*-collections even if he seems to have distrusted them.[26]

---

24 I give the main points here and will go into more detail in my commentary on the "Fragments in Physical Doxography" (FHS&G 225–245). See also the plausible hypothesis in Mansfeld (1990) and (1992).

25 Mansfeld (1992), 64–65 (and n. 10) with reference to Mansfeld (1990), 3065ff.

26 He mentions *anagraphais historikais* (*In Phys.* 28.33–34) in the same context. These 'lists'

Since his share in our knowledge for Theophrastus' *doxographica* amounts to 12 passages from his commentaries (mostly on Aristotle's *Physics*), it is debatable whether these texts relied on a *doxai*-collection like the *Physikai Doxai* or whether the FHS&G classification is misleading us into thinking that Simplicius used two types of works. While I have already highlighted four texts as certainly belonging to the *Physics*, we will do well to keep in mind that the texts from Simplicius in the section "*Doxographica physica*" in FHS&G are part of the section for Theophrastus' *Physics* in that collection (texts 137–245 FHS&G). Thus, it is reasonable to consider references in that subsection as informing us about Theophrastus' doxographical activities *within* his discussions of physics. This is the main reason why we must avoid conflating his *Physics* and his *doxai*-collection. This important distinction warns us that the subsection in FHS&G risks misleading its readers insofar as it suggests that Simplicius extracted his knowledge of early Greek philosophers from a doxographical work.

More material supporting the existence of the *doxai*-collection comes from outside of Simplicius' works; the following six passages are taken to be the evidence for the work *Physikai Doxai*.[27] They are mostly from the second and third centuries CE and can be briefly listed here:

(1) Taurus *apud* Philoponus, *Against Proclus* vi 8.27 [= **241A** FHS&G = *Phys.op.* 11 Diels[DG]] who mentions Theophrastus and a work on opinions in natural philosophy in which objections are stated (Καὶ Θεόφραστος μέντοι ἐν τῷ **περὶ τῶν φυσικῶν δόξων** κατὰ Πλάτωνά φησι γενητὸν τὸν κόσμον καὶ οὕτω ποιεῖται **τὰς ἐνστάσεις**.)

(2) Galen, *In Hippocr. de nat. hom.* XV 25 Kühn [= **231** FHS&G = *Phys.op.* Fr. 5a Diels[DG]; partial quote, *DG* 482.20–483.4]. καὶ Θεόφραστος δ' ἂν **ἐν ταῖς τῶν φυσικῶν δοξῶν ἐπιτομαῖς** τὴν Ξενοφάνους δόξαν, εἴπερ οὕτως εἶχεν, ἐγεγράφει. Καὶ σοι πάρεστιν, εἰ χαίροις τῇ περὶ τούτων ἱστορίᾳ, τὰς τοῦ Θεοφράστου βίβλους ἀναγνώναι, καθ' ἃς τὴν **ἐπιτομὴν** ἐποιήσατο **περὶ φυσικῶν δόξων**. (This reference to the "summaries" of the opinions of natural philosophers may not be a book title but an indication of the content of a work, since *epitome* can be either a summary *tout court* which follows a work in the original order or a compressed version of a work in its essential content).[28]

---

may belong to the doxographical tradition, which (in a sense) started with Theophrastus, see Mansfeld (1990); Baltussen (2000), 234–245, then became stock and trade of later traditions. For Simplicius' dual access to early views see Baltussen (2008), 56–57.

27 See Mansfeld (1990), (1992) for a detailed analysis.

28 Is it or is it not an extract from the *Physikai doxai*? Cf. Mansfeld-Runia (1997), 183 with n. 203. In addition, Mansfeld (1993), 361 suggests that the remark on Xenophanes implies that Theophrastus had used the original poem. On the double occurrence of the title

(3) Diogenes Laertius 9. 21 [= part of 227 FHS&G = *Phys.op.* Fr. 6a (first text) Diels^DG^] may well refer to the same "summary" but he also makes an interesting comment about the *physika*: Θεόφραστος ἐν τῇ ἐπιτομῇ Ἀναξαιμάνδρου φησὶν ἀκοῦσαι—… καὶ τὸν νοῦν καὶ τὴν ψυχὴν ταὐτὸν εἶναι, καθὰ μέμνηται καὶ **Θεόφραστος ἐν τοῖς Φυσικοῖς πάντων σχεδὸν ἐκτιθέμενος τὰ δόγματα**

(4) Alexander of Aphrodisias' commentary on Aristotle's *Metaphysics* p. 31.7–16 Hayduck [= **227C** FHS&G = *Phys.op.* Fr. 6 Diels^DG^], which (*pace* Diels) was not part of the so-called *Phys.op.*: Περὶ Παρμενίδου καὶ τῆς δόξης αὐτοῦ καὶ Θεόφραστος ἐν τῷ πρώτῳ περὶ τῶν φυσικῶν οὕτως λέγει (cf. DL, V 46 περὶ φύσεως α′β′γ′ or περὶ φυσικῶν α′-ιη′ 18 books[29]).

(5) Diogenes Laertius IX 21–22 (*Phys.op.* Fr. 6a Diels^DG^ [cf. nr. 3 above] reference to the *Physika*, perhaps a case of "self-quotation"[30]).

(6) Plutarch, *Against Colotes* 1115A, itself a polemical work [= **245** FHS&G],[31] writes: Θεοφράστου δὲ τὰ πρὸς τοὺς φυσικούς, which has a parallel at DL, V 46 (πρὸς τοὺς φυσικούς α′).

Note how many of these works mention claims that express disagreement with Plato (which explains why several Platonists among the sources aim to neutralise these objections), and two expressly indicate that this work on opinions on nature contained *objections*. This strongly suggests that Theophrastus' work in the domain of natural philosophy created a collection which represented a summary statement of earlier views combined with objections: in Plutarch both πρὸς and ἐπεναντιούμενοι may indicate that he refers to a book not by title but by its most typical content, counterarguments. In Diogenes Laertius several works refer to "objections" and "refutations", whether against individuals or groups.[32] The other work is referred to in these circles as *Physika*. These further titles on earlier individual philosophers raise the possibility that Theophrastus organised his work in several preparatory stages, in which individual evaluations preceded the larger work organised by themes, before he wrote his work on nature into which he could integrate such notes as were appropriate.

To the extent that we needed to discuss them, these materials have, I hope, clarified the scholarly debate about the two major works presumed most relev-

(compare DL, V 46 with V 48) and an *epitomē* in more than one book, see Mansfeld (1993), nn. 8–9 and Sollenberger (1985), 26, 35.

29 But presumably not DL V 46 φυσικῶν α′β′γ′δ′ε′ς′ζ′η′ (8 books), or περὶ φυσικῶν ἐπιτομῆς α′β′.

30 On self-quotation, see Gottschalk (1967) (Review of Steinmetz), Whittaker (1987) and Baltussen (2019).

31 On this Plutarchan work, see now Kechagia (2011).

32 Cf. the titles in DL: πρὸς Ἀναξαγόραν α′ (V 42); ἐνστασέων α′β′γ′ (V 43); πρὸς τοὺς ἐξ Ἀκαδημείας α′ (V 49).

ant to Theophrastus' activities as natural philosopher, the one dealing directly with the study of nature, the other with previous investigations into nature. We also reclaimed one of the texts in *Doxographi* for his *Physics*. Remarkably, much of the evidence appears in the second and third centuries (two Platonists, one Platonising physician), yet Simplicius shows little awareness of them. We can safely assume there was a large work (separate from physics), in which Theophrastus critically reviewed the opinions of previous philosophers on nature.[33] The titles in DL for the latter would suggest it was an extensive work, which *could* accommodate a work like *De sensibus*, if the *Physikai doxai* would be about *collecting* views and articulating *objections* and *refutations*. How Simplicius' evidence compares to this material can now be considered.

3.2 *Theophrastus'* Physika

In line with Aristotle's systematic thought and writings about nature, Theophrastus followed the same approach in outline, enquiring into the fundamental concepts of nature. I say "in outline", since he could in his teaching and research use the work of Aristotle as his starting point and foundation, that is, assume its existence as a given and thus build on it rather than create a "copy". The discussion on natural philosophy was still developing and Aristotle's impressive output was not perfect or final. As I have shown elsewhere, both he and Eudemus were seen by Simplicius as helpful in how they clarified and expanded on the treatises of their founder.[34]

The reconstruction of the first five books for Theophrastus' physics was already undertaken by L. Philippson (1831), who pointed to those passages in the ancient commentators (mostly Simplicius) where references to these books can be found. Some of Philippson's remarks (1831, 83–85) on this issue seem to have gone unnoticed (except, so far as I know, in Regenbogen (1940), col. 1396), which is especially significant when seen against the importance of the reconstruction initiated by Brandis (1860), and developed by Usener (1858) and Diels (*DG* 1879), in which confusion over the titles blurred the issues (above, section 2.2).[35] From the references to physical principles and motion (κίνησις)

33 Some smaller works are also listed in DL, V 42 (Anaxagoras), V 43 (Democritus, Diogenes, Empedocles), V 44 (Metrodorus), V 49 (Democritus), and previous note. Compare the Appendix nos. 19–20.

34 Baltussen (2008), ch. 3.

35 Regenbogen's suggestions (*RE*, 7 col. 1397) were taken up and enlarged by Steinmetz (1964), 10–13, 334 ff.

the evidence points to treatment in books I–III.[36] His writings on the soul and perception seem to have constituted books IV–V (see below). This is still the common view today.

The evidence in Diogenes Laertius' list of works confirms what we have seen so far: Theophrastus wrote many works on nature (many feature in Steinmetz' discussion), and some titles refer to a *Physika* in 8 books (DL, V 46, cf. Regenb. 1396.3–4), while there are also titles that are pointing to collections of opinions on natural philosophy (DL lists in V 48). Spread over these various sections, DL offers five titles related to natural enquiry.[37] This kind of evidence is regarded as reliable insofar as they go back to Hellenistic sources. But their variety does complicate matters.

Explicit references to books I, III and V are all found in Simplicius and Priscian. But there are no references to a second and a fourth book, the existence of which we must infer from the information on books I, III and V. There is one other source, and the full list looks as follows:

Book	Title	Simplicius, *In Phys./Cat.*	FHS&G
I	(1) ἐν ἀρχῇ τῶν ἑαυτοῦ φυσικῶν	*In Phys.* 9.7 (principles)	144B
I	(2) ἐν τῷ πρώτῳ φυσικῶν	*In Phys.* 20.19 (principles; senses)	143
I	(3) ἐν τῷ πρώτῳ τῶν φυσικῶν	*In Cat.* 435.25 (*kinēsis*)	153A
I	(4) Θ. ἐν τῷ πρώτῳ τῶν ἑαυτοῦ φυσικῶν	*In Phys.* 860.19–20 (motion)	153C
III	(5) Θ. ἐν τῷ α' περὶ κινήσεως ταὐτὰ περὶ τούτου δοξάζων φαίνεται	*In Phys.* 107.12–16[38]	155C
III	(6) ἐν τῷ πρώτῳ τῶν περὶ κινήσεως	*In Phys.* 964.30–31	271
III	(7) ἐν τῷ πρώτῳ τῶν περὶ κινήσεως	*In Phys.* 986.5 ff.	156B
III	(8) ἐν τρίτῳ τῶν φυσικῶν ἢ περὶ οὐρανοῦ	*In Phys.* 1236.1	176
IV?	(9) ἐν τῷ δευτέρῳ τῶν περὶ κινήσεως	*In Phys.* 413.1	153B

In addition, we have the following citations in Themistius and Priscian:

36 The references, preserved mostly in Simplicius (cf. Fr. 137 nos. 1–3 FHS&G), will be discussed below. See also Mai-Lan Boureau in this volume (@@45 with n. 25).

37 5. 45 περὶ φύσεως α'β'γ' (p. 373, line 177 Dorandi); 5. 46 περὶ φυσικῶν α'-ιη' (p. 373, line 178 Dorandi) and περὶ φυσικῶν ἐπιτομῆς α'β' (p. 373, line 179 Dorandi); 5. 48 φυσικῶν δοξῶν α'-ις' (p. 376, line 248 Dorandi), and φυσικῶν ⟨δοξῶν⟩ ἐπιτομῆς α' (p. 376, line 249 Dorandi) [Dorandi (2013), 376 (*in apparatu*) signals that Meursius added ⟨δοξῶν⟩].

38 Diels ad loc. (in app.) refers to "Themist. [*In Phys.*] p. 386.16" (ed. Spengel, 1866).

V	(10) ἐστὶ δ' ἐν τῷ **πεμπτῷ** τῶν φυσικῶν, δευτέρῳ τῶν περὶ ψυχῆς	Them. *Paraphrasis in De an.* 108.11, Heinze (CAG 5.3)	307A
V	(11) ... ἀλλ' ἐπὶ τὰ ἑξῆς ἴωμεν, ἀπ' ἄλλης ἡμῖν ἀρχῆς τὰ λοιπὰ τοῦ **πεμπτοῦ** βιβλίου ἐπεξεργαζόμενοι	Prisc. *Metaphrasis* 22.33 ff. (on imagination) Bywater (CAG suppl. I.2, p. 22.34)	297

The last two passages in the list are important sources from scholarly commentators and offer precious details: Priscian, *Metaphr.* 22.33 ff. (Bywater) refers to the fifth book when expounding Theophrastus' views on the soul and *phantasia*. For Priscian's sources we can use his *Solutiones ad Chosroem* 42.3–7 Bywater (CAG, Suppl. I.2);[39] Themistius' mention of "the fifth book of the *Physics*" confirms a second title "or the second of the *On the soul*".

The added titles (e.g., ἢ περὶ οὐρανοῦ) in several sources may not be original, but if they were, it could mean that the complete work had been composed as a combination of monograph-style treatises. Whether Theophrastus put them together himself or a later editor is impossible to say.

The list shows that we know a little more about its structure and the books' contents. In his *Physics* commentary Simplicius provides intriguing information about a discussion in the tradition regarding the overall structure of Aristotle's *Physics* (cf. 157 FHS&G). He reports that "those from the Peripatos" (so presumably also Theophrastus) divided the *Physics* into two main parts (*In Phys.* 923.7–16): they regarded books I–V as dealing with principles—a view Simplicius, writing in the sixth century, also attributed to the second-century CE Peripatetic Adrastus (*In Phys.* 6.4–10)—while books VI–VIII were on "motion" (*kinēsis* in the generic sense of "change"). This book division was a matter of some dispute in the later tradition, and Simplicius knows of two views: the division of distinct books was already established and the *Physics* was divided by some as "five books *On natural principles*, and three *On motion*" (*In Phys.* 6. 5–10 Adrastus; cf. 801.13–16 "Aristotle and the associates of Aristotle", 1358.8–9), while others chose a four-four division (see e.g. *In Phys.* 802.8–11, "Porphyry ... wisely reports on the division of the eight books, namely that all call the ⟨first⟩ five [books] *Physics*, the ⟨next⟩ three *On Motion*, [but that] he himself still says that the four books from the fifth to the eighth com-

39 On his independent status as a source see Huby (1985), 166 with n. 3. Regenbogen (1940), 1399.45–55 mentions Brandis' speculations on the 'last three books' (if they existed): "über den Inhalt der drei letzten Bücher der Physik Vermutungen bei Brandis III 290 ff. [with reference to Priscian's *Solutiones ad Chosroem*. 3, 62, 7 ff.]". This section is partly based on Baltussen (2016a).

ing after constitute the treatise on motion and is entitled idiosyncratically *On Motion*"[40]). This debate in the later Platonist tradition illustrates the difficulty for later readers to assess which broader structure Aristotle himself intended this work to have (if he had one), and consequently, which one Theophrastus had preferred. But Simplicius cites a correspondence between Eudemus and Theophrastus to claim that the latter "regards the fifth book too as part of the *Physics*" (157.9–10 FHS&G). As we will see shortly, this claim does not align well with the speculation that Theophrastus made the third book of *Physics* "on motion" and that it was also entitled "on the heaven" (see p. 24 with n. 43 on Regenbogen).

Whereto from here? The foregoing overview of materials informing us about broad structure, possible titles or distinct books clearly illustrates the difficulty of establishing a complete picture of the content and structure. The most secure references in Simplicius to Theophrastus' work are those when he mentions "his *own* physics" (twice), first at the very start of the *Physics* commentary (p. 9, line 7) and next in the commentary to book V (p. 860, line 19). Others are established from the context, e.g.:

(a) *In Phys.* 22.29 φησιν ὁ Θεόφραστος ἑτέρας εἶναι μᾶλλον ἢ περὶ φύσεως ἱστορίας τὴν μνήμην τῆς τούτου δόξης (i.e. Parmenides) [above section 2.2, text 6]

(b) *In Phys.* 70.16–18 (after citing Porphyry, he mentions Melissus' *On Nature* which suggests natural philosophy "by proxy") ὁ Μέλισσος καὶ τὴν ἐπιγραφὴν οὕτως ἐποιήσατο τοῦ συγγράμματος περὶ φύσεως ἢ περὶ τοῦ ὄντος (cf. 71.9–10); and

(c) *In Phys.* 1236.1 διαιρεῖ ὁ Θεόφραστος ἐν τριτῷ τῶν φυσικῶν ἤ περὶ οὐρανοῦ τὰ γινόμενα οὕτως

Because no references to any books beyond the fifth have survived, the total number of books is uncertain and the evidence from the catalogue of works in Diogenes Laertius is only partially helpful in this respect: as we saw, it may have been a work in 3 (short version?), 8 (similar to Aristotle) or 18 books (DL, V 46, but this is a stretch since not all works fit in), next to a work of 16 books on physical opinions (DL, V 48), depending on what we take the different titles to mean. And as to the remark in DL, IX 22 that Theophrastus expounded "nearly all the views of matters in nature", this "nearly" could be his choice of word reflecting a personal style;[41] then again, perhaps his *physiologia* comprised fewer books.[42] If we accept Simplicius' comment quoted above (p. 22) on the book division, we could speculate that he discussed principles of physics in books I–V and

40 See also McKirahan (2000), n. 558 ad loc.

41 But some would consider this source problematic: DL, IX 22 sits in the doxography on Heraclitus. Cf. Gottschalk (1967), 19 with his note 3.

42 On φυσιολογία cf. 142 FHS&G [= Fr. 51 W].

motion or change in books VI–VIII (and these last three books could be listed separately in DL, V 46 as περὶ κινήσεως α′, β′, γ′).[43]

One more detail seems relevant. Steinmetz (1964, 11) appears to agree with Regenbogen's "Vermutung" ([*sic*], *RE* Suppl. 7, 1396.6) that books I–III which deal with "Raum und Zeit" would be known also under the title περὶ κινήσεως ("on motion") α′β′γ′, in 3 books. I would note, however, that Regenbogen (*RE* suppl. 7, col. 1397.31–37; cf. 1396.3–6) only offers a "conjecture" here, when he asks the *question* whether the entry mentioning 3 books [i.e. DL, V 44] is identical to *Phys.* I–III, whereas DL, V 49 mentions a περὶ κινήσεως in 2 books. Gottschalk (1967, 18) also speaks of "Regenbogen's conjecture", but Regenbogen's speculative question is very non-commital (or at least tentative): "Die Frage muß wenigstens gestellt werden, ob nicht π. κινήσεως α′ identisch sei mit Physik α′, so dass die Buchgruppe π. κινήσεως α′β′γ′ der Physik α′β′γ′ entsprechen würde, wobei dann π. κινήσεως γ′ = π. οὐρανοῦ wäre" (1397.32–37). Regenbogen's speculation, expressed with prudent restraint in 1940, was a good intuition: that book III was περὶ οὐρανοῦ, *On the heaven*, seems confirmed by Simplicius *In Phys.* 8.5, 126.1–9 (= text 176 FHS&G).[44]

In sum, while a full attribution of subjects to the individual books (esp. VI–VIII) of the *Physics* remains difficult, a plausible case can be made for a work in 8 books, which due to their themes could have been listed separately.[45] In his valuable study Steinmetz decided to reconstruct the *Physika* in 18 books, attempting to include all the works on nature (titles and surviving works). Steinmetz' reconstruction of the *Physics* has been criticised, mostly because the inclusion of such materials resulted in a forced construct, while other elements simply would not allow for a smooth fit. As we saw (above p. 16), his comments concerning the evidence (pp. 11–13) were not always accurate.

---

43 We cannot easily explain DL V 49, περὶ κινήσεως α′β′, i.e., a work on motion in two books alongside the mention of a work in three books. One could argue, as is implied in Barbotin (1954), 52, following Regenbogen (1940), 1397.32–37, that the reference to a περὶ κινήσεως in two books indicates the books which were 'on motion' in general terms, whereas the third treated a specific theme and could be referred to with an alternative title (περὶ οὐρανοῦ, 'on the heaven'). While not fully satisfactory, it is perhaps supported by the connection Plutarch makes between Aristotle's 'on the heaven' and Theophrastus' arguments against the *phusikoi* (above, section 2.2, text 6).

44 That *De caelo* was 'on motion' was not acceptable to Alexander, who declared it to deal with the cosmos, as Simplicius' opening comments in his commentary on that work show (*In DC* 1.1–2; 10–12).

45 On the number of books, see Steinmetz (1964), 11–13, 335ff.; id. (1968), 247 [on bks. 1–5]; Regenbogen (1940), col. 1396; Fortenbaugh in FHS&G (1992), 277.

4 Does Simplicius Offer Further Details on the *Physika*?

We have not yet covered all the evidence, although a comprehensive treatment cannot be undertaken here.[46] It is worth pointing out that, thanks to the scholars Ilsetraut Hadot, Phillippe Hoffmann, Pantelis Golitsis as well as the translators of the London Translation Project, led by Richard Sorabji and now Michael Griffin, and my own contributions, our knowledge of Simplicius' philosophical exegesis and methodology has improved considerably over the past decades. His core methodological principles, many of which are based on instances of self-characterisation by Simplicius as a writer and commentator, are as follows:

(i) exegesis is a spiritual exercise (for instance at *In Enchiridion Epicteti* 194.56–60 Hadot).[47]

(ii) he articulates the required qualities of an exegete while he lists the eighth introductory question at *In Cat.* 7.24–32.

(iii) he offers a definition of *hypomnēmata* as works of Aristotle at *In Cat.* 4.15ff. in contradistinction to *syntagmata* (ibid. 4.20–22).

These are useful core notions for us to keep in mind when we consider how he assessed and uses his sources.[48] Such an analysis involves citation habits, selection of sources, and general aims of ancient scholarship by the sixth century CE (to some extent their "doxographical" method). But the over-arching spiritual or religious aspect present (see n. 47) should not be ignored—one which he articulates most clearly at the end of the commentary on the *De caelo*. As to citation habits, it is a well-known fact that his earlier books of the *In Phys.* contain more references to previous thinkers than the later ones. This will mirror Aristotle's approach in the parts commented on (*Phys.* A, *Metaph.* A etc.), but other factors may play a role (e.g., the importance of first principles, or an intrinsic connection between the frequency of citing and their occurrence in different contexts of the commentary later).

While Gottschalk already criticised Steinmetz' reconstruction as failing to accommodate all the evidence, Mansfeld has pointed out that the earliest suggestions about which passage referred to the *Physics* and which to the "enquiry"

---

46 There are a further 21 texts from *In Phys.*, *In DC*, *In Cat.* in FHS&G 139–223 ('Physics') which could not be incorporated here (see my Appendix).

47 Cf. Hadot-Hoffmann, *In Cat.* 8–9, n. 22: "La composition d'un commentaire est donc pour Simplicius une sorte d'exercice de méditation par écrit". Cf. ibid. n. 25–26.

48 There are many other patterns and 'rules of thumb' we see in his work, e.g., comparative method (the verb used is παραβάλλω, e.g., Simplicius, *In Phys.* 717.23, after Eudemus Fr. 86 W.).

into physical opinions were off track: since most of the citations in Simplicius refer to the *Physics* the burden of proof is on those who want to regard them as coming from another work. Some cross-contamination from the *doxai*-collection is of course possible.

The references to the first book of Theophrastus' *Physics* (Simplicius, *In Phys.* 964.29–965.6, ἐν τῷ πρώτῳ τῶν περὶ κινήσεως; cf. ibid. 986.5–7[49]) and the second book of a work with a (possible) title "On motion", raise the question whether it is likely that Simplicius cites a book under two different titles (on different occasions). The overall pattern of quoting rather seems to suggest that Simplicius sometimes mentions a second title in conjunction with the number-plus-overall title.[50]

Only one explicit testimony has been taken to support the assumption that Simplicius had access to Theophrastus' doxographical work. It can be found at *In Phys.* 115.11–13 Diels (234 FHS&G), but as it turns out, he probably acquires the materials via Alexander:[51]

> τὸν Παρμενίδου λόγον, ὡς ὁ Ἀλέξανδρος ἱστορεῖ, ὁ μὲν Θεόφραστος οὕτως ἐκτίθεται ἐν τῷ πρώτῳ τῆς φυσικῆς ἱστορίας, … (Theophrastus sets out the account of Parmenides, as Alexander reports, as follows in the first (book) of his enquiry into nature …)

But in this instance, this title of the work, ἡ φυσικὴ ἱστορία, refers to the *Physika* and is probably the same work as the "*naturalis auditus*" in Priscian *Solutiones ad Chosroem*, prooem. (= *Suppl. Arist.* 1.2, p. 42.5).[52] Unfortunately, we cannot retrieve any confirmation from the surviving work of Alexander on this point. Out of the 65 name-references to Theophrastus by Alexander, only the text I mentioned earlier (above section 2.2, text 4) was wrongly regarded a doxographical passage by Diels. This is unfortunate because this kind of reference does not even have to come from a *doxographical* work. Additional questions

---

49 This *locus* not mentioned in Barbotin (1954) who aimed to collect the evidence regarding the writings on soul.

50 Compare references to Aristotle's works: ἐν ἀρχῇ τοῦ πρώτου βιβλίου (*in de Caelo* 614.32), ἐν ἀρχῇ τοῦ τρίτου βιβλίου λόγων (*In Phys.* 9.6, cf. 519.12), ἐν ἀρχῇ τοῦ τετάρτου βιβλίου (*In Phys.* 868.5), ἐν ἀρχῇ τοῦ πέμπτου βιβλίου (*In Phys.* 1037.22), ἀρχόμενου τοῦ ἕκτου βιβλίου (*In Phys.* 868.7), ὥς καὶ ἐν ἀρχῇ τῆς φυσικῆς εἴρηται ἀποκροάσεως (*in DA.* 96.31), ἐν ἀρχῇ τῆς περὶ οὐρανοῦ (*In Phys.* 451.21), ἐν ἀρχῇ τῶν περὶ τοῦ κενοῦ λόγων (*In Phys.* 693.5), ἐν ἀρχῇ δὲ τῆς πραγματείας (*In DC* 461.9), ἐν ἀρχῇ τοῦδε τοῦ βιβλίου (*In Phys.* 924.22; 1250.38; 1327.9). On the accuracy of such phrases, see Appendix and Baltussen (2002a) and (2015b).

51 He cites Eudemus' version of Parmenides' words next, lines 12–13.

52 On the question of the title see also Gottschalk (1967), 18 ff.

can be raised about the fact that Simplicius does not mention this, while he relies very frequently on Alexander. But Menn has recently argued that the view that Simplicius' commentary is an expurgated version of Alexander's should be abandoned mainly because Simplicius' commentaries are much longer than what we know about Alexander's commentaries.[53]

As a last step we should move on to examining more closely Simplicius' comments on the structure and content. We should note that in the three main commentaries, the *In Cat.* and *In DC* only contain one reference each to the *Physics* of Theophrastus (*In Cat.* 435.25; *In DC* 564.24—see also my Appendix). As one would expect, we should investigate the *Physics* commentary for clues about Theophrastus' work. In general, Simplicius makes claims about the broader ideas underlying the study of nature in Aristotle's *Physics*, but alongside these notes he often mentions how Theophrastus and Eudemus responded to their teacher's work *in their own works on nature*. We have already seen the macro-structure of the *Physika*. I have suggested that the *Physika* of Theophrastus seems to resonate with much of Aristotle's *Physika* as we know it (at least for books I–V), as well as his wider study of natural phenomena. The surviving titles in DL (137 FHS&G) and the references in Simplicius support our suspicions about the similarities. Simplicius sees Theophrastus as one authority around Aristotle's philosophy but also as a valuable authority in his own right as this extract from Simplicius shows (*In Phys.* 184a16–b14 [p. 18.29–34 Diels] = 142 FHS&G):

> (i) Plato was right to call natural science "an account of probabilities" [εἰκοτολογίαν (cf. τὸν εἰκότα μῦθον, *Tim.* 29d)]; and (ii) Aristotle too bears witness with him, wanting demonstration, *in the proper sense*, to be from principles, which are immediate and credible in themselves, and from explanations which are explanations *in the proper sense* and naturally prior. (iii) But natural science is not for this reason to be scorned; rather, we should be satisfied with what is in accordance with our nature and capacity, *as Theophrastus thinks too*.

No doubt we may infer that Theophrastus started out in his first book by determining what the first principles are: Simplicius mentions him several

53 In the editors' Preface in Menn (2022b), summarising Menn (2022a), 11. The editors suggest that Menn regards it is as a "meta-commentary on Alexander" [cf. Menn (2022a), 12–14], a term which has received different descriptions. On meta-commentary, compare Barnes (1992) in his review of Hadot.

times in the first book of the *Physics* commentary on Aristotle.[54] In the first text to be discussed (144B FHS&G) we see that he has spotted a small omission in Aristotle's observations. The second text deals with a discussion of place in which Theophrastus raised objections against Aristotle. The third passage reviews a quotation regarding what comes into being and the role of entelechy. I selected these three passages to illustrate this, withholding other materials for another occasion.

Simplicius' offers several (supposed) quotations from Theophrastus and discusses them critically.

> [**Example 1**] In 144A–B FHS&G we find texts from Philoponus and Simplicius discussing that the principles of nature (ἀρχαί τῶν φυσικῶν).[55] Simplicius reports that Theophrastus also offered proof (ἀπέδειξε), saying that "there are principles of natural things" (a comment paralleled in Philoponus' commentary on *Physics* = 144A FHS&G); Theophrastus stated in his own work *On nature* the whole syllogism, thinking that some explanation was also needed for the minor premise, that natural science involves principles and causes and elements".

The two texts I paraphrase here again show his eagerness to be explicit about the theses and arguments put forward (formalised into "syllogismos" in Philoponus). In addition, this reference so early in the commentary (*In Phys.* 9.7, ὁ μέντοι Θεόφραστος ἐν ἀρχῇ τῶν ἑαυτοῦ φυσικῶν) is presented by Simplicius as a counterpart to a quote from Aristotle. Presumably Theophrastus is quoted with approval of having confirmed this point, which makes it a worthy addition to the discussion.

Another passage that Simplicius discussed shows Theophrastus was not uncritical towards Aristotle's ideas and claims.[56] In his "Excursus on place and time", Simplicius provided us with valuable information on Theophrastus' evaluative comments on place. Theophrastus raised pertinent objections to Aristotle's definition of natural place as the inner limit of the surrounding place (*On the Heaven* 310b3; *Physics* 214b12–215a11). What he saw as problematic is the idea that place has some kind of dynamic power which acts as a final cause

54 See Baltussen (2015b).

55 Menn (2022a), 'Introduction', comments on the value of comparing the two.

56 See also his discussion of place and Theophrastus' puzzles raised about Aristotle's notion of place at *In Phys.* 639.12–20 (= 147 FHS&G). Cf. Algra (1995), Sorabji (1992b), Baltussen (2016b), 36.

and makes things go in their place.[57] One reason for such a move is that the vacuum alone could not explain this (as atomists claimed). This is intended "to make place *explain* (i.e. be the *aitia* or cause of) the natural movements of the elements".[58] Theophrastus thought that this is problematic for several reasons: he thought that place may not be a surface; that place may not be immobile, and so on. Such puzzles signal quite clearly that Aristotle's view was considered unsatisfactory (Fr. 149 FHS&G, quoted by Simplicius):

> [**Example 2**] "Theophrastus also seems to have held that notion in his *Physics* on place, in which he raises a question in his preamble to his account: 'may it not be that place is not something that exists on its own, but is spoken according to the arrangement and position of bodies, with reference to their natures and capacities? And similarly in the case of animals and plants and in general all things with a differentiated structure, whether living or not, provided that their nature involves a shape. And for this reason, each thing is said to be in its own space by virtue of having its proper ordering'." (Simplicius, *In Phys.* 639.13–22).[59]

This passage and several other indications suggest that Simplicius took up this discussion in his commentary on the *Physics* in the early sixth century CE and clearly read Theophrastus' *Physics* alongside Aristotle's (604.5–6 Diels: "Theophrastus in his *Physics* presents puzzles ..."). Those who argue that Simplicius took all the Theophrastean material from Alexander (see n. 44 and Appendix) will have to explain (away) certain evidence that goes against such a claim. One could add that Simplicius did not mention that Theophrastus' *Physics* was lost or at risk of becoming rare (as he did for instance for Parmenides' poem, admittedly a far rarer work, at *In Phys.* 144.25–28). From this passage, he developed a view which clearly goes beyond the Theophrastean comments.[60] His exploratory queries are converted into the suggestion—articulated in more positive terms—that we should move away from the idea of a place having a power,

---

57 Sorabji (1992b), 56.

58 Sorabji (1988), 186.

59 καὶ γάρ καὶ Θεόφραστος ἐν τοῖς φυσικοῖς φαίνεται τὴν ἔννοιαν ταύτην ἐσχηκὼς περὶ τόπου, ἐν οἷς φησιν ὡς ἐν ἀπορίᾳ προάγων τὸν λόγον· "μήποτε οὐκ ἔστι καθ' αὑτὸν οὐσία τις τόπος, ἀλλὰ τῇ τάξει καὶ θέσει τῶν σωμάτων λέγεται κατὰ τὰς φύσεις καὶ δυνάμεις, ὁμοίως δὲ καὶ ἐπὶ τῶν ζῴων καὶ φυτῶν καὶ ὅλως τῶν ἀνομοιομερῶν εἴτε ἐμψύχων εἴτε ἀψύχων, ἔμμορφον δὲ τὴν φύσιν ἐχόντων. καὶ γὰρ τούτων τάξις τι καὶ θέσις τῶν μερῶν ἐστι πρὸς τὴν ὅλην οὐσίαν. διὸ καὶ ἕκαστον ἐν τῇ αὑτοῦ χώρᾳ λέγεται τῷ ἔχειν τὴν οἰκείαν τάξιν, ἐπεὶ καὶ τῶν τοῦ σώματος μερῶν ἕκαστον ἐπιποθήσειεν ἂν καὶ ἀπαιτήσειε τὴν ἑαυτοῦ χώραν καὶ θέσιν".

60 On this approach see Sorabji (1992b).

but rather should think about things as organisms, where the *arrangement* or ordering (*taxis*) is responsible for the place of the whole and its parts. To understand this reading of Aristotle and Theophrastus, we need to remain aware that Simplicius espoused a type of Platonism that combined doctrines taken from Plato, Aristotle, Stoicism and Pythagoreanism (i.e., ensouled matter).

Finally, at *In Phys.* 1236.1–9 (to Arist. *Phys.* VIII 5), Simplicius alludes to Theophrastus' third book which he gives an additional thematic title "on the heavens", illustrating the link between the discussion of the heavens and physics towards the end of the work. A further interesting point made here is his mention of a distinction among the things that "have come to be". I will quote the passage in a longer version than I did earlier, since this will show how the comment contains useful doctrinal information:

> [**Example 3**] "In the third book of his *Physics*, or *On the heavens*, Theophrastus divides (διαιρεῖ) what comes into being as follows: 'either they are generated by something similar', he says, 'as man by man, and hot by the hot, or by a contrary, as we observe with thunderbolts and flashes of lightning, coming into being of this fire, in the air: it is generated by cold, gathering the heat within them into one and igniting it. Or thirdly, by what is in entelechy in general, as is the **weal**: for it is generated by a **whip** that is in entelechy, but it is no longer similar or contrary to what is generated. And what is generated by the sun, he says, is also generated by an entelechy: for it too is neither similar nor contrary to what is generated by it'."[61] [transl. Muller, slightly modified]

The passage shows Theophrastus commenting by paraphrase and reframing. That we have a quotation here is guaranteed by Simplicius' commonly used 'he says' (twice, φησίν, lines 2 and 8) and the phrase 'as follows' (οὕτως, 1236.2). The passage echoes some of Aristotle's wording but also clearly expands on it. Aristotle's discussion in book VIII has been on movers and motion, and a little before this lemma commented on by Simplicius (257b6 ff.) he starts a "fresh discussion" on how self-movers have the source of motion within themselves,

61 διαιρεῖ ὁ Θεόφραστος ἐν τριτῷ τῶν φυσικῶν ἤ περὶ οὐρανοῦ τὰ γινόμενα οὕτως: "(i) ἢ γὰρ ὑπὸ ὁμοίου γίνεται, φησίν, ὡς ἄνθρωπος ὑπὸ ἀνθρώπου καὶ θερμὸν ὑπὸ θερμοῦ, ἢ (ii) ὑπὸ ἐναντίου, ὡς ὁρῶμεν τοὺς κεραυνοὺς καὶ τὰς ἀστραπάς· ὑπὸ γὰρ ψυχρότητος ἡ τούτου τοῦ πυρὸς ἐν τῷ ἀέρι γένεσις ἀθροίζοντος εἰς ἓν τὸ ἐν αὐτοῖς θερμὸν καὶ ἐκπυροῦντος. ἢ (iii) τρίτον ὑπὸ ἐντελεχείᾳ ὅλως ὄντος, ὡς καὶ ὁ μώλωψ· ὑπὸ γὰρ ἐντελεχείᾳ οὔσης τῆς μάστιγος γίνεται, οὔτε δὲ ὁμοίας ἔτι οὔτε ἐναντίας τῷ γινομένῳ. καὶ τὰ ὑπὸ τοῦ ἡλίου δὲ, φησί, γίνομενα ὑπὸ ἐντελεχείας γίνεται· καί γάρ αὐτὸς οὔτε ὅμοιος οὔτε ἐναντίος τοῖς γινομένοις ὑπ᾽ αὐτοῦ".

and when he comes to potential movement, he brings in the example of hot and cold that Theophrastus is talking about. Aristotle is especially interested in demonstrating that anything that moves something else, can itself not be moved except by itself (the conclusion at 259a4–9 is clear). But it is Simplicius who uses the example of the whip and weal (1235.38ff.), not Aristotle. It is worth noting that his example of the whip may well come from Alexander.[62] Moreover, the Byzantine scholia containing fragments from Alexander's *Physics*, newly discovered by Marwan Rashed in two Paris manuscripts,[63] suggest that, in this case, Simplicius may have had access to Theophrastus via Alexander, or perhaps to both, thus combining Theophrastus' physics and Alexander's comment on Theophrastus' physics.

With the verb διαιρεῖ Theophrastus creates a *diairesis*—usually a dichotomy to distinguish things from each other and "divide" them (the verb's basic meaning) into sub-groups under a common concept. Theophrastus' remarks on two types of things coming to be, and whether the cause and effect are the same (in kind) seems not closely related to Aristotle's discussion on self-movers. Simplicius goes on to describe an objection by Alexander, who does think that what comes to be is generated by an actuality (or, in the analogy, by a blow in the case of a whip). Simplicius accepts this point, next offering his own objection. It signals that Simplicius can stand up to the much-admired successor of Aristotle, even though Simplicius often introduces Theophrastus' views to support a particular point of either Aristotle or of himself.

5 Some Conclusions

My discussion (though far from being comprehensive) has reviewed evidence and challenges regarding Theophrastus' *Physika* in Simplicius and led us to various sources and his writings on nature. The question about his *Physika* and its relation to a doxographical work arises from the materials these two works seem to share about the early thinkers but also from the fact that some references have been wrongly attributed to the *doxai*-collection while they in fact relate to the *Physics* (§2). A brief inventory of the evidence for the *doxai*-collection shows a striking confluence of texts under the Roman empire (second and third c. CE), while it is also clear that we should keep Simplicius more closely connected to the *Physics* (§3). These points still offer sev-

62 I owe this reference to Claire Louguet (Université de Lille).

63 Rashed (2011). On this passage see: pp. 550–551, no. 628.

eral insights into Simplicius' value in reconstructing Theophrastus' physics. My main goal was twofold, to show (i) that Simplicius offers helpful evidence for the structure and method of Theophrastus' *Physics*; (ii) that the passages containing doctrinal discussions are also useful in how they reflect both Theophrastus' views and the Platonist respect for Theophrastus in clarifying Aristotle. Both these points exemplify significant features of Simplicius' role and clarify his access to, and use of, Theophrastus in his commentary on Aristotle's *Physics*.

The specific insights we have gained so far can be summed up in three broad points: *firstly*, Simplicius' role for the *Physics* of Theophrastus is important both for its use of his work in commenting on Aristotle and for Theophrastus' absence in the commentary at certain times. Given the generous quotations and exegetical comments on the Peripatetic's work, the modern scholar can refine what we know about the second scholarch's work. Simplicius gives us secure clues as to content, structure and book division.[64]

Secondly, the question as to whether Simplicius had direct access to the work can now be answered with greater confidence: while we cannot include all materials in a general conclusion, his citation methods we have exemplified offer sufficient evidence for many direct points of access (more material is included in the Appendix). A striking feature of Simplicius' citation technique is the fact that whenever he quotes from Theophrastus, he very regularly also cites (and quotes) from the second-century commentator Alexander of Aphrodisias. Alexander was a hugely influential exegete on Aristotle. This collocation of the two authors raised the intriguing question whether Simplicius had access to both directly, or only to Alexander (whose name appears more than 700 times in Simplicius' commentaries[65]), and to Theophrastus only *via* Alexander (though in the latter cases we might expect him to say "Alexander says that Theophrastus says ..." as happens in other instances). The regular appearance of both in the same context raises interesting questions about the motives for such an approach. A compromise would be to assume that at times he uses both Theophrastus and Alexander, but that this does not preclude direct access to Theophrastus' works on nature, as well as access via authors other than Alexander (as my Appendix shows). Perry (1983: 252) believed that Diels' argument that Simplicius cites all Theophrastus via Alexander is convincing (with reference to DG 102–118, esp. 112–113). The evidence brought together in

---

64 A by-product of our survey will be that cases in which Theophrastus is said to "report"/ἱστορεῖ, we can exclude them from his own views and place them under the doxographical items, see Appendix, nos. 7–10.

65 Based on a TLG search.

my Appendix offers proof against this claim, given that Simplicius regularly references Theophrastus' *own writings* (often with mention of the book number) when quoting or paraphrasing him. To assume that all these came to him via Alexander would require a fuller argument for his role as the sole source.

And *thirdly*, the structure of Theophrastus' *Physika*, broadly comparable to Aristotle's *Physics*, seems to indicate a loyal stance towards his teacher in his own educational and research activities, yet does not also indicate ideological loyalty in all respects. While some statements in later sources suggest that he follows Aristotle (e.g., Boethius[66]) we have good evidence that a critical stance to his master was part of his approach (the example of place). Even in the Latin rhetorical tradition we find Quintilian stating *dissentire ab eo non timide solet* (694 FHS&G), which leaves little doubt as to Theophrastus' habit of disagreeing with his master—a habit we also see in the constant aporetic questioning of Aristotelian claims.

Clearly more work needs to be done to review *all* the references in Simplicius and other texts related to *Physics*, to test further the role of Alexander, and to answer questions surrounding Simplicius' approach and assess in full his value for a reconstruction of the lost *Physika*.[67]

Appendix: Theophrastus in Simplicius' Commentaries

A selection of passages in which Simplicius cites Theophrastus from direct access (*Physics*), or via authors other than Alexander.

Commentary on Aristotle's Categories **(3 of 7 passages from TLG / *CAG* VIII).**

1. 346.16 (refers to τὸ συμβεβηκός)
2. 402.20 (refers to change, μεταβολή)
3. 415.15 (refers to *Topika*)

Commentary on Aristotle's On the Heavens **(11 passages from TLG / *CAG* VII) transl. Mueller, Hankinson.**

4. 1.8 "Theophrastus in **his *On the Heaven*"** [brought in as a "witness"]

66 "Theophrastus, as is his custom in other works as well, when he is dealing with matters similar to those which have been dealt with by Aristotle … *touches lightly* on those matters which he knows have been said by Aristotle earlier but *follows up more diligently other things not dealt with by Aristotle*" (72A FHS&G = Boethius, *in de interpretatione* 1 Introd.)

67 I am grateful to the organisers and editors for their invitation. I also thank Claire Louguet (Lille) for the feedback I received on my paper.

5. 504.6 Simplicius quotes Sosigenes who mentions Aristotle and Theophrastus [= 165D FHS&G] (cf. G.L. Irby-Massy & P. Keyser, *Greek Science in the Hellenistic Era*, 2002, 80–81)
6. 553.4 **Theophrastus** in the **first book of his *Prior Analytics***
7. 564.24 **Theophrastus … in his *Physics*** recounts (on Democritus)
8. 576.14 **Theophrastus** ἱστορεῖ [not his own view ≈ 7] 564.24
9. 602.5 **Theophrastus** ἱστορεῖ on flame from the eye (*ekkrisis*) [not his own view] but cf. DL, V 46 περὶ ἐκκρίσεως α′ (p. 374, line 204 Dorandi)
10. 641.7 **Theophrastus** ἱστορεῖ [not his own view (Democritus mentioned) = 238 FHS&G (*in apparatu*)]
11. 700.6 "and Theophrastus in **his work *On the Generation of the Elements*** and Posidonius the Stoic taking from them using it widely" (καὶ ὁ Θεόφραστος ἐν τῷ περὶ τῆς τῶν στοιχείων γενέσεως καὶ Ποσειδώνιος ὁ Στωικὸς παρὰ τούτων λαβὼν πανταχοῦ χρῆται). Not in DL [nor in Wimmer] but perhaps the same as DL, V 23 (Aristotle) περὶ στοιχείων α′ β′ γ′

***Commentary on Aristotle's Physics* (23/26 from TLG occurrences/ *CAG* vols. IX–X).**

12. **9.7** (principles of natural things) (transl. Menn 2022a) "However, **Theophrastus at the start of his own *Physics* demonstrated even this**, saying 'But that there are principles of natural things is clear from the fact that natural bodies are compound, and that every compound has as principles the things out of which it is composed'."
13. **20.19** (natural principles and the senses) (transl. Menn 2022a) "… to put it in general terms, we must track down the truth about natural principles [starting] from the senses and sensible things, being persuaded also by Theophrastus who, investigating this **in the first book of his *Physics***, wrote as follows: 'since it is not possible to talk about even one thing without motion (for all things of nature are in motion), nor to talk about the things around the centre [of the cosmos] without alteration and qualitative change, with a view to these things and speaking about them it is not possible to do without sensation, but rather we must try to consider [these things] beginning from this [i.e., perception], either grasping the phenomena in themselves or [starting] from them, if perchance there are any principles stronger than them and prior to these'."
14. **107.12–16** [transl. Huby 2011, see her notes 26, 33, 36] (on Aristotle's criticism of Melissus that there is a starting point of everything): [a little before Simplicius cites Porphyry on the example of freezing, a standard example of events that come about "all at once"] "And **Theophrastus in the first book of his *On motion*** (περὶ κινήσεως) is clearly thinking the

same on this point. He says this: 'about the motion of the moving thing, and the [passive] motion of the thing that has been moved, one ought to say to—and this also applies to bodies and their alteration—that not always is the half first, but sometimes [it happens] **all at once**.' For all these reasons, it does not seem to have been a satisfactory assumption that everything that comes to be has a starting point with regard to the thing; so that the additional assumption which says: 'but what comes to be a starting point', is not true, and the proposition that comes from the conversion and says: 'what does not come to be, does not have a starting point', seems to be faulty, not only in the form of the conversion that comes from the antecedent, but in that it is also false itself, if again one were to take the starting point of the thing, and not of time."

15. **115.10–** (reference to his *Research concerning Nature*, reported *via* Alexander, followed by a quotation from Eudemus' *Physics*, where 115.21–25 = 118.11–13 = 121.13–16) "The argument of Parmenides, as Alexander reports, Theophrastus set around like this in the first book of **his *Research concerning Nature*** (ἱστορία φυσική, cf. no. 22, 149.32 and no. 23, 154.17 below): what is other than being is not: but also being is used in one sense only: therefore being is one. I cannot say if **Eudemus** discussed this anywhere else, as clearly as this, but **in his *Physics***, he wrote the following about Parmenides, and perhaps possible to derive the stated conclusion. 'Parmenides does not appear to prove that being as one, not even if anyone were to agree with him, that being is used in only one way, unless as in the 'what' category of each thing as man of men'."
16. **134.11–12**: "And as Theophrastus proposed: 'what is other than being, is not. What is not, is nothing'."
17. **149.32–150.4** [transl. C.C.W. Taylor] "**In his *Enquiry*** Theophrastus mentions **rarefaction and condensation** only in the case of Anaximenes, but it is clear that the others too [who belong to this first class] made use of **rarity and density**, since Aristotle says in general about all that 'they generate the rest by density and rarity making many things' (187a17–18) out of the one matter".[68]
18. **154.14–23** (**= 228B** FHS&G) [transl. C.C.W. Taylor] "Theophrastus assimilates Anaxagoras and Anaximander, understanding Anaxagoras' words as

68 N.B. Porphyry and Nicolaus of Damascus are cited earlier (149.17–18) as having attributed the unlimited to Diogenes of Apollonius (which is incorrect, given that we know from the fragments and Simplicius 149.7–8, 151.20–21 that he held that "the primal stuff was air", Taylor n. 13).

amounting to saying that the substrate is a single nature (μίαν αὐτὸν). **He writes as follows in his inquiry into nature**: if people were to understand him, that way, he would appear to be making the material principles, infinite, as has been said, but giving change incoming to be a single cause."

19. 166.17 Theophrastus in his second book on Anaxagoras ...
20. **412.35** Theophrastus in the **second book of his *On Motion*** ... [cf. also nos. 14, 25]
21. **412.31–413.4** [transl. J.O. Urmson 2002] (after a brief discussion of categories and change, Simplicius cautiously speaks up against Aristotle and uses Eudemus and Theophrastus to support this) "I know it seems rash to speak out contrarily to Aristotle, but until we are able to recognise accurately the explanation of this sort of classification, which he makes, we are sufficiently encouraged by the agreement of Eudemus quoted with regard to time [Fr. 59 Wimmer; cf. 411.17], and **still more by that of Theophrastus**, who clearly views change and transformation as being in all the categories. At any rate, he says in the **second book of his *On Motion*** that 'it is more appropriate as we say, and as is the case, that the actuality of the potentially changeable qua changeable should **reside in each of the other categories**, such as substance, quality, quantity, travel. For alteration, growth, travel, coming to be, and their opposites are so'."
22. **566.18–21 "It should be known that both Theophrastus and Eudemus presuppose** the immobility of place also in their assumptions. It will become clear why Aristotle left this out when he explains the difference between place and a vessel."
23. **604.5–11** Simplicius asks "Where is place properly?" (604.2) and brings in Theophrastus next: "It should be known that **Theophrastus also in his *Physics*** [= 146 FHS&G] raised difficulties (**ἐν τοῖς Φυσικοῖς ἀπορεῖ**) about the account of place given by Aristotle, ..."
24. **606.33–35** "Now, **Theophrastus and Eudemus themselves accept the immobility of place** also as a postulate, but Aristotle makes an addition to the postulate saying 'so that place is the immediate unmoved boundary of the container [*Physics*, 212a20]'." [same position as in 566.18–21]
25. **612.5–10** "How could a body and anything incorporeal be equal? For equality is found in quantities, moreover, in homogeneous quantities, as lines with lines surfaces, with surfaces and bodies with bodies. Therefore, place is a body if it is an interval. If it is a body, it is, and it is either mobile or immobile. But if mobile it must necessarily change its place so that place will over again require replaced. But that is impossible, as **Theophrastus** and, indeed, Aristotle agree" (ὡς καὶ Θεοφράστῳ δοκεῖ καὶ Ἀριστοτέλει μέντοι.) [= 148 FHS&G].

26. **639.13–22** [= 149 FHS&G, cf. Sorabji, Sourcebook, vol. 2, T13(e), pp. 235–237] "Now that I have set out this concept of place so far as I am able, and now that I have set out the difficulties raised against it, and adduced the resolution of the difficulties, I wish to show that the concept is neither novel nor unknown to famous philosophers. Thus, **Theophrastus appears to have held this position in his *Physics*, where he says as dealing with a problem:** 'perhaps, place is not a substance in itself, but is predicated in relation to the order and position of bodies, etc.'"
27. **700.11–21** "By the ingenious proofs brought forward by Aristotle that we have discussed, he showed that the existence of time is unclear since the arguments, for its nonexistence are so persuasive. He next says that one will find both the definition of time and its nature equally unclear, if one tries to examine them in the life of tradition, as its existence seemed in the discussions which we previously went through. So its definition is unclear also, given some say that time is the motion and revolution of the hall light Plato, according to you, Dimas, defrosted, and Alexander, some say that it is the fear of the heavens itself, as those Pythagoreans are reputed to say, who perhaps, misunderstood Archytas who said generally that time was an interval in the nature of the whole …".
28. **788.34–789.2** (twice) "We have already stated the view of Plato and Aristotle about time. **Theophrastus** and Eudemus, the colleagues of Aristotle, seem to have believed and taught the same view of time. But Strato of Lampsacus criticised the definition of time given by Aristotle and his colleagues, and although he was himself a pupil of **Theophrastus**, who followed Aristotle on almost everything, he followed a newer path." (n. 41: 151B FHS&G).
29. **860.19; 27**; [transl. Baltussen] (*In Phys.* V 2 on motion mentions "Eudemus' own *Physics*" at 860.9 as paraphrasing Aristotle, with a quotation, παραφράζων καὶ εἰπὼν); **at 860.19 Theophrastus** is quoted **(φησι γράφων οὕτως) from "the first book of his *Physics*". He writes:** "concerning change/motion it is not difficult to provide the general and common account, since it is a kind of incomplete activity of potential being … **A little further along Theophrastus says** 'we must investigate as to changes/motions, whether some are generations, others more like realisations'."
30. 923.10; 16; *In Phys.* VI 1: Simplicius opens with comments about the Peripatetic habit of organising books by giving them letters, noting that therefore the sixth book was probably marked with "zeta". He also notes that the first 5 books were called *Physika*, the next three *On Motions/Change*. At lines 15–16 he mentions that **Theophrastus regarded (νομίζει) book V as part of *Physika*.** Simplicius further argues why the next three books are

on change, adducing *De caelo* I as supporting evidence (Diels suggests *in app.* that this refers to DC I 5 272a28).

31. 964.29–965.5, *In Phys.* VI 4: **Theophrastus in the first book of his *On Change/Motion*** (Q): “the appetites and desires and bodily drives are motions and have their starting point from that, but however many decisions and thoughts” [NB A little later he quotes from Strato’s *On Motion* at 956.10].
32. 986.5 ff. Theophrastus “in the first book of his On Change/Motion” (and quotation).
33. 998.13 “**Theophrastus** seems to puzzle over the question whether all change is in time, suspecting perhaps changes from darkness to light, whenever, after a lamp has been brought into a home, the house is suddenly filled with light in a split second”.
34. 1236.1 “**Theophrastus** divides in the **third book of the *Physika*** (that is, his *On the Heaven*) …”

CHAPTER 2

Was Theophrastus in Assos? New Insights on Theophrastus' *De Igne* and the Debates about the Elements in the Academy

Mai-Lan Boureau

While genealogical approaches are sometimes dismissed as overly convenient shortcuts that evade systematic analysis,[1] this can scarcely be said of Konrad Gaiser's hypothesis in his 1985 study *Theophrastus in Assos: Zur Entwicklung der Naturwissenschaft zwischen Akademie und Peripatos*. For, at first glance, his attempt to associate Theophrastus' *De igne* with the period between 347 and 345 BCE, when Aristotle and other leading Academics took refuge at the court of Hermias of Atarneus appears nothing short of paradoxical. The seemingly prosaic concerns addressed by the treatise, such as why "the belly melts coins",[2] seem to be at odds with the intellectual character ascribed to this phase of Aristotelian thought in Werner Jaeger's seminal book *Aristoteles: Grundlegung einer Geschichte seiner Entwicklung*.[3] For Jaeger, the Assos period was one of intellectual flourishing, characterised by the discussion of abstract principles and their connection to the physical realm. This atmosphere stands in stark contrast to the widespread interpretation of Theophrastus' *De igne* as a rejection of metaphysical reasoning in favour of a purely physical system, an exegetical tradition from which Gaiser decisively departs.[4] The present article sets aside broader questions about the validity of the genealogical approach, focusing instead on

1 The most systematic presentation of this view is found in Rapp (2019). In Boureau (Forthcoming), I argue that while this method has sometimes been applied in an insufficiently rigorous way, genealogical approaches can be employed with greater nuance and need not be incompatible with a systematic interpretation of Aristotle's philosophical thought.

2 Theophrastus, *De igne*, § 45.

3 I am quoting here from Robinson's 1948 English translation of this book, which was revised and expanded by the author.

4 This interpretation is spelled out in Battegazzore (1984), 48: "La caratteristica saliente del *De igne* risulta essere l'assorbimento di ogni traccia metafisica in un sistema esclusivamente fisico e concreto: i quattro tradizionali elementi non sono più considerati i principi ultimi e teorici della realtà, ma corpi (σώματα) percettibili e tangibili: è evitata accuratamente la parola στοιχεῖον, sostituita dal termine ἁπλᾶ (con il quale Aristotele aveva qualificato gli elementi in ambito fisico)."

 DOI:10.1163/9789004739314_004

whether there are any heuristic benefits to be gained from drawing a connection between Theophrastus' *De igne* and Aristotle's time in Assos. According to Jaeger's division of Aristotle's philosophical career, the Assos period is typically linked to an internal Academic debate about the elementary principles. If we follow Jaeger's reconstruction, the lost dialogue *De philosophia*,[5] Book A of the *Metaphysics*[6] and the first two books of *De caelo*[7] can be attributed to Aristotle's time in Assos, which belongs to "the most productive epoch in Aristotle's life".[8] Within this framework, it is possible to identify three paradigmatic features:

1. Aristotelianism was not yet fully formed, but emerged from a complex relationship with Platonism. At this stage, the break had not yet been fully consummated through the establishment of a rival school explicitly acknowledging this rupture.[9] The *Elegy to Eudemus*, which includes a tribute to Plato, encapsulates the ambivalence of this period.
2. The focus was on critiquing Platonic ideas rather than elaborating a new philosophical school.[10] For this reason, Aristotle devoted special attention to Plato's critique of the Presocratics and assigned a prominent place to a

---

5 Jaeger (1948), 124.

6 Jaeger (1948), 128 and 172.

7 See Jaeger (1948), 306 n. 1: "We must therefore suppose that the work *On the Heaven* mainly arose, or that the first draft of it was sketched out, during his middle period, and that revisions, some of them drastic, took place during his later years." See also Jaeger (1948), 306: "It was written after the dialogue *On Philosophy*, and therefore at the earliest one or two years after Plato's death. It was probably not much later than this, however, for the whole point of view is that of the later Academy."

8 Jaeger (1948), 124.

9 See Jaeger (1948), 124–125: "At the same time, everything points to the conclusion that his close affinity with Plato and Plato's preoccupations continued undisturbed. He went on living and teaching among Plato's pupils. As we have seen, his departure from the school at Athens was in no sense a break with the Academic community as such [...]. We discover that between the early, dogmatically Platonic, stage of his development, and the final form of his thought in its maturity, there was a period of transition whose nature can be definitely ascertained in many particulars, a period when he was criticizing, rearranging, and detaching himself, a period, previously wholly overlooked, which was clearly distinct from the final form of his philosophy, although it reveals the entelechy of the latter in all essential points."

10 Jaeger (1948), 172: "Thus the original form of the criticism presupposes a group of *Platonic* philosophers, for whom Aristotle once more sums up, in a rapid survey, all those objections to the doctrine of the now dead master that had occupied the Academy in the course of the years, in order to infer the necessity for a complete reorganization of Platonism on the basis of these criticisms." See also Jaeger (1948), 173, for the composition of this group, which included "Plato's best-known disciples, men who were either objective enough to listen to an opponent's reasons, like Xenocrates, or themselves full of doubt about Plato's doctrine, as Erastus, Coriscus, and their convert Hermias, seem to have been."

discussion of the *Timaeus*.[11] The latter work was read as Plato's attempt to propose a new type of physics that addressed the shortcomings of Presocratic enquiries into nature, which purported to explain phenomena without reference to the soul.

3. The critique of Platonic physics formed part of a broader reflection on the principles and their connection with sensible phenomena. The study of these principles did not necessarily take the form of a theology or an attempt to arrive at knowledge of divine beings, but rather presented itself as the essential stage in a new type of *theoria*.

The Assos period thus emerges as a time of intellectual flourishing, in which the evolution of Peripatetic thought was shaped more by common lines of investigation than by direct opposition to Academic doctrines. Although it is still uncertain whether Theophrastus was actually present at Assos, recent scholarship has reassessed his *Metaphysics* in a way that is entirely consistent with the intellectual climate of Assos, sometimes even explicitly mentioning Assos.[12] Rather than viewing Theophrastus' *Metaphysics* as a post-Aristotelian work or merely a critique of Book Λ of Aristotle's *Metaphysics*, scholars now interpret it as a witness to the collective enquiries of the early Peripatos that addressed questions inherited from the Academy.[13] These studies suggest that Theophrastus' engagement with Aristotle in his *Metaphysics* reflects a background of shared enquiry and debate about the nature of the prime mover,

---

11 While Jaeger (1948), 303, acknowledges that the enquiry into the eternity of the world in the first book of *De caelo* and Aristotle's *De philosophia* constitutes "a running polemic against Plato's *Timaeus*, which is explicitly mentioned", he places greater emphasis on the internal discussions within the Academy concerning physical matters than on direct engagement with the *Timaeus* itself. Jaeger (1948), 300: "nevertheless, the disposition of the books *On the Heaven* does at any rate enable us to observe how Aristotle's cosmology arose out of Plato's. We know the latter directly only from the *Timaeus*, behind which lie the far-reaching Pythagorean speculations of the school, and it is therefore very important that this work of Aristotle's allows us a glimpse of the discussions that went on about the subject in the Academy." However, it seems to me that the discussion of the treatise was more influential than Jaeger suggests.

12 See Gutas (2010), 8: "These considerations make it certainly possible, and very much probable, that the Essay may have been written in Assos, giving us 347 as the earliest date of composition. Accordingly, it can be maintained with as much certainty as is possible in such cases that Theophrastus's *On First Principles* was written in the period between 347 and 334."

13 See, e.g., Gutas (2010), 31: "Furthermore, if one keeps the date of composition of the Essay in mind, and that it was written at the time when Aristotle himself was grappling with the problems discussed in it and had composed book Lambda, definitely, and possibly also Alpha Meizon and Beta on the very subject, it becomes apparent how naturally Theophrastus's Essay fits in their context and continues the discussion initiated in them."

thus highlighting the existence of an ongoing philosophical dialogue, rather than the assertion of a dogmatic alternative.[14] Such a philosophical endeavour appears wholly consonant with the intellectual spirit of Assos.

That said, to what extent can this perspective be extended to the study of Theophrastus' *De igne*? A pivotal development in addressing this question has been the reassessment of *De igne*'s supposedly anti-Aristotelian outlook. The traditional interpretation, established by scholars such as Eduard Zeller, emphasised the fundamental compatibility between Theophrastus' and Aristotle's physical theories, with Zeller highlighting their shared views about the nature of fire.[15] This claim received support from Otto Regenbogen in his *Pauly-Wissowa* entry, in which he argued that the originality of *De igne* did not lie in its theses, many of which could be traced back to Aristotle. This interpretation was famously challenged by Peter Steinmetz, whose 1964 monograph emphasised the independence and originality of Theophrastus' physics. Steinmetz argued that the divergence from Aristotelian doctrine was, in reality, more significant, crediting Theophrastus with a methodological shift and pointing to a decisive break with core Aristotelian principles—most notably, through what he interprets as a rejection of the fifth element and a turn toward atomistic theory.[16] More recent scholarship has, however, reassessed the extent of this divergence, concluding that, despite its critical stance, *De igne* remains deeply embedded in Aristotelian thought, and potentially even in its treatment of aether.[17]

14 Lefebvre (2015), 37–38: "Théophraste avait manifestement une problématique propre sur la question des principes; son orientation générale est différente de celle d'Aristote, même si elle utilise en partie une même matière."

15 Zeller (1963) II/1, 832–833: "In der Lehre von den Elementen, welcher die uns erhaltene Abhandlung über das Feuer angehört, hielt Theophrast zwar die aristotelische Grundlage fest, aber doch fand er auch hier Schwierigkeiten."

16 Steinmetz's arguments did not, however, result in a lasting consensus. For example, in his 1967 review, Hans Gottschalk underscored the inadequacy of the evidence supporting Steinmetz's claims. In contrast, James Longrigg found that Steinmetz "argue[d] persuasively" with regard to the differences between Aristotle's and Theophrastus' conceptions of fire and elemental bodies (Longrigg 1966, 179).

17 The issue remains a matter of ongoing debate. For instance, Marlein van Raalte (2010), 48, argues that it is "very unlikely" that "Theophrastus adhered to the Aristotelian concept of aether". However, as István Bodnár (2002b), 76, points out, even if we grant that Theophrastus considered the matter of the heavens to be "fire", it would be "fire in an etiolated sense only", merely "a more mundane variant of the Aristotelian aether". Similarly, David Lefebvre (2017), 61 notes that Aristotle himself may at some point have accepted that the heavens were filled with a kind of "fire" possessing properties distinct from ordinary fire. He also contends (Lefebvre 2017, 76) that there is no major doctrinal divergence between Theophrastus and Aristotle concerning the nature of fire.

My own view is that rather than reflecting a turning away from enquiries into first principles on the part of Theophrastus, *De igne* was deeply shaped by the methods and research traditions of the Academy. In the absence of new historical evidence, this paper does not focus on whether Theophrastus was actually present at Assos, but instead examines the extent to which *De igne* reflects the intellectual framework of Assos. This is not a matter of merely historical interest, since it offers new insights into the methods and dynamics of the early Peripatetic school and its complex relationship with the Academy.[18] The specific claim I make here is that *De igne* bears the imprint of three distinct influences: (1) Plato's *Timaeus*;[19] (2) Aristotle's critique of Plato's account of fire and heat, in which he sought to address its alleged explanatory limitations and internal contradictions; and (3) the challenges that emerged from Aristotle's own attempts to resolve these issues, particularly when it came to reconciling the properties of fire with a coherent and well-founded theory of the elements. On this hypothesis, *De igne* is informed by the philosophical problems connected to the nature and priority of the elements within a closed system that is not sustained by transcendent principles rather than by Theophrastus' supposed lack of interest in higher principles. The question is: if an element is considered ontologically prior, how can it simultaneously possess different qualities? Theophrastus' original solution to this problem opens up a new perspective on the difficulties Aristotle faced in conceptualising the relationship between heat and fire, as well as in articulating the different definitions of fire. While it is bold to suggest that, because Theophrastus' solution avoids some of the problems faced by Aristotle, his characterisation of fire was intended specifically to circumvent them, many methodological questions are explicitly raised by Theophrastus himself.

In order to bring out the coherence of *De igne*—a work frequently considered an unstructured compilation of notes[20]—I will begin by reconstructing

18 Jaeger (1948), 172: "The bereaved school is now standing at a decisive turning-point of its career. Outside of Athens, which he very soon left, Aristotle was surrounded by such a group of Platonists, after Plato's death, only in Assos, and then never again."

19 On Theophrastus' acquaintance with the *Timaeus*, see the hypothesis advanced by Han Baltussen (2000), 131–132, to the effect that Theophrastus' remarks in the *De sensibus* are based not on the full text but partly on a summary. See also Baltussen (2003), 56, which introduces a further assumption that Theophrastus "used a summary version as a working manuscript for personal use only alongside the 'complete' *Timaeus* text".

20 See, e.g., Coutant (1971), IX: "It seems probable that the work is a collection of lecture notes rather than an organized presentation in the sense that the works on botany are. First and foremost, the work does not have any central idea pervading it but is rather a medley of sub-topics occurring disjointedly and with an occasional reprise." See also Battegazzore

its Academic and Aristotelian background. I will then examine how Theophrastus' characterisation of fire navigates challenges arising in both traditions. Finally, I will argue that although Theophrastus' account successfully explains phenomena such as heat, it does so at the expense of considerations about the nature of the elements. I will thus provide a new perspective on Theophrastus' apparent reluctance to conduct a comprehensive enquiry into the first principles and reframe his focus on fiery phenomena, arguing that it is not a manifestation of his interest in trivial observations, but the outcome of a deeper reflection on physical theory. I suggest that *De igne* reflects Theophrastus' recognition of an implicit tension between explaining observable phenomena and establishing the coherence of a self-contained physical system, arguing that his work prioritises the former—a methodological choice that sets him apart from Aristotle.

1 The Enquiry into First Principles and Theophrastus' Study of Fire

Gaiser's hypothesis that Theophrastus was part of the Assos circle rested on two heterogeneous arguments. The first was a novel interpretation of § 46 of Theophrastus' *De igne*, particularly the phrase ὁ δὲ ἐν κύκλῳ λίθος, which Gaiser linked to a well-known stone from Assos also mentioned in Pliny the Elder's *Natural History* (II 211 and XXXVI 131).[21] Gaiser argued that this stone, described as "being around us" (ἐν κύκλῳ), provided explicit evidence that § 46 contained a reference to Assos,[22] that Theophrastus had travelled there with other former members of the Academy and that he wrote *De igne* during his stay. Despite the impressive convergence of philological, historical and geological considerations, most scholars have found this evidence too tenuous to substantiate Gaiser's claim that *De igne* was composed entirely during Theophrastus' time in Assos.[23] Gaiser's second argument is more directly relevant to the topic of this paper, which views "Assos" as a method rather than a specific period. In addition to his analysis of the stone of Assos, Gaiser emphasised that an enquiry into first principles underlies *De igne* as a "problem horizon",[24] particularly in the first ten paragraphs of the text. On two occasions, the introduction

(1984), 46: "Va peraltro subito precisato che il De igne, quale a noi è pervenuto, non è una trattazione organica, e non aspira a nessuna completezza, ma è viziato da un'intrinseca desultorietà: il fatto stesso che i riferimenti e le allusioni alle teorie di precedenti filosofi siano sporadici e sfumati."

21 Gaiser (1985), 28–35.

22 Gaiser (1985), 32–35.

23 See, e.g., Engberg-Pederson (1987).

24 Gaiser (1985), 41.

to *De igne* explicitly refers to an enquiry into higher (μείζω) and first principles, namely in paragraphs 7 and 10.[25] This suggests that the study of fire is closely linked to another form of *theoria* that is concerned with principles and causes—an interpretation that has been widely acknowledged by commentators on *De igne* at least since Zeller. A connection between the study of fire and the enquiry into first principles is established already in the opening sentence of the treatise, albeit without explicitly defining the framework within which the discussion unfolds: "The nature of fire has the most peculiar powers among the simple bodies."[26]

Although the treatise appears to assume the existence of simple bodies, this premise seems to be more of a commonly accepted idea than a demonstrated thesis. From the outset, then, we are confronted with an important ambiguity regarding the status and scope of *De igne*: does the enquiry start from the first principles and move towards the study of fire, or does it proceed from the study of fire towards the first principles? If we accept the latter hypothesis, the investigation of first principles would serve merely as the background against which the main argument takes shape. By contrast, the former hypothesis—namely that the study of fire is undertaken within the context of a broader enquiry into the first principles of nature—appears more consistent with the research conducted in Assos. Moreover, this reading is supported by the phrasing and rhythm of the opening sentence, which underscores the singularity of fire among the so-called "simple bodies". The discussion of simple bodies fits naturally within an Academic research programme, as outlined in the *Protrepticus*, which established a direct relationship of dependence between knowledge of the whole and knowledge of the first principles:

> For whether it is fire or air or number or any other natures that are the causes and principles of other things, if we are ignorant of them we cannot know any of the other things; for how could one recognise speech if one did not know the syllables, or know these if one knew none of the letters (μηδὲν τῶν στοιχείων)?[27]

---

25 Theophrastus, *De igne* § 7, 3–4: ἀλλὰ γὰρ ταῦτα ἔοικεν εἰς μείζω τινὰ σκέψιν ἐκφέρειν ἡμᾶς τῶν ὑποκειμένων, ἣ ζητεῖ τὰς πρώτας αἰτίας. See also § 10, 1–2: ἀφέντας δ' οὖν τὰ μείζω καὶ τὰ πρότερον πειρατέον ὑπὲρ τῶν ἐλαττόνων λέγειν ἀρξαμένους ἀπὸ τῆς γενέσεως αὐτοῦ καὶ φθορᾶς.

26 Theophrastus, *De igne* § 1, 1, transl. Van Raalte: Ἡ τοῦ πυρὸς φύσις ἰδιαιτάτας ἔχει δυνάμεις τῶν ἁπλῶν.

27 Aristotle, *Protrepticus* (Iamblichus, *Protrepticus* VI, 39, 15–19 = Fr. 52 Rose = Fr. 5 Ross, Walter = Fr. 36 Düring), transl. Barnes & Lawrence: Εἴτε γὰρ πῦρ εἴτ' ἀὴρ εἴτ' ἀριθμὸς εἴτ' ἄλλαι τινὲς φύσεις αἰτίαι καὶ πρῶται τῶν ἄλλων, ἀδύνατον τῶν ἄλλων τι γιγνώσκειν ἐκείνας ἀγνοοῦντας· πῶς γὰρ ἄν τις ἢ λόγον γνωρίζοι συλλαβὰς ἀγνοῶν, ἢ ταύτας ἐπίσταιτο μηδὲν τῶν στοιχείων εἰδώς.

In this passage, fire is mentioned as a possible first principle alongside other natural bodies, such as air. While the focus on physical bodies might suggest an Ionian perspective that seeks to identify which physical body underlies all things, the inclusion of number as a potential first principle reveals the Academic framework of the text. This debate about first principles is echoed in Plato's *Timaeus*:

> We must, in fact, consider in itself the nature of fire and water, air and earth, before the generation of the Heaven, and their condition before the Heaven was. For to this day no one has explained their generation, but we speak as if men knew what fire and each of the others is, positing them as original principles, elements (as it were, letters) of the universe; whereas one who has ever so little intelligence should not rank them in this analogy even so low as syllables.[28]

The parallel with the *Protrepticus* is striking: both texts implicitly compare hierarchically arranged principles to letters and syllables, and they both consider fire as a possible primary explanatory principle. However, the *Timaeus* goes on to reject this idea, suggesting that fire cannot serve as a first principle, as it is not even comparable to a syllable, let alone a letter. The implied reason for this is its inherent obscurity, as indicated by the remark: "as if men knew what fire and each of the others is", echoing *Timaeus* 49B ("it is hard to say, with respect to any one of these, which we ought to call really water rather than fire").[29] While the *Timaeus* explicitly dismisses the notion that fire could be a first principle, this discussion suggests that the hypothesis was debated, indicating that Theophrastus' emphasis on the specificity of fire in *De igne* extends beyond a mere Heraclitean reference. Furthermore, the *Timaeus* conducts its enquiry into the first principles within a generative framework that emphasises the question of the origin of the elements and the broader issue of how all things could emerge from them.

Theophrastus' *Metaphysics* testifies to both his interest in enquiries into first principles and his engagement with the *Timaeus*:

---

28 Plato, *Timaeus* 48b–c, transl. Cornford: τὴν δὴ πρὸ τῆς οὐρανοῦ γενέσεως πυρὸς ὕδατός τε καὶ ἀέρος καὶ γῆς φύσιν θεατέον αὐτὴν καὶ τὰ πρὸ τούτου πάθη· νῦν γὰρ οὐδείς πω γένεσιν αὐτῶν μεμήνυκεν, ἀλλ' ὡς εἰδόσιν πῦρ ὅτι ποτέ ἐστιν καὶ ἕκαστον αὐτῶν λέγομεν ἀρχὰς αὐτὰ τιθέμενοι στοιχεῖα τοῦ παντός, προσῆκον αὐτοῖς οὐδ' ἂν ὡς ἐν συλλαβῆς εἴδεσιν μόνον εἰκότως ὑπὸ τοῦ καὶ βραχὺ φρονοῦντος ἀπεικασθῆναι.

29 Plato, *Timaeus* 49b, transl. Cornford: τούτων γὰρ εἰπεῖν ἕκαστον ὁποῖον ὄντως ὕδωρ χρὴ λέγειν μᾶλλον ἢ πῦρ.

> Perhaps one might raise the *aporia*, how and of what sort one should at all suppose the first principles to be: whether shapeless and, as it were, potent (as do those who posit fire and earth), or already shaped, on the ground that it is necessary most of all for these to be determined, as he says in the *Timaeus*.[30]

Here, Theophrastus contrasts two possible approaches to defining first principles: we can do so either through shapes, as in the *Timaeus*—where triangles serve as the fundamental constituents of the elements—or through powers (δυναμικάς), such as the capacity to heat or dry. The reference to fire and earth suggests a link to their primary qualities, which emphasises a fundamental methodological question: should the first principles be defined through forms or qualities? This was also a pressing issue for Aristotle, whose *De caelo* and *De generatione et corruptione* offer alternative responses. In *De caelo*, the first principles are defined through their motions—with "form" being understood not as geometric shape, but rather in the broader Aristotelian sense of substantial form.[31] Conversely, *De generatione et corruptione* approaches first principles through their primary qualities, constructing the physical system of bodies through a combinatory framework involving four fundamental qualities. Theophrastus' *De igne* navigates this choice by weighing the primacy of heat—understood as a power—against fire as a simple body:

> For the heat and the power of heat are a principle. For heat is more common, reaches further, and exerts more influence on qualitative change and natural generation, whereas fire, as has been said, is productive of itself, but destructive of other things in most cases. From this it will also be clear that the nature of fire is different from the nature of heat.[32]

30 Theophrastus, *Metaphysics* 14, 6b23–27, transl. Gutas: Πῶς δέ ποτε χρὴ καὶ ποίας τὰς ἀρχὰς ὑποθέσθαι, τάχ᾽ ἂν ἀπορήσειέν τις, πότερον ἀμόρφους καὶ οἷον δυναμικάς, ὥσπερ ὅσοι πῦρ καὶ γῆν, ἢ μεμορφωμένας, ὡς μάλιστα δέον ταύτας ὡρίσθαι, καθάπερ ἐν τῷ Τιμαίῳ φησίν·

31 Aristotle, *De caelo* IV 3, 310a31–b1, transl. Stocks: "Now, that which produces upward and downward movement is that which produces weight and lightness, and that which is moved is that which is potentially heavy and light, and the movement of each body to its own place is motion towards its own form."

32 Theophrastus, *De igne* § 6, transl. Van Raalte: τὸ γὰρ θερμὸν καὶ ἡ τούτου δύναμις ἀρχή. κοινότερον γὰρ τοῦτο καὶ ἐπὶ πλείω διήκει καὶ κυριώτερον ἀλλοιώσεως καὶ φυσικῆς γενέσεως, τὸ δὲ πῦρ, ὡς εἴρηται, γεννητικὸν μὲν αὑτοῦ, φθαρτικὸν δὲ ὡς ἐπὶ πᾶν τῶν ἄλλων. ὅθεν καὶ δῆλον ὡς ἑτέρα τις φύσις πυρὸς καὶ θερμοῦ.

As a general rule, Theophrastus' account of fire for the most part echoes Aristotelian theses. For example, as already highlighted by Eduard Zeller,[33] § 26 of *De igne* states that "fire is hot and dry", directly paralleling Aristotle's definition of fire in *De generatione et corruptione* II 3, 330b3. Moreover, in the opening lines of *De igne*, Theophrastus connects his description of fire as a self-generating body to Aristotle's thesis that a smaller fire is consumed by a larger one:

> Fire, however, is of such a nature as to generate and destroy itself, the lesser fire to generate the greater, and the greater fire to destroy the lesser.[34]

While the thesis that the smaller fire generates the larger one directly follows from Theophrastus' definition of fire as a burning process, the second part of the sentence, which states that a smaller fire is destroyed by a larger fire, reflects a core Aristotelian principle. Not only is this thesis consistently repeated and sometimes attributed to earlier thinkers,[35] but it also plays a crucial role in Aristotle's theory of the preservation of life, particularly in relation to the need for a cooling principle.[36] Although Theophrastus largely upholds Aristotelian theses from the outset of *De igne*, the way in which he frames his enquiry reflects his originality as a thinker, a point noted by Regenbogen, who examined the specific ways in which these questions are posed and interconnected.[37] In addition to assessing the primacy of heat or fire as a first principle, the opening of *De igne* raises the question of how fire can be considered a first principle if it always requires fuel. Theophrastus also explores the relationship of fire to the heat of the sun and the warmth found in living beings, an issue likewise addressed by Aristotle:

---

33 Zeller (1963) II/1, 833.

34 Theophrastus, *De igne* § 1, 3–4, transl. Van Raalte: τὸ δὲ πῦρ γεννᾶν καὶ φθείρειν πέφυκεν αὑτό, γεννᾶν μὲν τὸ ἔλαττον τὸ πλέον, φθείρειν δὲ τὸ πλέον τὸ ἔλαττον.

35 Aristotle, *De generatione et corruptione* I 7.

36 On the significance of this principle in Aristotle's philosophy of life, see Rashed (2005), 133 n. 1. For a comparison of the role of this thesis in Aristotle's and Theophrastus' accounts of fire, see Federspiel (2017), 346–347. Notably, Federspiel called attention to Fr. 345 FHS&G, drawn from *On Fainting*, which illustrates the integration of this thesis into Theophrastus' philosophy of life: "fainting is a loss or chilling of the warmth in the region where respiration occurs. This is brought about either by cold or even by warmth itself; for a larger fire extinguishes a smaller one."

37 Regenbogen (1940), 1420: "Mit Recht betont Gercke 31 gegenüber Zeller II 2^3, 832 ff. die eindringende Selbständigkeit dieser Fragestellungen gegenüber Aristoteles."

> This is not fire nor any such force, but it is the breath included in the semen and the foam-like, and the natural principle in the breath, being analogous to the element of the stars. Hence, whereas fire generates no animal and we do not find any living thing forming in either solids or liquids under the influence of fire, the heat of the sun and that of animals does generate them. Not only is this true of the heat that works through the semen, but whatever other residue of the animal nature there may be, this also has still a vital principle in it. From such considerations it is clear that the heat in animals neither is fire nor derives its origin from fire.[38]

However, *De igne* does not clearly establish a connection between fire and heat, or between fire and the sun—it raises these issues without offering a definitive resolution.[39] Consequently, even if one acknowledges that a shared enquiry into the physical first principles within the Academy formed part of the intellectual backdrop, Theophrastus deliberately distances himself from this line of questioning from the outset. While the frequent references to first causes in *De igne* suggest that such an enquiry is acknowledged as a possible avenue of research, the treatise itself remains primarily concerned with the empirical manifestations of fire—a topic that Theophrastus explicitly differentiates from "an inquiry into first causes":

> But this seems to lead us off (ἐκφέρειν) into an inquiry of greater scope than the present one, an inquiry into first causes.[40]

Thus, while Theophrastus acknowledges the relevance of studying the first causes, he consistently sets this pursuit aside, considering it to lie beyond the scope of his enquiry.[41] Nonetheless, I argue that the Academy's approach to

38 Aristotle, *De generatione animalium* III 4, 736b35–737a7, transl. Platt: τοῦτο δ' οὐ πῦρ οὐδὲ τοιαύτη δύναμίς ἐστιν ἀλλὰ τὸ ἐμπεριλαμβανόμενον ἐν τῷ σπέρματι καὶ ἐν τῷ ἀφρώδει πνεῦμα καὶ ἡ ἐν τῷ πνεύματι φύσις, ἀνάλογον οὖσα τῷ τῶν ἄστρων στοιχείῳ. διὸ πῦρ μὲν οὐθὲν γεννᾷ ζῷον, οὐδὲ φαίνεται συνιστάμενον ἐν πυρουμένοις οὔτ' ἐν ὑγροῖς οὔτ' ἐν ξηροῖς οὐθέν· ἡ δὲ τοῦ ἡλίου θερμότης καὶ ἡ τῶν ζῴων οὐ μόνον ἡ διὰ τοῦ σπέρματος, ἀλλὰ κἄν τι περίττωμα τύχῃ τῆς φύσεως ὂν ἕτερον, ὅμως ἔχει καὶ τοῦτο ζωτικὴν ἀρχήν. ὅτι μὲν οὖν ἡ ἐν τοῖς ζῴοις θερμότης οὔτε πῦρ οὔτε ἀπὸ πυρὸς ἔχει τὴν ἀρχὴν ἐκ τῶν τοιούτων ἐστὶ φανερόν.

39 On Theophrastus' circular aporetic approach to fire and heat in *De igne*, and the philosophical gain it yields, see Lefebvre (2017), 79–83.

40 Theophrastus, *De igne* §7, 3–4, transl. Coutant: ἀλλὰ γὰρ ταῦτα ἔοικεν εἰς μείζω τινὰ σκέψιν ἐκφέρειν ἡμᾶς τῶν ὑποκειμένων, ἣ ζητεῖ τὰς πρώτας αἰτίας.

41 These observations derive in part from Fabienne Baghdassarian's response to my paper. Any errors are my own.

first causes is not simply dismissed in §7, but continues to shape *De igne* throughout, particularly in Theophrastus' characterisation of fire as a self-generating body. While this characterisation does not establish fire as a primary principle—although it is notable that no other simple body holds a stronger claim—it adeptly navigates the challenges posed by both Plato's and Aristotle's accounts of elemental fire.

2 Theophrastus' Original Characterisation of Fire and Its Philosophical Stakes

The aim of this section is to highlight the Platonic and Aristotelian background of Theophrastus' characterisation of fire, in order to shed light on its philosophical stakes. This Academic background is reflected not only in the theses that are defended in *De igne*, but also in the specific *aporiai* it seems to address. For example, Gaiser traces the three forms of fire that appear in *De igne*—flame (φλόξ), ember (ἄνθραξ) and light (φῶς)—back to a Platonic tradition,[42] as they are mentioned in the *Timaeus* to illustrate the "several varieties" (πολλὰ γένη) of fire:

> Next we must observe that there are several varieties of fire: flame; that effluence from flame which does not burn but gives light to the eyes; and what is left of fire in glowing embers when flame is quenched.[43]

Likely drawing on the same tradition, Aristotle also refers to light, flame and charcoal as three species of fire in *Topics* V 5.[44] While continuing to acknowledge this tripartition, Theophrastus notably invokes it to underline the need for a substrate:

42 Gaiser (1985), 51.

43 *Timaeus* 58c–d, transl. Cornford: Μετὰ δὴ ταῦτα δεῖ νοεῖν ὅτι πυρός τε γένη πολλὰ γέγονεν, οἷον φλὸξ τό τε ἀπὸ τῆς φλογὸς ἀπιόν, ὃ κάει μὲν οὔ, φῶς δὲ τοῖς ὄμμασιν παρέχει, τό τε φλογὸς ἀποσβεσθείσης ἐν τοῖς διαπύροις καταλειπόμενον αὐτοῦ.

44 *Topics* V 5, 134b25–30, transl. W.A. Pickard-Cambridge: "Sometimes, too, a man may even add the word 'specifically', and still make a mistake. For the things in question should all be of one species, whenever the word 'specifically' is added; and in some cases this does not occur, as it does not, in fact, in the case of fire. For fire is not all of one species; for live coals (ἄνθραξ) and flame (φλόξ) and light (φῶς) are each of them fire, but are of different species."

> Fire is in a substrate—which in any case is clear to our perception of the world around us, whether light should or should not be reckoned to be included in it. For if light also [should be considered as a kind of fire], it is obvious that this exists in air and water in any case; if not, the fire of flame and of charcoal in any case are in a substrate.[45]

Fire is thus described as a dynamic process, characterised by its unique ability to generate itself—a δύναμις that sets it apart from the other simple bodies:

> The nature of fire has the most peculiar powers among the simple bodies. For air, water, and earth perform their natural changes only into one another, while none is itself generative of itself; fire, however, is of such a nature as to generate and destroy itself, the lesser fire to generate the greater, and the greater fire to destroy the lesser.[46]

Theophrastus' description notably avoids the two alternatives outlined in his *Metaphysics* (6b23–27, see *supra*): i.e. defining a first principle either by (i) assigning it a specific geometric shape, as in the *Timaeus*, or by (ii) attributing to it a distinctive power, such as the capacity to heat, a move that can lead to conceiving it as an excess of heat (ὑπερβολὴ θερμότητος or θερμοῦ), as is the case in several Aristotelian texts.[47] While the ability to generate itself is referred to as a δύναμις in the text quoted above, it differs fundamentally from a power like heating. Rather than a capacity that affects other things, it is an active, self-sustaining principle, one that defines fire not by its effects but by its unique mode of generation:

> In short, everything that is burning is, in a sense, in generation, comparable with motion, for which reason it somehow is destroyed as it is created, and with the exhaustion of the fuel, it itself is done away with. The ancients were referring to this when they said that fire always seeks

45 Theophrastus, *De igne* § 3, 1–5, transl. Van Raalte: μεγίστη δὲ αὕτη διαφορὰ δόξειεν ἄν· τὰ μὲν γὰρ καθ' αὑτὰ καὶ οὐδὲν ἐν ὑποκειμένῳ, πλὴν ὅσα μικτά, τὸ δὲ πῦρ ἐν ὑποκειμένῳ, ὅ γε τῇ περὶ ἡμᾶς αἰσθήσει φανερόν, εἴτ' ἀριθμητέον εἴτε μὴ ἀριθμητέον εἰς τὸ αὐτὸ τὸ φῶς. εἰ μὲν γὰρ καὶ τὸ φῶς, φανερὸν ὡς ἐν ἀέρι γε τοῦτο καὶ ὕδατι. εἰ δὲ μή, ⟨τὸ⟩ πῦρ τό γε τῆς φλογὸς καὶ τοῦ ἄνθρακος ⟨ἐν⟩ ὑποκειμένῳ.

46 Theophrastus, *De igne* § 1, 1–4, transl. Van Raalte: Ἡ τοῦ πυρὸς φύσις ἰδιαιτάτας ἔχει δυνάμεις τῶν ἁπλῶν. ἀὴρ μὲν γὰρ καὶ ὕδωρ καὶ γῆ τὰς εἰς ἄλληλα μόνον ποιοῦνται μεταβολὰς φυσικάς, αὐτὸ δὲ αὑτοῦ γεννητικὸν οὐδέν· τὸ δὲ πῦρ γεννᾶν καὶ φθείρειν πέφυκεν αὑτό, γεννᾶν μὲν τὸ ἔλαττον τὸ πλέον, φθείρειν δὲ τὸ πλέον τὸ ἔλαττον.

47 Aristotle, *De generatione et corruptione* II 3, 330b25 and *Meteorology* I 3, 340b23.

> nourishment, being unable to persist without fuel. Hence it seems foolish to speak of fire as a first substance and original element if it cannot exist without fuel. For in that case it is not elemental nor prior to the substrate and the fuel, unless there is some such substance in the first heavenly sphere itself which is pure, unmixed heat. In that case it would no longer burn. But burning is the essence of fire, unless indeed there are several different kinds of fire, one pure and unmixed, an element, the other located in the earth's sphere, mixed, and always in generation.[48]

The emphasis on generation in Theophrastus' text is noteworthy. As mentioned above (p. 46) in connection with the *Timaeus*, Academic enquiries into first principles adopted a generative framework. Similarly, in *De caelo* III 1, which I have argued has a substantial Academic background,[49] Aristotle explicitly connects the enquiry into simple bodies with the study of generation. Theophrastus takes a different approach, however, defining fire through its capacity to burn—a process that is intrinsically tied to generation. Burning is not conceived of as a heating process, but as an act of transformation (i.e. turning fuel into fire). Fire's ability to burn is presented as its essential nature: "In that case, it would no longer burn. But burning is the essence of fire" (οὕτω δὲ οὐκ ἂν ἔτι καίοι. πυρὸς δ' αὕτη φύσις). This capacity is explicitly compared to the act of generation: "Everything that is burning is, in a sense, in generation, comparable with motion" (ἀεὶ καὶ πᾶν καιόμενόν τι καὶ ὥσπερ ἐν γενέσει καθάπερ ἡ κίνησις). Thus, while generation plays a crucial role in Aristotle's and Plato's accounts of the elements, its function in Theophrastus' theory of fire is distinct. Rather than being integrated into an elemental theory—where the goal is to explain how all things arise from first bodies—generation in *De igne* describes the very nature of fire itself.

The thesis that fire is able to generate itself was originally advanced by Aristotle, for example in the *Parva naturalia*:

---

48 Theophrastus, *De igne* § 3, 9–11, § 4, 1–9, transl. Coutant: ἁπλῶς δ' ἀεὶ καὶ πᾶν καιόμενόν τι καὶ ὥσπερ ἐν γενέσει καθάπερ ἡ κίνησις, διὸ καὶ γινόμενον φθείρεταί πως καὶ ἅμα τῷ ὑπολείπειν τὸ καυστὸν καὶ αὐτὸ συναπόλλυται. τοῦτο γὰρ ἦν καὶ τὸ παρὰ τῶν παλαιῶν λεγόμενον, ὅτι τροφὴν ἀεὶ ζητεῖ τὸ πῦρ ὡς οὐκ ἐνδεχόμενον αὐτὸ διαμένειν ἄνευ τῆς ὕλης. ⟨διὸ⟩ καὶ ἄτοπον φαίνεται πρῶτον αὐτὸ λέγειν καὶ οἷον ἀρχήν, εἰ μὴ οἷόν τ' εἶναι χωρὶς ὕλης. οὔτε γὰρ ἁπλοῦν οὕτω γε οὔτε πρότερον τοῦ ὑποκειμένου καὶ τῆς ὕλης, εἰ μή τις ἐν αὐτῇ τῇ πρώτῃ σφαίρᾳ τοιαύτη φύσις, ὥστε ἄμικτον εἶναι θερμότητα καὶ καθαράν. οὕτω δὲ οὐκ ἂν ἔτι καίοι. πυρὸς δ' αὕτη φύσις, πλὴν εἰ ἄρα γε πλείους καὶ διάφοροι καὶ ἡ μὲν πρώτη καθαρὰ καὶ ἄμικτος, ἡ δὲ περὶ τὴν τῆς γῆς σφαῖραν μεμιγμένη καὶ ἀεὶ κατὰ γένεσιν.

49 Boureau (2022).

> But in other cases the result is exhaustion—when the heat accumulates excessively owing to lack of respiration and of refrigeration. For the heat, accumulating in great quantity, quickly uses up its nutriment and consumes it all before more is sent up by exhalation. Hence not only is a smaller fire readily put out by a larger one, but of itself the lamp's flame is consumed when inserted in a large blaze, just as is the case with any other combustible. The reason is that the nutriment in the flame is seized by the larger one before fresh fuel can be added, for *fire is ever coming into being and flowing like a river*, but so speedily as to elude observation. Clearly, therefore, if the bodily heat must be conserved (as is necessary if life is to continue), there must be some way of cooling the heat resident in the source of warmth.[50]

In this passage, the ability of fire to continuously generate itself, as long as fuel is available, is compared to the flow of a river, a likely allusion to Heraclitus. However, this characteristic of fire is mentioned in other Aristotelian contexts, which shows that it is integrated into Aristotle's account of fire.[51] Finally, Theophrastus' characterisation of fire seems to take into account an objection raised by Aristotle to Plato's attribution of a pyramidal shape to fire (and Democritus' attribution of a spherical one):

> Again, combustion of a body produces fire, and fire is a sphere or a pyramid. The body, then, is turned into spheres or pyramids. Let us grant that these figures may reasonably be supposed to cut and break up bodies; still it remains quite inexplicable that a pyramid must needs produce pyramids or a sphere spheres. One might as well postulate that a knife or a saw divides things into knives or saws.[52]

---

50 Aristotle, *De juventute et senectute* 5, 469b28–470a7, transl. G.R.T. Ross: ὁτὲ δὲ μαραίνεσθαι συμβαίνει, πλείονος ἀθροιζομένου θερμοῦ διὰ τὸ μὴ ἀναπνεῖν μηδὲ καταψύχεσθαι· ταχὺ γὰρ καὶ οὕτω καταναλίσκει τὴν τροφὴν πολὺ συναθροιζόμενον τὸ θερμόν, καὶ φθάνει καταναλίσκον πρὶν ἐπιστῆναι τὴν ἀναθυμίασιν. διόπερ οὐ μόνον μαραίνεται τὸ ἔλαττον παρὰ τὸ πλεῖον πῦρ, ἀλλὰ καὶ αὐτὴ καθ' αὑτὴν ἡ τοῦ λύχνου φλὸξ ἐντιθεμένη πλείονι φλογὶ κατακαίεται, καθάπερ ὁτιοῦν ἄλλο τῶν καυστῶν. αἴτιον δ' ὅτι τὴν μὲν οὖσαν ἐν τῇ φλογὶ τροφὴν φθάνει λαμβάνουσα ἡ μείζων φλὸξ πρὶν ἐπελθεῖν ἑτέραν, τὸ **δὲ πῦρ ἀεὶ διατελεῖ γινόμενον καὶ ῥέον ὥσπερ ποταμός**, ἀλλὰ λανθάνει διὰ τὸ τάχος. δῆλον τοίνυν ὡς εἴπερ δεῖ σῴζεσθαι τὸ θερμόν (τοῦτο δ' ἀναγκαῖον, εἴπερ μέλλει ζῆν), δεῖ γίνεσθαί τινα τοῦ θερμοῦ τοῦ ἐν τῇ ἀρχῇ κατάψυξιν.

51 For example, Aristotle, *De caelo* III 8, 307a24–26, and *De anima* II 4, 416a9–12. It is worth noting that the passage from *De anima* is commonly interpreted as addressing Heraclitus, with the expression *δοκεῖ τισιν* further suggesting that the idea that fire has a self-generating power may have been particularly emphasised by Heraclitus.

52 Aristotle, *De caelo* III 8, 307a24–31, transl. Stocks: Ἔτι εἰ τὸ καιόμενον πυροῦται, τὸ δὲ πῦρ

In this passage, Aristotle once again associates combustion with fire, as his objection is fundamentally based on the combustion process. He rejects the notion that the nature of fire is determined by geometric shapes, arguing that while fire generates more fire through combustion, solids do not produce more solids. By emphasising the centrality of the combustion process without assigning fire a specific geometric form, Theophrastus seems to acknowledge and incorporate this Aristotelian critique. It is noteworthy that, in the Aristotelian texts, the ability of fire to generate itself is presented as a mere property of fire and not its essence. Theophrastus' innovation thus lies not in introducing a new thesis about fire, but in shifting the emphasis—making self-generation the main characteristic of fire.

My claim here is that this new characterisation allows Theophrastus to circumvent the challenges Aristotle faces in defining fire, which arise from his not relying, as Plato does, on an extrinsic principle, such as a geometric figure. This approach exemplifies a philosophical pattern that may be characteristic of Theophrastus' works in Assos, at least as it has been reconstructed by Jaeger and Gaiser: criticising Plato from a Peripatetic perspective, while skilfully avoiding the difficulties Aristotle encountered in reforming Academic theories. In this specific case, Aristotle's rejection of the geometric determination of primary bodies forces him to develop an alternative framework to account for the definitional properties of elemental bodies. However, the abandonment of extrinsic principles when defining a simple body gives rise to at least two significant challenges within Aristotle's account:

1. Articulating the relationship between fire's various proper predicates and its essence. Connecting what is merely proper to fire with what constitutes its essential definition poses a genuine difficulty for Aristotle. Moreover, this difficulty highlights the paradoxical notion of primacy, as an element is supposed to be a prior concept and therefore should not rely on a principle of articulation to conciliate heterogeneous properties. A related issue pertains to the determination of fire's essence, as opposed to its proper predicates. This includes questions such as whether fire is defined by its motion—specifically, its natural upward movement—or by its essential primary qualities, such as dryness and heat.[53]

ἐστι σφαῖρα ἢ πυραμίς, ἀνάγκη τὸ καιόμενον γίγνεσθαι σφαίρας ἢ πυραμίδας. Τὸ μὲν οὖν τέμνειν καὶ διαιρεῖν ἔστω κατὰ λόγον συμβαῖνον τῷ σχήματι· τὸ δ' ἐξ ἀνάγκης τὴν πυραμίδα ποιεῖν πυραμίδας ἢ τὴν σφαῖραν σφαίρας παντελῶς ἄλογον, καὶ ὅμοιον ὥςπερ εἴ τις ἀξιοίη τὴν μάχαιραν εἰς μαχαίρας διαιρεῖν ἢ τὸν πρίονα εἰς πρίονας.

53 As previously noted, Theophrastus acknowledges both of these characteristics, though he does not treat either as constitutive of fire's essence.

2. Drawing a distinction between truly elemental fire and other forms of fire. Establishing the nature of the relationship between elemental fire, as defined in its pure, theoretical form, and empirical fire proves challenging, regardless of its various manifestations in sensory experience.

These two issues come to the fore in following passage from *De generatione et corruptione*:

> In fact, however, fire and air, and each of the bodies we have mentioned, are not simple, but combined. The simple bodies are indeed similar in nature to them, but not identical with them. Thus the simple body corresponding to fire is fire-like, not fire; that which corresponds to air is air-like; and so on with the rest of them. But fire is an excess of heat, just as ice is an excess of cold.[54]

In merely suggesting a resemblance between simple (pure) bodies and composite bodies (i.e. those commonly encountered), Aristotle remains ambiguous. The relationship between familiar elements, such as fire or earth, and their pure forms remains unclear. How do simple bodies manifest within the mixed bodies we perceive? If terrestrial fire shares the form of pure fire, is the difference between them solely due to the fact that the former is mixed with other elements? If so, what role does the substance with which it is mixed play in determining the form of the composite body? As a result, the connection between first principles and terrestrial bodies remains elusive within Aristotle's account.

The contrast between pure fire and another form of fire that is *almost* fire is displayed in another Aristotelian text, *Meteorology* I 3, albeit in a somewhat different form. Here, the distinction is no longer between empirical fire and pure, elemental fire, but between two kinds of fire that are not commonly encountered in everyday experience:

> From another point of view we may look [at the motion] as follows. The body that lies below the circular motion of the heavens is, in a way, matter, and is potentially hot, cold, dry, moist, and possessed of whatever other qualities are derived from these. But it actually acquires or retains one of these in virtue of motion or rest, the cause and principle of which has

54 Aristotle, *De generatione et corruptione* II 3, 330b21–26, transl. Joachim: Οὐκ ἔστι δὲ τὸ πῦρ καὶ ὁ ἀὴρ καὶ ἕκαστον τῶν εἰρημένων ἁπλοῦν, ἀλλὰ μικτόν. Τὰ δ' ἁπλᾶ τοιαῦτα μέν ἐστιν, οὐ μέντοι ταὐτά, οἷον εἴ τι τῷ πυρὶ ὅμοιον, πυροειδές, οὐ πῦρ, καὶ τὸ τῷ ἀέρι ἀεροειδές· ὁμοίως δὲ κἀπὶ τῶν ἄλλων. Τὸ δὲ πῦρ ἐστιν ὑπερβολὴ θερμότητος, ὥσπερ καὶ κρύσταλλος ψυχρότητος·

> already been explained. So at the centre and round it we get earth and water, the heaviest and coldest elements, by themselves; round them and contiguous with them, air and what we commonly call fire. It is not really fire, for fire is an excess of heat and a sort of ebullition.[55]

In this passage, Aristotle contrasts true fire—defined, as in *De generatione et corruptione*, as an "excess of heat" (ὑπερβολὴ θερμοῦ)—with another form of fire, described as "what we commonly call fire", which, strictly speaking, is not genuine fire. Unlike the earlier distinction between empirical and elemental fire, this contrast opposes fire, as defined by its intense heat, to a different theoretical conception, grounded in the natural motions of simple bodies. This so-called "fire", referred to as the ὑπέκκαυμα in the Peripatetic tradition—which Theophrastus never considered a viable candidate for the elemental form of fire[56]—is characterised by its position as the highest-moving body relative to other elements. Located just beneath the celestial sphere, it is hot,[57] but not excessively hot *per se* and therefore does not qualify as true fire within the framework Aristotle lays out in the *Meteorology*. While it can be indirectly ignited by the motion of the heavens, it lacks the intrinsic ability to burn.

Aristotle diverges here from Theophrastus' approach by attributing the excessive heat of this "fire" to the influence of the aether. The only connection between true fire and the ὑπέκκαυμα lies in their states of heat: the former is excessively hot in actuality, as it is defined as an excess of heat, while the latter is only potentially so. However, neither serves as the principle of the other, nor can they be traced back to a common underlying principle that is distinct from themselves. Notably, this ambiguity surrounding the definition of fire does not extend to other elements such as earth and water, which are both the heaviest—as demonstrated by their natural downward motion—and the coldest elements.

The difficulty facing Aristotle illustrates the architectonic challenge of establishing a criterion for identifying an element as such and reveals the antinomic tensions inherent in the notion of elements. For example, fire exists both within

55 Aristotle, *Meteorologica* I 3, 340b14–23, transl. Webster: δεῖ δὲ νοεῖν οὕτως καὶ ἐντεῦθεν ἀρξαμένους. τὸ γὰρ ὑπὸ τὴν ἄνω περιφορὰν σῶμα οἷον ὕλη τις οὖσα καὶ δυνάμει θερμὴ καὶ ψυχρὰ καὶ ξηρὰ καὶ ὑγρά, καὶ ὅσα ἄλλα τούτοις ἀκολουθεῖ πάθη, γίγνεται τοιαύτη καὶ ἔστιν ὑπὸ κινήσεως καὶ ἀκινησίας, ἧς τὴν αἰτίαν καὶ τὴν ἀρχὴν εἰρήκαμεν πρότερον. ἐπὶ μὲν οὖν τοῦ μέσου καὶ περὶ τὸ μέσον τὸ βαρύτατόν ἐστιν καὶ ψυχρότατον ἀποκεκριμένον, γῆ καὶ ὕδωρ· περὶ δὲ ταῦτα καὶ ἐχόμενα τούτων, ἀήρ τε καὶ ὃ διὰ συνήθειαν καλοῦμεν πῦρ, οὐκ ἔστι δὲ πῦρ· ὑπερβολὴ γὰρ θερμοῦ καὶ οἷον ζέσις ἐστὶ τὸ πῦρ.

56 On this issue see Bodnár (2002b), 86–87.

57 Aristotle, *Meteorologica* I 3, 340b26–27.

the physical system, as empirical fire involved in transformations, and at its boundaries, serving to close causal explanations and mark out the limits of the physical system. This raises the question of how we can reconcile the different types of fire and their distinct functions as outlined by Aristotle. Moreover, what principle makes it possible to articulate Aristotle's various definitions of fire? Within the framework of *De caelo*, natural motion defines the nature of bodies. It seems that motion functions as the *explicans* rather than the *explicandum*: it constitutes the form of the elemental body, as suggested in *De caelo* IV 3.[58] That said, how can we reconcile fire's lightness with its heat without resorting to an arbitrary or forced connection? This difficulty appears to have been recognised by Aristotle himself, as indicated in the following passage from *Parts of Animals*:

> Now there are three degrees of composition; and of these the first in order, as all will allow, is composition out of what some call the elements, such as earth, air, water, fire. Perhaps, however, it would be more accurate to say composition out of the elementary forces; nor indeed out of all of these, as said elsewhere in previous treatises. For wet and dry, hot and cold, form the material of all composite bodies; and all other differences are secondary to these, such differences, that is, as heaviness or lightness, density or rarity, roughness or smoothness, and any other such properties of bodies as there may be.[59]

Although Aristotle claims that lightness and heaviness can be derived from moisture, dryness, heat and cold, he does not spell out the details of this derivation. This omission underscores a real challenge in the articulation of his framework, as he merely juxtaposes lightness and heaviness with heat and dryness in the case of fire, without providing a clear explanation of their interrelation. Within the Aristotelian corpus itself, we thus encounter several definitions of fire that do not seem to refer to the same entity.[60]

---

58 Aristotle, *De caelo* IV 3, 310a32–b24.

59 Aristotle, *De partibus animalium* II 1, 646a12–20, transl. Ogle: Τριῶν δ' οὐσῶν τῶν συνθέσεων πρώτην μὲν ἄν τις θείη τὴν ἐκ τῶν καλουμένων ὑπό τινων στοιχείων, οἷον γῆς ἀέρος ὕδατος πυρός. Ἔτι δὲ βέλτιον ἴσως ἐκ τῶν δυνάμεων λέγειν, καὶ τούτων οὐκ ἐξ ἁπασῶν, ἀλλ' ὥσπερ ἐν ἑτέροις εἴρηται καὶ πρότερον. Ὑγρὸν γὰρ καὶ ξηρὸν καὶ θερμὸν καὶ ψυχρὸν ὕλη τῶν συνθέτων σωμάτων ἐστίν· αἱ δ' ἄλλαι διαφοραὶ ταύταις ἀκολουθοῦσιν, οἷον βάρος καὶ κουφότης καὶ πυκνότης καὶ μανότης καὶ τραχύτης καὶ λειότης καὶ τἆλλα τὰ τοιαῦτα πάθη τῶν σωμάτων.

60 E.g. Paparazzo (2023) attempts to reconcile three distinct conceptions of fire in *Meteorologica* I–III by associating them with different theoretical perspectives: fire as an element, as a form of exhalation, and as a combustion process.

Conversely, Theophrastus minimises the gap between phenomenal fire and elemental fire,[61] thereby sidestepping the challenge of explicitly connecting them. Moreover, by rejecting a definition of fire based on a single quality, such as upward motion or heat, Theophrastus offers a seemingly clearer, albeit initially paradoxical, solution. Nonetheless, he does not abandon the qualities traditionally ascribed to fire in the Peripatetic tradition—such as heat[62]—nor does he abandon the notion of natural motion, as is evident in his critique of Democritus' account of flame:

> But the view of those who speak of the motion of the air which encounters the flame is quite ridiculous. The fire is self-moving and it imparts motion to the air rather than receives it from the impulse of the air, since when the weather is calm, the pointed shape of the flame is no less in evidence.[63]

This passage should not be interpreted as a departure from the standard Aristotelian thesis that fire possesses an upward motion. Theophrastus is not asserting that fire is genuinely *self-moving*, which would contradict Aristotle's careful efforts in *De caelo* IV and *Physics* VIII 4 to avoid such a claim. Rather, his argument targets the view that fire's upward motion is merely accidental, resulting from its being pushed by air. Theophrastus instead suggests that fire is moved by fire itself, in the sense that it grows and propagates. Ultimately, while heat and natural motion remain closely associated with fire in Theophrastus' account, neither is presented as its defining essence. It is precisely this distinction between essence and observable qualities that enables Theophrastus to articulate a coherent explanation of these characteristics, as the final section of this paper will demonstrate.

---

61 This observation has been made, for example, by Van Raalte (2010), 49, in her commentary on *De igne* §§1–10. However, it has often been interpreted in terms of Theophrastus rejecting metaphysical concerns, such as first principles, in favour of focusing on the everyday manifestations of fire (see the interpretations previously noted by Steinmetz 1964 and Battegazzore 1984).

62 On the relationship between elemental heat and fire in Theophrastus' preserved works and fragments, see the discussion in Sharples (1998), 113–116.

63 Theophrastus, *De igne* §53, 1–3, transl. Coutant: γελοία δὲ ὅλως καὶ τούτοις ἡ τοῦ προσπίπτοντος ἀέρος κίνησις. ὑφ' αὐτοῦ γὰρ κινεῖται τὸ πῦρ καὶ μᾶλλον κινεῖ τὸν ἀέρα ἢ ὑπ' ἐκείνου τῷ διωθεῖν, ἐπεὶ καὶ εὐδίας οὔσης οὐδὲν ἧττον ἡ ὀξύτης τῆς φλογός.

3 Explaining Phenomena vs. Establishing Priority: A Theophrastean Dilemma?

Although heat is consistently associated with fire in Theophrastus' *De igne*,[64] the precise relationship between them remains elusive. In order to clarify this connection, it would be necessary to resolve the problem posed in § 5, namely to establish the nature of the link between the sun (potentially conceived as elemental fire) and empirical fire, understood as a burning process.[65] Simply identifying the sun as a form of fire would be insufficient, as it would still be necessary to reconcile celestial, non-burning fire with empirical, combustible fire under a single concept. This problem remains unsolved, as evidenced by the ongoing scholarly debate on whether *De igne* endorses Aristotle's notion of aether (see *supra*). Nevertheless, Theophrastus' account of fire offers a satisfactory and systematic explanation of heat, effectively addressing some of the major difficulties faced by both the Aristotelian and the Platonic accounts. My aim in this section is to demonstrate that Theophrastus' emphasis on explaining fiery phenomena and heat does not merely reflect an empirical focus, but rather a methodological engagement with the fundamental tension between fire's role as an element within a physical system and its observable effect, namely heat production. To that end, I will begin by examining the methodological criticism made by the Peripatetics of Plato's explanation of heat in the *Timaeus*, before examining the challenges arising from Aristotle's attempts to link heat and his definition of elemental fire. Against this background, I will argue that Theophrastus' decision to leave the elemental status of fire unresolved enables him to develop a coherent account of heat, one that strategically draws on Platonic insights while avoiding the difficulties inherent in Aristotle's approach.

At first glance, Plato's account of fire successfully reconciles its determination as tetrahedron with its primary quality: the production of heat. Of the regular solids, the tetrahedron possesses the sharpest angles and the fewest faces, making it the most suitable candidate for fire:

64 As István Bodnár (2002b), 86, puts it: "Theophrastus' discussion always links heat and the quality of hotness to processual and self-subsistent forms of fire. Heat apparently is a causally efficacious property (or even *the* causally efficacious property) of fire; θερμόν and θερμότης feature practically in all the arguments of the second, larger part of the treatise."

65 On this issue, see once again Lefebvre (2017), 79–83.

> Accordingly we shall preserve the probability of our account, if we assign this figure to earth; and of the remainder the least mobile to water, the most mobile to fire, and the intermediate figure to air. Again, we shall assign the smallest body to fire, the largest to water, and the intermediate to air; and again the body with the sharpest angles to fire, the next to air, the third to water. Now, taking all these figures, the one with the fewest faces (pyramid) must be the most mobile, since it has the sharpest cutting edges and the sharpest points in every direction, and moreover the lightest, as being composed of the smallest number of similar parts.[66]

Aristotle criticises this theory by presenting a general criterion for assessing the validity of a physical theory. A robust theory must be both coherent and comprehensive, offering explanations that extend beyond isolated phenomena. It should not offer *ad hoc* justifications—e.g. accounting for heat while failing to explain cold—but instead present a unified framework capable of addressing the full range of natural phenomena:

> In addition, since hot and cold are contrary powers, it is impossible to allot any shape to the cold. For the shape given must be the contrary of that given to the hot, but there is no contrariety between figures. That is why they have all left the cold out, though properly either all or none should have their distinguishing figures.[67]

Aristotle's point is not that Plato fails to provide a theory of cold; such a theory is explicitly presented in *Timaeus* 62a. Rather, he takes aim at the inconsistency of Plato's account of heat and cold, and more generally the use of shapes to determine simple bodies and physical powers. According to Aristotle, shapes cannot account for physical powers, as long as such powers possess contrar-

66 Plato, *Timaeus* 56a–b, transl. Cornford: διὸ γῇ μὲν τοῦτο ἀπονέμοντες τὸν εἰκότα λόγον διασῴζομεν, ὕδατι δ’ αὖ τῶν λοιπῶν τὸ δυσκινητότατον εἶδος, τὸ δ’ εὐκινητότατον πυρί, τὸ δὲ μέσον ἀέρι· καὶ τὸ μὲν σμικρότατον σῶμα πυρί, τὸ δ’ αὖ μέγιστον ὕδατι, τὸ δὲ μέσον ἀέρι· καὶ τὸ μὲν ὀξύτατον αὖ πυρί, τὸ δὲ δεύτερον ἀέρι, τὸ δὲ τρίτον ὕδατι. ταῦτ’ οὖν δὴ πάντα, τὸ μὲν ἔχον ὀλιγίστας βάσεις εὐκινητότατον ἀνάγκη πεφυκέναι, τμητικώτατόν τε καὶ ὀξύτατον ὂν πάντῃ πάντων, ἔτι τε ἐλαφρότατον, ἐξ ὀλιγίστων συνεστὸς τῶν αὐτῶν μερῶν.

67 Aristotle, *De caelo* III 8, 307b5–10, transl. Stocks: Πρὸς δὲ τούτοις, ἐπεὶ τὸ θερμὸν καὶ τὸ ψυχρὸν ἐναντία τῇ δυνάμει, ἀδύνατον ἀποδοῦναι τῷ ψυχρῷ σχῆμά τι· δεῖ γὰρ ἐναντίον εἶναι τὸ ἀποδιδόμενον, οὐθὲν δ’ ἐναντίον ἐστὶ σχῆμα σχήματι. Διὸ καὶ πάντες ἀπολελοίπασι τοῦτο· καίτοι προσῆκεν ἢ πάντα ἀφορίσαι σχήμασιν ἢ μηδέν.

ies.[68] To justify his claim, Aristotle invokes an irreconcilable contrast between shapes and physical powers. Physical powers inherently involve opposition, as the same matter can be affected in opposite ways—e.g. being either warmed or cooled. However, no shape can stand in opposition to another in the same way. Therefore, no shape can adequately explain a physical power. Remarkably, a passage from Theophrastus' *De sensibus* closely echoes Aristotle's criticism:

> Yet this is also absurd. First of all, he gives no uniform account of all things, not even of those that belong to the same class. For he describes heat in terms of figure, but he has not given a like account of cold.[69]

Theophrastus similarly attacks the lack of coherence in Plato's attribution of regular solids to simple bodies, pointing out that "no uniform account of all things" is provided. However, the strength of Plato's approach lies in its ability to unify the determination of fire with the production of heat. In contrast, Aristotle's account of heat struggles to establish a clear connection between heat as an observable phenomenon and its role as a defining characteristic of fire:

> In some of the bodies which are called hot the heat is derived from without, while in others it belongs to the bodies themselves; and it makes a most important difference whether the heat has the former or the latter origin. For one of them comes close to being hot accidentally and not in its own right—as if, finding that some man in a fever was a musician, one were to say that musicians are hotter than healthy men. Of that which is hot per se and that which is hot per accidens, the former is the slower to cool, while not rarely the latter is the hotter to the touch. The former again is the more burning of the two—flame, for instance, as compared with boiling water—while the latter, as the boiling water, which is hot per accidens, is the more heating to the touch. From all this it is clear that it is no simple matter to decide which of two bodies is the hotter. For the first may be the hotter in one sense, the second the hotter in another. Indeed in some of these cases it is impossible to say simply even whether a thing

68 Aristotle, *De caelo* III 8, 307b6: ἐπεὶ τὸ θερμὸν καὶ τὸ ψυχρὸν ἐναντία τῇ δυνάμει.

69 Theophrastus, *De sensibus* 87, 1–3, transl. Stratton (slightly modified): Ἄτοπον δὲ καὶ τοῦτο· πρῶτον μὲν τὸ μὴ πάντα ὁμοίως ἀποδοῦναι μηδὲ ὅσα τοῦ αὐτοῦ γένους. ὁρίσας γὰρ τὸ θερμὸν σχήματι τὸ ψυχρὸν οὐχ ὡσαύτως ἀπέδωκεν.

> is hot or not. For the actual substratum may not itself be hot, but may be hot when coupled with heat as an attribute, as would be the case if one attached a single name to hot water or hot iron. It is after this manner that blood is hot.[70]

This passage attempts to draw a distinction between heat *by nature* and heat *by accident*, contrasting a flame, which is inherently hot, with boiling water. Although this distinction arises in a biological context and aims to explain how blood is hot, it can be situated within a broader framework, namely Aristotle's attempt to connect heat to the elemental nature and defining principle of fire. However, this approach raises at least two significant challenges. The first challenge is that the distinction in question does not have a clear experiential basis, a difficulty that is acknowledged by Aristotle ("from all this it is clear that it is no simple matter to decide which of two bodies is the hotter"). Perception, for example, does not allow us to determine whether something is hot by nature or by accident, as boiling water "is the more heating to the touch". The second challenge is the lack of an explanation for why something that is hot *by nature* should retain its heat longer than something that is hot by accident, highlighting the absence of a comprehensive theory of heat. In contrast, Theophrastus' theory of heat provides a consistent and systematic framework that accounts for a wide range of phenomena:

> Thus we may lay all these and like phenomena to the fineness or the coarseness of the fuel. For example, the flame from reeds heats flesh and reeds are light, compact because they are continuous. A fire that is fine can penetrate, and the heat is made possible by the continuous admix-

70 Aristotle, *De partibus animalium* II 2, 648b35–649a17, transl. Ogle: Πρὸς δὲ τούτοις τῶν λεγομένων θερμῶν τὰ μὲν ἀλλοτρίαν ἔχει τὴν θερμότητα τὰ δ' οἰκείαν, διαφέρει δὲ τὸ θερμὸν εἶναι οὕτως ἢ ἐκείνως πλεῖστον· ἐγγὺς γὰρ τοῦ κατὰ συμβεβηκὸς εἶναι θερμὸν ἀλλὰ μὴ καθ' αὑτὸ θάτερον αὐτῶν, ὥσπερ ἂν εἴ τις λέγοι, εἰ συμβεβηκὸς εἴη τῷ πυρέττοντι εἶναι μουσικῷ, τὸν μουσικὸν εἶναι θερμότερον ἢ τὸν μεθ' ὑγιείας θερμόν. Ἐπεὶ δ' ἐστὶ τὸ μὲν καθ' αὑτὸ θερμὸν τὸ δὲ κατὰ συμβεβηκός, ψύχεται μὲν βραδύτερον τὸ καθ' αὑτό, θερμαίνει δὲ μᾶλλον πολλάκις τὴν αἴσθησιν τὸ κατὰ συμβεβηκός· καὶ πάλιν καίει μὲν μᾶλλον τὸ καθ' αὑτὸ θερμόν, οἷον ἡ φλὸξ τοῦ ὕδατος τοῦ ζέοντος, θερμαίνει δὲ κατὰ τὴν ἁφὴν τὸ ζέον μᾶλλον, τὸ κατὰ συμβεβηκὸς θερμόν. Ὥστε φανερὸν ὅτι τὸ κρῖναι δυοῖν πότερον θερμότερον οὐχ ἁπλοῦν· ὡδὶ μὲν γὰρ τόδε ἔσται θερμότερον, ὡδὶ δὲ θάτερον. Ἔνια δὲ τῶν τοιούτων οὐδ' ἔστιν ἁπλῶς εἰπεῖν ὅτι θερμὸν ἢ μὴ θερμόν. Ὃ μὲν γάρ ποτε τυγχάνει ὂν τὸ ὑποκείμενον, οὐ θερμόν, συνδυαζόμενον δὲ θερμόν, οἷον εἴ τις θεῖτο ὄνομα ὕδατι ἢ σιδήρῳ θερμῷ. Τοῦτον γὰρ τὸν τρόπον τὸ αἷμα θερμόν ἐστιν.

> ture of the fuel. The flame of reeds then is in generable preferable to the heat of charcoal; its output is both greater and more concentrated.[71]

In Theophrastus' theory, much like in the *Timaeus*, the production of heat is consistently explained as a phenomenon of penetration (διαδυτικόν), which is grounded in the fineness (τὸ λεπτόν) of fire's parts. However, while Theophrastus' position appears to draw inspiration from Platonic physics, it simultaneously takes into account Peripatetic criticisms of Plato's methodology. Thus, Theophrastus deliberately avoids attributing a specific geometric form to fire. Moreover, while the fineness of its parts accounts for the intensity of heat, it does not define the essence of fire itself. By conceptualising fire not as a particular kind of matter but as a generative process—an entity that sustains itself through the consumption of fuel—Theophrastus circumvents the problems that the Peripatetic tradition identified in Plato's model. Here, the intensity of heat, which is determined by the subtlety of fire's parts, is directly linked to the nature of the burning substrate, such as reed or charcoal, rather than being seen as an intrinsic property of fire. This approach offers a coherent and non-arbitrary explanation that is closely tied to Theophrastus' characterisation of fire as a self-generating body. By distinguishing heat from the essential definition of fire, while still acknowledging that it is a proper attribute, Theophrastus offers a consistent and unified account of heat as a physical phenomenon across its various manifestations. Notably, he differentiates the heat caused by the sun and the heat found in animal bodies from the heat of fire, while appealing in both cases to the same underlying principle, namely the fineness of parts:

> Now the heat which is natural to animate bodies, being infused into a comparatively large number of creatures and in a rather special way becomes in a sense alive and capable of generating similar creatures. Even more so the heat from the sun. It is able to create animals and plants. It is mixed, to be sure with air (occurring rather in it),[72] but by reason of its

71 Theophrastus, *De igne* § 33, 1–6, transl. Coutant: πάντων δὲ τούτων καὶ τῶν τοιούτων αἰτιάσαιτ' ἄν τις τὴν λεπτότητα καὶ τὴν παχύτητα· θερμαίνει τε γὰρ τὴν σάρκα καὶ τὸ ὕδωρ ἡ τῶν καλάμων τάχιστα διὰ τὴν λεπτότητα καὶ πυκνότητα (λεπτὴ μὲν γὰρ ὅτι κοῦφος, πυκνὴ δὲ ὅτι συνεχής· τὸ δὲ λεπτὸν διαδυτικόν, καὶ τὸ θερμαίνεσθαι διὰ τὴν συνεχῆ κατάμιξιν) ἥ τε φλὸξ ὅλως τῶν ἀνθράκων μᾶλλον διὰ τοῦτο, καὶ ἅμα πλείων ἡ ἀπορροὴ ⟨καὶ⟩ ἀθροωτέρα.

72 Steinmetz (1964), 135, considers the words between brackets as a gloss, added in order to bring Theophrastus' text into line with the Aristotelian orthodoxy.

> gentleness and fineness has a certain appropriateness for generating life; it is not, like the heat of fire, hard and caustic.[73]

In a manner reminiscent of Aristotle's mention of *proper heat* (θερμότητα οἰκείαν) or being *hot in its own right* (καθ' αὑτό),[74] Theophrastus' text focuses on the inherent heat of animals. However, unlike in Aristotle's account, this is not distinguished from accidental heat. Instead, the divide lies between the generative heat (γόνιμος) of the sun and living beings and the destructive heat of fire. Theophrastus explains how these differing effects arise, attributing them to the greater fineness of the particles involved, in contrast to Aristotle, who could not concretely explain the supposed effects of his own distinction, such as differences in cooling time.

In a similar vein, Theophrastus is able to consistently account for the upward motion of fire:

> If fire moves in a kind of rush or flow, and the result for all flowing things is that they diminish over time unless replenished from another source, then it is necessary that, as it extends into a larger space, it becomes narrower while maintaining continuity. Thus, this could also explain the shape of the flame: the base is the strongest and most abundant, and from there it diminishes continuously, weakening as it ascends, though it preserves its continuity.[75]

This passage explains not only the shape of the flame, but also its motion. As fire "extends into a larger space", it does not remain confined to its place, but its behaviour resembles a continuous flow. This echoes the description of fire quoted above in Aristotle's *De juventute et senectute*. However, within Theophrastus' framework, the upward motion of fire is not conceived of as a movement of actualisation toward its form, nor does it play a role in defining fire as an element, as it does in Aristotle's account in *De caelo* IV. Instead, Theophrastus' explanation aligns with his characterisation of fire as self-generating: fire originates from its core and diminishes as it moves away from its fuel source.

73 Theophrastus, *De igne* § 44, transl. Coutant.

74 See *supra* (p. 61–62), Aristotle, *De partibus animalium* II 2, 648b35–649a17.

75 Theophrastus, *De igne* § 54, transl. Coutant: εἰ δ' ἐν φορᾷ τινι καὶ ῥύσει τὸ πῦρ, ἀεὶ δὲ τῶν ῥεόντων εἰς ἔλαττον ἡ ἀποτελεύτησις μὴ γινομένης ἄλλοθεν ἐπιρροῆς. (ἀναγκαῖον εἰς τὸν πλείω τόπον διατεινόμενον καὶ τὸ συνεχὲς διατηροῦν ἀποστενοῦσθαι μᾶλλον.) ὥστε τοῦτο ἂν εἴη καὶ τῆς φλογὸς αἴτιον εἰς τὸν σχηματισμόν. ἡ γὰρ ἀρχὴ κάτωθεν ἰσχυροτάτη καὶ πλείστη, μετὰ δὲ ταῦτα ἐλάττων ἀεὶ καὶ ἀπολήγουσα ἀσθενεστέρα, διασώζουσα δὲ τὸ συνεχές.

The comparison with Aristotle's approach to heat and upward motion suggests that Theophrastus' ability to articulate a consistent account of fiery phenomena stems precisely from his decision not to fully engage with the elemental status of fire. Instead, he adopts a distinctly analytical approach, explaining the heat and upward motion of fire in terms of combustion, which he treats as the primary factor responsible for both. By refraining from making heat the defining characteristic of fire or establishing fire as the fundamental principle of heat, Theophrastus develops a coherent explanatory framework that extends beyond fire to encompass other forms of heat.

However, the precise role of combustion in the definition of fire remains ambiguous, beyond being described as one of its "most peculiar powers" (§ 1, ἰδιαιτάτας δυνάμεις) in the case of terrestrial fire. Moreover, since Theophrastus does not rule out the possibility of a celestial, non-burning form of fire, combustion cannot even be considered a proper attribute of fire. This point is further reinforced in § 4, where the essence of fire is left undetermined, suggesting that Theophrastus deliberately avoids committing to a rigid definition of fire within a structured elemental framework.[76]

4 Conclusion

This paper has sought to challenge the view that Theophrastus' *De igne* is the work of a resigned empiricist who had abandoned the study of principles, while also rejecting the comparison that has occasionally been made of Theophrastus to a modern scientist constructing a physical theory through systematic observation and experimentation. Instead, *De igne* reveals itself as a work that is shaped on a deep level by the philosophical dialogue between the Platonic and Aristotelian traditions, particularly with regard to the challenges posed by the enquiry into the elements. While Theophrastus appears to set this issue aside at the outset, it remains a persistent undercurrent throughout the treatise, informing his approach to fire and heat. The aporetic method evident in *De igne* offers a productive way to apply the genealogical approach and assess its heuristic value in a novel manner. Since Theophrastus does not elaborate a dogmatic theory and, as has been noted, none of his core claims about fire are entirely new but were already upheld by Aristotle, the Assos reference does not help trace the evolution of specific theses. Rather, it highlights a shared set

76 See Lloyd (1973), 10: "Similarly *On Fire* hits on a fundamental objection to Aristotle's doctrine on that subject. Yet Theophrastus attempts no new element theory, nor even a definite statement of what fire, in particular, is."

of questions and problems directly connected to the Academy. The effect is to recast Theophrastus not as a reader of Aristotle, commenting on an established corpus, but rather as a colleague independently drawing on the same Academic tradition to address the same challenges, such as the tension between providing a limited but consistent explanation of observable phenomena and engaging with the paradoxes bearing on the nature of the elements. In this regard, *De igne* appears not so much "post-Aristotelian", as it is sometimes labelled (e.g. in the subtitle of Coutant's edition), but rather as *para*-Aristotelian.[77]

77 This paper was written during my FWO Junior Postdoctoral Fellowship at KU Leuven (File no. 12Z1822N) and also forms part of a project funded by the European Research Council (ERC) under the European Union's Horizon 2020 research and innovation programme (Grant agreement no. 885273). The initial version of this paper was presented at a conference at the University of Lyon 3 in November 2022, organised by Gweltaz Guyomarc'h and Alessio Santoro. I am grateful to the organisers and participants for the engaging discussions, and especially to my respondent, Fabienne Baghdassarian, for her insightful and stimulating comments. Finally, I wish to thank the participants of the doctoral colloquium of the De Wulf-Mansion Centre at KU Leuven for the valuable feedback they gave on a later version of the paper.

CHAPTER 3

Theophrastus on Odorous Exhalations and the Nature of Perception

Robert Roreitner

Perception was an important theme for Theophrastus. He is the author of *De sensibus*, which offers a detailed critical analysis of the views on the nature of perception ascribed to Aristotle's predecessors. The sixth book of *De causis plantarum* is dedicated to explaining phenomena connected to flavours and odours, especially the naturally produced ones. In *De odoribus* we find a similar investigation focusing primarily on artificially produced odours.[1] Theophrastus is also credited with other thematically related writings that are now lost, such as *On Vision* in four books.[2] And he wrote a book covering the topics of (at least) Aristotle's *De anima* II 5–III 5 that is referred to by our sources as either Theophrastus' *De anima* II or his *Physics* V.

What can we learn from these sources about Theophrastus' views on perception? How far did he follow Aristotle? Did he take positions on some of the notoriously difficult questions in Aristotle's account? Did he introduce some important innovations? Relatively little attention has been paid to these questions, apparently because our resources for answering them are far from optimal. As on most other issues, we have evidence suggesting that Theophrastus took over the central tenets of Aristotle's theory of perception. But, more interestingly, we also have evidence (in *CP* VI and *Od.*) suggesting that he went beyond, and seemingly against, what Aristotle said—especially with regard to how odours are mediated and perceived. He ascribed an important role to what he calls "exhalations" and what he seems to conceive of as streams of tiny particles continuously emitted by odorous objects and transported by the air to the perceivers' olfactory organs. There has been some discussion of exactly how radical Theophrastus' innovation was in this respect, how

1 Thompson (1941), 16–20, argues that some version of *Od.* (treating mainly of odours produced by human art) was originally (a part of) the lost eighth book of *CP* (following book VI, which is about naturally produced flavours and odours and, presumably, book VII on flavours produced by human art). Cf. the programme laid down at *CP* VI 3.3. Cf. Wöhrle (1988) and Reger (2005).

2 DL V 49.

far he shifted from Aristotle and whether he ended up being consistent with himself or not.[3]

The question of consistency becomes more urgent as Theophrastus moves closer to Aristotle's own discussion of smell—and perception in general—in *DA* II 5–12 and *De sensu*. This is the case with Theophrastus' *De sensibus* (containing, among other things, an extensive polemic against effluence accounts of perception)[4] and especially with his *De anima* II. The latter should be our key piece of evidence, but unfortunately our only source of information about it is indirect and not very reliable. Unlike in the case of passages concerning *nous*, where we have at least two sources that can be checked against each other (yielding a reasonable amount of certainty that we possess several verbatim quotes from Theophrastus), for perception (and *phantasia*) we are limited to the exclusive testimony of Priscian of Lydia's *Metaphrasis in Theophrastum* (6th century CE). The author openly acknowledges his allegiance to the Iamblichean Neoplatonic doctrine of the soul and perception as the lens through which he tries to extract and absorb what Theophrastus "adds beyond that which Aristotle has handed down" and what Theophrastus "offers by raising difficulties".[5] Priscian provides us with 22 pages of quotes, paraphrases and, especially, his own reflections and solutions concerning perception, all integrated within a single flow of thought, where distinguishing the individual elements with any certainty becomes very difficult and often impossible.[6] Combined with the terse and aporetic nature of Theophrastus' discussion (that, we are told, was "compressed and concise in style" but contained "many difficulties (ἀπορίαι), many analyses, and many solutions"),[7] the task of establishing something firm about Theophrastus' views on perception may seem hopeless. Still, a comparison with the later section of *Metaphrasis* (25–37) dealing with *nous* shows that Priscian *is* sometimes quoting Theophrastus and that he is doing so quite reli-

3 See Sharples (1985); Sedley (1985); Eigler and Wöhrle (1993), 16–18; Schorlemmer (2022), 71–82, and (Forthcoming).

4 For an analysis of *De sensibus* as "an integral part of the same Peripatetic dialectic as the *DA*", see Johansen (2019) (quoted from p. 228). Cf. also Mansfeld (1996).

5 Priscian, *Metaphr.* 7.17–23. For Priscian's account, see Steel (1997), 105–137 (with Steel 1978, 142–154, for Priscian's strategy in general), and Kalderon (2017). Cf. also Perkams (2005) and De Haas (2010).

6 Huby (1999) (commenting on Fortenbaugh et al. 1992, 12–38 = 273–296 FHS&G) is the most detailed existing assessment of what may, anyhow, come from Theophrastus (cf. Huby 1997 for *Metaphrasis* 1–22 as a whole). The question is helpfully revisited by Caston (2019) with important new insights and what I find to be a more solid selection of relevant passages.

7 Themistius, *In DA* 108.8–10.

ably.[8] Hence, one cannot simply dismiss *Metaphrasis* 1–22 as unreliable and stick to other texts.[9] If we are interested in Theophrastus' views on the very nature of perception (rather than particular phenomena of smell), it is likely that Priscian's *Metaphrasis* contains key pieces of evidence—the task is to tell them apart from Priscian's own philosophical outbursts with a reasonable degree of certainty and to reconstruct at least some general insights from what must have been quite an elusive text, even in its full form. That is, admittedly, a very difficult task, and so it is hardly surprising that only two attempts of this kind have been undertaken—and that they arrived at diametrically opposed conclusions.[10]

This paper reassesses the existing evidence and offers a somewhat different overall reconstruction of Theophrastus' views on perception. My diagnosis is that with both main *foci* of scholarly discussion, the existing approaches tend toward one extreme or the other, whereas Theophrastus' account occupies an interesting middle position. Theophrastus' account of odorous exhalations has either been treated as articulating an entirely different theory of smell than that held by Aristotle or been ignored, as if Theophrastus had nothing to add to Aristotle's explanation of how we smell things. The only other option seemed to be that Theophrastus was inconsistent and followed Aristotle in some contexts, while radically diverging from him in others. In section 1, I argue that Theophrastus' account of odorous exhalations does constitute an important innovation with respect to Aristotle, but that it was intended to be integrated within Aristotle's general account of perception—and of smell in particular. What Theophrastus does is to introduce a novel, complementary causal factor which enhances the explanatory force of the Aristotelian account, allowing it to explain phenomena that could not be explained without this additional factor. In the rest of the paper, I turn to our evidence for Theophrastus' views on the

8 Compare especially Priscian, *Metaphr.* 27.8–14 with Themistius, *In DA* 108.1–6; but also Priscian, *Metaphr.* 25.28–29 with Themistius, *In DA* 107.34–35; Priscian, *Metaphr.* 26.1–2 with Themistius, *In DA* 108.6–7; and Priscian, *Metaphr.* 28.20–22 with Themistius, *In DA* 108.15–17.

9 For such a tendency, see Schorlemmer (2022), 71–79 subsuming everything we find in *Metaphrasis* under "Wahrnehmung nach Priskian" and contrasting it with "Wahrnehmung in Theophrasts eigenen Worten", which is reconstructed solely on the basis of *CP* VI and *Od.* One danger in this approach is that it ends up ascribing the status of a general theory to something that was intended as a much more specific enquiry operating within a very different general framework. More on this below.

10 See Laks (2002) with the revised English version in Laks (2019) (arguing for a spiritualist reading of Theophrastus) and Caston (2019) (arguing for a materialist reading of Theophrastus along the lines of Caston 2005; cf. Caston 2020). In what follows I refer to Laks (2019)—rather than Laks (2002)—as the latest version of Laks' interpretation.

nature of mediation and perception. In section 2, I discuss difficulties he seems to have raised against competing accounts of mediation. In sections 3 and 4, I provide evidence against the idea that Theophrastus endorsed a spiritualist account of mediation and perception, as well as offering a reconstruction of his views on mediation as the transference of a thoroughly physical and perceiver-independent affection. At the same time, in section 4, I question the existing materialist interpretation of our fragmentary evidence for Theophrastus' views on the nature of perception, offering an alternative that is at least equally consistent with the sources and that better agrees with the account of mediation defended in section 3.[11]

1 Odorous Exhalations and the Aristotelian Account of Smell

The idea that in his explanations of olfactory phenomena in *CP* VI and *Od.* Theophrastus might have embraced an effluence theory of smell sounds surprising, not only because such a theory sharply conflicts with Aristotle's views, but also because Theophrastus himself launches an extended attack on effluence theories of perception in *Sens.* 12–22. It has been suggested that in his argument against effluences in *Sens.* 20 Theophrastus in fact *limits* his target to hearing and sight, to the exclusion of smell.[12] At 505.6–10, he argues that if (1) effluence causes wasting away (εἰ ἡ φθίσις διὰ τὴν ἀπορροήν) and if (2) odours come about by means of effluence (συμβαίνει δὲ καὶ τὰς ὀσμὰς ἀπορροῇ γίνεσθαι), then those objects with the strongest odours would perish most quickly—which according to Theophrastus conflicts with phenomena: in fact, the most odorous things tend to last longest. David Sedley translates the first premise as "if wasting away is due to effluence" and interprets it as articulating a universal claim about wasting away being due to effluences. This claim is easy to challenge—why should there not be other kinds of wasting away, due, say, to excessive excretion or lack of nourishment?—[13] which allows for the possibility that Theophrastus in fact wanted to retain as true the second premise endorsing an effluence account of smell.

11 The questions bearing on the nature of mediation and perception naturally draw one's attention back to much-discussed passages in Aristotle. However, in this paper, I refrain from discussing Aristotle's views in their own terms and limit myself to evaluating the fragmentary Theophrastean evidence. For the former, I refer the reader to Roreitner (2025).

12 Sedley (1985).

13 Cf. *Sens.* 20, 505.5–6.

It is doubtful, though, that the first premise is intended as a universal claim about wasting away. The implication from wasting away to effluences is irrelevant to the argument. What is needed is the opposite implication, namely that if there are effluences from object *x*, then *x* is being wasted away.[14] While there are likely to be cases of wasting away that do not involve effluences, there are no effluences that would not involve at least some amount of wasting away of the underlying material. This is a sound common-sensical premise that would be hard to deny and that is exactly what one needs to ground the conclusion of Theophrastus' argument—a conclusion that conflicts with the phenomena. If this analysis is correct, then in *Sens.* 20 Theophrastus must mean to deny the second premise (i.e. that smelling comes about by means of effluences) as his proper target. Hence, this passage of *De sensibus* increases, rather than reduces, the tension with Theophrastus' apparent adoption of effluence account of smell in other treatises.[15]

This tension comes primarily from *De causis plantarum* VI and *De odoribus*, where Theophrastus repeatedly speaks of exhalations (usually ἀποπνοαί) coming out of fragrant objects, which conspicuously resemble Empedoclean effluences (ἀπορροαί). Some scholars have concluded that the tension between these passages and Theophrastus' other works is hard, if not impossible, to resolve.[16] Others have gone even further, seeing this as evidence that Theophrastus abandoned Aristotle's account of how odours are mediated and perceived.[17] The truth is, though, that in none of the relevant passages from *CP* VI is the precise role of the exhalations and their contribution to mediation of odours and to smelling itself entirely clear, and none shows that exhalations are intended as an alternative to Aristotle's account of the perceptual object's acting on the medium and, through it, on the sense organ. The most suggestive text

14 That is clearly how the premise is read by Stratton (1917) ("if effluence involves a loss of substance"). The translation by Ierodiakonou and Bydén (Forthcoming) ("if decay is caused by the effluence") is more suggestive of Sedley's understanding.

15 At *Sens.* 90, 526.25–26, Theophrastus treats the idea that odour is a kind of effluence as something more or less agreed upon (σχεδὸν ὁμολογεῖται). If this means, as it seems, that smelling always and essentially involves effluences and that this is how smelling is to be explained, then Theophrastus is obviously excluding Aristotle from the agreement, and so there is no reason to assume that he is silently including himself. Perhaps the idea can also be given a more relaxed interpretation, according to which it claims only that some cases of smelling somehow involve effluences (cf. Eigler and Wöhrle 1993, 16). Even Aristotle and Theophrastus could agree on this point.

16 See Sharples (1985), 204: "How far Theophrastus abandons or compromises an Aristotelian account of the transmission of odours is a question that seems difficult to resolve". Cf., e.g., Stratton (1917), 39, or Gottschalk (1967), 25.

17 Besides Sedley (1985), see now Schorlemmer (2022), 76–79, and (Forthcoming).

is *De odoribus* 45, where exhalations are not only unmistakably described in terms of tiny material particles coming out of flowers; the author is, moreover, offering an explanatory account in terms of fine and light particles of rose perfume penetrating and filling up "passages" (πόροι) associated with the organ of smell.[18] That sounds very much like an effluence account of smell—but only at first sight.

In fact, what is explained by the case of fine and light particles filling up the passages is *not* smelling, but, on the contrary, a certain incapacity to smell—an inability to discriminate odours. Theophrastus relates the story of perfumers who, when the customer is unwilling to buy their rose perfume, scent him with it, "so that he is unable to smell the scents of other perfumes" (πρὸς τὸ μὴ αἰσθάνεσθαι τὰ παρὰ τῶν ἄλλων, *Od.* 45.7). Moreover, he provides an explanation for this phenomenon: due to its fineness and lightness, the rose perfume "penetrates and fills up the passages, so that the sense, being entirely taken up and filled with it becomes incapable of discriminating" (διικνεῖται καὶ συμπληροῖ τοὺς πόρους, ὥσθ' ἡ αἴσθησις κατειλημμένη καὶ πλήρης οὖσα κρίνειν ἀδυνατεῖ, *Od.* 45.9–11). In his translation, Arthur Hort weakens the last claim by adding "unable to judge *the others*", as if the sense of smell in this condition were still able to "judge" the odour of the rose perfume—and that seems to be how the passage is understood by those who see in it an effluence account of smell.[19] But there is no justification for this addition.[20] While the cunning perfumer may believe that his trick condemns the poor customer to continuously smell his rose perfume, Theophrastus analyses what happens in terms of the customer being rendered unable to genuinely smell anything whatsoever: the customer's sense of smell becomes unable to κρίνειν—period.[21] In the context of percep-

18 Schorlemmer (Forthcoming) argues that πόροι are not passages, but rather "tiny caves or holes" and that "Theophrastus would have explained the transition from one perceived smell to another by extrusion of the first particles by the next". But, to my knowledge, there is no good evidence for this understanding of πόροι in relation to smell, while there is at least one solid piece of evidence for the idea of πόροι as passages, namely *CP* VI 9.3.6–9, which speaks of an ἀτμός sent from the mouth "through the passages" (διὰ τῶν πόρων) to the sense of smell (εἰς τὴν ὄσφρησιν). Theophrastus' talk of olfactory passages finds support in Aristotle's answer to the question of why humans need inhalation in order to smell at *DA* II 9, 421b32–422a3. The reason is that inhalation uncovers a covering "because the veins and passages (πόροι) are dilating" (cf. *Sens.* 5, 444b20–28, and Priscian, *Metaphr.* 16.17–18).

19 See, e.g., Schorlemmer (2022), 77: "Diese Geruchspartikel hätten unterschiedliche Formen, womit er [d.h. Theophrastus] einerseits erklärt, warum ein Rosenduft, wenn er mit einem anderen Geruch zusammen auftrete, anstelle des anderen wahrgenommen werde."

20 Cf. Eigler and Wöhrle (1993), who duly translate it "so daß die Sinnesempfindung … jegliches Unterscheidungsvermögen verliert" or Baltussen (2015a), 44.

21 The case under consideration represents one of three situations κωλύοντες τὴν κρίσιν (*Od.* 46.2).

tion, κρίνειν means primarily to discriminate, to tell things apart,[22] and one can hardly discriminate *x*—that is, tell *x* apart from other qualities in a given range—when one is unable to discriminate anything else in that range. The fact that Theophrastus invokes olfactory passages and their being filled with the exhalations from rose perfume is perfectly in line with this Aristotelian analysis of perception as discrimination: what the exhalations explain is not a genuine case of smelling (i.e. olfactory discrimination), but rather a case of disabling the sense of smell.[23] This situation thus resembles cases in which a sense is disabled by an extreme perceptual experience as analysed by Aristotle, for instance, in *DA* II 10: having tasted something very bitter or being sick, everything appears (φαίνεται) bitter to the person, so that she is unable to perceive anything.[24] The fact that exhalations of rose perfume can have a similar effect on the sense of smell does not imply that smelling must or indeed can be realised by means of exhalations fitting into passages in the perceiver's organs, any more than standard cases of perception involve extreme affections on Aristotle's account.[25]

Similar conclusions can be drawn about Theophrastus' explanation of the fact that some flowers, such as violets, smell better at a greater distance in *CP* VI 17.1: this is because these plants also emit other exhalations coming from

22 Cf. Ebert (1983).

23 The proposed interpretation of *Od.* 45–46 is, I believe, perfectly compatible with what follows in *Od.* 47. Here Theophrastus explains why, when it is added to a mixed perfume, the rose scent initially crowds out the other scents, but soon vanishes. The *explanans* here is the power of the rose scent to quickly disperse, in the form of exhalations, throughout the surrounding air. This, I take it, allows it to act on the olfactory organs first, before anything else can do so. Although Theophrastus does not repeat here the claim that, when filling the passages, the rose scent makes the sense of smell incapable of discrimination, everything he says is perfectly compatible with such an account: presumably, one genuinely smells the rose perfume only for a certain period of time, namely until one's olfactory passages are filled with it, which prevents one from smelling anything, including the rose perfume itself.

24 *DA* II 10, 422b5–10 (cf. II 12, 424a14–15, 28–32, II 2, 426a30–b8, III 4, 429a31–b3; *Insomn.* 2, 459b13–18). See also Theophrastus, *Sens.* 19, 504.30–505.1: "when an echo is present in the ears, a flavour in the sense of taste or an odour in the sense of smell, they all become blunted, and the fuller they are of what is like [their objects], the more blunted they are."

25 With the support of the Alchemies of Scent laboratory, we ran a little experiment with Sean Coughlin in Prague, with the goal of recreating the experience of the deceived customer. One suggestive outcome was that it is not enough to be exposed to the rose perfume once. If it is to "destroy" other odours, a bit of it needs to be applied to the customer's body, ideally near the nose (that is apparently the point of "scenting" the customer). When done like this, the trick seems to be quite effective. While the text of *Od.* 45–46 may read as if the olfactory passages remain filled up after being briefly exposed to the rose perfume, it seems more likely to me now that what Theophrastus has in mind is a situation where the customer is exposed to it *continually*, so that the passages are continually being filled up.

other parts than the flower, which apparently block the passages and make it impossible for the exhalations from the flower to come through, or rather act through properly.[26] This is only possible from a greater distance, because the disturbing exhalations cannot travel far due to their earthly nature—they fall down. Due to its inability to act through the obstacles or to overpower other odours acting on the olfactory organ, the odour of violets is described as being "weak". Nothing here suggests that exhalations have anything to do with explaining the smelling of violet itself, or indeed that Theophrastus is interested in this general phenomenon. Rather he seems to be focusing on a much more specific problem.[27]

An additional reason to think twice before assuming that Theophrastus' talk of odorous exhalations was intended as an alternative to Aristotle's account is that Theophrastus' criticism of effluence theories is not limited to *De sensibus*. We also find a similar criticism in *De causis plantarum* VI. In VI 5.4, Theophrastus discusses why non-human animals, although they often have a sharper sense of smell, do not perceive certain odours—namely those that, in *Sens.* 5, Aristotle calls pleasurable in virtue of themselves (cf. *CP* VI 5.2; *Od.* 4). Theophrastus mentions the possibility of explaining this difference in terms of a difference in passages (πόροι)—say, by claiming that large animals have large passages and small animals small ones, which determines what odours they can and cannot perceive. In his response, Theophrastus not only rejects this simpleminded contrast between the structure of olfactory organs in large and small animals, but claims more generally that "perception does take place in passages; rather, one would better account for the affection and the lack of it (in which pleasure and pain consist) with respect to a disposition and a mixture [on the side of the perceiver]".[28] This strongly suggests that whatever the

26 Cf. *Od.* 12; *Probl.* XII 2, 906a30–35, XII 4, 906b36–907a4, XII 9, 907a24–28. Hieronymus of Rhodes (Fr. 53 W) seems to incorporate this observation within a general effluence account of smell and perception—but precisely there he appears to go beyond Theophrastus.

27 In *CP* VI 17.2, Theophrastus contrasts the "strong" odours which, on the contrary, are unpleasant when not tempered in some way—usually by mixing the odorous stuff with something else (Theophrastus' favourite example is the myrrh which is soaked in hydromel or sweet wine, so that it smells pleasant and sweet upon burning—cf. *Od.* 12–13). The idea of explaining smell in terms of exhalations is a far cry from Theophrastus' concept of "mixed odours" here.

28 Οὐδ' ἐν τοῖς πόροις ἡ αἴσθησις ἀλλὰ μᾶλλον πρὸς τὴν διάθεσιν καὶ τὴν κρᾶσιν τό τε πάθος ἀποδοίη καὶ τὴν ἀπάθειαν ἐν οἷς τὸ ἡδὺ καὶ τὸ λυπηρόν (*CP* VI 5.4.9–12). This is the very first allusion to odorous effluences and passages within *CP* VI. For both the objection and the alternative, see also Theophrastus, *Sens.* 34–35. Cf. Aristotle, *Sens.* 5, 444a33–b2.

role of the passages that are occasionally invoked in Theophrastus' discussion of smell may be, it is *not* to explain the basic fact that animals perceive odours.

That said, Theophrastus' talk of odorous exhalations and passages seems to mark a genuine advance with respect to Aristotle, insofar as it allows Theophrastus to address questions and explain phenomena that Aristotle never even formulates in the extant writings and that would be rather difficult, if not impossible, to address solely on the basis of Aristotle's teachings that have come down to us. One striking example is the question of why the odours of some things are only smelled while being eaten, which is addressed in CP VI 9.3.[29] Here an exhalation (ἀτμός) and its passing through passages (πόροι) seems crucial for the odour to be registered. Still, it is neither claimed nor implied that smelling itself is to be explained in terms of exhalations and passages or that every smelling involves exhalations.[30] In fact, the latter (and so *a fortiori* the former) seems to be denied by Theophrastus' explanation in CP VI 14.11 of why the rose's fragrance is retained when the rose is being dried *but does not carry far*—as it does when the rose is fresh. This is, Theophrastus claims, because drying roses become hard and "emit no exhalation" (οὐ δίδωσιν ἀναπνοήν, CP VI 14.11.7).[31] Obviously emitting an exhalation is not even a necessary condition for being smelled (otherwise dry roses that do not emit an exhalation could not be smelled—but they can, as is the case with many other fragrant dry plants), and so *a fortiori* it cannot figure in the general account of smell.[32]

29 Cf. *Od.* 12.

30 The πόροι here only seem to play the role of connecting the two senses located close to each other (cf. VI 9.1.4–5).

31 Indeed, throughout CP VI 14.9–11, exhalations are emitted by fresh flowers with sufficient humidity and only until they dry up. Theophrastus' point is to distinguish between those plants that, after exhalations have ceased because they are already dry, lose their fragrance (their fragrance is "lost by evaporation" (διαπνεῖται)) and those that do not—roses being an example of the latter class, together with melilots (cf. *Od.* 34). While this is relatively rare for flowers, there are many other types of plants like the one mentioned by Theophrastus. Indeed, many herbaceous plants (such as fenugreek) only become fragrant when dried (so their fragrance is emphatically *not* "lost by evaporation"). In many cases getting rid of exhalations, which is facilitated by ventilation, is the condition for becoming fragrant, see, e.g., CP VI 14.8.1–4, VI 16.4, VI 16.6. Cf. also CP VI 19.3 on the difference between water and oil: while water "evaporates" (διαπνεῖ) the odours, oil retains them.

32 That is not to deny that most flowers, according to Theophrastus, can only be smelled for as long as they are emitting exhalations (thanks to Richard Schorlemmer for stressing this). The point is that other things, and even some flowers, can be smelled without exhalations, and so even in other flowers exhalations cannot be a part of the account of how they are smelled (they can, at most, explain why flowers are smelled from afar, how the experience of their odour varies across distance or why they cease to be smelled after some time).

All this suggests the following conclusion: Theophrastus' use of exhalations in explaining various olfactory phenomena is intended as an innovation *within*—rather than *against*—the general theoretical framework set out by Aristotle in *DA* II 9 and *Sens.* 5. This is confirmed by the fact that at the outset of *CP* VI Theophrastus clearly recalls Aristotle's account of flavours and odours from *Sens.* 4–5:

> The nature of each kind [i.e. flavour and odour] has been distinguished elsewhere: both are somehow mixed according to a ratio, and while flavour is admixture of the dry and earthy in the wet or percolation of the wet through the dry by means of the hot (for it doesn't really matter [which description we choose]), odour is [some kind of presence] of the dry component of the flavour to the transparent (for that is the common nature of air and water). (Theophrastus, *CP* VI 1.1.4–10)[33]

This reads like a fair, albeit somewhat casual, summary of Aristotle's account in *Sens.* 4–5, especially 441b15–21 and 442b27–443a3.[34] When Theophrastus says that it does not matter whether we speak of percolation (as Aristotle does at *Sens.* 4, 441b17–19) or admixture, this seems to simply reflect the fact that Aristotle is elsewhere willing to speak of flavours as something *admixed* to the wet (see *DA* II 10, 422a13–14). The only difference Theophrastus cares to draw here between flavour and odour is that while the former is limited to the wet, the latter is present in both water and air. The mode in which odour is present in the transparent is not specified. From the context, it would seem to be some kind of mixture (as in Einarson's translation), which sounds somewhat surprising in comparison with Aristotle,[35] or a kind of percolation (διήθησις, as supplied by Amigues). However, Theophrastus says clearly that he is not interested, at the moment, in specifying whether we should understand flavour on the basis of an admixture or a percolation, and he does not give a name to how odours are present in the transparent at all; hence, even a neutral translation like Thompson's ("odour is produced by the dry component of the liquid flavour

33 Ἡ μὲν οὖν φύσις ποία τις ἑκατέρου τοῦ γένους ἐν ἄλλοις ἀφώρισται καὶ ὅτι μικτά πως ἄμφω κατὰ λόγον ἐστί. Χυμὸς μὲν ἡ τοῦ ξηροῦ καὶ γεώδους τῷ ὑγρῷ ἐναπόμιξις ἢ ἡ τοῦ ξηροῦ διὰ τοῦ ὑγροῦ διήθησις ὑπὸ θερμοῦ· διαφέρει δ' ἴσως οὐδέν. Ὀσμὴ δὲ τοῦ ἐν χυμῷ ξηροῦ ἐν τῷ διαφανεῖ· τοῦτο γὰρ κοινὸν ἀέρος καὶ ὕδατος.

34 Cf. 443b6–8, 445a13–14.

35 For the contrast between flavour as something mixed with the wet and odour as something merely acting on the medium, see Johansen (1997), 237–242, and (1996), or Johnstone (2012), 167–170.

in that which is transparent") can perhaps be justified.[36] The main point is that, for present purposes, Theophrastus is *not* interested in emphasising the difference between taste as a contact sense and smell as a distal sense—a difference that often remains unarticulated in Aristotle's own texts.[37] But not emphasising is not the same as undermining, and if Theophrastus wanted to do the latter, we would expect him to be explicit about it, especially since he seems to be doing nothing more here than recalling the shared Peripatetic background.[38]

If Theophrastus accepts this background, how do his innovative explanations in terms of exhalations fit in? It has been suggested that he abandons this framework later in *CP* VI, when he comes to plant fragrances,[39] but that is nowhere implied in the text.[40] Moreover, we have seen that, even later, Theophrastus is far from accepting exhalations as an essential or even necessary ingredient of smelling (he denies the claim that smelling should take place in passages, and he explicitly mentions dry flowers, and more generally dry plants, that do not emit exhalations and yet can be smelled). This suggests that he at least thinks of his explanatory accounts of involving exhalations as fitting into the broadly Aristotelian conception of flavours and odours recalled in *CP* VI 1.1.

It is not that hard to understand how this can be so. If odour is the effect exercised by the flavoured dry on air or water (cf. *Sens*. 5, 443a6–8, 445a13–14), then there are at least two options for explaining how the odour of a flower can reach me through the air so that I register it: either the flavoured dry stuff acts on the air *immediately where the flower is located* (as must happen in the case of dry roses and melilots), or the flower emits flavoured exhalations which are carried by the air to my vicinity—or perhaps even inside my olfactory system—while presumably acting on the air *all along the way*. This idea is supported by observable phenomena (the fragrances of flowers may be accompanied by a visible cloud of tiny particles) and more importantly by the explanatory power of the hypothesis: once adopted, it helps to explain numerous phenomena that would otherwise be hard to account for, including the fact that things that taste

36 Cf. *Sens*. 5, 443a7: ἡ ἐν ὑγρῷ τοῦ ἐγχύμου ξηροῦ φύσις ὀσμή.

37 Cf., e.g., *Sens*. 5, 443b12–14. This makes sense given that the difference between contact and distal senses is anyway relativised in *DA* II 7 and 11.

38 Cf. Thompson (1941), 70–71, who argues that the back-reference is to Aristotle's *De sensu* and *De anima*. Theophrastus speaks of "a kind of mixing" in the medium of smell also in *Physics* V (see Priscian, *Metaphr*. 14.10–12 quoted below in fn. 73), but here he makes it clear that he does not have genuine Aristotelian mixture in mind, but rather a specific kind of being affected (πάσχειν) by the odorous object. Presumably, he is speaking of "a kind of mixing" in place of Aristotle's "washing" and "cleansing".

39 See Schorlemmer (2022), 79.

40 Indeed, the framework seems to be reaffirmed at *CP* VI 14.7.6–9.

worse often smell better[41] or that things (supposedly) have a stronger odour in the morning and a weaker one at noon.[42] Theophrastus thus has very good reasons to hypothesise that exhalations play a role in how we do—or do not—perceive certain odours, especially the fragrances of flowers, without any need to reject or transform the general account of smell developed by Aristotle—something that Theophrastus shows no inclination to do.[43]

In undertaking this task, Theophrastus is likely to have felt encouraged by Aristotle's analysis of odours beneficial to human health in *Sens.* 5.[44] Some of Aristotle's formulations here sound as if he were already speaking of a kind of exhalation: some odours are carried up (ἀναφερόμεναι) to the brain on account of the lightness (κουφότης) of the heat in them (*Sens.* 5, 444a21–24).[45] In any case, Aristotle is absolutely clear that this process, which is directly connected to breathing, is not essential to smelling as such: smelling is conceived of as a unitary phenomenon across the animal kingdom, independently of whether a particular animal breathes or not and whether it lives on land or in the sea (cf. *Sens.* 5, 444b15–28, with *DA* II 7, 419a35–b3, and II 9, 421b13–26). As such, it must have a unitary explanation—as dictated by the principle of commensurate universality (*APo* I 4–5). This theoretical commitment is taken over by Theophrastus in the above-quoted passage from *CP* VI 1.1, where he follows Aristotle in identifying the medium of smell not exclusively with air, but rather with what is transparent—i.e. both air and water (cf. *Sens.* 5, 442b29–30). And Theophrastus retains this commitment consistently throughout his writings.[46]

---

41 See *CP* VI 16.8, where the answer is given in terms of exhalations: the unperfected flavours emit more exhalations, which enables them to affect the sense of smell more forcefully, regardless of whether the odour is agreeable or disagreeable. This is usually the case with wild fruits in comparison with cultivated ones. The situation with flowers is more complicated, as revealed by *CP* VI 20. For the more general question of why some things smell good but taste bad, see Katerina Ierodiakonou's contribution to this volume.

42 *CP* VI 17.2.5–8: the explanation is not spelled out, but from the context it seems that exhalations play some role. Cf. VI 17.3.1–6 for the differences between the seasons of the year.

43 There is thus nothing "tortuous" (Sharples 1985, 199) about the idea that flowers emit exhalations that act on the mediating air.

44 For a discussion of the distinction between the two kinds of odours and how they are perceived, see Lautner (Forthcoming). Cf. *Od.* 35–36.

45 See also the notion of ἀναθυμίασις at *Sens.* 5, 445a25–27. Cf. *Od.* 49–50 (with Eigler and Wöhrle 1993, 72–73) or *CP* VI 16.8–17.2.

46 See Theophrastus, *Sens.* 21–22, and Priscian, *Metaphr.* 16.13–14, who quotes Theophrastus as taking for granted (like Aristotle) that smelling can occur without breathing. Cf. *CP* VI 5.5.6, where Theophrastus considers the possibility that odours exhaling from chasms and caves are special in only being accessible to breathing animals, implying that most odours are not limited in this way.

Aristotle presumably discusses the specifics of human smell in *Sens.* 5 in order to explain the alleged effects on health and at least a part of the pleasure we take in odours in their own right: these are phenomena pertaining to smell that, however, cannot be explained on the basis of the general account of smell provided in *DA* II 9. Similarly, I suggest that, throughout *CP* VI, Theophrastus is invoking exhalations to explain phenomena pertaining to smell that cannot be explained on the basis of the general account of smell provided by Aristotle. In this way, he enhances the explanatory power of the Aristotelian account of perception, which he seems to treat as a common intellectual heritage. He advances the collective Peripatetic project by explaining: how odours can be mediated at a vast distance; how distance determines what we do and do not smell with regard to each given object; why most flowers are only fragrant while fresh; why wild fruits are often more fragrant than cultivated ones and, in general, why things with unperfected flavours tend to be more fragrant; how some odours—or even non-odorous exhalations—can "block" other odours or even temporarily disable the perceptual faculty; how the strength of odours depends on the time of the day and of the year; and so on and so forth. None of these phenomena can be directly explained on the basis of Aristotle's *De anima* and *De sensu* alone. At the same time, none of Theophrastus' explanations implies any change to the general account. Rather they all seem to be developed within the framework that we know from Aristotle's foundational texts. If Theophrastus emphasises this dependence less than we would like him to do, it is presumably because he simply sees it as obvious.

2 Confronting Effluence and Other Competing Theories in *Physics* V?

As we turn, with this in mind, to the meagre testimonies we possess of Theophrastus' discussion of perception in *Physics* V, it is interesting to observe that the discussion of effluence accounts of perception plays a more important role here than it does in Aristotle's *DA* II 5–12.[47] Theophrastus is reported to have raised the following difficulty for both effluence and extramission theories of vision:

47 Aristotle only mentions effluences once in passing at *DA* II 10, 422a14–15 (and once in connection with the nature of light at II 7, 418b13–15); cf. *Sens.* 5, 443b1–2. Ps.-Simplicius, *In DA*, omits any discussion of perceptual effluences (he only mentions them in passing at 155.24–25 when paraphrasing *DA* II 10, 422a14–15).

> And since he [i.e. Theophrastus] is clearly wanting the transparent to be moved by colours, and extending the problem about how the movement comes about, whether by some effluence, or whether by one thing acting and the other being affected, also to those who do not make seeing happen by means of the transparent, but from colours themselves being borne in, and again to those who suppose that the perceptive organ sends [something] out, it is right for us to enquire further also if in fact the transparent is moved, and what is its movement. (Priscian, *Metaphr.* 11.14–20)[48]

And a bit earlier he is reported to have considered (and ruled out) an effluence account of light:[49]

> But if light is incorporeal, why, he says, is it that it occurs in the presence of fire or, in general, body? For it would appear to be a kind of bodily effluence and a body. (Priscian, *Metaphr.* 9.31–33)[50]

The effluence account of light alluded to here was rejected by Theophrastus. If we can trust Priscian, the basis for this rejection was a comparison with heating: when something is heated, it receives the form of heat, not hot particles—and similarly when something is illuminated.[51]

In the first text quoted above, the effluence and extramission theories of vision are confronted with a dilemma that is not entirely transparent from

48 Ἐπεὶ δὲ φαίνεται κινεῖσθαι τὸ διαφανὲς ὑπὸ τῶν χρωμάτων βουλόμενος, καὶ τὴν ἀπορίαν τὴν πῶς κινεῖται, πότερον ἀπορροῇ τινι ἢ τῷ τὰ μὲν ποιεῖν τὰ δὲ πάσχειν, διατείνειν καὶ ἐπὶ τοὺς μὴ διὰ τοῦ διαφανοῦς ποιοῦντας τὸ ὁρᾶν ἀλλ' ἀπ' αὐτῶν ἐπιφερομένων τῶν χρωμάτων, καὶ ἔτι ἐπὶ τοὺς ἐκπέμπεσθαι τὸ αἰσθητήριον ὑπολαμβάνοντας, δίκαιον ἡμᾶς ἐπιζητῆσαι ἔτι, εἰ τῷ ὄντι κινεῖται τὸ διαφανὲς καὶ τίνα κίνησιν. Here and elsewhere in this paper I use the translation by Huby (1997), with modifications.

49 cf. *DA* II 7, 418b13–15.

50 Ἀλλ' εἰ ἀσώματον τὸ φῶς, διὰ τί, φησί, πυρὸς ἢ ὅλως σώματος παρουσίᾳ γίνεται; δόξειε γὰρ ἂν ἀπορροή τις εἶναι σωματικὴ καὶ σῶμα. Cf. *Metaphr.* 8.1–9 (where Theophrastus is reported to have considered—and rejected—the possibility that the transparent is a body present in the air or water) and 5.22–24 with 13.25–31 (where effluence accounts are rejected without any explicit reference to Theophrastus).

51 ὃ δὴ ἐπιλυόμενος, ληπτέον, φησίν, οὐ τοῦτον τὸν τρόπον ἀλλ' ὡς πέφυκεν, ἐκεῖνο πάντως δηλῶν ὡς εἰδητική τις ἐστὶν ἡ τοῦ φωτὸς ἐνέργεια· ἐπεὶ καὶ ἡ θερμότης ἀπ' εἴδους (*Metaphr.* 9.33–10.1). The parallel with heat was already introduced at 8.27–9.9, apparently as Theophrastus' own (cf. Huby 1999, 54–55). Moreover, the account at 9.33–10.1 contradicts Priscian's account at 9.12–30 (see especially 9.19–20), suggesting that here, too, the comparison comes from Theophrastus.

Priscian's report. The thought may be related to Theophrastus' objections against Empedocles in *Sens.* 13. He may have meant to press on the effluence (and extramission) theories the question of whether perception is caused by the effluence (or extramission) itself or rather by its acting on the perceptive organ (or its being affected by the perceptual object). The former leads to difficulties such as having to explain why the perceiver perceives the external object rather than the other way round (*Sens.* 13, 503.4–6) or why the perceptive organ does not perceive itself (*Sens.* 13, 503.6–9). If the latter is right, one can further ask whether it is not better, at least for some sense modalities, to dispense with effluences (not to mention extramission) altogether and limit the account to acting and being affected.

Be that as it may, what is interesting about Priscian's testimonies is the need Theophrastus seems to have felt to address the effluence accounts—and to explain why Aristotle's action/affection account is superior to them. This interest may have been motivated by discussions in the post-Aristotelian Peripatos which was certainly more sympathetic to ideas akin to effluence accounts than Aristotle was.[52] Moreover, given Theophrastus' own use of exhalations in explaining various phenomena associated with smell (as discussed above), one can appreciate his concern to distance himself from the idea that effluences should explain smelling itself and perception more generally.

This polemical context can also shed some light on a worry that occupies several pages of Priscian's discussion. This worry is not explicitly ascribed to Theophrastus, but it is followed by φημὶ τοίνυν—one of the typical marks of switch from Theophrastus' queries to Priscian's responses:[53]

52 We find these tendencies attested in *De coloribus* (see, e.g., 793b33–794a2: "all [perceived] hues represent a threefold mixture of light, that through which the light appears, such as water or air, and underlying colours from which the light is reflected"), *Problemata* (e.g. VII 7, 887a15–21, XII 5, 907a8–12, XIII 4, 908a20–21), and Aëtius, *Placita* 4.13.3 (= 64 Sharples) referring to Strato: "Strato claims that colours travel from the bodies and give their colours to the intermediate air" (Στράτων χρώματά φησιν ἀπὸ τῶν σωμάτων φέρεσθαι συγχρῴζοντ' αὐτοῖς τὸν μεταξὺ ἀέρα).

53 Cf. Caston (2019), 194–195. Unlike Fortenbaugh et al. (1992), Caston lists this passage among Priscian's quotations/paraphrases of Theophrastus, but without commenting on it. Priscian's response at 12.17–28 (which is further elaborated on at a later point) sharply contrasts in both language and doctrine with the formulation of the query, and it is based on a denial that the medium is really affected by the perceptual object (rather it is perfected by it)—something that clearly conflicts with Theophrastus' commitments as attested elsewhere. The question is echoed in Ps.-Simplicius, *In DA* 136.26–28, followed by a reference to the author's "epitome of Theophrastus' *Physics*" which is not unlikely to be our text (cf. also Priscian, *Metaphr.* 17.21–24). A similar objection is found in Plotinus, *Enn.* IV 5.2.50–56. But I do not think it should be taken as a specifically Platonist objection. A

> Is it then that they [i.e. perceptual objects] move the transparent, and it moves the sight, as if it were given shape by the colours? But if the transparent had been affected in this way we would perceive *it* and not the colour [out there]; but if we perceive the colour [out there], this would be what also moves our sight. But as for the transparent—the transparent, when illuminated, is also of itself a thing which moves sight, not by means of something else and not in the way that it is moved by colours—what then is its movement? So that we may know also how the things seen themselves move sight through the medium. (Priscian, *Metaphr.* 12.10–16)[54]

The concern here is not with effluence accounts as such, but rather with an account of mediation that assumes that the medium takes over the quality of the perceptual object.[55] If the medium acts on the perceiver on the basis of having taken over the quality of the perceptual object, then there is nothing to explain why the perceiver should perceive the distant object rather than the contiguous medium (or the closest part of it).[56] Or, indeed (one could continue), why should the perceiver perceive anything at all—if it is true that one cannot perceive what is put directly on the organ without something mediating the affection?[57] Now, although this worry does not concern effluence theories, it can be easily extended to them: if what the distant object does is to send effluences, then there is nothing to explain why the perceiver should perceive the distant object rather than the effluence which has reached her—and indeed how she could perceive anything at all. In fact, the worry seems to be directed against non-effluence accounts of mediation that too closely resemble efflu-

---

structurally similar consideration (directed against effluence accounts) is found, e.g., in Alexander, *In Sens.* 58.16–22, and *Mant.* 11, 136.6–11, and it is obvious how Alexander's own account of mediation seeks to disarm this kind of objection (see, e.g., *In Sens.* 59.10–15; *Mant.* 15). As far as I can see, it is not unlikely that the difficulty (implicit in Aristotle) was first articulated by Theophrastus.

54 Ἆρα οὖν αὐτὰ μὲν τὸ διαφανές, τοῦτο δὲ κινεῖ τὴν ὄψιν, οἷον μορφωθὲν αὐτὸ ὑπὸ τῶν χρωμάτων; ἀλλ' οὕτω τοῦ διαφανοῦς ἂν πεπονθότος ᾐσθανόμεθα καὶ οὐ τοῦ χρώματος· εἰ δὲ τοῦ χρώματος, τοῦτο ἂν καὶ τὴν ὄψιν κινοίη. ἀλλὰ τοῦ διαφανοῦς—τὸ δὲ διαφανὲς καὶ καθ' αὑτὸ κινητικὸν τῆς ὄψεως, ὅταν φωτίζηται, οὐ δι' ἑτέρου καὶ ὡς ὑπὸ τῶν χρωμάτων κινηθέν—τίς οὖν ἡ κίνησις; ἵνα καὶ ὅπως αὐτὰ τὰ ὁρώμενα διὰ τοῦ μεταξὺ κινεῖ τὴν ὄψιν.

55 Cf. Aristotle, *DA* II 10, 422a13–14.

56 The contrast is not drawn exactly in these terms (which may be due to Priscian's imprecise rendering), but this must be what is intended. The idea that the perceiver would perceive the medium rather than the colour only makes sense if under "colour" here we understand the colour *of the object out there*.

57 See *DA* II 7, 419a12–30; II 9, 421b16–17.

ence accounts. Potential targets include Strato and the author of *De coloribus* (if it was written by someone else):[58] if what colours do is to give themselves to the intermediate air, or to mix with the colour of the air, then what we really see—if anything—can only be the resulting colour of the intermediate air, not the colour of the distant object itself. Under perfect conditions, the colour of the air can happen to be exactly the same as the colour of the distant object. However, it does not come to be perceived as a colour *of that object*, but—if anything—as a colour *of the contiguous air*. How it is possible to get from here to perceiving colours of distant objects remains unexplained.

Now if Theophrastus did indeed raise this kind of dilemma (with the indicated target in mind), how did he intend to answer it? And how did he intend to answer it in the specific case of smell? Due to exhalations and the relatively slow process of affection, odours obviously have more independence from their sources than sounds or affections caused by coloured objects. But that just seems to be the way it is—and Theophrastus deserves credit for not only respecting this fact, but also attempting to explain various manifestations of it. The proper fragrance of a violet flower, for instance, can only be smelled from a distance: this fragrance that we perceive is not, like the colour of the violet, something that sticks to the distant object. Rather, it literally fills up the space between it and us, and indeed our olfactory organs, mixing with other exhalations along the way. Still, the way in which the flavoured dry particles act, via the air, on our organ must, presumably, be in principle the same as the way in which distant odorous objects without exhalations (like dry roses) act on it. As suggested in the preceding section, the bottom line for Theophrastus is that exhalations are neither a part of what mediation and perception are, nor a necessary condition for them. What needs to be explained then is how a distant perceptual object can act and be perceived *through* a medium rather than turning the medium into another perceptual object that acts on the perceiver and is perceived by them in its own right.

3 The Nature of Mediation

Thus, if mediation and perception are, according to Theophrastus, not defined by means of effluences or exhalations and can take place without them, how are they defined and how "material" are they? After reporting Theophrastus'

58 See fn. 52 above. For two different views on the authorship of *De coloribus*, see Wöhrle (1999), 45–51, and Gottschalk (1964). For the account contained in *De coloribus* (and its connection to Strato), see also Richard Schorlemmer's contribution to this volume.

question (which is closely related to the above-mentioned dilemma) as to "how [in mediation and perception] like is affected by like, air by external air and the wet by the wet [outside]" (15.29–31), Priscian offers a tentative solution, introducing the notion of the transsonant (διηχές) and the transodorant (δίοσμον).[59] The following passage spells out the kind of answer to Theophrastus' query these notions are supposed to provide:[60]

> And it is not that the like is affected by the like, but the potential by the actual: not indeed as air is affected by air, but as the potentially transsonant by the actually transsonic. (Priscian, *Metaphr.* 16.3–6)[61]

There is a spiritualist interpretation of this answer: instead of a genuine alteration turning the air, say, from odourless to stinky (or from wet to dry), what happens in the medium is a mere "phenomenal" change, which consists in the fact that someone perceives through it, and nothing else. Under the spiritualist interpretation of mediation, the change in question only takes place in the presence of perceivers, which turns it into a truly extraordinary kind of "change".

The first thing to observe here is that Priscian himself has a tendency to read Aristotle in spiritualist terms—for instance, when he claims that neither media nor perceptive organs are really *affected* by perceptual objects, but rather are *perfected* by them.[62] But this is a point made *against* Theophrastus, so if it implies anything at all, it is that Theophrastus did *not* endorse spiritualism in the sense in which Priscian does. Moreover, the above-quoted response, as such, by no means implies spiritualism. In mediating sounds, for instance, the

59 The notion of transsonance was already introduced earlier in an answer to the query about why the distal senses should need anything beyond their objects: "But, someone will say, as flavour is the proper perceptual object of taste, and sound of hearing, needing nothing else from outside for the organ of sense to be moved, unless indeed someone should speak of the transsonant, as colour needs the transparent [...]" (ἀλλ', ἐρεῖ τις, ὡς ὁ μὲν χυμὸς ἴδιον αἰσθητὸν τῆς γεύσεως καὶ ὁ ψόφος τῆς ἀκοῆς οὐδενὸς ἔξωθεν εἰς τὸ κινεῖσθαι τὸ αἰσθητήριον δεόμενος, εἰ μὴ ἄρα τοῦ διηχοῦς τις εἴποι, ὡς τὸ χρῶμα τοῦ διαφανοῦς [...], *Metaphr.* 10.17–20).

60 For an analysis of the entire section in which this passage comes up, which seems to inextricably combine quotes, paraphrases and Priscian's own reflections, see Huby (1999), 42–46.

61 καὶ οὐ τὸ ὅμοιον ὑπὸ τοῦ ὁμοίου πάσχει, ἀλλὰ τὸ δυνάμει ὑπὸ τοῦ ἐνεργείᾳ· οὐ μὴν ὡς ὁ ἀὴρ ὑπὸ τοῦ ἀέρος, ἀλλ' ὡς δυνάμει διηχὲς ὑπὸ τοῦ ἐνεργείᾳ διηχητικοῦ.

62 See *Metaphr.* 12.17–28 (in response to 12.10–16 and extended to hearing at 12.32–13.4) or 14.12–15 (in response to 14.10–12). Cf. 5.24–29, 6.27–7.1, 9.12–30 (about light) and 14.6–8. Another point suggesting spiritualism is Priscian's claim that the perceptive organ is not moved before it is active and would not be moved at all without being active at 2.4–5.

air can be affected in a perfectly physical, perceiver-independent, wave-like manner, meaning that a motion passes through it without leaving any persisting trace in the air itself. It is conceivable that mediation in general consists in changes being *passed through* the medium, so that the medium does not retain any trace of them—but without turning these changes into merely spiritual changes. For instance, the drying involved in the mediation of odours may well have literal drying or heating effects on the perceiver, as seems to be presupposed in Aristotle's account of human smell in *Sens.* 5, without normally leaving that effect in the mediating air: in standard cases, the air is not itself turned into a dryer, warmer or odorous air, because what it does is precisely to *mediate* the drying which "passes through" it (διικνεῖσθαι).[63]

The terms "transsonant" and "transodorant" may or may not come from Theophrastus.[64] André Laks seems to go too far in also ascribing to Theophrastus the metaphor of a lever (μοχλός) that comes up in this context (i.e. that of the question of why we see distant objects rather than contiguous air) in Ps.-Simplicius (*In DA* 136.20–29) with a reference to the author's "epitome of Theophrastus' *Physics*".[65] But the main point is that the notion of the transsonant and the transodorant, and even the metaphor of a lever, may well faithfully capture Theophrastus' original reply, without committing him to spiritualism.[66] The idea, I take it, is *not* that the medium undergoes a merely phenomenal change as something is perceived through it. Rather, it is *being affected* and *changed* in a perfectly physical, perceiver-independent way (like the beam of a lever when it is pushed down). However, if the medium is to duly play its mediating role,

---

63 For Theophrastus' use of διικνεῖσθαι for mediation, see *Metaphr.* 14.16–17 and 17.14–15.

64 Besides the two passages in the *Metaphrasis* (10.17–20, 15.31–16.9), the notions are ascribed to Theophrastus by Philoponus, *In DA* 354.12–16 (for Philoponus, see Ellis 1990, 295–299). Themistius, *In DA* 62.31–32, ascribes the two notions to anonymous "exegetes". The notion of the transodorant is also mentioned by Alexander, *In Sens.* 88.18–89.7, without any ascription (cf. *Quaest.* I 2, 7.10–11). The report of Arius Didymus in Stobaeus, *Anth.* 484.15–21 shows that the notion of the transsonant certainly predates Alexander. Whether the two notions can be traced back to Theophrastus has been a thorny issue. Baltussen (2000), 86 n. 67, and Schorlemmer (Forthcoming) argue against the ascription. Huby (1999), 51–52, 57–58, Laks (2019), 176–177 n. 34, and Johansen (2019), 234, defend it. Caston (2019), 195 n. 20, wisely remains agnostic.

65 Laks (2019), 181–183.

66 The two notions were probably introduced as a response to Aristotle's complaint at *DA* II 7, 419a32–b1, that there is no existing name to capture the nature common to air and water that is responsible for mediating odours parallel to the notion of "transparent" (cf. *Sens.* 5, 442b29–443a2). Surprisingly, Aristotle is satisfied here with identifying air as what mediates sounds, but it is easy to find out that water mediates sounds, too (cf. *DA* II 8, 418b13), and so the very same question arises for sounds as well.

the affection must never end up *having changed* the medium's proper qualities; if that happened, it would mark a failure to mediate, turning the alleged medium of perception into something perceptible in its own terms. In *De causis plantarum* VI, Theophrastus aptly describes what makes the air a good medium for smell: unless the odours are very strong (presumably), the air "is not able to retain them, but only to send them through" (οὐδ' ὁ ἀὴρ δύναται κατέχειν ἀλλὰ διαπέμψαι μόνον).[67]

There is an interesting passage in Priscian's *Metaphrasis* where he attempts to integrate ideas that at least *prima facie* challenge spiritualism. One explanation is that some of these ideas come from Theophrastus and what Priscian does is to try to show that he is able to take them on board.[68] After offering what reads like a spiritualist response (12.17–13.4) to the above-quoted question as to why we perceive distant objects rather than contiguous media—rephrased as "what then the transparent contributes to vision?" at 12.29—Priscian asks again "What then?" (13.4–5). He then goes on to elaborate on the three distal sense modalities one by one—recognising in each case, after all, a passive process (πάθημα καὶ κίνησις) or a passive disposition (διάθεσις παθητική) in the medium. In the case of hearing (13.5–9) and seeing (13.9–16), he is able to argue that this process or disposition is to be distinguished from the activity of mediation and perception proper.[69] In the case of smelling (and tasting and touching), though, the situation is more difficult for Priscian, and here is what he has to say:

> (i) But with regard to the object of smell and smelling, the affection which occurs in between is clearer, since even when the object smelt has gone away the odour remains, not only through the particles which have flowed off, but also because the air has been affected in some way, just as also it has been warmed when it is between that which is warming and that which has been warmed, and as the air is made sweet which is between the tongue and the honey. (ii) And both perception and the perceptual objects are even here in accordance with the form, but since both the perceptive organ and the perceptual object are in these cases more involved in matter, the affection in the air as well as in the percept-

67 *CP* VI 19.3.12–13. The odour "passes through it without fastening on it" (διϊὸν οὐχ ἅπτεται, VI 19.4.8–9).

68 The passage is not included by either Fortenbaugh et al. (1992) or Caston (2019).

69 For hearing, see also *Metaphr.* 14.25–15.1.

> ive organ becomes more forcefully, and thus more clearly, apparent in that they can come to share in the perceptible form. (Priscian, *Metaphr.* 13.16–25)[70]

The idea seems to be that we can understand the way in which the transodorant medium is affected when we observe the aftereffects of strongly odorous objects. When such an object is removed, not only do the emitted particles remain in the air (an echo of Theophrastus' exhalations); moreover, the air itself takes over the odour of the object that has been removed (it comes to share in the perceptible form). The latter observation, which is reminiscent of *DA* II 12, 424b17–19, is then extended to the hot and the sweet—that is, the objects of the two remaining (namely contact) sense modalities.[71]

What is the point of this observation? It is presumably not that, in each case of smelling, tasting and touching, the (external or internal) medium first comes itself to have the same quality as the perceptual object. Allowing this would make any response to the original dilemma impossible (we would always smell the odour of the contiguous air, taste the flavour of our own tongue and feel the warmth of our own body—or rather we would not perceive anything at all). Instead, the point seems to be that we can understand the nature of the affection and change passing through the media by observing what happens when the affection is too strong: the air itself becomes odorous, my hand warm and I come to have an aftertaste on my tongue.[72] What this is supposed to show, I take it, is that in normal cases of perception, too, the affection transmitted through the medium is exactly of this kind, and if it were strong enough it could have precisely these kinds of effects. It is precisely because the affection does *not*

70 (i) Ἐπὶ δὲ τοῦ ὀσφραντοῦ καὶ τῆς ὀσφρήσεως ἐναργέστερον τὸ συμβαῖνον ἐν τῷ μεταξὺ πάθημα, ὡς καὶ ἀπελθόντος τοῦ ὀσφραντοῦ ἐμμένειν τὴν ὀσμὴν οὐ διὰ τὰ ἀπορρεύσαντα μόρια μόνον ἀλλὰ καὶ αὐτοῦ πεπονθότος τι τοῦ ἀέρος, ὥσπερ καὶ θερμανθέντος ὅταν μεταξὺ τοῦ θερμαίνοντός τε καὶ τοῦ θερμαινομένου γίνηται, καὶ γλυκαζομένου τοῦ ἀέρος τοῦ μεταξὺ τῆς γλώττης καὶ τοῦ μέλιτος. (ii) καὶ κατὰ τὸ εἶδος μὲν κἀνταῦθα ἥ τε αἴσθησις καὶ τὸ αἰσθητόν, ἀλλὰ διὰ τὸ ἐνυλότερον καὶ τοῦ αἰσθητηρίου καὶ τοῦ αἰσθητοῦ ἐν τούτοις εἴδους σφοδρότερον καὶ διὰ τοῦτο ἐναργέστερόν τε τοῖς αἰσθητηρίοις καὶ ἐν τῷ ἀέρι, ἵνα μετάσχῃ τοῦ αἰσθητοῦ εἴδους, διαφαίνεται τὸ πάθος.

71 Cf. *Metaphr.* 14.22–25, where Priscian reports that on the issue of internal media of touch and taste, Theophrastus basically agrees with Aristotle without raising significant questions (cf. 5.19–20). The idea of sweetened air is surprising—it may have originated in Priscian's misunderstanding of Theophrastus who may have rather had in mind the internal media of touch and taste.

72 If we stick to Priscian's rendering, the consideration in the case of flavours and tangible qualities would rather be that from the results of their acting on non-perceptive bodies, we can infer the nature of their acting on perceivers' bodies (cf. *DA* II 12, 424b12–13). The upshot would be the same.

have these effects—the medium is being affected (πάσχον) by the object, but it passes the affection along before it can be completed in it (πεπονθός)—that the medium functions as a medium rather than becoming a perceptual object of its own.[73] The notion of a "transodorant" function captures this idea rather nicely, without necessarily having any spiritualist implications: at any rate, what happens in the medium is a perfectly physical process which occurs independently of whether a perceiver is present or not.

It is not clear what Priscian means when he says that, even here, both perception and the perceptual object "are in accordance with the form". Apparently, he wishes to preserve a spiritualist account of mediation proper and dissociate it from the physical changes in the medium (as in the case of seeing and hearing), but without really knowing how to do so.[74] There is, of course, nothing that rules out the possibility that Priscian himself created this difficulty. But it seems more likely that he is struggling with something he found in Theophrastus—attempting, not very successfully, to integrate Theophrastus' insights within his own account, as he often does.[75] If so, then at least the first part of the above-quoted passage—namely (i)—can shed some light on Theophrastus' account of mediation.[76]

73 We never learn anything more precise about the exact mechanism of this affection—either from Theophrastus' botanical works or from Priscian's reports—than what Aristotle lays down in DA II 9 and *Sens.* 5. A little later (*Metaphr.* 14.10–12), Theophrastus is quoted as saying: ἡ δὲ ὄσφρησις, φησί, διὰ τοῦ ἀέρος ὥσπερ ἀναμιγνυμένου πως ἔοικε καὶ πάσχοντος, ἡ δὲ ἀκοὴ σχηματιζομένου. But the mention of mixture here (echoing CP VI 1.1) should not be overinterpreted (as already noted in fn. 38). It may be just another expression for Aristotle's "washing" and "cleansing" (*Sens.* 5, 443a1–2, 445a13–14), and Theophrastus' καὶ πάσχοντος suggests that he understands it at any rate as a way of being affected rather than a mixing strictly speaking (cf. Huby (1999), 38–39). What is potentially relevant here is the present tense of πάσχειν: the medium is *being affected*—without (in successful cases) this affection ever coming to be completed, i.e. the medium *having been affected* (πεπονθός).

74 Compare modern spiritualist accounts of mediation (including that of odours) in Burnyeat (1992), (1995) and (2001) and especially Johansen (1997), 116–147. Cf. Broadie (1993).

75 Cf. Priscian's reaction at *Metaphr.* 14.12–15 to Theophrastus' claim (at 14.10–12) that the mediation of odours involves affection.

76 As also suggested by the Theophrastean idea of emitted (flavoured dry) particles there. Another interesting passage is *Metaphr.* 18.1–7, with a discussion of DA II 10, 422a34–b2. The question is why Aristotle insists that the tongue must be neither moist nor incapable of being moistened. Someone's suggestion (φαίη ἄν τις) against Aristotle is reported: it would seem that what matters is only that the tongue is not flavoured. This suggestion is then countered by affirming that since flavours are perceived by contact (ἁπτικῶς), the organ must be affected by the moist, too. Thus, perceiving flavours essentially involves moistening of the tongue, which appears, again, to go against the core idea of spiritualism. This passage is not included by Fortenbaugh et al. (1992) or Caston (2019), but see Huby (1999), 79, for a suggestion that the argument may come from Theophrastus. This

4 The Nature of Perception

If the preceding considerations are on the right track, it remains to be asked what kind of alternative to spiritualism Theophrastus might have embraced in his understanding of perception. I have argued that he did not endorse an effluence account (in which effluences would be necessary, or even essential, to perceiving). And everyone agrees that Theophrastus did not endorse literalism, as has been ascribed to Aristotle by some recent interpreters.[77] The very first testimony from Priscian seems to rule this out, as does a related passage in *De sensibus* (both are quoted below). The only remaining option currently on the table is the idea that Theophrastus conceived the transmission of perceptual information on the transduction model, as was originally proposed by Victor Caston for Aristotle.[78] Before evaluating this suggestion and asking whether Theophrastus' remarks may be understood in yet another way, we need to look at the key texts, starting with the very first preserved passage from Priscian's *Metaphrasis*:

> His [i.e. Theophrastus'] next target is concerned with perception. Since Aristotle wants the perceptive organs, when moved by the perceptual objects, to become like those objects by being affected, he asks what the becoming like consists in. For with perceptive organs, and even more with the soul, the capacity to become like [the perceptual object] in colour and flavours and sound and shape seems absurd. So he claims, too, that the becoming like occurs with regard to the forms and λόγοι without the matter. (Priscian, *Metaphr.* 1.3–8)[79]

The question about the kind of assimilation involved in perception naturally arises for any reader of DA II 5, where Aristotle concludes that "that which

suggestion can be supported by drawing a contrast with Ps.-Simplicius, *In DA* 157.11–29, where the author argues that the tongue needs to *be moist* and seems to leave no room for it being *moistened* when tasting (cf. also the continuation at 18.7–10, where Priscian hastily attempts to disarm the argument by recalling Aristotle and Theophrastus' view that the tongue is, in fact, not the proper organ of taste).

77 See, e.g., Slakey (1961); Sorabji (1974); Sorabji (1992a); Everson (1997).

78 See Caston (2005) and (2020) for Aristotle, and Caston (2019) for Theophrastus.

79 Περὶ αἰσθήσεως αὐτῷ ὁ σκοπὸς ἐφεξῆς. ἐπεὶ δὲ ὑπὸ τῶν αἰσθητῶν κινούμενα τὰ αἰσθητήρια ἐξομοιοῦσθαι τοῖς αἰσθητοῖς τῷ πάσχειν βούλεται ὁ Ἀριστοτέλης, ζητεῖ τίς ἡ ὁμοίωσις. καὶ γὰρ ἐπὶ τῶν αἰσθητηρίων καὶ ἔτι μᾶλλον ἐπὶ τῆς ψυχῆς ἄτοπον φαίνεται τὸ χρώματι καὶ χυμοῖς καὶ ψόφῳ καὶ μορφῇ ἐξομοιοῦσθαι δυνατόν. λέγει μὲν οὖν καὶ αὐτὸς κατὰ τὰ εἴδη καὶ τοὺς λόγους ἄνευ τῆς ὕλης γίνεσθαι τὴν ἐξομοίωσιν.

can perceive (τὸ αἰσθητικόν) [...] is being affected as being unlike, but once it has been affected it is assimilated and is like it [i.e. the perceptual object]" (418a4–6). In what sense is the perceiver like the perceptual object that she is perceiving? One reason why this is a difficult question is that basic principles of Aristotle's natural philosophy (spelled out most clearly in *GC* I 7) seem to imply that as soon as the perceiver *is like* the perceptual object, she can no longer *be affected* by it (for only unlike things can act upon each other). So how can she retain perceptual contact with it? A related concern has been raised above with respect to media: if a medium were to take over the quality it is supposed to mediate, it would thereby become unable to fulfil its function. In the case of media, this worry can be resolved by saying that they are never really assimilated (if they were, we would—at best—perceive them and not the distant objects). But this answer is not available in the case of perceivers (and their organs), since these *are* clearly assimilated somehow according to *DA* II 5—hence the urgency of Theophrastus' question about the mode of their assimilation.

The same kind of question is raised more fully and eloquently in *Sens.* 19 as a part of Theophrastus' polemic with the Empedoclean account of perception:

> It is particularly difficult to determine the affections of sight, but as far as other senses are concerned, how do we discriminate by the like? The problem is that "like" is indeterminate. For [we do] not [discriminate] sound by a sound, odour by an odour or, in general, quality by a quality of the same kind; rather, it would be more, so to speak, by contraries [that we discriminate]. For the sense has to be set before [the external perceptual objects] as being impassive (ἀπαθῆ γὰρ δεῖ τὴν αἴσθησιν προσάγειν). When, on the other hand, an echo is present in the ears, a flavour in the sense of taste or an odour in the sense of smell, they all become blunted, and the fuller they are of what is like [their objects], the more blunted they are—unless one draws distinctions about these. (Theophrastus, *Sens.* 19, 504.25–505.2)[80]

The demand for ἀπάθεια on the part of αἴσθησις here has most often been understood as requiring that the sense be devoid of any modifications prior to being

80 ἀλλὰ τὰ μὲν τῆς ὄψεως πάθη χαλεπώτερον ἔσται διελεῖν. τὰ δὲ περὶ τὰς ἄλλας αἰσθήσεις πῶς κρίνωμεν τῷ ὁμοίῳ; τὸ γὰρ ὅμοιον ἀόριστον. οὔτε γὰρ ψόφῳ τὸν ψόφον οὔτ' ὀσμῇ τὴν ὀσμὴν οὔτε τοῖς ἄλλοις τοῖς ὁμογενέσιν, ἀλλὰ μᾶλλον ὡς εἰπεῖν τοῖς ἐναντίοις. ἀπαθῆ γὰρ δεῖ τὴν αἴσθησιν προσάγειν· ἤχου δὲ ἐνόντος ἐν ὠσὶν ἢ χυλῶν ἐν γεύσει καὶ ὀσμῆς ἐν ὀσφρήσει κωφότεραι πᾶσαι γίνονται μᾶλλον ὅσῳ ἂν πλήρεις ὦσι τῶν ὁμοίων, εἰ μή τις λεχθείη περὶ τούτων διορισμός.

affected.[81] This neutrality is certainly insisted on, but it cannot, I think, be all that Theophrastus has in mind, given that his point about the impossibility of discriminating sound by a sound surely applies equally well to the situation in which the perceiver has already been affected by a sounding object: not even then can there be, for instance, "a sound" in the perceiver by which she would supposedly discriminate sound, for this inhering sound would blunt her hearing and make any perceptual contact with the external sound impossible. Thus, it is necessary for the perceiver to be not just "without modifications" prior to being affected, but also "impassive" in the sense of being able to resist being modified in a way that would turn her organs into perceptual objects similar to those acting upon her. This consideration seems to make any assimilation account impossible—unless, Theophrastus adds, one draws distinctions with regard to sound, odour, and other perceptual qualities. That would, effectively, mean drawing distinctions about the meaning of likeness.

Presumably, the distinctions Theophrastus is calling for are supposed to explain in what sense something like an odour can (and must) come to be present in the perceiver while she is perceiving and in what sense it cannot—that is, in what sense the latter is and in what sense it is not *like* the former. The above-quoted passage from Priscian closes by reporting Theophrastus' understanding of Aristotle's answer to this question: the perceivers are assimilated with respect to forms and λόγοι without the matter. We can be reasonably sure that this is Theophrastus' own formulation, for Priscian later repeats it verbatim (*Metaphr.* 3.27–29) before introducing his own qualifications. And the claim is not so surprising. Theophrastus is basically just saying that in order to fully understand the assimilation model introduced in *DA* II 5, we need to wait until II 12, where the kind of likeness is finally specified: perception is "the capacity to receive perceptual forms without the matter" (424a18–19) and the perceiver "is affected […] with respect to λόγος" (a23–24). So far, so good. But how exactly should we understand the specification? This is, of course, a notoriously difficult question, and Priscian does not give us many hints as to how Theophrastus may have answered it.

What is reasonably clear, as noted above, is that Theophrastus would not accept the literalist reading of *DA* II 12, 424a17–24, which takes the idea of forms without matter as referring to nothing more than the fact that perceptive organs do not receive material particles coming from perceptual objects, but are only changed qualitatively. It is not just that Theophrastus *does*, as we

81 See, e.g., Caston (2019), 198, and Johansen (2019), 242. Along the same lines, Ierodiakonou and Bydén (Forthcoming) translate ἀπαθῆ as "in an unaffected condition". Strangely enough, George M. Stratton renders ἀπαθῆ as "in a passive state".

saw, speak about material exhalations being received by the organ of smell. Also and more importantly, the way in which he raises his question about the mode of assimilation seems to exclude the idea of a literal likeness being established in the organ: that would, apparently, be precisely the kind of likeness that makes our senses imperceptive, rather than making them perceive anything. It has been argued that, since Theophrastus cannot be referring to a literalist understanding of likeness, he must have a spiritualist assimilation in mind.[82] But we have seen that Theophrastus could not have endorsed a spiritualist understanding of mediation; and if media are affected by perceptual objects in a perfectly physical way—independently of whether a perceiver is present or not—then the perceivers can hardly be affected by the media in a merely spiritual way.

Victor Caston proposes that what happens in the perceptive organs, according to Theophrastus, is a case of transduction: the same ratio that defines the perceptual quality in question (as, say, a ratio of white and black) comes to be realised in the organ in a different pair of contraries (say, the contraries of viscous and runny). Caston has one key piece of evidence for this proposal, namely the following passage from Priscian:

> In what way, then, is λόγος the cause of perception? For λόγος lies in the mixture of the several and [or: that is] in the relation of them to one another. He solves this himself by placing λόγος not in the relation of the elements [to one another], but in that of the perceptive organ to the perceptual objects. (Priscian, *Metaphr.* 21.4–8)[83]

We can be reasonably sure that both the question and the answer here come from Theophrastus, since Priscian goes on, in his customary way, to polemicise against what he finds unacceptable in Theophrastus' answer, arguing that instead of identifying the relevant meaning of λόγος as that of relation (σχέσις) of the sense organ to its object, we should understand it as pertaining to the "living substance" (οὐσία ζωτική) from which the λόγος is "projected" (προβάλλεσθαι) in perception. Caston understands Theophrastus' "relation of the perceptive organ to the perceptual objects" in terms of an equation between the λόγος defining the perceptual object and the λόγος present in the perceptive organ: as *a* is to *b*, so *c* is to *d*, which is the point of transduction. However,

82 Laks (2019).

83 Πῶς οὖν ὁ λόγος τῆς αἰσθήσεως αἴτιος; ὁ γὰρ λόγος ἐν τῇ τῶν πλειόνων μίξει καὶ ἐν τῇ πρὸς ἄλληλα αὐτῶν σχέσει. ἢ τοῦτο καὶ αὐτὸς ἐπιλύεται, οὐκ ἐν τῇ τῶν στοιχείων ἀλλὰ ἐν τῇ τοῦ αἰσθητηρίου πρὸς τὰ αἰσθητὰ σχέσει ἀφορίζων τὸν λόγον.

this is hard to find in the text: instead of two λόγοι being equal, it speaks of a single λόγος consisting in a certain relation of the organ to the perceptual objects.[84] Moreover, the idea of two λόγοι seems to spoil the contrast between a relation of elements *to each other* on the one hand and a relation of the perceptive organ *to perceptual objects* on the other: if "elements" are understood in a sufficiently broad sense, then the λόγος realised in the viscous and the runny *is* a λόγος of elements, and thus provides no alternative.[85] Furthermore, the transduction interpretation implies that the relation in question (i.e. of equality) is perfectly symmetrical, whereas Theophrastus' formulation suggests that *unlike* in the case of the elements, where λόγος consists in their relation to *each other*, the relevant notion of λόγος consists in an asymmetrical relation of the perceptive organ to perceptual objects—which is as it should be, because the former perceives the latter and not the other way round.[86]

It seems more plausible to assume that only one λόγος is in play and that this λόγος is not to be understood in terms of a ratio of any elements or qualities inherent to the perceptive organ, but rather in terms of a relation of the perceptive organ to the perceptual objects. The context suggests that not being a λόγος of the elements is intended as equivalent to being a λόγος received "without the matter". There is, admittedly, very little to draw on in reconstructing Theophrastus' understanding of this notoriously tricky phrase, but the evidence at least seems to be perfectly compatible with the following interpretation. The assimilation involved in perceiving consists in receiving the form of the quality alone—that is, receiving the form without it coming to be enmattered in the organ (not even in a transduced way). Rather, the matter of the received quality remains outside—in the perceived object. Presumably, the form is nothing other than the λόγος—that is, the ratio defining the quality in question. The point is precisely that this λόγος comes to be present in the organ

84 Cf. *Metaphr.* 19.30–31, 20.23–25.

85 Caston also refers to Theophrastus, *Sens.* 32, 508.19–20 (ἐν συμμετρίᾳ τινὶ καὶ κράσει πρὸς τὸ αἰσθητὸν ἡ αἴσθησις). But the idea of two λόγοι being equal is not articulated here either.

86 One can also wonder, since—at least in the case of odours, flavours, and tangible qualities—what occurs when the medium or the organ is too forcefully affected is that it turns to have an odour, flavour or, say, heat of its own (as we saw above), how it could happen that in standard cases of perception, the organ does not become odorous, flavoured or hot, but acquires quite different qualities that only share the same ratio. Moreover, if what a quality *F* does to the organ is to make it *G*, then as soon as the organ has become *G*, it is not obvious how it could be affected by the quality *F* any further, and thus how it could continue to perceive it: it would seem that the organ once again becomes a *perceptible* object in its own terms rather than *perceiving* anything. It is not obvious why being *G* should make the organ any less blunted, in Theophrastus' terms, than being *F*.

not as a ratio of the organ's own constituents, but in the form of a relation that the organ establishes—during the time when the perceiving occurs—to the perceived object.[87]

But is this not just a regression to spiritualism? Not really. We have seen that the way in which the medium is affected and acts on the perceptive organ has nothing spiritual to it: the affection that mediates odours is such that it can end up turning the medium into an odorous object; the only reason why the medium remains neutral is that it immediately passes the affection along. It is this physical affection that arrives at the perceptual organ, and so the way in which the organ is being affected is certainly not merely spiritual. The point is simply that this affection cannot result in the organ acquiring the quality in question *as its own*—any more than the medium does. The only way in which the organ can be assimilated to the perceived object is purely formal: i.e. the form comes to be present in it while its matter remains outside. But, unlike in the case of spiritualism, this assimilation results from a perfectly physical kind of action exercised by the perceptual object, via the medium, on the perceptive organ.

How exactly it is that this way of being affected yields the requisite result (the form being received without its matter) is a different question, for which we find almost no clues in the testimonies we possess. However, one negative clue may perhaps be found in the very first passage from Priscian's *Metaphrasis* quoted above (1.3–8). It has been argued that the idea that literal assimilation is even more absurd in the case of the soul than in the case of the perceptive organ is Priscian's interpolation, since the concern with the soul's impassivity is typical of later Neoplatonist approaches and would make little sense in Theophrastus' case.[88] That, however, is far from obvious. In his so-called *Metaphysics*, Theophrastus comes close to accepting the view that perception is a motion of the soul itself[89]—just as Aristotle seems to accept it in *Physics* VII 3 (248a6–9) and *De somno* 1 (454a7).[90] And in his *On Motion* Theophrastus is

87 Does this mean that λόγος loses the meaning of "ratio" and acquires the vaguer meaning of "relation" (σχέσις) instead? I do not think so. The λόγος in question will be the ratio defining the perceived quality. The point is that it will not be present in the perceptive organ as a ratio of its own constituents, but exactly as the defining ratio of the quality of the perceptual object *out there*. (This, I take it, is possible due to the neutrality of the organ—both before and during perception.) In this way, the λόγος constitutes a relation of the perceiver to the perceived object.

88 See Caston (2019), 197.

89 ἐπεὶ καὶ αἱ αἰσθήσεις καίπερ ἐν τῷ πάσχειν οὖσαι δι' ἑτέρων ὅμως ἐν ψυχῇ γίνονται (*Metaph.* II 8 = 8a Gutas, 5b5–7).

90 The idea reflects Plato's view as articulated in *Theaet.* 186b11–c2; *Phlb.* 33d2–7, 34a3–5; or *Tim.* 43c–44c, 45c–d, 67b.

reported to have claimed that κρίσεις καὶ θεωρίαι are motions which "have their origin, their activity, and their end in the soul itself".[91] I am inclined to think that this claim in *On Motion* is made specifically for "theoretical discriminations"—that is, the acts of thinking as analysed in Aristotle's *De anima* III 4–5—and not for acts of perceptual discrimination.[92] Irrespective of that, it would not at all be surprising if Theophrastus were concerned in *Physics* V with the question as to whether or not the soul itself is affected by and assimilated to perceptual objects: following Plato and certain texts of Aristotle, he at least flirted with the idea elsewhere and he insists in *On Motion* that theoretical thinking, at any rate, *is* a motion of the soul itself. It would thus make perfect sense for him to emphasise (as Aristotle does in *DA* I 3–4) that the soul itself cannot be moved by, let alone assimilated to, perceptual objects acting on the perceptive organs.[93] One motivation for this emphasis comes from the fact that, in Aristotle's *De anima* II 5, the assimilation account is applied to τὸ αἰσθητικόν which often (but not always) means the perceptive part of the soul. If what Priscian says about the soul at *Metaphr*. 1.5–7 is a genuine report, then Theophrastus seems to have explicitly excluded an interpretation that can be traced back to antiquity and has also enjoyed some popularity in recent discussion of Aristotle.[94]

∴

Although perception was clearly an important topic for Theophrastus, the sources we possess are regrettably limited. I have argued that they allow us nevertheless to establish some conclusions with a reasonable degree of likelihood and to make educated guesses about other aspects of Theophrastus' view. There is no good reason to think that Theophrastus abandoned Aristotle's general account of perception, or his account of smell, in favour of a quasi-Empedoclean effluence theory. What he did was rather to introduce odorous exhalations as an additional, complementary causal factor allowing him to explain a significantly broader range of olfactory phenomena than would be possible on the basis of *De anima* and *De sensu* alone. Furthermore, there are

---

91 See Simplicius, *In Phys*. 964.29–965.6 = 271 FHS&G.

92 More on this in Roreitner (2024), 438–444. For a similar view, see, e.g., Sharples (2006), 171.

93 Cf. Aristotle, *DA* I 5, 410a10–11; III 8, 431b28–29.

94 It can be traced back to Themistius, followed later by, among others, Thomas Aquinas (for a reconstruction, see Roreitner 2023). Recently this view has been most forcefully defended by Lorenz (2007) (cf., e.g., Charles 2000, 111, Johansen (2012), 166, and Carter (2018), 53–58).

good reasons to think that the way perceptual objects act on the media and on the perceptive organs is not, according to Theophrastus, merely spiritual. Rather he seems to have conceived the role of the medium in terms of transferring perfectly physical (i.e. perceiver-independent) affections, but without ever having these affections completed in itself—that is, without itself acquiring the quality in question as a quality of its own. The same clearly holds for the perceptive organs, at least as far as literal assimilation is concerned: Theophrastus seems to have explicitly excluded the idea that when perceiving red colour, the eye-jelly itself comes to be red. Venturing beyond these findings seems impossible without introducing some amount of speculation. I have argued that nothing in our testimonies really suggests that Theophrastus understood the assimilation involved in perceiving along the lines of transduction. They are more suggestive, overall, of the idea that what the physical (i.e. perceiver-independent) acting of the perceptual object (via the medium) on the perceptive organ results in is a reception of the *form* of that object alone, while its matter remains outside: the form of the perceived object comes to be present in the ensouled organ (rather than in the soul itself) without becoming enmattered in it in any way. This is apparently what, for Theophrastus, perception is.[95]

---

95 This paper was presented at the "Methods in Aristotelian Tradition" conference in Olomouc (2024). Many thanks to the audience for the helpful discussion. I am most grateful to those who read and commented upon earlier versions: Mike Arsenault, Sean Coughlin, Katerina Ierodiakonou, Richard Schorlemmer, Luca Torrente and, in particular, Gweltaz Guyomarc'h, not only for his feedback, but also for kindly inviting me to contribute to the present volume. The work was supported by the Czech Science Foundation (project No. 25–16298S).

CHAPTER 4

Why Do Flowers Smell Good but Taste Bad? Theophrastus' Reply

Katerina Ierodiakonou

1 Introduction

In his psychological treatises *De anima* and *De sensu* Aristotle talks about the senses of taste and smell in close sequence. In *De anima* II 9 and II 10 he gives an account first of how smell and then of how taste perceive their respective objects, namely odours and flavours, while in *De sensu* 4 and 5 he treats these sense objects in reverse order, first flavours and then odours. Although the sense of smell is, according to Aristotle (*DA* II 9, 421a7–16), less accurate than taste—in fact, smell is regarded by him as the worst sense human beings have both in comparison to the other four senses and in comparison to the sense of smell other animals have—, flavours and odours are said to exhibit a close affinity to the extent that they are considered as "almost the same affection" (*Sens.* 4, 440b30: σχεδὸν … τὸ αὐτὸ πάθος).[1] For Aristotle conceives of flavour as the affection produced by something dry being immersed in what is moist (*Sens.* 4, 441b22–23), and he conceives of odour as the affection produced by the already flavoured moisture in air and water alike (*Sens.* 5, 442b30–31).[2] Moreover, Aristotle (*Sens.* 5, 443a9–22) corroborates the close connection between taste and smell by invoking our everyday experience that flavourless things—for instance, stones and gold—are also odourless. And it would have been strange if the ancients did not notice that, when we have a bad cold and we lose our sense of smell, our food tastes bland.

1 The Aristotelian commentators follow Aristotle in pointing out that the sense of smell is the worst sense human beings have (Alexander, *DA* 51.8–10; 51.25–52.11; Themistius, *In DA* 67.26–68.4; Philoponus, *In DA* 386.4–12; 'Simplicius', *In DA* 150.35–152.26), and all of them discuss, too, the close affinity between flavours and odours (Alexander, *DA* 51.13–25; Themistius, *in DA* 68.4–9; Philoponus, *In DA* 389.23–390.2; 'Simplicius', *In DA* 153.10–18).

2 On Aristotle's accounts of flavour and odour, see Johansen (1997), 230–237; Johnstone (2012), 164–170; Ierodiakonou (Forthcoming (b)).

 | DOI:10.1163/9789004739314_006

The close affinity between flavours and odours is further stressed by Aristotle when, in *De sensu* 5, he introduces the analogy between the different kinds of flavour and the different kinds of odour:[3]

T1: Aristotle *De sensu* 5, 443b3–16

It is not unclear, therefore, that the moist which is in either breath or water can draw upon flavoured dryness and be affected by it; for air, too, has a moist nature. Further, if dry stuff which is, as it were, washed out acts in a similar way within liquids and air, it is evident that odours must be analogous to flavours (δεῖ ἀνάλογον εἶναι τὰς ὀσμὰς τοῖς χυμοῖς). And this, in fact, happens in some cases (ἐπ' ἐνίων συμβέβηκεν). For odours are pungent (δριμεῖαι), sweet (γλυκεῖαι), tangy (αὐστηραί), sour (στρυφναί), and fatty (λιπαραί), and one might say rotten things (σαπράς) are analogous to bitter ones (πικροῖς).[4] That is why, just as bitter things are hard to drink, rotten things are hard to breathe in. It is clear, therefore, that odour in air and water is the very thing which flavour is in water. And for this reason cold and freezing dull flavours and make odours disappear; for cooling and freezing make the heat, which is the moving and creative force, disappear.

However, what particularly surprises us, in this passage, is Aristotle's statement that the correspondence between flavours and odours holds "in some cases"; and we find a similar parenthetical remark in *De anima* II 9:

T2: Aristotle, *De anima* II 9, 421a26–b3

Just as some flavours are sweet and some are bitter, so it is with odours. (But some things possess an analogous odour and flavour—I mean, for example, a sweet odour and a sweet flavour—while others possess a contrary one (τὰ δὲ τοὐναντίον).) Similarly an odour may be pungent, tangy, acidic (ὀξεῖα), or fatty. But because, as we said [i.e. 421a7–16], odours can-

3 All translations from Aristotle's psychological treatises are based on Miller (2018), with some changes.

4 Needless to say, most Greek terms denoting different kinds of flavour or odour are difficult to translate, since we do not know with certainty to what exactly they principally refer. For instance, the term "στρυφνός", which is used both for characterising apples (Athenaeus, *Deipn.* II 56.33) as well as wine (Galen, *De vict. att.* 100.4), is translated here as "sour", but it is also sometimes translated as "astringent"; "αὐστηρός", which is used for water (Plato, *Phlb.* 61c7) and for wine (Aristotle, *Top.* III 3, 111a5), is translated as "tangy" but also as "tart" or "grating"; "δριμύς", which is used for thyme (Aristotle, *DA* II 9, 421b2) and for oregano (Athenaeus, *Deipn.* II 77.31), is translated as "pungent" but also as "harsh".

> not be sharply distinguished the way flavours can, they have acquired names from these objects in virtue of their similarity: the sweet odour of saffron or honey and the pungent odour of thyme and similar objects, and the same for other odours.

So, although the close affinity between flavours and odours is presented, here, as the reason why we use the same terms both for the different kinds of flavour and for the different kinds of odour, Aristotle also implies that the analogy between flavours and odours sometimes breaks down. Which are, then, these cases that Aristotle has in mind?

The Aristotelian commentators of the *De anima* notice Aristotle's reluctance to generalise in an absolute way this analogy, and they consider it helpful to add examples in order to justify his reserved claim. For instance, Themistius (*in* DA 68.18–25) refers to the case of honey in which the sweet flavour coincides (συντρέχει) with the sweet odour, just like Aristotle does, and he also mentions that in the case of most perfumes their odour is sweet while their flavour is bitter.[5] John Philoponus (*In* DA 389.28–33) adds garlic and leek to the examples given by Aristotle of things that have an odour consonant (σύμφωνον) with their flavour—that is, sweet in the cases of saffron and honey, and pungent in the cases of thyme, garlic, and leek—, and he further gives two examples of things whose bitter flavour is not consonant (ἀσύμφωνον) with their sweet odour, namely aloe and storax.[6] Finally, Priscian ('Simplicius', *In* DA 153.35–154.1)[7] repeats Aristotle's example that the flavour of honey coincides (συντρεχούσης) with its odour, but he does not agree that saffron has a sweet flavour and thus presents it as an example of something whose flavour is in discord (διαφωνεῖ) with its sweet odour.[8]

---

5 πλὴν ὅπερ εἶπον, οὐκ εὐποροῦντες τῶν οἰκείων ὀνομάτων ἐπὶ τῆς ὀσμῆς ἀπὸ τῆς γεύσεως αὐτὰ μεταφέρομεν, καὶ ὥσπερ τῶν χυμῶν τοὺς μὲν γλυκεῖς, τοὺς δὲ πικροὺς ὀνομάζομεν, οὕτω καὶ τῶν ὀσμῶν τὰς μὲν γλυκείας, τὰς δὲ πικράς. ἐνίοτε μὲν οὖν ἐπὶ τῶν αἰσθήσεων ἀμφοτέρων συντρέχει καὶ τὰ ὀνόματα καὶ τὰ πράγματα, καὶ ὁ γλυκὺς τῇ γεύσει χυμὸς καὶ τῇ ὀσμῇ γλυκύς, ὥσπερ τοῦ μέλιτος· πολλὰ δὲ ἡδέα μὲν τῇ ὀσμῇ, πικρὰ δὲ τῇ γεύσει, ὥσπερ τὰ πολλὰ τῶν μύρων.

6 τῶν δὲ γευστῶν φησιν ἅμα καὶ ὀσφραντῶν τὰ μὲν σύμφωνον ἔχειν τὸν χυμὸν τῇ ὀσμῇ, οἷον κρόκος τε καὶ μέλι (γλυκεῖα γὰρ ἐπ' ἀμφοτέρων λέγεται ἡ ὀσμή, ὥσπερ καὶ ὁ χυμός· ὁμοίως τὸ θύμον, σκόροδον, πράσον δριμὺ καὶ τῇ ὀσφρήσει καὶ τῇ γεύσει), τὰ δὲ ἀσύμφωνον, οἷον ὁ ἀλόας τὴν μὲν γεῦσιν πικρὰν ἔχει, γλυκεῖαν δὲ τὴν ὀσμήν, ὁμοίως δὲ καὶ ὁ στύραξ.

7 On the identity of the author of this commentary, see the introduction in Steel (1997) and Hadot (2002).

8 ⟨ἡ μὲν γλυκεῖα κρόκου καὶ μέλιτος⟩, ἐπὶ μὲν τοῦ μέλιτος συντρεχούσης τῇ γλυκείᾳ ὀσμῇ τῆς γεύσεως, οὐχὶ δὲ καὶ ἐπὶ τοῦ κρόκου, ἀλλὰ διαφωνεῖ πρὸς τὴν ὀσμὴν γλυκεῖαν οὖσαν οὐ τοιαύτη ὑπάρχουσα ἡ γεῦσις.

Hence, Aristotle and his commentators seem to have been well aware of the fact that sometimes the objects of taste and smell are not analogous but contrary. How did they try to account for such a phenomenon? My aim, here, is to look into Aristotle's suggested explanations of the possible discrepancy between flavours and odours, in order to compare them with Theophrastus' views on this issue and to figure out how they both influenced later authors.

2 Aristotle's Reply

Let me begin with Aristotle's position. In *De sensu* 5, right after his claim that there is an analogy between the different kinds of flavour and the different kinds of odour, Aristotle makes a distinction between two kinds of odour: the first kind includes subkinds that are analogous to the different kinds of flavour, while the second kind includes subkinds that are independent of the different kinds of flavour:[9]

> T3: Aristotle, *De sensu* 5, 443b16–444a8
>
> There are two kinds (εἴδη) of odour. For it is not the case, as some say, that there are no kinds of odour, because there are. But it is necessary to distinguish in what way there are and in what way there are not. One kind of odour is classified in line with flavours, as we have said [i.e. 443b7–8], and is coincidentally pleasant or unpleasant. For because flavours are properties of what nourishes, odours are pleasant to animals which experience appetite but not pleasant when they are full and not in need of anything; nor are they pleasant to those animals which do not find the food that has the odours pleasant. So, these odours, as we have said, involve pleasure and pain coincidentally, and hence they are also common (κοιναί τῶν ζῴων) to all animals.
>
> But odours of the other kind are pleasant in virtue of themselves, for example, those of flowers. For they do not provide any inducement, great or small, toward food, nor do they contribute anything to appetite, but instead the contrary; for what Strattis said when mocking Euripides is true: "When you cook lentils don't pour on perfume!" Those who in fact mix such potent ingredients into drinks compel our pleasures by habituation, until a pleasure arises from two senses as it might from one. This

9 Aristotle, unfortunately, uses the term "εἴδη" both when he talks about the two kinds of odour and when he talks about the different subkinds belonging to the two kinds of odour, and this practice sometimes results in making the relevant passages somewhat difficult to follow.

> second kind of odour is peculiar to human beings (ἴδιον ἀνθρώπου), while the first, which is classified in line with flavours, belongs also to the other animals, as we said before; and the first kind, because it involves pleasure coincidentally, is divided into subkinds (εἴδη) corresponding to the flavours, whereas the second kind is not, because it has a pleasant or unpleasant nature in virtue of itself.

Aristotle begins, in this passage, by airing his objection to the view that there are different kinds of flavour but no kinds of odour, a view that was indeed put forward by his predecessors, namely by Democritus and Plato. For, if we are to trust Theophrastus' report (*Od.* 64; *Sens.* 82), Democritus introduced seven different kinds of flavour, but he neglected the sense of smell and failed to distinguish kinds of odour. And Plato, in the *Timaeus* (66d1–67a6; see also, Theophrastus, *Sens.* 84–85), defended the view that we cannot classify different kinds of odour; we can only adopt, he said, a rough scale between the extremes of the pleasant (ἡδύ) and the unpleasant (λυπηρόν) odour. Aristotle, on the other hand, makes the distinction between odours that are said to be pleasant or unpleasant coincidentally (κατὰ συμβεβηκός)—that is, insofar as they are related to things that have a pleasant or an unpleasant flavour—and odours that are said to be pleasant or unpleasant in virtue of themselves (καθ' αὑτάς).[10]

Aristotle explains that all animals experience the odour of something that tastes good as pleasant, whereas they experience the odour of something that tastes bad as unpleasant. He also maintains that the same thing that tastes and smells good when animals are hungry, no longer tastes and smells good when they are full. In Aristotle's view, therefore, human and non-human animals experience odours as pleasant or unpleasant depending on the nutritive value of the corresponding flavours; and these are the coincidentally pleasant or unpleasant odours. Moreover, Aristotle adds, human beings may judge odours as pleasant or unpleasant in virtue of themselves, and thus independently of flavours. In other words, human beings find enjoyable or repelling odours that have nothing to do with food; for example, they enjoy the odours of flowers and are repelled by the odours of filth or staleness. In fact, it is this Aristotelian distinction between two kinds of odour—the odours that are pleasant or unpleasant coincidentally and those that are pleasant or unpleasant in vir-

10 It is worth noting that Aristotle, in this passage, does not take into consideration that animals use their sense of smell in order to find mates, a fact which he observes in his zoological treatises; for instance, in *HA* VI 18, 572b9–13, he says explicitly that each stallion recognises the mares that form his company by their odour (I owe this reference to Sophia Connell).

tue of themselves—that explains, I think, the occasional discord between flavours and odours, to which Aristotle alludes in the *De sensu* (T1) as well as in the *De anima* (T2). Since human beings are able to disassociate the odours of perceptible objects from their corresponding flavours, it is reasonable to expect that sometimes they may experience the odour of something as pleasant while they experience its flavour as unpleasant; for instance, to use the commentators' examples already mentioned, it should not be surprising that perfumes, aloe, and storax taste bad but smell good.

Most importantly, Aristotle does not only claim that human beings experience two kinds of odour; he also tries to understand why human beings are unique among animals in this respect, and he offers in *De sensu* 5 two types of explanation. First, he gives a physiological explanation: the human brain, he states, in comparison with that of other animals, is disproportionately large, moist, and cold; having such a brain does not make our sense of smell more accurate but at least contributes to making it more refined, in the sense that we are thereby able to appreciate odours completely unrelated to food. Second, Aristotle presents a teleological explanation: human beings are meant to appreciate certain odours completely unrelated to food, because such odours have been generated in order to restore the equilibrium between heat and coldness in the region around the human brain, and this proves beneficial to our health:[11]

> T4: Aristotle, *De sensu* 5, 444a8–b2
>
> The reason why the second kind of odour is peculiar to human beings involves the condition of the region around the brain. For the brain is cold by nature, and the blood in the vessels around it is thin and pure and easily cooled (which is also why the exhalation rising from nutriment is cooled by this region and produces sickly discharges). Hence, this kind of odour has been generated for human beings to support their health (τοῖς ἀνθρώποις πρὸς βοήθειαν ὑγιείας γέγονε); indeed it has no function (ἔργον) other than this. And it obviously performs it; for food, though pleasurable, whether it is dry or moist, is often noxious, while the fragrance result-

11 Aristotle seems to follow, in this regard, a Platonic tradition: In the *Timaeus* (67a1–6), Plato claimed that pleasant odours bring the body back to its natural state. Moreover, in the ninth book of the *Republic* (584b4–7), he referred to odours as examples of pure pleasure, and in the *Philebus* (51b3–7; 51e1–5) pure odours are treated as less divine than other sense objects but are also said to be pleasant in virtue of themselves (I owe these references to Xanthippe Bourlogianni and Claire Louguet). For Plato's account of the sense of smell, see Ierodiakonou (Forthcoming (a)).

ing from the odour which is pleasant in itself is almost always beneficial (ὠφέλιμος) to them whatever condition they are in. And for this reason smelling takes place through breathing—not in all animals but in human beings and some animals with blood such as those with four legs and those which partake to a greater extent in the nature of air. For when the odours are carried upward to the brain due to the lightness of the hot stuff in them, this region of the body is in a healthier condition; for the power of odour is by nature hot …

The second kind of odour is peculiar to human beings because they possess the largest and moistest brains relative to their size; for this is also why human beings are almost (ὡς εἰπεῖν) the only animals that perceive and enjoy the odours of flowers and similar things. For the heat and change produced by these things is commensurate (σύμμετρος) to the excess (ὑπερβολὴν) of moistness and coldness in this region of the body.

I have not been able to trace any early medical sources on which Aristotle may have depended for these claims about the condition of our brain and the effect of pleasant odours on the human health. Still, it is worth quoting a passage from Galen's treatise *De instrumento odoratus*, which comes quite close to what Aristotle says:[12]

T5: Galen, *De instrumento odoratus* 62.1–12 Kollesch (= II 884.1–16 Kühn); transl. B.S. Eastwood, with changes

That intense odours at once cause pains in the head, make it stuffy and heavy, sometimes even bringing on delirium, is also a sign that some of the material itself reaches the brain. For the brain could not be affected to so violent and acute a degree just on the basis of a similar quality [i.e. cool and wet]. Similarly it is now to be expected, when the brain suffers as a result of heatstroke, that the odour of roses is healthful. Whereas nothing less than roses, and also inhaled cold air, helps, so on the other hand very warm air rather clearly is deleterious. And when in fact the whole respiration is carried on by the brain in order to preserve the correct amount

12 In addition, it is worth quoting a passage in Plutarch's *De Iside et Osiride* (383D), in which we find a reference to Aristotle's view about the beneficial effect of pleasant odours to the human health, in the context of a discussion about related practices performed by the Egyptians (I owe this reference to Sean Coughlin): Ἀριστοτέλης δέ φησι καὶ μύρων καὶ ἀνθέων καὶ λειμώνων εὐώδεις ἀποπνοίας οὐκ ἔλαττον ἔχειν τοῦ πρὸς ἡδονὴν τὸ πρὸς ὑγίειαν, ψυχρὸν ὄντα φύσει καὶ παγετώδη τὸν ἐγκέφαλον ἠρέμα τῇ θερμότητι καὶ λειότητι διαχεούσας. εἰ δὲ καὶ τὴν σμύρναν παρ' Αἰγυπτίοις σὰλ καλοῦσιν, ἐξερμηνευθὲν δὲ τοῦτο μάλιστα φράζει τῆς λη‍ρήσεως ἐκσκορπισμόν, ἔστιν ἣν καὶ τοῦτο μαρτυρίαν τῷ λόγῳ τῆς αἰτίας δίδωσιν.

> of heat in the heart, it is quite reasonable for the brain to provide this equilibrium (τὴν συμμετρίαν) first for itself, since by nature it is most susceptible of being injured through excessive (ἀμετρότερον) heating up or cooling down.

So, for Galen, too, some odours are beneficial to the health of human beings. Contrary to Aristotle, however, Galen claims that the sweet odour of roses helps those suffering from a heatstroke, because it cools the human brain and hence restores the equilibrium of its heat and coldness.[13] But leaving aside Galen's dissent from the Aristotelian doctrine as well as his own general theory of smell,[14] let us next investigate the way Aristotle's followers and commentators account for the exceptions of the correspondence between the senses of taste and smell.

3 Theophrastus' Reply

The Aristotelian who showed special interest in theories of perception and, in particular, in the senses of taste and smell is, undoubtedly, Theophrastus. Among his surviving writings, we find the short treatise *De sensibus*, in which he presents and criticises the views of Aristotle's predecessors on the function and objects of taste and smell, but we also find many interesting passages about flavours and odours both in the sixth book of his botanical work *De causis plantarum* and in his small treatise on odours, the *De odoribus*.[15] Indeed, when reading Theophrastus' works, one often gets the impression that his treatment of issues such as the production of flavours and odours as well as the classification of their different kinds is much more thorough than the relevant passages in Aristotle's psychological treatises. It is indicative that, when Galen discusses

---

13 On the beneficial effect of the sweet odour of roses, see also Galen, *De instr. odor.* 40.6–8 Kollesch (= II 862.12–16 Kühn); 48.1–4 Kollesch (= II 870.3–6 Kühn). Furthermore, according to Galen (*SMT* XI 537.5–539.8; 557.10–564.9 Kühn), rose juices blended with olive oil have a moderately cooling capacity (I owe this reference to John Wilkins); for the use of roses as ingredients in the production of medical drugs, see Squillace (2020), 181–183.

14 For Galen's views on the senses of taste and smell, see Siegel (1970), 140–173; for his dependence on Plato's and Aristotle's theories of smell, see Eastwood (1981); and for his anatomical description of the olfactory apparatus, see Smith (1971), 81–84.

15 Diogenes Laertius (V 44) includes in his list of Theophrastus' treatises the *De odoribus* in one book but also lists *De causis plantarum* in eight books, although we have it in six books. It has been argued that some version of the *De odoribus* was originally part of the eighth book of *De causis plantarum*; see Thompson (1941), 10–20; Wörle (1988); Sharples (1995), 133; Reger (2005), 257.

Plato's treatment of the senses of taste and smell in the *Timaeus*, he compares them to Theophrastus' rather than to Aristotle's account of these senses. In particular, in his work *De simplicium medicamentorum temperamentis ac facultatibus* (XI 450.16–452.2 Kühn), Galen presents the Platonic list of seven kinds of flavour and then attributes to Theophrastus the addition of a further kind, namely the fatty flavour, even though Aristotle, too, talks about this kind of flavour in the *De sensu* (4, 442a12–25).

Theophrastus actually seems to have been engaged in a more systematic study of the senses of smell and taste, as well as of their respective objects, than his illustrious teacher and collaborator did; for instance, concerning the generation of the different kinds of flavour, Theophrastus clearly has a more precise idea with regard to the role played by the different degrees of dryness and moisture. Although Aristotle (*DA* II 10, 422b10–14; *Sens.* 4, 442a12–25) lists between the sweet and the bitter flavour five, or sometimes six, other kinds of flavour—fatty, pungent, tangy, sour, acidic, and salty—, without specifying the criteria for the order of this list, Theophrastus claims that what is dry has a flavour closer to the sweet flavour—for instance, the pungent flavour—, whereas what is moist has a flavour closer to the bitter flavour—for instance, the acidic flavour:[16]

> T6: Theophrastus, *De causis plantarum* VI 11.1
>
> Another point that we must endeavour to settle as relevant to flavours is this: Which, what is dry or what is moist, prevails more in each of the kinds of flavour? Thus the acidic is believed to belong more to what is moist, and so the tangy; whereas the pungent is believed to belong more to what is dry, and so the sweet (at all events what is moist gets sweeter as it thickens).

Moreover, Theophrastus supports this view by many concrete examples; not only something gets sweeter as it thickens, i.e. as it loses its moisture, he is also reported to have argued that unripe fruits, which are sour, become less sour and more acidic when ripe, due to the increase of their moisture (Galen, *SMT* XI 660.5–10;[17] see also, *SMT* XI 636.16–638.9 Kühn).

---

16 All translations from Theophrastus' *De causis plantarum* are based on Einarson and Link (1990) and those from the *De odoribus* on Hort (1916), with changes.

17 ὡς γὰρ καὶ πρόσθεν εἴρηται, σχεδὸν ἅπαντες οἱ τῶν δένδρων καρποὶ κατὰ τὴν πρώτην γένεσιν, ὡς καὶ Θεόφραστος ἔλεγε, στρυφνοὶ γευομένοις εἰσὶν, εἶτ' ἐπειδὰν ὑγρότητος πληρῶνται, προσλαμβάνουσι μὲν ὀξύτητος, ἀφαιροῦσι δὲ τῆς στρυφνότητος.

Concerning now the relation between the senses of taste and smell, Theophrastus (*CP* VI 5.3; 17.10; *Od.* 4), just like Aristotle, points out that, although human smell is less accurate than taste, there is a close correspondence between them; and this, he also adds, we observe both in the fact that flavourless things are also odourless (*Od.* 67) as well as in the analogy between the seven kinds of flavour and the seven kinds of odour (*CP* VI 4.1–2). But although Aristotle does not doubt the applicability of the terms used for the different kinds of flavour to those of odour, Theophrastus expresses the worry that, since odours are more difficult to distinguish than flavours, there are occasions on which the terms used for flavours may not be appropriate for odours:

> T7: Theophrastus, *De odoribus* 1–2
>
> Of odours some are, as it were, indistinct and insipid, as is the case with flavours, while some have a distinct character. And these characters appear to correspond to those of flavours, yet they have not in all cases the same names—as we said in a former treatise—nor, on the whole, are they marked off from one another by such specific differences as are flavours; rather, one may say that, in generic character, some things have a good and some a bad odour (τὰ μὲν εὔοσμα τὰ δὲ κάκοσμα). But the various kinds of good or bad odour, although they exhibit considerable differences, have not received further distinguishing names, marking off one particular kind of sweetness or of bitterness from another; we speak of an odour as pungent, powerful (ἰσχυρά), mild (μαλακή), sweet, or heavy (βαρεῖα), though some of these descriptions apply to bad-smelling things as well as to those with a good odour.

Most importantly for our purposes, Theophrastus states that the cases in which the analogy between flavours and odours breaks down are many, thus stressing more sharply than Aristotle the possibility of a discrepancy between the senses of taste and smell; for example, wines with a rich and powerful odour sometimes do not have a pleasant flavour, and the same applies in the cases of perfumes, incense, coronary plants, and other aromatic substances—as he himself remarks:

> T8: Theophrastus, *De causis plantarum* VI 16.6–7
>
> Again, wines with a rich (πολύοσμοι) and powerful odour are not mild when tasted, which shows that good flavour (εὐστομίας) does not coexist with rich and powerful odour. This holds also of perfumes, incense, coronary plants, and other aromatic substances: all such things are bitter and of poor flavour (δύσχυλα), as are the almonds too. Again mountain wines

have in some cases a fragrance, but they are harsh (σκληροί) and not good in flavour (οὐκ εὔχυλοι).

Also, the reverse is said to be the case; in particular, Theophrastus gives the example of the carob and the Phoenician cedar, which both have a sweet flavour but produce an unpleasant odour:

T9: Theophrastus, *De odoribus* 5
And, as was said above [i.e. *De odor.* 1], things which have a good odour are generally of poor, sour, or somewhat bitter (ὑπόπικρα) flavour. Again some things which have a good flavour have also a bad odour (κακώδη), such as the carob, which is sweet (this is true of some regions, if not of all). Also, the Phoenician cedar, though it is sweet, when chewed produces a sort of bad odour but makes the urine fragrant (εὐῶδες).

There are even some passages about plants in which Theophrastus makes the sweeping remark that all fragrant things have a bad flavour; he even goes as far as to say that the bitter flavour is "the origin" (ἀρχήν) of fragrance:

T10: Theophrastus, *De causis plantarum* VI 9.4
It seems that of the two opposites, namely the sweet and the bitter, the sweet is the origin (as it were) of good flavour, whereas the bitter is the origin of good odour; and in a way the bitter is to a greater extent the origin of good odour, since it is hard to find anything fragrant that is not bitter, but many non-sweet things have good flavour (these being more or less the things that are also fragrant when tasted and eaten). Sweetness on the other hand has rarely any odour at all, least of all a good one. All this suggests that the sweet and the fragrant do not mix; and yet both are products of concoction. But of this problem later [i.e. *CP* VI 16.1–8].

To account for such discrepancies, it is not surprising that Theophrastus follows Aristotle and makes a distinction between coincidental odours and odours in themselves:

T11: Theophrastus, *De odoribus* 5
Now the odour of some things which have a good odour resides in things which are used for food—for instance, that of stone-fruits, pears, and apples—, the odour of which is sweet even if one does not eat them; indeed it may be said to be sweeter in that case. However, to make a general distinction, some odours are in themselves (καθ' αὑτάς), while others

are coincidental (κατὰ συμβεβηκός); those of juices and things used for food are coincidental, those of flowers are in themselves.

And similarly to Aristotle, Theophrastus (*Od.* 4) also states that the different kinds of odour that are unrelated to food are experienced by human beings, whereas other animals do not seem to appreciate a pleasant odour in virtue of itself; he even adds that there are some animals—for instance, vultures and beetles—which seem to be annoyed by pleasant odours. But Theophrastus, who had a particular interest in doing research on non-human animals, is more careful than Aristotle in this respect and makes sure to also point out that the temperament of animals in question should always be taken into consideration as well as their capacity of smell. There is actually evidence that he considers it rather questionable whether non-human animals, in general, do not enjoy pleasant odours for their own sake (see also, *CP* VI 5.5):

T12: Theophrastus, *De causis plantarum* VI 5.1–2

In fact, among animals other than human beings there are either none at all or very few indeed that seek out good odour for its own sake (τὴν εὐωδίαν διῶκον αὐτῆς χάριν); if animals do seek it out, they do so only coincidentally, when their food has that odour, for their desire is relative to food and it is the food that they want. In the second place, if there is in fact such an animal, the matter is not evident (οὐ σαφές), at least to us. So the panther is reported to have a pleasant odour for other animals in its own right (this being why it hunts them by lying down, luring them by this odour); but it has no noticeable fragrance to us.[18]

The interesting question, of course, is how Theophrastus explains the phenomenon of the discord between flavours and odours. Does he follow Aristotle in giving two types of explanation, namely a physiological and a teleological? It is telling, I think, that he does not give a teleological explanation. That is to say, he does not claim that only human beings have the capacity to experience odours of the second kind, i.e. odours in virtue of themselves, because their function is simply to restore human health; on the contrary, he seems to doubt such an explanation:

18 On the panther's odour, see also Aristotle, *HA* IX 6, 612a12–15; [Aristotle], *Probl.* XIII 4, 907b35–37.

T13: Theophrastus, *De causis plantarum* VI 5.5

This occurs also in human beings, for the same odours are not agreeable (προσφιλεῖς) to all of us. Uniformity occurs rather in bad and heavy odours, some of which affect all alike, as the odours exhaling from chasms and caves that are lethal to all who draw near. Even if these odours affect only the animals that breathe, yet this point is clear, as it was with flavours, that with odours too there are some that are appropriate to different kinds of animal, and that fragrance not associated with food is agreeable to few or none. Whether it is also beneficial (συμφέρον) to some animal is still more uncertain (ἀδηλότερον).

Is this, then, another occasion on which Theophrastus diverts from Aristotle's teleology?

The subject of Theophrastus' take on Aristotle's teleology has often been raised in contemporary scholarship.[19] It has been remarked that, in the ninth chapter of his *Metaphysics*, Theophrastus discusses passages from the Aristotelian treatises in which the Stagirite seems to have struggled to explain certain phenomena in terms of his teleological principle, when it may have been better for him to admit that such explanations are simply inappropriate. For instance, Theophrastus (*Metaph.* 9, 10a22–11a5) claims that natural phenomena such as extremely dry and moist seasons, the breasts in males, the large size of the deer's horns, and the insects that only live for a day seem to violate Aristotle's doctrine that all things in nature are for the sake of something (τοῦ πάνθ' ἕνεκά του). He thus proposes that such instances should urge us instead not to posit teleological reasons in all cases without qualification, and to acknowledge that there may be limits to the determination of causes. Nevertheless, scholars have also stressed that such statements by Theophrastus do not imply that he completely fails to recognise the role of necessity in nature nor that he rejects all teleological explanations, even though it is true that in his surviving works there are few such examples. Therefore, his stance in this issue has been interpreted, rightly I think, as showing nothing but caution towards an indiscriminate application of the principle that nature does nothing in vain. It is most likely, then, that his uncommitted position regarding the teleological explanation of the beneficial role of pleasant odours should be understood as another case of such a cautious stance.

---

19 See e.g. Lennox (1985); Repici (1990) and (2013), 226–229; Van Raalte (1993), 491–492; Gourinat (2015).

As to the physiological explanation Theophrastus offers for the sweet odours that are produced by bitter flavours, or for the reverse case, it does not resemble the Aristotelian account which is found in the *De sensu* and involves the condition of the human brain and the region around it. To explain, for instance, why a fruit tastes sweeter the less fragrant it is, Theophrastus says the following (see also, *CP* VI 16.1–3):

> T14: Theophrastus, *De causis plantarum* VI 14.5
>
> But no sweet fruit (one may say) is fragrant, or it is so only slightly. Thus the sweet apple is of all apples the least fragrant, and the sweeter it is, the less its fragrance. (Yet at the same time the sweet apple too has a certain vinous character in its flavour.)
>
> The reason for the absence of odour is that the sweet flavour is thicker and too earthy (παχύτερος καὶ γεωδέστερος) (especially when mingled with the pericarpion), whereas odour belongs to a finer and drier flavour (λεπτοτέρου καὶ ξηροτέρου) and to things that evaporate more readily. These matters will be treated in greater detail later.

And this is, in fact, what he says later about the reverse cases, that is, when unpleasant flavours produce pleasant odours:

> T15: Theophrastus, *De causis plantarum* VI 16.8
>
> It appears that the other answer is yielded as well: fragrance lies rather (as it were) in the unperfected (ἀτελέσιν) and privative (στερητικοῖς) flavours, inasmuch as fragrant things are some of them bitter, some pungent, some faint (ἀσθενῆ), some sour, some with another unpleasant flavour. This is not perhaps unreasonable, since there is more exhalation of a sort (ἀποπνοή τις) from these unperfected flavours. Hence if the exhalation carries off particles that are well-tempered (εὔκρατα),[20] the odour is commensurate (σύμμετρον) to the sense of smell; whereas if it takes particles that run to excess (ὑπερβάλλοντα), the odour becomes disagreeable (δυσχερῆ) and heavy. And it would not appear that odours of this type are a mere matter of privation (ἐν στερήσει), but possess a nature of their own, inasmuch as the effects are in some cases extreme.[21]

20 For Theophrastus' use of the adjective εὔκρατος in a similar context, in which though it primarily denotes quality than quantity, see *HP* I 4.1: Πλείω μὲν γὰρ δοκεῖ τὰ ἄγρια φέρειν, ὥσπερ ἀχρὰς κότινος, καλλίω δὲ τὰ ἥμερα, καὶ τοὺς χυλοὺς δὲ αὐτοὺς γλυκυτέρους καὶ ἡδίους καὶ τὸ ὅλον ὡς εἰπεῖν εὐκράτους μᾶλλον.

21 Theophrastus questions, here, Aristotle's view that bitter odours should be considered

> Fragrance, then, for this reason lies more in non-sweet and poor-flavoured things (so to speak).

Hence, according to Theophrastus, it is the way odours are produced from the flavoured moisture in air or in water that seems to be mainly responsible for the discord between the taste of bitter apples, for instance, and their fragrance: if the particles emitted by the bitter apple are well-tempered, i.e. not in excess, and thus commensurate to the sense of smell, the odour is pleasant.

It is also worth noting that, in this context, Theophrastus presupposes a theory of effluences that postulates particles being emitted by perceptible objects and reaching the perceiver's organ of smell; and interestingly enough, we find him subscribing to the same effluence theory in *De odoribus* 3 and in more passages from the *De causis plantarum* (e.g. VI 9.3; 14.11; 17.1). In the *De sensibus* (20–22; see also, Priscian, *Metaphr.* 13.16–27; 14.10–15), on the other hand, Theophrastus seems to argue against the general theory of effluences that Empedocles and Plato put forward and defended. Does this mean that he is inconsistent in this respect? Or, does he perhaps recognise that there is something particular in the case of the sense of smell, in contrast to the other senses, that requires the assumption of effluences? It may not be accidental, after all, that the same ambivalence seems to characterise Aristotle's views in the *De sensu*; for he clearly rejects the effluence theory of his predecessors (5, 443a2–8; a21–b2), whereas he appears to make use of such a theory especially when it comes to the sense of smell (5, 444a21–24; 445a25–27).[22]

4 Theophrastus' Legacy

But whether or not Aristotle's and Theophrastus' theories of sense perception are inconsistent, at least Theophrastus offered a more worked out physiological explanation, than the one suggested by Aristotle, of the reason why the analogy between flavours and odours does not hold in certain cases. Furthermore, it seems that Theophrastus' reasoning as well as his general treatment of the

---

as privative, just as he questions earlier on that bitter flavours are privative (*CP* VI 4.2); for Theophrastus' dissent from Aristotle's account of different kinds of flavour, see Ierodiakonou (Forthcoming (b)).

22 On whether or not Theophrastus' account of perception is inconsistent, see Sharples (1985), 193–197; Sedley (1985), 205–206; Eigler and Wörle (1993), 16–18; Baltussen (2015a), 43; Schorlemmer (2022), 71–82; Roreitner's contribution to this volume.

objects of taste and smell had an influence on later ancient authors, who also tried to give plausible descriptions of the function of these senses.

For instance, in his rather long comment on the coincidental odours and those unrelated to food, Alexander of Aphrodisias subscribes to Aristotle's teleological explanation and elaborates on it (*In Sens.* 99.9–15),[23] but he also adopts Theophrastus' criteria of what makes an odour sweet or bitter. For Alexander remarks that the odours of the second kind, which are not related to flavours, are more pleasant the hotter they are and more unpleasant the more moist they are; he even illustrates this with the example of lands and things that are more fragrant in hot and dry areas than in cold and moist ones:

> T16: Alexander, *In Sens.* 98.23–99.9; transl. A. Towey, with changes
>
> For there is no function (ἔργον) for odour of this sort other than to heat and be dispersed through the places around the brain which are too cold and to bring them to what is commensurate (εἰς τὸ σύμμετρον ἄγειν). For odours of this kind are by their own nature fragrant and pleasant. For they are dry and distinguished by the hot which is mild and pleasant, and the more they have been overpowered by the hot the more fragrant they are. For generally there is fragrance in them because of the ripening of the moist in them by the hot. It is at any rate for this reason that things in hot and dry places are more fragrant, and land which is hot and dry is fragrant when it first admits rain because of the heat which is present in it. For at that time the flavours from it are fine and dry. Odours which are unpleasant and troublesome (ἀηδεῖς καὶ μοχθηραί) are generated because of the cold and unripeness of the moist which is present in that from which they are generated.

Galen, too, is influenced by the Theophrastean views on taste and smell, especially in his previously mentioned work, *De simplicium medicamentorum temperamentis ac facultatibus*, in which he talks about the odours unrelated to food and, without even mentioning Aristotle, explicitly refers to Theophrastus as someone who studiously examined the phenomenon of the roses that smell good and taste bad (*SMT* XI 697.16–698.4 Kühn).[24] However, Galen's own

23 Alexander also gives, in this passage, an example of the fact that odours can be beneficial to human beings, when he claims that pleasant odours soothe the pain caused by pestilential diseases (ἐν ταῖς λοιμώδεσιν).

24 ἐπὶ μέντοι τῶν ἡδίστων, οἷά περ καὶ τὰ ῥόδα, διαφέρουσί τε καὶ πάμπολυ διεστήκασιν. οὐ μόνον γὰρ οὐχ ἡδέα γευομένοις ἐστὶν, ἀλλὰ καὶ πικρότητος οὐκ ὀλίγον ἐμφαίνει. καίτοι τόγε πικρὸν οὐδέποτ' ἦν εὐῶδες οὐδ' ὡς φάρμακον, οὐδ' ὡς τροφὴ ἡμῖν ἡδύ. τίς οὖν ἡ τῆς διαφωνίας ταύτης αἰτία, διὸ καὶ Θεόφραστος μὲν ἐζήτησεν, ἄμεινον δ' ἴσως καὶ ἡμᾶς ὅσα γιγνώσκομεν ὑπὲρ αὐτῆς εἰπεῖν.

account somewhat differs from Theophrastus', since Galen introduces the role of the *pneuma* in the brain ventricles with which the odours should be compatible in order for them to be experienced as pleasant (*SMT* XI 698.15–699.5 Kühn).[25] In spite of this, I think that Galen's explanation of the discord between flavours and odours is along the same lines as Theophrastus' doctrine; it is not a teleological explanation and it also depends on how the different qualities of the elemental particles—hot and cold, dry and wet—affect our senses of taste and smell. For Galen (*SMT* XI 701.14–702.11 Kühn)[26] explicitly states that, from the three constituent elements of roses, namely earth, air, and water, it is the coldness of the water alone that produces their sweet smell, whereas all other three elements are responsible for its bitter flavour.[27] This is, I think, an elaboration of the type of explanation given by Theophrastus. That is to say, Theophrastus implies that apples have a bitter or a sweet odour depending on the quantity of particles emitted by them, while Galen clearly suggests that the rose's flavour is due to the admixture of constituent elements that are other than the ones responsible for its odour.[28]

---

25 οὐ μὴν εὐθὺς καὶ πᾶν ὀσμῶδες ἡδὺ, διότι μηδ' οἰκεῖόν ἐστιν ἅπαν τοῦ κατὰ τὰς κοιλίας τοῦ ἐγκεφάλου πνεύματος. ὥσπερ γὰρ τῶν προσπιπτόντων τῇ γλώττῃ χυμῶν οἱ μὲν οἰκειότατοι γλυκεῖς ἦσαν, οἱ δ' οὐκ οἰκεῖοι πολλὰς ἐκέκτηντο διαφορὰς, οὕτω καὶ τῶν ἀτμῶν οἱ μὲν οἰκεῖοι τῷ κατὰ τὸν ἐγκέφαλον πνεύματι φίλιοί τ' εἰσὶ καὶ ἡδεῖς, οἱ δ' οὐκ οἰκεῖοι διαφέρουσι μὲν ἀλλήλων οὐκ ὀλίγαις διαφοραῖς.

26 ἐν δὲ τῇ τοῦ ῥόδου φύσει τὸ μέν ἐστι στρυφνὸν γευομένοις, ὅπερ ἀναγκαῖον εἶναι γεῶδές τε καὶ παχυμερὲς καὶ ψυχρὸν, ἕτερον δὲ πικρὸν, ὃ δὴ λεπτομερές τ' ἐστὶ καὶ θερμὸν, καὶ τρίτον ἐπὶ τούτοις τὸ ὑδατῶδες, ψυχρὸν μὲν ἐξ ἀνάγκης, ἀλλ' ἐν τῇ κατὰ τὸ λεπτομερές τε καὶ παχυμερὲς ἀντιθέσει μέσον πως ὑπάρχον τῶν ἄκρων, οὗ τῇ μίξει καὶ τὸ στρυφνὸν καὶ τὸ πικρὸν οὐκ ἔστιν ἄκρως τοιοῦτον ἑκάτερον οἷόν περ λέγεται· πάσας γὰρ ἐκλύει τὰς ποιότητας ἡ τοῦ ὑδατώδους ἐπιμιξία. τοῦτ' οὖν αὐτὸ τὸ ὑδατῶδες ἔοικέ πως ἤδη τε καὶ κατειργάσθαι καὶ λελεπτύνθαι καὶ τεθερμάνθαι, καὶ διὰ τοῦτο ῥᾳδίως εἰς ἀτμοὺς ἄγεσθαι καὶ ταύτῃ τὸ ῥόδον ὀσμῶδές τε εἶναι καὶ ξηραίνεσθαι τάχιστα. καὶ ταῦθ' ὑπὲρ αὐτοῦ πάντα διὰ μὲν τῆς γεύσεως ἔνεστι τεκμήρασθαι, διὰ δὲ τῆς ὀσμῆς οὐχ οἷόν τε τῷ μήθ' ὁμοίως ἀτμίζειν ἅπαντ' αὐτοῦ τὰ μόρια μήθ' ὁμοίως κινεῖν τὴν αἴσθησιν.

27 On Galen's explanation of why things that smell sweet often taste bad, see also Sharples (1985), 196–197.

28 Theophrastus, in fact, analysed the ingredients of the rose perfume (ῥόδινον): it is obtained from the flowers of the rose (*Od.* 27) together with certain spices—for instance, gingergrass and sweet-flag—and a considerable amount of salt (*Od.* 25) as well as with some plant-based colouring substances (*Od.* 31, 33). He also claimed that it is due to the salt that rose perfume is used in medicine for curing earaches and strangury (*Od.* 35) and for alleviating headaches caused by other fragrances (*Od.* 50). So, although Theophrastus did not explain the sweet odour of roses and their bitter flavour in terms of their different elemental particles, he did realise that the different constituents of an admixture produce different effects.

Therefore, according to my suggestion, it is the same principle that lies behind the physiological explanation, proposed both by Theophrastus and Galen, of the observed discrepancies between flavours and odours. To put it briefly, the elemental particles that are emitted by the perceptible object and produce its odour differ in some respect from those that produce its flavour. And this is actually the principle behind Priscian's explanation, at least as much as it is possible to assess from his extremely concise remark in his commentary on Aristotle's *De anima*. For Priscian, too, does not talk of a teleological explanation; rather, he claims that the discord we experience between flavours and odours is caused by the fact that, during the process of the production of an odour, i.e. when the flavoured moisture affects the air or the water, the odour generated sometimes may be similar to the flavour but sometimes may change to its contrary:

> T17: 'Simplicius', *In DA* 153.27–34; transl. C. Steel, with changes
>
> The names "sweet", "bitter" and the like have been transferred from flavours to odours. But not idly—for why not from objects of touch or sounds? Rather it is because, as was said, there is an affinity (συγγένεια) of odours to flavours also with relation to their bearers, because of the solution of the flavoured in the compounded moisture of air and water. Why then does a sweet odour not always follow (ἀκολουθεῖ) on a sweet flavour? It is because sometimes the flavoured matter, as it dissolves, changes its character (παρατρέπεται), and the change of character (παρατροπή) turns it into something of the same kind or into just the privation of the bearer.

In other words, the elemental particles responsible for the flavour of something tasteable are not always the same as those responsible for its odour; the production of the odour may require processes that result in significant changes of the constituents involved in the production of the corresponding flavour.

5 Conclusion

To conclude: Aristotle constructed a general theory of perception and tried to apply it in the case of the five specific senses, marking their differences as well as their affinities. But he also observed cases in which his blanket doctrine did not conform with his everyday experience. If the senses of taste and smell are closely related, as he thought, how could one explain the fact that flowers have a sweet odour but a bitter flavour? Aristotle's treatment of such phenomena did not seem adequate to his followers. Among them, Theophrastus was the one who undertook to closely study the relevant experiential evidence and to suggest a more detailed explanation, which in certain respects even doubted some of the Aristotelian principles. Those who later engaged themselves in settling the same issue acknowledged the Theophrastean contribution, although they had to adapt it to their own doctrinal predilections. Hence, there was nothing remarkable for the ancients regarding the fact that roses smell good but taste bad; it was just a spinoff phenomenon accounted for in various ways by ancient theories of perception introduced and defended by different philosophers, with the result that it remained a point of contention for a long period of time in antiquity.[29]

---

29 I would like to thank all those who attended my talks on this topic in Geneva, Lyon, London, and Athens—and especially, Robert Roreitner and Alessio Santoro—for their constructive comments which helped me to improve my first draft.

CHAPTER 5

Eudemus' *Physics* on Cosmic Issues: Frs. 111–123b W, on *Physics* VIII

István Bodnár

What we know about Eudemus' *Physics* is in very large part transmitted to us through Simplicius' *Commentary on Aristotle's Physics*. This is important to keep in mind. It means that the bits and pieces of text we have are always quoted by Simplicius in a particular context, with some specific argumentative goal. We do have also Simplicius' overall characterization of this work.[1] Where e.g. Eudemus skips a whole book, and does not comment on Book VII, Simplicius writes

> And Eudemus, up to this having followed practically all the major points of the treatise passed over this book as superfluous and went over to what is in the last book. (1036.13–15; Fr. 109 W, transl. Stork modified)[2]

We can read these lines as an assurance, and also as a warning. An assurance, because Simplicius asserts that Eudemus' *Physics* covered much the same ground as Aristotle's, and the constant flow of comments Simplicius makes should be expected to highlight the major differences of the two treatments. But it is also a warning. Simplicius' straight out claim that Eudemus "passed over this book as superfluous," in view of what is in store in Book VIII, may go back to something in Eudemus. But it just as well might be Simplicius' evaluation about the book, proposed as a guess about Eudemus' motivations for the decision to leave this book aside. Taking also for granted that Eudemus' silence rests on such a decision in the first place.[3]

Accordingly, we should exercise due caution. The first issue to keep in mind is that Eudemus' *Physics* takes as its starting point the Aristotelian text. But it is not a commentary, not even a paraphrase. The cases where Eudemus fills gaps

1 See Sharples (2002b) and Stork (Draft).

2 καὶ ὅ γε Εὔδημος μέχρι τοῦδε σχεδὸν τοῖς ὅλοις τῆς πραγματείας κεφαλαίοις παρακολουθήσας τοῦτο παρελθὼν ὡς περιττὸν ἐπὶ τὰ ἐν τῷ τελευταίῳ βιβλίῳ μετῆλθεν. (This is Stork's text resting on the reading of manuscripts. Diels reads τοῖς ὅλης τῆς πραγματείας κεφαλαίοις)

3 For a similar assessment see Baltussen (2002b), 130 n. 14.

in Aristotle's arguments suggest that the aim is to produce in four or five books a somewhat streamlined version, which is somewhat augmented nevertheless.

1

A prime example for such an intervention is Simplicius' following report:

> Eudemus reproaches Anaxagoras *not only because* he says that motion, after not having existed, began at a certain moment, *but also because* he omitted to talk about whether it endures or will cease, although this is not obvious.
>
> "For what is there," he says, "to prevent the intellect from deciding to stop all things at a given moment, as that fellow said [that it decided] to move [them]?"
>
> But Eudemus also reprehends Anaxagoras on the following point:
>
> "How is it possible for a privation to be prior to its opposing possession? If, then, rest is the privation of motion, it could not exist prior to motion." (1185.9–15; Fr. 111 W, transl. Chase modified)[4]

In this case there is no reason to doubt Simplicius' testimony, that Eudemus included Aristotle's criticism of Anaxagoras, and reinforced it with an additional point. The additional point, that by Anaxagoras' lights his Intellect should just as well be in a position to stop the world process as it could launch it foregrounds the demiurgic quality of the operation of the intellect. This is also reinforced by the terminology Eudemus deploys. The intellect is portrayed as making a decision to stop motion in the world, just the same way as he started the motion which led to the formation of the world.

Wehrli already indicated that with this move Eudemus taps into an argument Aristotle raised against the special status of eternal, divine entities in Plato's *Timaeus*. These—just like the cosmos itself—are generated by the Demiurge. And "anything that is bound is liable to being undone." Nevertheless as the Demiurge continues, addressing the lesser gods "Still, you will not be undone nor will death be your portion, since you have received the guar-

4 Ὁ δὲ Εὔδημος μέμφεται τῷ Ἀναξαγόρᾳ οὐ μόνον, ὅτι μὴ πρότερον οὖσαν ἄρξασθαί ποτε λέγει τὴν κίνησιν, ἀλλ᾽ ὅτι καὶ περὶ τοῦ διαμένειν ἢ λήξειν ποτὲ παρέλιπεν εἰπεῖν, καίπερ οὐκ ὄντος φανεροῦ. "τί γὰρ κωλύει, φησί, δόξαι ποτὲ τῷ νῷ στῆσαι πάντα χρήματα, καθάπερ ἐκεῖνος εἶπεν κινῆσαι."
καὶ τοῦτο δὲ αἰτιᾶται τοῦ Ἀναξαγόρου ὁ Εὔδημος· "πῶς ἐνδέχεται στέρησίν τινα προτέραν εἶναι τῆς ἀντικειμένης ἕξεως· εἰ οὖν ἡ ἠρεμία στέρησις κινήσεώς ἐστιν, οὐκ ἂν εἴη πρὸ τῆς κινήσεως."

antee of my will—a greater, more sovereign bond than those with which you were bound when you came to be." (41b)

Aristotle rejects this option briskly (in *De caelo* I 10) relying on the very same principle which Plato formulated: "for everything that comes into being is evidently such that it perishes." (277b20–21) The difference is that he speaks only about the entities which are produced by coming into being, and does not allow for any exceptions on the basis of the specific qualities or intentions of the producer. Eudemus' criticism of Anaxagoras, interestingly, is couched in such terms. Anaxagoras' account of the beginning of the world-process refers to the activity of the Intellect as an agent and claims that it is in the power of the very same agent to reverse course and halt the world-process—it only needs to decide to do so.[5]

Eudemus' second objection rests on construing the two stages—the precosmic one, without the motion imparted by the Intellect, and the ensuing cosmic one—in terms of being at rest and then being in motion, and then identifies these with states of privation (being deprived of motion)[6] and possession. After this Eudemus goes on to claim that privation cannot precede possession.[7] This most probably is connected to an understanding of privation and possession that is present in Aristotle's *Categories* 10, i.e. that possession never can come after a state of privation (13a31–37). Such an understanding of this short line of criticism against Anaxagoras is not unproblematic. For two reasons. First it is not clear whether, and why these restrictions of the *Categories* would hold about motion and rest. Furthermore, if the consideration of *Categories* 10 can be utilised, then they will cut not only against Anaxagoras, but also against Empedocles. If the absence of motion from the cosmos is privation, opposed to the presence of motion as possession, and privation cannot precede possession, Empedocles' account of a world where states of motion and rest come alternating after one another could be just as problematic as Anaxagoras'.

5 Note that the Demiurge's guarantee is not just some arbitrary *fiat*. Instead, it rests on the goodness of the maker, and the excellence of the product, as "only one who is evil would consent to the undoing of what has been well fitted together and is in fine condition." (41b) Eudemus' suggestion about Anaxagoras, however, need not take any such buttressing of indestructibility. Anaxagoras' Intellect was famously criticised by Socrates as insensitive to the goodness of the outcome, nor does it act out of goodness.

6 Aristotle in *Physics* V 2, 226b14–16 submits that "[rest] will be privation [of motion] in that which is capable of receiving [motion]."

7 Note that Fr. 92 W also attests interest on Eudemus' part in the issue of how change between privation and possession can be included among the different types of changes (μεταβολαί) distinguished by Aristotle in *Physics* V.1.

Eudemus, however, might find a way to exempt Empedocles. For this he needs to take the periods of rest in the Empedoclean cycle not as some inert states, as privations of motion, but as flowing from the nature of the entities responsible for the cosmic process. Perhaps this is exactly the point of what we read in Fr. 110 W. Aristotle in 252a5–10 submits that proposing that the alternating periods of motion and rest are just natural for the entities involved, is insufficient. Eudemus' remark—"Eudemus takes immobility to be in the reign of Love, throughout the *Sphairos*, when everything has been mixed together"[8]—apparently provides a causally relevant factor, Love, responsible for this immobility, which will be intercepted by the onslaught of the other causal factor, Strife. This need not be an objection to Aristotle. It may only specify the position of the state of rest within the Empedoclean cosmic cycle, perhaps also suggesting that after the complete domination of Strife and before the start of the influence of Love there is no state of rest. Strife, then, as a source of shaking things up by its nature will always produce motion. The account remains fictitious, but it does not need to involve the claim that lack of motion fully preceded possession of motion.

Although it does not touch upon these considerations, I wonder whether the lines Simplicius gives after the single line of testimony I adduced from Fr. 110 W, are added by him or go back to Eudemus. These lines provide some textual evidence for Eudemus' claim about the Sphairos, quoting Empedocles B27 lines 1 and 3–4. The difference this might make is one of style plus authorial intent. If these lines go back to Eudemus, he will be engaged in enlarging the evidentiary basis of Aristotle's claims, otherwise he will just add a terse one liner of explication.

Fr. 113 W raises similar concerns. The reference to Eudemus occurs in the last lines of Simplicius' comments on chapter 3 about the possible distributions of motion and rest in the world. These comments are about the last section of the chapter, where Aristotle summarised the argumentation of the chapter and set out what option remained to be tackled in what follows. The options as set up at the beginning of the chapter (253a24–30)[9] were: either (1) everything is always in motion, or (2) everything is always at rest,[10] or (3) some are in motion and some are at rest, and in this latter case either (3.1) the items that move are

8 Εὔδημος δὲ τὴν ἀκινησίαν ἐν τῇ τῆς Φιλίας ἐπικρατείᾳ κατὰ τὸν Σφαῖρον ἐκδέχεται, ἐπειδὰν ἅπαντα συγκριθῇ. (1183.27–28; Fr. 110 W)

9 This list is repeated at 254a16–22, in a slightly different order. Moreover option (3.1) is missing there, and is supplied by Ross in his editions.

10 Actually, this should be a non-starter: chapter 1 has already established that "there never was and never will be a time when there was not or will not be motion …" (252b5–6). Nev-

always in motion, and the items at rest are always at rest, or (3.2) everything is of such a nature that they can partake both in motion and rest, or—what Aristotle calls, both there and at the end of the chapter, the remaining option (3.3) that some of the things are always without motion, some are always in motion, and some partake both in motion and rest. Simplicius in closing his comments on the chapter points out that the way the "remaining option" is formulated is somewhat hasty. Here two further options should also be included: (3.4) that in addition to the items which take part both in motion and rest there are only such further entities that are always in motion, and (3.5) that the additional entities are all always at rest. Interestingly, Simplicius claims that Eudemus also included these further options in his *Physics*:

> One should note that it would have been possible to produce by division two further sections (which would divide things in two, some of them going either way and being at times in motion and at times at rest, the rest being either always in motion or always at rest). Eudemus at any rate, him also pretty much paraphrasing Aristotle, assumes also these sections when he concisely states: "either all of them go either way or some do and the rest are either in motion or at rest." (1206.23–29; Fr. 113 W, transl. Share modified)[11]

It is quite hard to assess whether one should accept Simplicius' claim here. This is not about the quote. The issue is whether what Simplicius quotes should be taken to carry the meaning that he imputes to Eudemus.

For one thing Simplicius may have been deceived by the similarity, indeed, almost identity of the formulation of *his* additional two options ("the rest being either always in motion or always at rest," τῶν δὲ λοιπῶν ἢ ἀεὶ κινουμένων ἢ ἀεὶ ἠρεμούντων, 1206.26–27) and the wording Eudemus used ("the rest are [either] in motion or at rest," τὰ δὲ λοιπὰ κινεῖται ἢ ἠρεμεῖ, 1206.29) The only apparent difference between the two formulations is that Eudemus does not say that these entities are always in motion, or always at rest. But Eudemus can dispense with this qualification, because, just like Simplicius, he speaks about things

ertheless, in the argumentative flow it makes sense to keep this option, as a preparatory step for retaining *some* items that are always at rest.

11 χρὴ δὲ ἐφιστάνειν, ὅτι καὶ ἄλλα δύο τμήματα δυνατὸν ἦν ἐκ τῆς διαιρέσεως ἀναφαίνεσθαι διχῇ τέμνοντα τὰ ὄντα, τῶν μὲν ἐπαμφοτεριζόντων καὶ ποτὲ μὲν κινουμένων ποτὲ δὲ ἠρεμούντων, τῶν δὲ λοιπῶν ἢ ἀεὶ κινουμένων ἢ ἀεὶ ἠρεμούντων. καὶ ὅ γε Εὔδημος παραφράζων σχεδὸν καὶ αὐτὸς τὰ Ἀριστοτέλους, τίθησι καὶ ταῦτα τὰ τμήματα συντόμως λέγων· "ἢ πάντα ἐπαμφοτερίζει ἢ τινὰ αὐτῶν, τὰ δὲ λοιπὰ κινεῖται ἢ ἠρεμεῖ."

that do not "go either way", that do not switch from motion to rest, or from rest to motion. So there is no difference between Simplicius' and Eudemus' claim on this score. One may want to cling to the difference that Simplicius stresses the division between the two options by using a more stressed "either … or" (ἤ … ἤ) formulation instead of Eudemus' "[either] … or" (… ἤ). This, however, is again immaterial. No object is both in motion and at rest. The reason why Simplicius' wording can introduce the two additional options is that he has just formulated that that is the point, the intended meaning of his formulation. The difference is that we do not have anything similar from Eudemus' side, anything that would contrast what is meant by this formulation to option (3.3) as introduced by Aristotle.

The fact that Simplicius quotes only this concise statement from Eudemus, suggests that he did not present such a contrast between the two cases putatively introduced here and option (3.3). So Simplicius' claim could be that Eudemus, with this formulation, "assumes also these sections." (l. 28) This is a distinct possibility. Here this is an issue of how Eudemus' general claim, about those items that do not switch from motion to rest (and *v.v.*), distributes over the predicates that they are [always] in motion or [always] at rest. One option can be to take Eudemus to refer to the same two options that Simplicius as read in the context introduced: either all of them are always in motion, or all of them are always at rest. But as I have submitted, we seem to lack such a context in Eudemus, in *his Physics* this sentence appears to be a standalone claim. In that case, however, the natural way to parse the sentence is that all of these items are either in motion or at rest, that is to say, this group of entities divides into ones that are always in motion and ones that are always at rest. A claim that is treacherously close to Aristotle's option (3.3).

Simplicius, however, may have a point, because indeed, Eudemus' formulation does not only cover this Aristotelian option, but also allows for a setup in which there are no items always in motion, and also for one in which there are no items that are always at rest. The claim that the items that do not go either way are all of them either in motion or at rest is true also in case one of these latter options is not populated by any objects.

Simplicius then turns out to be vindicated after all. But the case is not really reassuring. The question remains whether Eudemus, in giving this succinct formulation, could have intended to refer to all possible distributions, including also these two extremely imbalanced ones, and could have expected his readers to take his formulation as also referring to these setups. My doubts about this rest on an argument from silence, and those are notoriously weak. But nevertheless, the fact remains: Simplicius only quotes this concise statement,[12] from

12 Eudemus' pursuit of finding concise formulations can be attested also elsewhere. See also

an author about whom he several times asserts that his *Physics* proceeds in the footsteps of Aristotle. And all this happens at the end of the comments on a chapter,[13] as part of a parting shot.

He adds after the quote from Eudemus, in his own voice: "But it is clear that if that threefold division [i.e. (3.3)] is proved correct, the present two-way divisions [i.e. (3.4) and (3.5)] are also refuted." (1206.29–31, transl. Shape modified) This already indicates that Simplicius looks forward to dispensing with the additional two options—but he does not inform us how Eudemus disposed of the same two additional options allegedly introduced by him. The most parsimonious supposition then is to think that Eudemus' formulation was not meant to introduce these options, which, however, as Simplicius realised could also be covered by Eudemus' formulation.

2

There is an additional fragment about a detail in Simplicius' commentary on chapter 3 (Fr. 112 W), one in chapter 4 (Fr. 114 W), six on chapter 5 (Frs. 115–120 W), one on chapter 6 (Fr. 121 W) and four on chapter 10 (Frs. 122a–123b W). The uneven spread of these fragments can reflect the presence of deviations from Aristotle's procedures, but we should also expect that it is at least as much influenced by Simplicius' interests.

The additional fragment on chapter 3 is Fr. 112 W. There Aristotle's observation against the universal presence of motion, that "stone grows neither harder nor softer," (*Physics* VIII 3, 253b30–31), quoted by Simplicius at 1200.3–5, is followed by the claim

> but, as Eudemus says, "nor is it the case that something flows from it each day, since over many years the efflux would be greater than the body from which it flows, and yet," he says, "many things, even small ones, are seen to endure, remaining untouched for a very long time." (1200.5–9, transl. Share modified)

Fr. 117 W, where—to the chagrin of Simplicius—Eudemus gives a joint formulation for two options in VIII 5 256a21–b3 which Simplicius is adamant to keep apart. Then next, in Fr. 118 W this joint formulation is applied in the discussion of the three plus one options of how a self-mover can be articulated into the mover and the moved entity.

13 Note that I do not claim that Simplicius had our chapter division. What I stress here is that our chapter 3 is a unit, setting out a well-demarcated task at the beginning, returning to that articulation in the last twenty-odd lines, and formulating in no ambiguous terms what remains to be done in what follows.

Two things are worth mentioning here. After Aristotle's claim about an instance of qualitative stability, Eudemus is quoted as citing cases of quantitative (and also qualitative) stability. This is a step of providing a more inclusive set of examples for Aristotle's claim. The other thing to note is that Eudemus here in an argument against the universal and eternal presence of motion—this time, of separation of parts from a body—formulates considerations similar to the arguments against Anaxagoras in *Physics* I.4.

The next fragment, Fr. 114 W, similar to Fr. 113 W, comes around the end of Simplicius' comments on chapter 4.[14] Having finished the comments on the chapter, Simplicius adds this fragment of Eudemus, and then concludes with some somewhat critical, and somewhat elucidatory remarks about Aristotle's claim that everything that moves is moved by something. In the course of which, however, the point of Fr. 114 W need not be critical:

> He [Aristotle] has proved here in a more physical manner than in the preceding book that everything that is moved is moved by something. Eudemus proves that that which is moved is moved by something also from the fact that relatives subsist together.
>
> "For as it is not reasonable," he says, "that that which is cut is not cut by something, and that which is dragged is not dragged by something, and in general that which is made is not made by something, so [it is] also [not reasonable] that that which is moved is not moved by something." (1219.37–1220.4, transl. Bodnár)[15]

Instead, Simplicius' wording suggests that Eudemus *also* indicated an additional different way, contrasted to Aristotle's, who proceeded in a way more appropriate to natural investigation (φυσικώτερον) than in the preceding book.[16] Eudemus' proof is again a dialectical one (although Simplicius does not mention this), starting out from considerations about relatives, again in line with the *Categories*. Eudemus' move certainly is a shortcut. And we need

14 Just as chapter 3 (see the considerations above), chapter 4 is also a well-demarcated argumentative unit.

15 φυσικώτερον δὲ νῦν ἔδειξεν, ὅτι πᾶν τὸ κινούμενον ὑπό τινος κινεῖται, ἤπερ ἐν τῷ πρὸ τούτου βιβλίῳ. ὁ δὲ Εὔδημος, ὅτι τὸ κινούμενον ὑπό τινος κινεῖται, δείκνυσι καὶ ἐκ τῆς τῶν πρός τι συνυποστάσεως· "ὡς γὰρ τὸ τεμνόμενον, φησίν, ἄλογον μὴ ὑπό τινος τέμνεσθαι καὶ τὸ ἑλκόμενον μὴ ὑπό τινος ἕλκεσθαι καὶ ὅλως τὸ ποιούμενον μὴ ὑπό τινος ποιεῖσθαι, οὕτω καὶ τὸ κινούμενον μὴ ὑπό τινος κινεῖσθαι".

16 Cf. Alexander's assessment of book VII, that there Aristotle used "more dialectical proofs" (λογικωτέραις … ταῖς ἀποδείξεσι, 1036.12–13).

not presume that Eudemus would have taken this as more than an initial move of physical enquiry. Simplicius' repeated quotes from Eudemus' *Physics* corresponding to chapter 5, on self-motion attest that such categorical considerations did not rule out that the same entity by his lights could in principle qualify as both the mover of a motion and the moved entity.

After this in Fr. 115 W, at the start of the following chapter, Eudemus sets out why the investigation about self-movers is the next step, instead of discussing whether what is moved is moved by something in motion or at rest (1220.28–29),[17]

> … but because, as Eudemus says, it seems to practically everybody that a thing that causes motion is also in motion itself, and given that, of things that cause motion and are at the same time in motion, some are self-moved, others moved from outside by something else (there being two things that are originative of motion: the unmoved and the self-moved), he [Aristotle] first introduces the self-moved and shows that among things that both cause motion and are in motion [themselves] there is something primary which causes motion without being moved from outside by something else … (1220.30–36, Fr. 115 W, transl. Share modified)[18]

Fr. 116 W, again a testimony about Eudemus tagged to the end of comments on a chunk of the text of the *Physics* (256a4–21) might seem problematic at first sight:

> Eudemus adds that among motions here [below] the motions associated with life appear to be causes and that the heavenly [motion] is of course [the cause] of the transformation of the elements. The latter too seems

17 Such an overarching interest, where Eudemus raises issues not tackled by Aristotle is also evidenced at the beginning of Fr. 118 W, where Simplicius attests that Eudemus noted that the self-mover was introduced by Plato.

18 διὰ δὲ τὸ πᾶσι σχεδόν, ὥς φησιν Εὔδημος δοκεῖν τὸ κινοῦν κινεῖσθαι καὶ αὐτό, τῶν δὲ κινούντων ἅμα καὶ κινουμένων τὸ μὲν αὐτοκίνητόν ἐστι, τὸ δὲ ὑπ᾽ ἄλλου ἔξωθεν κινεῖται, δύο ὄντων ἀρχικῶν τῆς κινήσεως τοῦ τε ἀκινήτου καὶ τοῦ αὐτοκινήτου, τὸ αὐτοκίνητον πρῶτον παραδίδωσι καὶ δείκνυσιν, ὅτι ἔστι τι πρῶτον ἐν τοῖς κινοῦσί τε καὶ κινουμένοις, ὃ κινεῖ μὴ ὑπ᾽ ἄλλου ἔξωθεν κινούμενον …

Note that the claim about the two things that are originative of motion, and the explicit reference to Aristotle may most probably go back to Simplicius rather than to Eudemus.

> in some sense to be associated with life, for the heaven is not moved by something else but by itself. (1222.17–20, Fr. 116 W, transl. Share modified)[19]

Accordingly, Eudemus added this summary characterization—we are not informed whether before adding this he charted the same argumentative trajectory as Aristotle. This is quite similar to what we found in Fr. 113 W: a concise, even terse formulation meant to include the options Aristotle enumerates (perhaps covering some additional ones, as Simplicius claims), and to what we shall see in Fr. 117 W too.

This is important to keep in mind, because it can allay our consternation about Eudemus apparently rejecting the unmoved mover, the mover of the heaven. This worry would be misplaced: Eudemus' short summary deploys two major considerations. First, on the basis of the argument of the previous chapter, he submits that self-movers should be conceived as fundamentally similar to animals that move themselves, which are characterised, first at 254b14–19 and then also at 255a5–11 in contradistinction to the motion of the elements. There, Aristotle even used the term "associated with life" (ζωτικόν) to describe the capacity for self-motion.

The other component of Aristotle's argumentation here that is summarised in Eudemus' pithy assertion is what we have at 256a13–21, the last lines of the section where Simplicius quotes these lines from Eudemus. There Aristotle—keeping in force all way through the presupposition that the investigation should be restricted to the domain of moved movers—in the first part, up until 256a19, rules out that the chain of moved movers could go back in an infinite sequence. Hence there has to be among them a first moved mover. The consequence of this is drawn in the closing lines of the section: "if then everything that is in motion is moved by something, and the first mover is moved but not by something else, it must be moved by itself" (256a19–21).

Accordingly, what Eudemus does may not amount to anything more than identifying this first moved mover as the heavens, about which Aristotle talked in general terms only.[20] We have no reason to suppose that he would resist the introduction of the unmoved mover. That will be reached by Aristotle's argumentation, and we actually have fragments 121–123b W, where Eudemus tackles some problems connected to the unmoved movers.

19 προστίθησι δὲ ὁ Εὔδημος, ὅτι καὶ τῶν ἐνταῦθα κινήσεων αἱ ζωτικαὶ φαίνονται κινήσεις αἴτιαι, καὶ ἡ οὐρανία τῆς τῶν στοιχείων δηλονότι μεταβολῆς. δοκεῖ δὲ καὶ αὕτη ζωτική πως εἶναι· οὐ γὰρ ὑπ' ἄλλου κινεῖται ὁ οὐρανός, ἀλλ' ὑφ' ἑαυτοῦ.

20 This move can also be compared to what Eudemus does in Fr. 119 W, where he replaces Aristotle's examples in which he sets out what would be required from an item so that it could qualify as a proper case of self-causation with ones that are more illuminating.

Up to this point we could not ascertain Eudemus' argumentative strategy with any confidence. Indeed, it was not pellucid how the additional points Simplicius mentioned were connected to the rest of what Eudemus had to say. Fr. 118 and Fr. 120 W indicate how Eudemus articulated the cases that have to be tackled. In both cases this is so because Eudemus set out a different line of argumentation from the one pursued by Aristotle.[21] What we see in these two cases is that in Fr. 118 W Eudemus stakes out the different options rigorously, and as Simplicius presents in his paraphrase, goes through them in a straightforward manner, to arrive at the core case to be discussed, which is "what has within itself something that causes motion but is unmoved according to itself, and another that is in motion." (1234.5–6, transl. Chase modified)[22]

In Fr. 118 W we see Eudemus' articulation of the different options in Simplicius' paraphrase. Then, Fr. 120 W is even more illuminating, as it professes to be an actual quote:

> Eudemus also divides this as follows: "the part causing motion," he says, "will either do so while being unmoved, or [will be] moved. But if it is unmoved, what we want has been proved: that what is self-moved is composed of what is moved and a mover that is unmoved. But if the part causing motion does so while itself being moved, since all that is moved is moved by something, this too will be moved by something. Accordingly, what moves it cannot be external to it, for the whole could no longer be said to be moved by itself, but by something else moving the part that moves it. Therefore, it will be moved by some one of the things within what is self-moved. Thus, the part causing motion will be moved either by one of its own parts, or by the thing moved. Again, this mover causes motion either while remaining unmoved—and the prime mover will be unmoved—, or while being moved. And again, either by a part of itself or by the thing moved."
>
> "By using reciprocal removal this way," says Eudemus, "at some point we will arrive at the prime movers: these were what caused motion not by some [part], but moved as wholes by means of wholes. For the reciprocal removal cannot be infinite, for they [the parts we have gone through] would not be in motion if there were no first item. But it has been proved that it is impossible for a whole to be moved by the whole of itself. There-

21 Although in the case of Fr. 118 W the articulation used by Eudemus is found later in a concise formulation also in Aristotle, at 257b28–30.

22 τὸ ἔχον ἐν αὑτῷ τὸ μέν τι κινοῦν ἀκίνητον ὂν καθ᾽ αὑτό, τὸ δὲ κινούμενον.

> fore, it is necessary for the prime mover to cause motion while remaining unmoved."
>
> This is how Eudemus carries out the division. (1237.10–27, transl. Chase modified)[23]

This longer quote shows a style of argumentation distinctly different from how Aristotle proceeds. As a first step it divided the self-moving entity into a part responsible for the motion, and another part that is moved by this part. The reason for this division is also stated at a later step: "it has been proved that it is impossible for a whole to be moved by the whole of itself." (lines 24 ff.), and as we have seen, Eudemus employed his more accessible, more convincing examples in his Fr. 119 W to substantiate the claim Aristotle makes in 257b2–6.

Once this division of the self-moved entity into mover and moved part is established, Eudemus sets up a dilemma: either the part that is effecting motion is unmoved—and then we have already divided the self-mover into an unmoved mover and a part moved by this unmoved component. Alternatively, this part is moved. But then it needs a mover, just as the original self-moving entity needed one. At this point, there are two options: the part effecting motion can be a self-mover, just as the self-moving entity that is under scrutiny was one. But there is also another option, which Eudemus also includes, for completeness's sake. The part in motion which is responsible for the motion of the self-moving entity may be moved by the other part, the part which is being moved.

What is not allowed at this point is that mover and moved entity move each other: Aristotle stresses that that is also ruled out by the considerations formulated in 257b2 ff., to which Eudemus Fr. 119 W corresponds.[24] Hence when we

---

23 διαιρεῖ δὲ καὶ τοῦτο ὁ Εὔδημος οὕτως· "τὸ γὰρ κινοῦν, φησί, μέρος ἢ ἀκίνητον ὂν κινήσει ἢ κινούμενον· ἀλλ᾽ εἰ μὲν ἀκίνητον, δέδεικται ὃ βουλόμεθα, ὅτι τὸ αὐτοκίνητον σύγκειται ἐκ κινουμένου καὶ ἀκινήτου τοῦ κινοῦντος. εἰ δὲ τὸ κινοῦν μέρος κινούμενον καὶ αὐτὸ κινεῖ, ἐπειδὴ πᾶν τὸ κινούμενον ὑπό τινος κινεῖται καὶ τοῦτο ὑπό τινος κινηθήσεται, οὐκοῦν ἔξωθεν μὲν οὐχ οἷόν τε εἶναι τὸ κινοῦν αὐτό· οὐ γὰρ ἂν ἔτι ὑφ᾽ ἑαυτοῦ λέγοιτο κινεῖσθαι τὸ ὅλον, ἀλλ᾽ ὑπ᾽ ἄλλου τοῦ τὸ κινοῦν αὐτὸ μέρος κινοῦντος· ὑπό τινος ἄρα τῶν ἐν τῷ αὐτοκινήτῳ κινήσεται. ἢ οὖν ὑπό τινος τῶν ἑαυτοῦ μορίων κινήσεται τὸ κινοῦν μέρος ἢ ὑπὸ τοῦ κινουμένου. καὶ τοῦτο πάλιν τὸ κινοῦν ἢ ἀκίνητον ὂν κινεῖ, καὶ ἔσται τὸ πρώτως κινοῦν ἀκίνητον, ἢ κινούμενον· καὶ πάλιν ἢ ὑπὸ μέρους ἑαυτοῦ ἢ ὑπὸ τοῦ κινουμένου.

ἀνταναιροῦντες δέ, φησὶν ὁ Εὔδημος, οὕτως ἥξομέν ποτε εἰς τὰ πρώτως κινοῦντα· ταῦτα δὲ ἦν, ἃ μὴ τινὶ ἐκίνει, ἀλλ᾽ ὅλα ὅλοις· οὐ γὰρ ἐνδέχεται τὴν ἀνταναίρεσιν ἄπειρον εἶναι· οὐ γὰρ ἂν κινοῖντο μὴ ὄντος πρώτου. δέδεικται δέ, ὅτι ὅλον ὑπὸ ὅλου ἀδύνατον κινεῖσθαι ἑαυτοῦ· ὥστε ἀνάγκη τὸ πρώτως κινοῦν ἀκίνητον ὂν κινεῖν." οὕτω μὲν οὖν ὁ Εὔδημος διαιρεῖ·

24 See also the more complex case discussed earlier at 256b27–257a14, where mover is moved with a motion that is of a different type from the motion that it effects. The discussion

have to find a mover for this moved entity, which is the mover of the item that moves it, we will have to specify either a part of this moved entity, or a part of the item that moves this. In case the item so specified is not unmoved, the same step needs to be reiterated. This procedure is then dubbed by Eudemus "reciprocal removal,"—the participle he uses is ἀνταναιροῦντες—with the proviso that it cannot go on indefinitely. It must stop at a final remove: the mover of the whole series of moved movers which is unmoved.

This proof is remarkable for its simplicity and its structural clarity. Moreover, it is quite unlike Aristotle's typical arguments that most often do not follow this kind of strict structure of introducing the consequences of previous steps in this order. Instead, what we find most often in Aristotle is that he introduces a statement and then provides the support for this statement in the form of a complex system of grounds in a web of γάρ clauses.[25] In contrast to this, Eudemus' argumentative is one of setting out different options, and then introducing consequences, then calling attention to what has already been established along the way, all this culminating in a proof of the thesis argued for, that the prime mover causes motion while remaining unmoved.

3

After all these we have arrived at the fragments about the unmoved movers of the celestial spheres. My aim here is to characterise Eudemus' approach in these cases without going into all the details.[26] First of these is Eudemus' Fr. 121 W, connected to the thorny issue of Aristotle's claims at 259b20–31. The fragment is just three words, that

> Next, he [Aristotle] proves that the primary mover "for each motion" (as Eudemus adds) must be unmoved both according to itself and also incidentally, and must also be eternal as the cause of eternal motion. (1262. 17–21, Fr. 121 W, transl. McKirahan modified)[27]

---

of this more complex case then leads to the same difficulty, that when the intermediate movers are set aside, we will have to arrive at a case that is identical to the one rejected at 257b2–6.

25 For detailed presentation of this contrast between Eudemus' and Aristotle's argumentative style see White (Forthcoming), who contrasts Eudemus' argumentative strategy to Aristotle's as presented in Netz (2001).

26 For these see Bodnár (2002a).

27 εἶτα ἐφεξῆς δείξας, ὅτι τὸ πρώτως κινοῦν "καθ' ἑκάστην κίνησιν", ὡς ὁ Εὔδημος προστίθησιν, ἀκίνητον εἶναι χρὴ καὶ καθ' αὑτὸ καὶ κατὰ συμβεβηκός, εἶναι δὲ καὶ ἀίδιον ὡς ἀιδίου κινήσεως αἴτιον.

The difficulty with interpreting what these three words inserted into the phrase at 259b23–24[28] are meant to achieve is that this very passage, at b28–31 closes with a claim that attributes some sort of incidental motion to some celestial entities or to their movers. Moreover, Simplicius' testimony comes at a slightly belated moment, after the discussion of the passage itself, in the introductory lines of the following passage of 259b32–260a10, summarising what has been established so far. One would think that Eudemus' introduction of these words could have, or even should have been indicated earlier, in Simplicius' discussion of the passage on pp. 1259–1261. Simplicius, however, has good reason to postpone this testimony about Eudemus. His discussion of the passage containing 259b23–24 concentrates on the contrast between unmoved movers that move *themselves* incidentally like souls and the prime mover that is not moved even incidentally: the former—unlike the latter—cannot produce eternal, continuous motion.

Then, when Simplicius discusses 259b28–31, where Aristotle submits that eternal, continuous motion can also be caused by the movers of the planetary spheres, or in the planetary spheres themselves (lines b29–31 are notoriously difficult to interpret, but this is immaterial for our present discussion), Simplicius comes up with a radical proposal. According to him the clause at 259b28–29, οὐκ ἔστιν δὲ τὸ αὐτὸ τὸ κινεῖσθαι κατὰ συμβεβηκὸς ὑφ' αὑτοῦ καὶ ὑφ' ἑτέρου cannot mean "Being moved incidentally by itself and by something else are different"—i.e. different in that motion caused by a mover that is incidentally moved by something else can be eternal and continuous. This is anathema to Simplicius. Nothing incidental can be eternal by his lights. Therefore, he claims that the proper understanding of the clause is "Being moved incidentally by itself and [being moved] by something else are different." Planetary spheres can perform, and their movers can cause eternal continuous motion even though these spheres are moved by something else, and their movers are under the causal influence of something else some way or another.

This means that Eudemus' intervention in the text is fully in line with Simplicius' intentions, but it comes at the wrong moment. Therefore, it makes perfect sense to mention it only after the fact, retrospectively in the summary of what has been achieved so far.

---

28 Note that the phrase in Simplicius' wording adds "both according to itself" (καὶ καθ' αὑτό) to what we find at 259b13–14. Bodnár (2002a), 179 takes it that Eudemus' additional words were inserted into an earlier occurrence of an almost identical phrase ("both without qualification and also incidentally", καὶ ἁπλῶς καὶ κατὰ συμβεβηκός) at 258b14–15. That may not make a huge doctrinal difference, but for reasons of sorting out the argumentative structure of Simplicius' discussion one should stick to the suggestion in Jaeger (1948), 366 n. 1.

This has some further repercussions on how we should take Fr. 121 W itself. It assigns a mover that is unmoved both according to itself and also incidentally to every celestial motion. But then it is highly unlikely that Eudemus would have subscribed to Simplicius' rather forced understanding of 259b28–29. There is quite a range of options how Eudemus' movers that are unmoved also incidentally can allow for yet some further entities that are incidentally moved.[29] If Eudemus opted for any one of them, it would be somewhat disingenuous on Simplicius' part to use Eudemus in the summary fashion he does after his quite non-standard proposal for 259b28–29. Hence my guess—which once again rests on an argument from silence—is that Eudemus' *Physics* cut the Gordian knot and went for a concise and disambiguated presentation of the proper account of celestial theory. Once Eudemus made up his mind that all the prime movers of the heaven had to be unmoved both according to themselves and also incidentally, there was no reason to include anything about incidental motions in the celestial domain, and so no such details were available for Simplicius to report about.

There is another quote from Eudemus which could have been used to argue that Eudemus exempts each and every prime mover from incidental motion. This is Fr. 123a W:

> "If the prime mover is without parts and does not touch what is moved, how is it related to it?" (1355.28–30, Fr. 123a W, transl. McKirahan modified)[30]

Simplicius suggests that this was raised as an introductory problem about the localization of the prime mover in VIII 10, starting out from the proof of the indivisibility of the prime mover at the beginning of the chapter. But—as Simplicius' immediately following lines testify—this can also be called upon as supporting the rejection of incidental motion of such an indivisible entity:

> How can what is without parts and does not touch what is moved but is unrelated to it and transcendent, be moved along with what is moved, so as to be moved incidentally? (1355.30–32, transl. McKirahan)[31]

29 See Bodnár (2014), 401 ff., with a chart of the available options on p. 405.

30 εἰ ἀμερές, φησίν, ἔστι τὸ πρώτως κινοῦν καὶ μὴ ἅπτεται τοῦ κινουμένου, πῶς ἔχει πρὸς αὐτό;

31 τὸ οὖν ἀμερὲς καὶ μὴ ἁπτόμενον τοῦ κινουμένου, ἀλλ' ἄσχετον πρὸς αὐτὸ καὶ ἐξῃρημένον, πῶς δύναται συγκινεῖσθαι τῷ κινουμένῳ, ὥστε κατὰ συμβεβηκὸς κινεῖσθαι;

Note, however, that the difficulty Eudemus raised is about the relationship between mover and the entity moved by this mover. Accordingly, the incidental motion Simplicius is entitled to speak about, that it might be imparted to the mover is an incidental motion *by itself*. In order to formulate considerations about incidental motion of celestial movers imparted to them *by something else*, one would need some further steps.[32] And indeed: Simplicius' remark here is much rather targeted against Alexander's putative localisation of the prime mover, than it is indicative of Eudemus' stance on the incidental motions of celestial movers.

4

Counting Fr. 123a W again, there are four more fragments on book VIII remaining. Three about what Eudemus proposed for the localisation of the prime mover, and a fourth one about the mode of operation of the indivisible, unmoved celestial mover. The interest of the three fragments about the localisation of the prime mover is not so much about the actual proposal they formulate, but the fact that in the first one, Fr. 122a W we get Alexander's testimony about the issue, but then Simplicius goes out of his way and checks the text of Eudemus' *Physics* available to him, and reports that there is no trace of that proposal in the text.[33] Instead, he gives two quotes from Eudemus, Frs. 123a and 122b W. The Aristotelian claim behind these proposals is that "the things nearest the mover move the quickest." (267b7–8) On the option which Alexander imputes to Eudemus the mover would be "on the largest circle through the poles" (1354.10, repeated at 1355.33),[34] whereas Eudemus' actual words quoted by Simplicius claim that "On a sphere the place around the poles moves most quickly." (1355.35–36)[35]

32 First among these is to argue that these considerations are not about the first celestial mover, the mover of the sphere of the fixed stars, but about every celestial mover. On this see Bodnár (2002a), 181–183.

33 Note that on quite a few occasions it is reasonable to suppose that Simplicius was prompted to give a summary statement about, or even a quote from Eudemus by following Alexander's commentary on the *Physics*. See e.g. Rashed (2011), 535–536, who makes a case that Frs. 115 and 116 W may originate from Alexander. The case of Fr. 123b W is all the more important, because it shows that on occasion Simplicius could exercise caution against a report he found in Alexander.

34 ἐν τῷ μεγίστῳ κύκλῳ … τῷ διὰ τῶν πόλων. Although it would be tempting to circumvent this testimony and render it as "on the celestial equator," as McKirahan does, this is a highly problematic move.

35 "τῆς δὲ σφαίρας ὁ μὲν τόπος ὁ περὶ τοὺς πόλους τάχιστα κινεῖται".

Both of these claims look problematic. One of them more than the other: Alexander's proposal can be made sense of by claiming that the celestial spheres have depth, and the motion induced by their unmoved mover is quickest at their outer periphery, hence the mover should be located there. Putting the mover on a meridian would lead to a setup in which the sphere's each disk parallel to its equator would reach one branch of the mover on each half turn, and then would be running on towards the mover's other branch. This way the interaction would not occur through some extended area, like in Alexander's proposal, who suggested that the mover should be located "on the entire periphery of the outer sphere."

In contrast to Alexander's testimony about the option Eudemus endorsed, the one Simplicius found (or thought to find) in Eudemus' *Physics* makes precious little sense. The only point to keep in mind is that the very claim that "on a sphere the place around the poles moves most quickly" is misleading or dead wrong. Dead wrong if linear velocity is meant—the part of a sphere with the quickest motion in this sense is its equator—and misleading if we talk about angular velocity. Each part of a sphere takes part in exactly the same circular motion, otherwise it would be torn apart. Therefore, if we do not reject the whole enterprise of locating the prime mover—interestingly, in view of the difficulty proffered in Fr. 123a W that is also a distinct possibility[36]—all we can ascribe to Eudemus is that similarly to what we see at Fr. 116 W, he can be taken to come up with some specific proposal for the localisation of the prime mover on the basis of the general characterisation Aristotle formulates. Moreover, as far as Simplicius is concerned the fact that he double-checked Alexander's testimony is highly reassuring. Even if we may have our doubts, we should take it that in this particular case Simplicius exercised proper care and due caution in reporting about Eudemus' *Physics*.

This leaves us with Fr. 123b W about the mode of operation of the prime mover. Here Eudemus returns to the problem he has already voiced in Fr. 123a W, about the relationship between an indivisible mover and the item moved by it:

> "What causes motion in respect of place," he declares, "seems to cause motion by pushing or pulling. And if not only in these ways, in any case it touches it either directly or through one or more intermediaries. But what is without parts cannot touch anything; for it does not have one part

36 For this option see Bodnár (2002a), 186.

> as a beginning, the other part as an extremity, whereas things that touch have their extremities together. How then can what is without parts cause motion?"
>
> He solves the difficulty, saying, "Some things cause motion being moved and others [cause motion] being at rest. Things that are moved cause motion by touching, while those that cause motion without being moved do so otherwise, though not all in the same way: the prime mover does not cause motion in the same way as the earth causes the ball thrown towards it to move upwards. For it [the prime mover] does not cause motion when there [already] was a prior motion, for [that way] it would no longer cause motion in the primary way. On the other hand, the earth, being at rest, will never cause motion primarily." (1357.19–29, Fr. 123b W, transl. McKirahan modified)[37]

Before introducing this quote Simplicius stressed that Eudemus goes beyond the issue Aristotle is discussing here. He does not address the difficulty whether a moved entity can be the cause of continuous motion. Rather he formulates the fundamental query how an indivisible entity can be causally responsible for local motion in the first place. As the quote suggests, Eudemus did not, and indeed, could not come up with a positive account. What he formulates, nevertheless, is a typology of contrast cases. Moved movers are responsible for the motion that they cause through touch. Indivisible entities—like e.g. points – cannot be in touch, neither with other indivisible entities, nor with something that is divisible. Hence there should be some other type of causal action. Aristotle's suggestion for this at the end of *De generatione et corruptione* I 6 is that some entities can be in one directional touch. Maybe Eudemus is also feeling his way towards such a solution. In any case, we do not hear from Simplicius about the solution of the difficulty. All we get is yet another distinction, between moved and unmoved movers. It turns out that not only moved movers operate through touch. Some unmoved movers can also be responsible through touch for the motion they cause. This is the case with those unmoved movers that cause motion in a derivative way, by being responsible for a motion that

37 "δοκεῖ γάρ, φησί, τὸ κινοῦν κατὰ τόπον ἢ ὠθοῦν ἢ ἕλκον κινεῖν· εἰ δὲ μὴ μόνον οὕτως, ἀλλ᾽ οὖν ἁπτόμενόν γε αὐτὸ ἢ διὰ ἄλλου ἢ δι᾽ ἑνὸς ἢ πλειόνων· τὸ δὲ ἀμερὲς οὐδενὸς ἐνδέχεται ἅψασθαι· οὐ γάρ ἐστιν αὐτοῦ τὸ μὲν ἀρχή, τὸ δὲ πέρας· τῶν δὲ ἁπτομένων τὰ πέρατα ἅμα. πῶς οὖν κινήσει τὸ ἀμερές;"

καὶ λύει τὴν ἀπορίαν λέγων ὅτι "τὰ μὲν κινούμενα κινεῖ τὰ δὲ ἠρεμοῦντα. καὶ τὰ μὲν κινούμενα κινεῖ ἁπτόμενα, ἃ δὲ κινεῖ μὴ κινούμενα, ἄλλως, οὐχ ὁμοίως δὲ πάντα. οὐ γὰρ ὡς ἡ γῆ τὴν σφαῖραν ῥιφθεῖσαν ἐπ᾽ αὐτὴν ἄνω ἐκίνει, οὕτω καὶ τὸ πρώτως κινῆσαν· οὐ γὰρ προγενομένης κινήσεως ἐκεῖνο κινεῖ· οὐ γὰρ ἂν ἔτι πρώτως κινοίη· ἡ δὲ γῆ οὐδέποτε ἠρεμοῦσα πρώτως κινήσει."

is consecutive on, and indeed causally dependent on a previous motion. As a result of this contrast, we can take away that the indivisible unmoved movers will need to be causally efficacious in yet a different way. How that is so, is not broached here. I do not think this should be the result of Simplicius' selective quoting procedure here. Instead, the most probable supposition is that Eudemus closed his *Physics* with some crucial distinctions aimed at showing how the prime mover cannot cause motion. Nevertheless, that causal interaction must be possible, this much has effectively been proved in the preceding considerations. But perhaps by Eudemus' lights the specific mode of operation which has been distinguished here in Fr. 123b W from these other options is such that it cannot be given an account here, because it may very well be beyond the purview of natural philosophy.

5

It is time to take stock. What we have found is quite in line with the standard characterisations of Eudemus in the literature. In Sharples' concise words "Eudemus follows Aristotle quite closely. Sometimes his exposition is more compressed than Aristotle's, sometimes he expands on it; often he will draw upon his knowledge of other parts of Aristotle's *Physics* or of other Aristotelian doctrines, and often he will seem to be striving after a more systematic exposition."[38] But, as Sharples also stresses, there is more. Sharples suggests that his discussion of the Eudemian testimonies and fragments on change, place and time enables us to assess Eudemus' methods and contributions. Here, in connection with what I indicated as "cosmic issues" we can have some grasp of what Eudemus took to be the task of natural philosophy in this domain, and how he sets about presenting these issues in the few cases where we have some evidence. And there is even more: Eudemus, as White (Forthcoming) argues, has a distinctive argumentative style, quite unlike that of Aristotle. In the context of this specific, much more rigorous argumentative style, quite a few of his changes—be they omissions, or abbreviations, or additions—take on distinct significance.

Sharples in the introductory pages of his paper asks whether we can take Eudemus to be engaged in the exercise of "'interpreting Aristotle through Aristotle' in order to elucidate Aristotle's thought and make it more systematic". In response to this André Laks pointed out to him that such a claim

38 Sharples (2002b), 124.

would involve the further one that he anticipated "the procedure of interpreting Homer through Homer developed in Hellenistic Alexandria." Sharples replies that drawing connections across different Aristotelian texts is a "natural response to the presence of common themes" in different places of the corpus, and refers to *Metaphysics* V as a prime example, which evidences that such an integrative enterprise was well underway already in the Peripatos.[39]

I think that, in view of what we have found here, we can take yet a further step. Eudemus does not stitch together bits and pieces found in Aristotle. What he is engaged in is an original way of covering the issues presented in Aristotle's *Physics*. A recapitulation, from someone who lives the life of research and enquiry as pursued in Aristotle's school. This is so, even though here he takes as his starting point a written document corresponding to Aristotle's *Physics*.*

39 Sharples (2002b), 110 with n. 9.

* This chapter is a substantially revised version of my original presentation, in light of the discussion at the conference in Lyon. I am especially grateful to Gweltaz Guyomarc'h for his thoughtful comments on the penultimate draft. Moreover I am indebted to Orna Harari for a constant flow of discussions about Alexander and Simplicius, and to Stephen White for sending me his forthcoming paper on Eudemus' distinctive argumentative style.

CHAPTER 6

Eudemus and Alexander of Aphrodisias against the Idea of an Infinite Universe

Robert Mayhew

Aristotle discusses the infinite or unlimited (τὸ ἄπειρον) in *Physics* III 4–8. Eudemus (ca. 370–300 BCE), a student and then colleague of Aristotle, dealt with this topic in the second book of his (lost) *Physics*, which was based on Aristotle's work of that name,[1] as did Alexander of Aphrodisias (2nd/3rd c. AD) in his (lost) commentary on the *Physics* (and elsewhere).[2] What follows is an account of Eudemus' contribution to or revisions of an objection levelled by Aristotle against an argument in favour of an idea they both rejected—that the universe is infinite—followed by different objections levelled by Alexander against this same idea. Simplicius is the source for most of the texts that I consider, apart from Aristotle.[3]

Early in his discussion of the infinite (or unlimited),[4] Aristotle lays out the five considerations that might seem to make plausible its *actual* existence (III 4, 203b15–25):[5]

1 We know it was in his second book from Simplicius, *In Phys.* III 4, 203a16–23, 459.24–26 (Fr. 62.4–6 W = 84.12–13 Stork). Throughout the paper, in citing Simplicius' commentary on the *Physics*, the page and line numbers refer to Diels (1882a). W refers to Wehrli (1955); Peter Stork's edition of the fragments of Eudemus is still forthcoming. Where I refer to Stork but not to Wehrli, it means the latter did not include that text.

2 See Rashed (2011) for the fragments of Alexander's commentary on books IV–VIII, recovered from Byzantine scholia.

3 Baltussen (2006), 40: "The amount and detail of Simplicius' information, illustrated in particular by the numerous references to individual books …, make it more than probable that he had a copy of Eudemus' work." For a list of these references, see Baltussen (2006), 3. On the nature of Eudemus' *Physics*, see Baltussen (2002b) & (2006) and Sharples (2002b). On Simplicius' use of Eudemus, and Eudemus' value to him, see Menn (2022), 19–23. Simplicius clearly had a copy of Alexander's commentary, and assumed his readers did as well (Menn (2022), 13). On Simplicius' use of Alexander, see Menn (2022, §§1–3).

4 In the context of *Phys.* III 4–8, I prefer 'the infinite' as a translation of τὸ ἄπειρον, though 'the unlimited' is possible as well—as in Urmson's translation of Simplicius' *In Phys.* III (2002). I would like to have translated the related terms ἄπειρος, πέρας, and περαίνω in such a way that the connection among them was always apparent (using in every case the root '-fin-'); but that was not possible, as πέρας is 'a limit' and περαίνω 'to limit'. See Menn (2022), 107.

5 The text of the *Physics* used is Ross (1936). All translations from the Greek are my own.

 | DOI:10.1163/9789004739314_008

> The belief in the existence of something infinite can come mainly from considering five things: (1) from time, for it is infinite, and (2) from the division in magnitudes, for mathematicians too make use of the infinite; further (3) by the [idea] that in this way only do generation and destruction not give out, if there is an infinite from which what comes to be is taken away; further (4) by the [idea] that what is limited always reaches a limit in relation to something, such that there must be no [ultimate] limit, if each thing must always reach a limit in relation to something else. But most of all and most authoritative, the one which creates the common *aporia* for everyone: (5) for by not giving out in thought, (a) number seems to be infinite as well as (b) mathematical magnitudes and (c) what is outside of the universe (*ouranos*).[6] Etc.[7]

Two comments, looking ahead to Eudemus' re-ordering of this presentation: the infinite division in magnitudes (2) could have been considered alongside the infinite addition (in thought) to numbers and mathematical magnitudes (5a & b); and, the idea that there is no giving out in thought regarding what is outside of the universe (5c) is related to the idea that every limit must itself be limited by something else, *ad infinitum* (4).[8]

---

6 I believe I am here justified in translating οὐρανός 'universe' (which in this context I prefer to 'heaven' or 'heavens'). First, Aristotle makes it clear in *De caelo* (I 9, 278b9–24) that οὐρανός is said in three ways, one of which is τὸ ὅλον καὶ τὸ πᾶν. Second, after quoting a line from *Physics* 4.4 (211a23–25), in which Aristotle twice refers to what is in the *ouranos* (ἐν τῷ οὐρανῷ), Simplicius writes (*In Phys.* 212b3–22, 594.34–595.6 [Fr. 100A.1–10 Stork, cf. 80 W]): "And it is clear that *ouranos* meant either the whole cosmos or 'the revolving all'; but he produced much unclarity in the words before us [sc. 212b3–22] by sometimes using the name 'the all' (τὸ πᾶν) and sometimes 'the *ouranos*'. […] But Eudemus does not even use the name 'the all'; and in the case of the *ouranos*, he is making an inquiry into the whole [sc. cosmos?] …." (καὶ δῆλον ὅτι οὐρανὸν ἢ τὸν ὅλον κόσμον ἔλεγεν ἢ τὸ κυκλοφορητικὸν πᾶν· πολλὴν δὲ τοῖς προκειμένοις ἀσάφειαν ἐνεποίησε τὸ ποτὲ μὲν τὸ πᾶν ποτὲ δὲ τὸν οὐρανὸν ὀνομάζειν· […] ὁ δὲ Εὔδημος τὸ μὲν πᾶν οὐδὲ ὀνομάζει· ἐπὶ δὲ τοῦ οὐρανοῦ τὴν ζήτησιν ποιεῖται τοῦ ὅλου ….) Earlier, in *Physics* III 4, 203a6–8, Aristotle claimed that the Pythagoreans (οἱ Πυθαγόρειοι) held that what is outside of the universe is infinite (τὸ ἔξω τοῦ οὐρανοῦ ἄπειρον).

7 τοῦ δ' εἶναί τι ἄπειρον ἡ πίστις ἐκ πέντε μάλιστ' ἂν συμβαίνοι σκοποῦσιν, (1) ἔκ τε τοῦ χρόνου, οὗτος γὰρ ἄπειρος, (2) καὶ ἐκ τῆς ἐν τοῖς μεγέθεσι διαιρέσεως, χρῶνται γὰρ καὶ οἱ μαθηματικοὶ τῷ ἀπείρῳ· ἔτι (3) τῷ οὕτως ἂν μόνως μὴ ὑπολείπειν γένεσιν καὶ φθοράν, εἰ ἄπειρον εἴη ὅθεν ἀφαιρεῖται τὸ γιγνόμενον· ἔτι (4) τῷ τὸ πεπερασμένον ἀεὶ πρός τι περαίνειν, ὥστε ἀνάγκη μηδὲν εἶναι πέρας, εἰ ἀεὶ περαίνειν ἀνάγκη ἕτερον πρὸς ἕτερον. μάλιστα δὲ καὶ κυριώτατον, ὃ τὴν κοινὴν ποιεῖ ἀπορίαν πᾶσι· (5) διὰ γὰρ τὸ ἐν τῇ νοήσει μὴ ὑπολείπειν καὶ ὁ ἀριθμὸς δοκεῖ ἄπειρος εἶναι καὶ τὰ μαθηματικὰ μεγέθη καὶ τὸ ἔξω τοῦ οὐρανοῦ. κτλ.

8 See Furley (1969), 92–96.

According to Simplicius, Eudemus believes there are rather *six* reasons, not five (*In Phys.* III 4, 203b15–30, 468.4–7; Fr. 64 W = 85.53–58 Stork):

> So Aristotle in this way claims that the belief in the existence of the infinite comes mainly from five reasons, whereas Eudemus claims it is owing to six reasons. Having said that the first of these is the division in the continuous *ad infinitum*,[9] he claims that the counterpart of this is the one theorizing about addition in the case of numbers.[10]

Earlier, Simplicius had commented (*In Phys.* III 4, 203b15–30, 466.7–8; Fr. 63 W = 85.1–3 Stork):[11]

> The second belief in the existence of the infinite is 'from the division in magnitudes', or as Eudemus claims, [from the division] in continuous things.[12]

So Aristotle's second point was apparently the first in Eudemus' presentation. And whereas Aristotle included the infinite addition of numbers under his fifth (5a), Eudemus claims it should be removed from that classification and placed alongside—but separate from—the argument from division in the continuous, making the total number of reasons given for believing in an actual infinite six. Assuming Eudemus did not change the order of the others (though nothing hangs on this), his presentation then would have been: (1) division in the continuous and its counterpart, (2) addition to numbers, followed by (3) time, (4) generation and destruction, (5) limits requiring limits, and (6) mathematical magnitudes and what is outside of the universe not giving out in thought. At least judging by the extant evidence, the focus of his final reason is what is supposedly outside of the universe.

Simplicius will later note that Aristotle *had* included the infinite addition of numbers (namely, under his fifth consideration);[13] and he is right in a sense. But

9 An alternative translation of τὴν ἐν τῷ συνεχεῖ διαίρεσιν ἐπ᾽ ἄπειρον is "the division in the continuous *for an infinite* [*time*]" (LSJ s.v. ἐπί C. II). Urmson's translation of ἐπ᾽ ἄπειρον is "without limit".

10 οὕτως μὲν οὖν ὁ Ἀριστοτέλης ἐκ πέντε μάλιστα τὴν πίστιν εἶναί φησι τοῦ εἶναι τὸ ἄπειρον, ὁ δὲ Εὔδημος δι᾽ αἰτίας ἕξ φησιν. ὧν πρώτην εἰπὼν τὴν ἐν τῷ συνεχεῖ διαίρεσιν ἐπ᾽ ἄπειρον γινομένην ταύτης ἀντίστροφόν φησι τὴν κατὰ πρόσθεσιν ἐν τοῖς ἀριθμοῖς θεωρουμένην.

11 See also *In Phys.* I 2, 185a14, 55.19–24; Fr. 140.7–12 W = 154.65–73 Stork.

12 δευτέρα δὲ πίστις τοῦ εἶναι τὸ ἄπειρον «ἐκ τῆς ἐν τοῖς μεγέθεσι διαιρέσεως», ἢ ὡς Εὔδημός φησιν, ἐν τοῖς συνεχέσι.

13 *In Phys.* III 4, 203b15–30, 468.7–10; Fr. 85.58–62 Stork: "But Aristotle did *not* pass over

Eudemus treats the addition of numbers in its own right, not as an example of something that does not give out in thought, elevating its status, so to speak, as a counterpart to division in the continuous. This arguably improves the organisation and presentation, as infinite division and addition do seem to go together, in the way described. In any case, I expect this difference between Aristotle and Eudemus is not conceptually important,[14] and it will not concern me further.

Before turning to what in some sense is the central text of this essay, I want to present Simplicius' summary of the reasoning behind Aristotle's fourth consideration that might seem to make plausible the actual existence of the infinite. This is worth doing, as the reasoning here is quite similar to the reasoning employed in that central text. Simplicius writes (*In Phys.* III 4, 203b15–30, 466.31–36):

> The fourth too is hard to face, that everything limited appears always to reach a limit at something. For if everything limited reaches a limit at something else that is outside it, that outside at which the limit is reached is either infinite [= unlimited] or limited. And if it is infinite [= unlimited], we have from this itself that the infinite exits. But if it is limited, like the earth, and it reaches a limit at something else, and this is *ad infinitum*, then the infinite exits.[15]

Shortly thereafter, he adds (*In Phys.* III 4, 203b15–30, 467.1–3):

> It is especially by relying on this argument, Alexander claims, that those around Epicurus say that the all [i.e. the universe] is infinite, because everything limited has something outside it limiting it.[16]

this reason, having placed it among the things considered infinite in thought, when he said that 'number too seems to be infinite, and mathematical magnitudes' it is clear can be divided *ad infinitum*." (οὐδὲ ταύτην δὲ τὴν αἰτίαν ὁ Ἀριστοτέλης παρῆκεν ἐν τοῖς διὰ τὴν ἐπίνοιαν νομιζομένοις ἀπείροις θεὶς αὐτήν, ὅτε ἔλεγε «καὶ ὁ ἀριθμὸς δοκεῖ ἄπειρος εἶναι καὶ τὰ μαθηματικὰ μεγέθη» διαιρεῖσθαι ἐπ' ἄπειρον δῆλον.)

14 Sharples (2002b), 114: "the difference, it seems, is one of arrangement rather than of doctrine."

15 τέταρτον δὲ καὶ δυσαντίβλεπτον τὸ φαίνεσθαι πᾶν τὸ πεπερασμένον ἀεὶ πρός τι περαῖνον. εἰ γὰρ πᾶν τὸ πεπερασμένον πρὸς ἄλλο τι ἔξωθεν αὐτοῦ ὂν περαίνει, τὸ ἔξωθεν ἐκεῖνο πρὸς ὃ περαίνει, ἢ ἄπειρόν ἐστιν ἢ πεπερασμένον. καὶ εἰ μὲν ἄπειρον, ἔχομεν αὐτόθεν ὅτι ἔστι τὸ ἄπειρον· εἰ δὲ πεπερασμένον οἷον ἡ γῆ, καὶ αὐτὸ πρὸς ἄλλο περαίνει, καὶ τοῦτο ἐπ' ἄπειρον. εἰ δὲ ἐπ' ἄπειρον, ἔστι τὸ ἄπειρον.

16 τούτῳ δὲ [μάλιστα] τῷ λόγῳ ὡς μάλιστα, ὥς φησιν ὁ Ἀλέξανδρος, οἱ περὶ Ἐπίκουρον πιστεύοντες ἄπειρον ἔλεγον εἶναι τὸ πᾶν, διότι πᾶν τὸ πεπερασμένον παρά τι περαινόμενον ἔξωθέν τι

Simplicius goes on to present Eudemus' elaboration on what is in Aristotle's presentation the fifth consideration.[17] It is one of the most famous source-texts for Eudemus (*In Phys.* III 4, 203b15–30, 467.26–35; Fr. 65 W = 85.33–45 Stork):[18]

> But Archytas,[19] as Eudemus claims, used to propound the argument in this way: Could I, coming to the extreme [edge] of the universe—that is, to the fixed [sc. sphere of the stars?][20]—extend my hand or staff into what is outside, or not? Now indeed, not being able to extend it is absurd; and if I extend it, what is outside will be either body or place. It'll make no difference, as we shall learn. So he will always go in the same manner to the limit always [i.e. in each case] assumed, and he will ask the same thing, and if there will always be somewhere else to [or 'something else into'] which the staff [extends], it is clear that it[21] is indeed infinite. And if it is a body, what was proposed has been proved; but if it is a place, and place is that in which a body is or potentially could be—and what is potential must be treated as [actually] existing in the case of eternal things—in this way too there would be infinite body and place.[22]

ἔχει. (οἱ περὶ Ἐπίκουρον likely means 'Epicurus and his followers'). See Epicurus, *Letter to Herodotus* (DL, X 41 [514–516 Dorandi]) and Lucretius I 958–983.

17 There is scholarly disagreement over whether this text is part of a reply to Aristotle's fourth or fifth consideration (or both). Simplicius saw it as a reply to the fifth, as the passage prior to the Archytas-text makes clear (467.5–25 = Fr. 85.5–32 Stork); and Huffman (2005), 545–546 argues (I believe successfully) for the fifth. Eudemus seems to treat them together, however; for in the Archytas-text he refers to continually reaching and then going beyond a limit. (In commenting on Aristotle's fourth consideration, Simplicius does not mention Eudemus: see *In Phys.* III 8, 208a11–14, 516.1–38.)

18 For discussion of this text, see Furley (1969, 92–94) and Huffman (2005), 541–550. On Archytas' argument and its afterlife, see Sorabji (1988), chs. 8 & 9 and Ierodiakonou (2011). On how much of this text might actually come from Archytas, and how much is Eudemus' paraphrase, see Huffman (2005), 542–544. In translating it, I was helped by Barnes (1982), 204, Urmson (2002), 87, and Huffman (2005), 541.

19 A fourth century BCE Pythagorean. He and Plato knew each other. See Huffman (2020).

20 Huffman brackets οἷον τῷ ἀπλανεῖ (lit. "that is, to what does not wander"), explaining that the expression τῷ ἀπλανεῖ "is not the idiom of the fourth century and almost surely does not belong to Archytas or Eudemus ... The use of οἷον ... points to an explanatory gloss" (2005), 542—from Simplicius or someone else? He translates the remaining words—τῷ ἐσχάτῳ ... οὐρανῷ—"at the outermost edge of the heaven," the purported gloss being "that is to say, at the fixed heaven."

21 That is, what is beyond the edge of the universe—beyond the fixed sphere of the stars.

22 Ἀρχύτας δέ, ὥς φησιν Εὔδημος, οὕτως ἠρώτα τὸν λόγον· «ἐν τῷ ἐσχάτῳ οἷον τῷ ἀπλανεῖ οὐρανῷ γενόμενος πότερον ἐκτείναιμι ἂν τὴν χεῖρα ἢ τὴν ῥάβδον εἰς τὸ ἔξω, ἢ οὔ; καὶ τὸ μὲν οὖν μὴ ἐκτεί-νειν ἄτοπον, εἰ δὲ ἐκτείνω, ἤτοι σῶμα ἢ τόπος τὸ ἐκτὸς ἔσται· διοίσει δὲ οὐδέν, ὡς μαθησόμεθα. ἀεὶ οὖν βαδιεῖται τὸν αὐτὸν τρόπον ἐπὶ τὸ ἀεὶ λαμβανόμενον πέρας, καὶ ταὐτὸν ἐρωτήσει, καὶ εἰ ἀεὶ

The Archytas-text is all that Simplicius has preserved of the views of Eudemus on what is in Aristotle's presentation the fifth consideration. Simplicius' immediate response to this not only introduces his provisional thoughts[23] on Archytas' argument, it also anticipates two fundamentally different ways of assessing the argument, one of them being Alexander's, as we shall see (*In Phys.* III 4, 203b15–30, 467.35–468.3; Fr. 85.45–52 Stork):

> Perhaps this argument will be bitterly puzzling for us as well, who are saying that there is nothing outside of the universe, as the cosmic body, whose limit is the universe, occupies the whole[24] of space. So if, having come to the back [or "surface"] of the universe, he would stretch out his hand, where would it be stretched out? For surely not into nothing, since none of the things that exist are in what does not exist. But nor will he be prevented from stretching out [sc. his hand]. For it is not possible to be prevented by nothing.[25]

"For surely not into nothing" etc. anticipates Alexander's response; "But nor will he be prevented" etc. is the position of Archytas. More on this *aporia* later.

In *Phys.* III 6–7, Aristotle (*inter alia*) offers a sophisticated defence of the concept of the *potentially* infinite, rejecting any conception of its actual existence.[26] In *Phys.* III 8, he ends his account of the infinite with a summary of his replies to the five considerations or reasons given in favour of its actual existence. As I am here solely concerned with no. 5, I present his reply to this one alone (208a14–19):[27]

ἕτερον ἔσται ἐφ᾽ ὃ ἡ ῥάβδος, δῆλον ὅτι καὶ ἄπειρον, καὶ εἰ μὲν σῶμα, δέδεικται τὸ προκείμενον· εἰ δὲ τόπος, ἔστι δὲ τόπος τὸ ἐν ᾧ σῶμά ἐστιν ἢ δύναιτ᾽ ἂν εἶναι, τὸ δὲ δυνάμει ὡς ὂν χρὴ τιθέναι ἐπὶ τῶν ἀιδίων, καὶ οὕτως ἂν εἴη σῶμα ἄπειρον καὶ τόπος.

23 The passage begins μήποτε ('perhaps'). On Simplicius' use of μήποτε, signalling his own views, see Menn (2022), 26–32.

24 Diels prints ὕλην, which Urmson (2002), 160 n. 247 considers a misprint, and translates ὅλην.

25 μήποτε δὲ οὗτος ὁ λόγος καὶ πρὸς ἡμᾶς ἀπορήσει δριμέως τοὺς λέγοντας μηδὲν εἶναι ἔξω τοῦ οὐρανοῦ, ὡς τοῦ κοσμικοῦ σώματος, οὗ πέρας ὁ οὐρανός, τὴν ὅλην χώραν κατειληφότος. εἰ οὖν ἐν τῷ νώτῳ τοῦ οὐρανοῦ γενόμενος ἐκτείνοι τὴν χεῖρα, ποῦ ἂν ἐκταθείη; οὐ γὰρ δὴ εἰς τὸ μηδέν· οὐδὲν γὰρ τῶν ὄντων ἐν τῷ μὴ ὄντι. ἀλλ᾽ οὐδὲ κωλυθήσεται ἐκτεῖναι· ὑπὸ γὰρ τοῦ μηδενὸς οὐ δυνατὸν κωλυθῆναι.

26 See e.g. Ross (1936), 367–369 & 554–561, Wagner (1967), 519–530, Coope (2012), and Cooper (2016).

27 As indicated earlier, Aristotle's nos. 4 & 5 are related, though Aristotle's reply to the fifth does not seem to include the fourth, and his reply to the fourth (which distinguishes being finite or limited and being in contact, 208a11–14) is not part of the context necessary for

> To trust in thought is absurd; for the excess and the deficiency are not in the thing, but in the thought. For one might think each of us is many times larger than he is, growing *ad infinitum*;[28] but it is not for this reason that someone is beyond *the city or* the particular size we are, because someone thinks it, but because he is [there, or that size]; and this [sc. the thought] is incidental.[29]

Before continuing, I have to say something about τοῦ ἄστεος … ἢ, 'the city or' (208a18), which I have italicised. Ross's apparatus reports that τοῦ ἄστεος … ἢ is the reading of all the manuscripts and is also found in the *Physics*-commentaries of Philoponus, Simplicius, and Themistius,[30] though according to a *scholium* on τοῦ ἄστεος in *Parisinus gr.* 1853 (ms. E), Eudemus supposedly offered the variant τοῦ ἀστέρος, 'beyond *the star*'.[31] Philoponus, however, rejects τοῦ ἄστεος … ἢ (*In Phys.* III 8, 208a14, *CAG* vol. XVI, 495.6–7 Vitelli):[32] "The most accurate of the copies [of the *Physics*] do not have 'of the city', but only 'beyond the particular size'" (τὰ ἀκριβέστερα δὲ τῶν ἀντιγράφων οὐκ ἔχει τὸ τοῦ ἄστεος, ἀλλὰ μόνον ἔξω τοῦ τηλικούτου μεγέθους). Diels, in his *Textgeschichte* of the *Physics*[33] follows Philoponus, and Ross follows Diels, thus bracketing τοῦ ἄστεος and ἢ. Ross comments: "Diels is very likely right in thinking the interpolation arises from some words in Eudemus' commentary …."[34]

As for the supposed Eudemus-variant ἀστέρος, Wehrli[35] is right that Eudemus' actual comment on Aristotle here (more on this shortly) indicates that we

understanding Eudemus' treatment of Aristotle's reply to the fifth. In *De caelo* I 5–9 as well, Aristotle discusses and rejects the idea of an infinite universe. Some of Simplicius' commentary on *De caelo* I 5–9 is relevant in the present context.

28 I translate εἰς ἄπειρον "*ad infinitum*," though Aristotle may have in mind "into an infinite (space)."

29 τὸ δὲ τῇ νοήσει πιστεύειν ἄτοπον· οὐ γὰρ ἐπὶ τοῦ πράγματος ἡ ὑπεροχὴ καὶ ἡ ἔλλειψις, ἀλλ' ἐπὶ τῆς νοήσεως. ἕκαστον γὰρ ἡμῶν νοήσειεν ἄν τις πολλαπλάσιον ἑαυτοῦ αὔξων εἰς ἄπειρον· ἀλλ' οὐ διὰ τοῦτο ἔξω *τοῦ ἄστεός* τίς ἐστιν *ἢ* τοῦ τηλικούτου μεγέθους ὃ ἔχομεν, ὅτι νοεῖ τις, ἀλλ' ὅτι ἔστι· τοῦτο δὲ συμβέβηκεν.

30 Ross's *app. crit.*, ad loc.: "τοῦ ἄστεος ΠPST: τοῦ ἀστέρος γρ. Eudemus: om. γρ. P., Diels" and "ἢ ΠPST: om. γρ. P., Diels."

31 The *scholium* in *Parisinus gr.* 1853, fol. 23ᵛ (Fr. 71 Werhli = 90B Stork): γρ. ὁ Εὔδημος ἔξω τοῦ ἀστέρος ἤγουν τῆς ἐν τοῖς ἄστροις φορᾶς.

32 The full context of this passage is provided in the Appendix below.

33 Diels (1882b), 39.

34 Ross (1936), 562. Ross continues: "such as οὐ διὰ τοῦτο Διάρης χιλίων σταδίων ἐστὶν ἢ τοῦ ἄστεος μείζων." We encounter words very much like these in Simplicius, *In Phys.* III 8, 208a14–19, 517.13–20, quoted below, p. 144.

35 Wehrli (1955), 102.

should reject it.[36] But even if τοῦ ἄστεος goes back to Eudemus and not to Aristotle, I thought it best to include these words in the text and translation (in italics). Without them, Aristotle's meaning is for instance "… that someone is larger than the particular size we are" (more literally, "beyond the particular size we have" [ἔξω τοῦ τηλικούτου μεγέθους ὃ ἔχομεν]). A point in favour of keeping ἔξω τοῦ ἄστεος, however, is that it parallels ἔξω τοῦ οὐρανοῦ in Aristotle's presentation of the five reasons (III 4, 203b25), better than ἔξω τοῦ τηλικούτου κτλ. alone.

If τοῦ ἄστεος … ἢ ("of the city or") is not original to Aristotle, then his reply is focused on mathematical magnitudes not giving out in thought, but not on numbers giving out in thought or on what is supposedly outside of the universe. Perhaps he believed that the same objection (namely, that the excess is in the thought alone) could easily be made in the other two cases as well and need not be spelled out in his lecture notes. If so, then τοῦ ἄστεος was originally a marginal gloss (arguably going back to Eudemus), which later made its way into the text. Or perhaps something has dropped out of the text, such that Aristotle's original, longer reply (which included references to the addition to numbers in thought, as well as to what is supposedly beyond the edge of the universe) can no longer be recovered. In any case, as we shall see—or as I shall argue—Eudemus will refer to the increase in magnitude in thought in answering Archytas' argument for an infinite universe. By contrast, Themistius takes Aristotle to be referring to two cases in which thinking something does not make it true (*In Phys.* III 8, 208a14–19, *CAG* vol. V 2, 100.29–101.5 Schenkl): I think of you (νοῶ σε) as outside of the city (which presumably may or may not be the case), and I think of you as tall as a mountain (which cannot be the case).[37] Similarly, Philoponus (*In Phys.* III 8, 208a14–19, 495.1–5) refers both to Socrates increasing in size without limit and to Socrates being outside of the city—even though he thought 'outside of the city' lacked the support of the most accurate manuscripts.[38]

Simplicius has a very positive appraisal of Aristotle's response to the fifth reason given in favour of its actual existence (*In Phys.* III 8, 208a14–19, 517.6–7): "He surely refutes even this argument concisely and clearly" (λύει δὴ καὶ τοῦτον τὸν λόγον συντόμως καὶ σαφῶς).[39] In support of Aristotle, he continues, contrast-

---

36 Moreover, if this did come from Eudemus, would he not have written ἔξω *τῶν ἀστέρων*?

37 See the Appendix below.

38 See the Appendix below.

39 Urmson (2002), 142 translates this "He refutes this argument concisely and clearly," ignoring the δή and καί, and so mutes Simplicius' enthusiasm. I translate καὶ τοῦτον τὸν λόγον "even this argument," taking it to refer to Archytas' argument, because of how 'bitterly puzzled' Simplicius was in his immediate reaction to it (as indicated above).

ing Eudemus' and Alexanders' interpretations of the text,[40] and this contrast gives us an indication of how the two answered Archytas' argument (*In Phys.* III 8, 208a14–19, 517.13–20; Fr. 70 W ≈ 90A.15–23 Stork):[41]

> So it is possible to imagine even a Diares[42] of a thousand stadia [in height] or bigger than the city, and [to imagine] the part of him rising above [or 'the excess part of him'] outside of the city. For I think it is better to interpret 'beyond the city' in this way, as Eudemus understood the [words] of his teacher—as [referring to] people imagining Diares as bigger than the city—and not as Alexander explains it, when we who are in the city think of ourselves outside of it *or* many times larger in magnitude, as though 'outside of the city' involves a difference according to place alone.[43]

The contrast is between thinking that ἔξω τοῦ ἄστεος means 'extending beyond the city' (Eudemus' interpretation, with the emphasis on size) and thinking that it means 'physically outside of the city' (Alexander's interpretation, with the emphasis on physical position).[44] Without emending the text, Simplicius is stating that Alexander's interpretation includes either our imaging ourselves not in the city or our imaging ourselves so large that we would occupy more space than the city, which I think weakens the contrast. There may well be a problem with the manuscripts' final ἤ ("*or* many times larger in magnitude"), as "many times larger in magnitude" works better as a description of the interpretation of *Eudemus* and *not* that of Alexander, whatever their different emphases

40 On Simplicius' practice of contrasting Eudemus and Alexander of Aphrodisias, and of favouring the former, see Menn (2022), 20. Rescigno (2004), 471 nevertheless believes Alexander is here indebted to Eudemus.

41 There is no reference to what is outside of the universe in this text, but I think that is certainly part of the context here and that "outside of the city" is meant as an analogy to it. This is Simplicius' view, as the opening (517.3–6 = 90A.1–5 Stork) and the closing (517.32–33 = 90A.40–41) of his discussion of *Phys.* III 8, 208a14–19 make clear.

42 In *De anima* II 6, Aristotle uses the son of Diares (Διάρους υἱός) as an example of the object of perception, and in perceiving his pale colour, we perceive the son of Diares incidentally (418a21). Perhaps Diares was a standard example, in all sorts of contexts, in the early Lyceum. He appears as an example in one other Eudemus-text: Simplicius, *In Phys.* IV 10, 218a31–b9, 702.4 (Fr. 83.18 W = 103.12 Stork). See Rescigno (2004), 473–474.

43 οὕτω δὲ καὶ Διάρη χιλίων σταδίων ἔστι φαντάζεσθαι ἢ τῆς πόλεως μείζω καὶ τὸ ὑπερέχον αὐτοῦ ἐκτὸς τῆς πόλεως. κάλλιον γὰρ οἶμαι τὸ ἔξω τοῦ ἄστεος οὕτως ἀκούειν, ὡς ὁ Εὔδημος ἐνόησε τὰ τοῦ καθηγεμόνος, ὡς μείζονα τῆς πόλεως τὸν Διάρη φανταζομένων, καὶ μὴ ὡς ὁ Ἀλέξανδρος ἐξηγεῖται, ὅταν ἡμᾶς αὐτοὺς ἔνδον ὄντας τοῦ ἄστεως ἔξω νοήσωμεν ἢ πολλαπλασίους κατὰ τὸ μέγεθος, ὡς τοῦ ἔξω τοῦ ἄστεος κατὰ τὸν τόπον μόνον ἔχοντος τὴν παραλλαγήν.

44 A less relevant contrast is between imagining *someone else* larger than the city (Eudemus) and imagining *oneself* larger than the city *or* outside it (Alexander).

(note the *κατὰ τὸν τόπον μόνον*). In fact, I think a case can be made for emending the text to read ⟨μᾶλλον⟩ ἤ ("think of ourselves being outside it ⟨rather⟩ than many times larger in magnitude"). However, a relevant fragment from Alexander's lost commentary on *De caelo*,[45] in Simplicius' commentary on that work, would seem to support instead replacing ἤ with καί, yielding: "when we who are in the city think of ourselves outside of it *and* many times larger in magnitude" (even though I still find ⟨μᾶλλον⟩ ἤ better fits the context). Simplicius reports (*In* DC I 9, 279a11–18; CAG vol. VII, 286.23–27 Heiberg = Fr. 91a.17–19 Rescigno):

> Alexander also reproaches thinking what was said [about an actual infinite] to be true, following the case of the hand in the imagination. For, he claims, we have imaginations of many even impossible things; indeed, each person might imagine himself outside of the city *and* many times larger in magnitude or much smaller and the size of a millet-seed.[46]

In any case, Eudemus' interpretation, which Simplicius prefers, is, I take it, that just because someone can imagine Diares a thousand stadia (between 150 and 200 kilometres) tall, and so beyond or above the city, does not mean that he could in fact be a thousand stadia tall. Not only does thinking not make it so, in some particular case; more fundamentally, conceivability is not the same as, nor does it imply, metaphysical possibility. And the fact that I might *in some sense* be able to imagine Diares going (or growing) beyond the edge of the universe would in no way alter the fact that it is impossible to go beyond the edge of the universe.[47]

Now I would have thought that Eudemus further believed that it *is* impossible to go outside of the universe or to extend one's hand beyond the edge of the universe, just as it is impossible for a human being to become a thousand stadia tall. Unfortunately, we are given little indication as to what arguments of his own, if any, he might have presented against the idea of something going beyond the edge of the universe, aside from the fact that thinking it does not make it possible. Much later in the commentary, and in a different context, Simplicius reports that Eudemus (whom he quotes) planned to investigate whether any-

45 Fr. 91a Rescigno = *In* DC 284.28–285.9 & 285.21–286.27 Heiberg. For the entire text, with Italian translation and commentary, see Rescigno (2004), 466–502.

46 διαβάλλει δὲ ὁ Ἀλέξανδρος καὶ τὸ τῇ φαντασίᾳ τῆς χειρὸς ἀκολουθοῦντας ἀληθὲς ἡγεῖσθαι τὸ λεγόμενον. πολλῶν γάρ, φησί, καὶ ἀδυνάτων φαντασίας ἴσχομεν· καὶ γὰρ ἕκαστος ἑαυτὸν ἔξω τοῦ ἄστεος φαντασθείη ἂν καὶ τὸ μέγεθος πολλαπλάσιον ἢ βραχύτατον καὶ κεγχριαῖον. See Rescigno (2004), 470–472.

47 Cf. Modrak (2002), 195–201.

thing was outside of the universe (*In Phys.* IV 5, 212b3–22, 595.10–11; Fr. 80.13–15 W = 100A.17–18 Stork):[48]

> But is [sc. the universe] itself in a place or not, or how could either [be the case]? For it is not in a place at all, *unless there is something outside*; for in that way it would be contained. *We shall investigate this.*[49]

By contrast, according to Simplicius, Alexander seems to interpret Aristotle as follows:[50] Just because I can imagine myself outside of the city I am in does not establish that I am in fact outside of that city. Thinking doesn't make it so. I can imagine myself in Lyon, say, far beyond my home in New Jersey, but that cognitive ability does not change the fact that I am now in New Jersey, and it has no probative value in determining where I am (or might possibly be?). In the same way, just because I can (or it might seem that I can) imagine myself beyond the edge of the universe, does not mean that I am in fact (or could in fact be) beyond the edge of the universe.

Simplicius, after contrasting the views of Eudemus and Alexander, a little later comments (*In Phys.* III 8, 208a14–19, 517.24–26; Fr. 90A.29–32 Stork):

> For the same thing being imagined to be such as to be in another place shows only that imagination alters things [i.e. the things being imagined], however it does not show that this is relevant to the infinite.[51]

Simplicius is implying that all that Alexander's interpretation of Aristotle's objection does is show that the ability to imagine myself beyond the edge of the universe does not demonstrate that it is possible for me to be outside the universe, though it does not rule that out either.

---

48 Simplicius is quoting Eudemus, whom he named a few lines earlier (595.3). See the following note.

49 αὐτὸς δὲ (sc. ὁ οὐρανὸς) πότερόν ἐστιν ἐν τόπῳ ἢ οὐκ ἔστιν ἢ πῶς ἑκάτερον; ὅλως μὲν γὰρ οὐκ ἔστιν ἐν τόπῳ, *εἰ μὴ ἔστι τι ἐκτός*· οὕτω γὰρ ἂν περιέχοιτο· *τοῦτο δὲ ἐπισκεψόμεθα*. Reading ὅλως with ms. E against ὅλος of ms. F ("it is not in a place *as a whole*"). In the same line in Themistius, *In Phys.* IV 5, 212a31–b3 (CAG V 2, 119.27–29 Schenkl = Fr. 100B.16–17 Stork), the manuscripts are split between ὅλως (M W B[2]) and ὅλον (B[1] L). See Algra (1995), 255–257 and Todd (2003), 86.

50 For a brief summary of Alexander's discussion of the infinite in his lost commentary on the *Physics*, which goes beyond the issues discussed in this essay, see Moraux (2001), 147–148.

51 τὸ γὰρ ἐν ἄλλῳ τόπῳ φαντασθῆναι τὸ αὐτὸ τοσοῦτον μόνον δείκνυσιν, ὅτι παραλλάττει τὰ πράγματα ἡ φαντασία, οὐ μέντοι ὡς ἐπὶ τοῦ ἀπείρου δείκνυσιν.

If his presentation of Alexander's commentary on the relevant Aristotle passage is accurate, then I think he is right to maintain that what Eudemus says is much more effective as an objection to Archytas' argument. For my ability to think of myself now in Lyon (or in China, where I have never been), rather than in New Jersey, would on its own not count against Archytas' argument.[52]

In the passage under discussion, Alexander (as presented by Simplicius) is attempting to explain the relevant text of Aristotle, and it may seem that he himself maintains merely that one should not conclude that the universe is infinite on the grounds that we can in some sense imagine ourselves in a place beyond it. But in another work, attributed to Alexander,[53] the author (and I think it *is* Alexander)[54] argues forcefully that it is impossible to go beyond the edge of the universe, *as there is nothing there*. *Quaestio* III 12, "That what exists is not infinite" (Ὅτι μὴ ἄπειρον τὸ ὄν), concludes (CAG sup. vol. II 2, 107.1–4 Bruns):[55]

> For how could someone extend something into nothing, or how could he come to be in what is wholly not existing? He would not get an impulse to extend one of his limbs into nothing in the first place; for such is the nature of what does not exist.[56]

Now I think Alexander is absolutely right, though I cannot fully defend that here (in part because that would require moving from the history of philosophy to philosophy *per se*). What Alexander is saying is that Archytas is guilty of *reifying* non-existence or nothing—treating it as a special kind of existence, into which something can extend, when in fact it is the *absence* of existence.[57] *Contra* Archytas, nothing or non-existence is not a place into which one can extend anything. Archytas (and his defenders) are in fact making the same mistake as

52 For a discussion of Alexander's response to Archytas, in a different context, see Harari (2020), 288–295.

53 Φυσικαὶ σχολικαὶ ἀπορίαι καὶ λύσεις (*Puzzles and solutions in natural philosophy*); its standard Latin title is simply *Quaestiones*.

54 This gets support from a passage in Simplicius' commentary on the *De caelo*, which I turn to shortly. See Todd (1984), 185 n. 1.

55 For a fuller discussion of this text, see Todd (1984).

56 πῶς γὰρ ἄν τις εἰς τὸ μηδὲν ἐκτείναι τι, ἢ πῶς ἂν ἐν τῷ μηδὲ ὅλως ὄντι γένοιτο; τὴν ἀρχὴν γὰρ οὐδ' ἂν ὁρμήν τι λάβοι εἰς τὸ ἐκτείνειν τι τῶν αὐτοῦ μερῶν ἐν τῷ μηδενί· τοιαύτη γὰρ ἡ τοῦ μὴ ὄντος φύσις.

57 See Rand (1990), 60–61, who calls this fallacy the Reification of the Zero. Archytas would no doubt claim that there is a difference between empty space and the absence of existence, but that is what in my view the Aristotelians are rejecting.

the White King in Lewis Carroll's *Through the Looking-Glass, and What Alice Found There* (1871). Waiting for his two messengers, the King asks Alice:

> "Just look down the road and tell me if you can see either of them."
> "I see nobody on the road," said Alice.
> "I only wish *I* had such eyes," the King remarked in a fretful tone. "To be able to see Nobody! And at such a distance too!"

The Aristotelians argue that the cosmos is finite and as such that there is nothing outside of the universe. The opposition takes this to imply that there is a vast area of empty space into which obviously one is able to go (just as the White King takes 'Nobody' to refer to a particular person). But it does not imply that at all. Moreover, as Alexander recognised, Archytas' argument is question-begging. At one point in his *De caelo* commentary, Simplicius presents the views of Alexander on this, again from Alexander's lost commentary on that work (*In DC* I 9, 279a11, 285.25–27 Heiberg = Fr. 91a.49–53 Rescigno):

> So, the assumption is absurd and assumes beforehand, in the imagination, what is being investigated, that there is something outside of the universe [i.e. the all], either void or solid.[58]

Carl Huffman does not think much of the Aristotelian answer to Archytas: "The problem with all of this ... is that this will only be convincing if Archytas ... accepts Aristotle's view that our normal assumptions about space do not apply at the edge of the cosmos ..., and there is no obvious reason why Archytas is compelled to do so."[59] But I disagree, and submit that one cannot in fact imagine anything going beyond the edge of the universe into non-being.

Moreover, Alexander is aware of such claims, and has an answer to them. In *Quaestio* III 12, he first paraphrases the kind of reasoning at play here (101.18–22 Bruns):[60]

58 ἄτοπος οὖν ἡ ὑπόθεσις καὶ τὸ ζητούμενον προλαμβάνουσα τῇ φαντασίᾳ ὅτι ἔστι τι τοῦ παντὸς ἐκτὸς ἢ κενὸν ἢ στερέμνιον.

59 Huffman (2005), 549. David Furley (1969), 93 too seems to think Archytas has common sense (our 'normal assumptions') on his side, against Aristotle: "The problem is not that one *can* imagine something beyond any given limit but that one *cannot help* imagining it."

60 In translating this and the following passage, I was helped by Todd (1984) and Sharples (1994).

> For we have not encountered any such limited thing, which has nothing beyond it; for such are all the things that are limited as parts [i.e. of a whole], which are the only things that we perceive; ... for outside of all the things that we perceive there is something.[61]

Later (p. 104.29–32) he responds to this line of reasoning:

> To these [i.e. the things we perceive] belongs the property of being seen beside another thing, not because they are limited, but because they are parts of the universe; [but this argument][62] thinks it right to transfer this [property] also to what is limited as a whole, when it can be attributed only to things that are limited as parts—not *qua* limited, but *qua* parts.[63]

If I understand him correctly, then the line of reasoning Alexander is refuting is a textbook example of what is now called the fallacy of composition—the fallacy of assuming that what is true of every part of a whole must be true of the whole as well. For it does not follow that because every object we have encountered *in* the universe has something beside it or outside of it that therefore the universe itself must have something beside it or outside of it. Being limited by something outside it is *not* an essential feature of being finite, which Aristotle maintains the universe is.

Of course, one could try to present other arguments in defence of an infinite body or void beyond the fixed sphere of the stars (an idea Aristotle argues against in *De caelo* I 9), but it would require something other than Archytas' thought experiment.

There is much more that one could say about all of this, and much more that Alexander does say about this in *Quaestio* III 12 and in the relevant 'fragment' from his lost commentary on *De caelo* in Simplicius' commentary on that work. Especially important is to understand more fully the Aristotelian view that "to trust in thought [i.e. in thought experiments] is absurd," and how this connects to the Aristotelian conception of possibility, and how they differ from the 'mod-

61 οὔτε γάρ τινι περιπεπτώχαμεν τοιούτῳ πεπερασμένῳ, οὗ μηδέν ἐστιν ἐπέχεινα· τοιαῦτα γὰρ πάντα τὰ ὡς μέρη πεπερασμένα, ὧν ἐσμεν αἰσθανόμενοι μόνων ...· ἁπάντων γὰρ ὧν αἰσθανόμεθα ἐχτός τι ὄν.

62 A subject of ἀξιοῖ must be supplied, and I follow Todd. Cf. "but the person who argues thus" (Sharples).

63 τούτοις γὰρ ὑπάρχοντος τοῦ παρ᾽ ἄλλο ὁρᾶσθαι, οὐ διὰ τὸ πεπεράνθαι, ἀλλὰ διὰ τὸ μέρεσιν τοῦ παντὸς εἶναι, ἀξιοῖ τοῦτο μεταφέρειν χαὶ ἐπὶ τὸ ὡς πᾶν πεπερασμένον, μόνοις τοῖς ὡς μέρεσι πεπερασμένοις τούτου παραχολουθεῖν δυναμένου, οὐχ ᾗ πεπερασμένα, ἀλλ ᾗ μέρη.

ern' conception of the connection between conceivability and various kinds of possibility.[64] But that is beyond the scope of this essay.

Instead, I will end by quoting Martin Heidegger, who arguably knows more about nothing than anyone. In *An Introduction to Metaphysics*, he writes:

> One cannot, in fact, talk about and deal with Nothing as if it were a thing, such as the rain out there, or a mountain, or any object at all; Nothing remains in principle inaccessible to all science. Whoever truly wants to talk of Nothing must necessarily become unscientific. But this is a great misfortune only if one believes that scientific thinking alone is the authentic, rigorous thinking, that it alone can and must be made the measure even of philosophical thinking. [...] True talk of Nothing always remains unfamiliar. It does not allow itself to be made common. It dissolves, to be sure, if one places it in the cheap acid of a merely logical cleverness.[65]

The Aristotelians, and especially Alexander in this case—maintaining that scientific thinking alone is the authentic, rigorous thinking—have dissolved the talk of an infinite nothingness outside of the universe in the powerful acid of logical reasoning.[66]

64 For instance, many of the views presented in Gendler & Hawthorne (2002). But to give one example, I assume that neither Aristotle nor Alexander would grant any legitimacy to Descartes' supposition in the *Meditations* that it is possible that "some malicious demon of the utmost power and cunning has employed all his energies in order to deceive me" (First Meditation [AT 22], Cottingham transl.). Rather they would claim that to trust in such a thought experiment is absurd. For a forceful statement of the view I am attributing to Aristotelians, see the passage from Themistius, presented below in the Appendix.

65 Fried and Polt translation (2014), 29. Heidegger (1953), 19–20: "Man kann in der Tat nicht über das Nichts reden und verhandeln, als sei dies ein Ding wie draußen der Regen oder ein Berg oder überhaupt irgend ein Gegenstand. Das Nichts bleibt grundsätzlich aller Wissenschaft unzugänglich. Wer vom Nichts wahrhaft reden will, muß notwendig unwissenschaftlich werden. Aber dies bleibt nur solange ein großes Unglück, als man der Meinung ist, wissenschaftliches Denken sei das einzige und eigentliche strenge Denken, es allein könne und müsse zum Maßstab auch des philosophischen Denkens gemacht werden. Die Sache liegt aber umgekehrt. [...] Das wahre Reden vom Nichts bleibt immer ungewöhnlich. Es läßt sich nicht gemein machen. Es zerrinnt freilich, wenn man es in die billige Säure eines nur logischen Scharfsinnes bringt."

66 This essay (a version of which was presented at the conference in Lyon in 2024) had its origins in one half of a paper ("Two studies on Eudemus' *Physics* 2–3") that I presented at the Project Theophrastus conference *Eudemus of Rhodes: Philosophy and Science in the Early Peripatos*, at the University of Rome in September 2015. I also presented versions of the 'infinite universe' half of that paper at work-in-progress workshops at Rutgers (2015)

Appendix

Themistius and Philoponus on *Phys.* III 8, 208a14–19 (τὸ τῇ νοήσει πιστεύειν ἄτοπον …)

Themistius, *In Phys.* III 8, 208a14–19, CAG vol. V 2, 100.29–101.5 Schenkl.

> But what is the necessity, by Zeus, if when we imagine in our thinking some excess of a magnitude [growing?] *ad infinitum*, that some such thing also exists among the [actual] things? For the [actual] things are not true in the way that notions are. Now, I *think* of you outside of the city; I *think* of you many times larger than a mountain. *Are* you, therefore, outside of the city? And *are* you, as Homer [says],[67] equal to the summit of a mountain? Indeed, it would be nice for it to be up to us to make true whatever we might wish; for then it would be sufficient merely to assume that we are stronger than Zeus. But not everything such as it is thought of also exists, rather few things are thought of such are they exist.[68]

Philoponus, *In Phys.* III 8, 208a14–19, CAG vol. XVI, 494.25–495.7 Vitelli.

> Neither does the *aporia* arising from thought have any necessity. "For the excess and the deficiency are not in the thing, but in the thought." Because thought thinks of some infinite magnitude, he claims, it is not also immediately necessary that such (a magnitude) exists. For neither, he claims, if you were to *think* of Socrates having increased infinitely, is it immediately necessary for Socrates to be that size and to exceed the three-cubit size that he has, because of the thought; nor if we were to think of him outside of the city, is it immediately necessary that he be outside; for the [actual] things do not follow the thought, but the thought—at least when it is correct—is added to the things. But the most accurate of the copies

---

and Seton Hall (2023). My thanks to the audience members at Rome, Rutgers, Seton Hall, and Lyon (and especially, on this last occasion, Alessio Santoro and Pavel Gregorić) for their helpful feedback.

67 *Iliad* II 456: οὔρεος ἐν κορυφῇς.

68 ἀλλὰ τίς ἀνάγκη, πρὸς Διός, εἰ νοοῦμεν ὑπεροχήν τινα ἐπ᾿ ἄπειρον φανταζόμενοι τοῦ μεγέθους, εἶναί τινα τοιαύτην καὶ ἐν τοῖς πράγμασιν; οὐ γὰρ τὰ πράγματα ἀληθῆ, ὡς ἔχει τὰ διανοήματα. νοῶ γοῦν σε ἔξω τοῦ ἄστεος· νοῶ σε πολλαπλασίονα τοῦ ὄρους. ἆρ᾿ οὖν ἔξω τοῦ ἄστεος εἶ; καὶ εἶ ἄρα καθ᾿ Ὅμηρον ἴσος ὄρεος κορυφῇ; χάριεν μέντ᾿ ἂν ἦν ἐφ᾿ ἡμῖν εἶναι τὸ ποιεῖν ἀληθές, ὅ τι περ ἂν βουλοίμεθα· ἐξήρκει γὰρ ὑπολαμβάνειν μόνον, ὅτι κρείττους ἐσμὲν τοῦ Διός. ἀλλ᾿ οὐ πάντα οἷα νοεῖται, καὶ ἔστιν, ἀλλ᾿ οἷα ἔστιν, ὀλίγα νοεῖται.

[of the *Physics*] do not have 'of the city', but only 'beyond the particular size,' that is, the one that human beings have.[69]

69 ὅτι οὐδὲ ἡ ἐκ τῆς νοήσεως ἀπορία ἔχει τι ἀναγκαῖον. «οὐ γὰρ ἐπὶ τοῦ πράγματος ἡ ὑπεροχὴ καὶ ἔλλειψις, ἀλλ᾽ ἐπὶ τῆς νοήσεως.» οὐ διὰ τὸ τὴν νόησιν, φησίν, ἄπειρόν τι μέγεθος ἐννοῆσαι, ἤδη καὶ ἀνάγκη τοιοῦτον εἶναι· οὐδὲ γάρ, φησίν, εἰ νοήσαις Σωκράτην ἄπειρον ηὐξημένον, ἤδη ἀνάγκη τὸν Σωκράτην τηλικοῦτον εἶναι καὶ ὑπερβάλλειν τὸ τρίπηχυ μέγεθος ὃ ἔχει, διὰ τὴν νόησιν· οὐδὲ ἂν νοήσωμεν αὐτὸν ἔξω τοῦ ἄστεος ὄντα, ἤδη ἀνάγκη αὐτὸν ἔξω εἶναι· οὐδὲ γὰρ τῇ νοήσει τὰ πράγματα ἕπεται, ἀλλὰ τοῖς πράγμασιν ἐπισυμβαίνει ἥ γε ὀρθῶς ἔχουσα νόησις. τὰ ἀκριβέστερα δὲ τῶν ἀντιγράφων οὐκ ἔχει τὸ «τοῦ ἄστεος», ἀλλὰ μόνον «ἔξω τοῦ τηλικούτου μεγέθους», τουτέστιν οὗ ἔχομεν οἱ ἄνθρωποι.

CHAPTER 7

Strato and the Peripatetic Background of Alexander's *On Time*

Frans A.J. de Haas

1 Introduction

The physics of Strato of Lampsacus (c. 335–270 BCE) has drawn considerable attention in modern scholarship.[1] If one thing has become clear it is how much of our knowledge of his philosophy rests entirely on assorted fragments from much later authors who use Strato's observations for their own agenda. If, for the purposes of this paper, we limit ourselves to the fragments on time (Fr. 31–37 Sharples, in Desclos & Fortenbaugh (2011)) we see Sextus Empiricus collecting opposing arguments to generate suspension of judgment, and, most of all, Simplicius presenting Strato's observations as deviating from Aristotle so as to generate stepping-stones to Plotinus (*Enn.* III.7)—with whom he also disagrees—to finally confirm Damascius' theory of time.[2] Yet in all likelihood—as the participants in the Lyon conference agreed—Strato's remarks were meant not so much to oppose or criticise Aristotle, but rather to draw attention to reasonable questions that require further elucidation so as to understand and strengthen the Aristotelian position on time. In this paper I shall provide more support for this approach.

Simplicius rebuts some of Strato's suggestions by means of Aristotelian arguments which may well rely on Alexander's commentary on *Phys.* IV 10–14, if not on Aristotle himself to whom Strato remains close. In Alexander's *On Time* 95.16 ff. (henceforth *Temp.*) we find a discussion of some of the observations that Simplicius lists as Strato's. This raises the question that will occupy us in the second part of this paper: what role do Strato's remarks play in the development of Alexander's own theory of time, which is believed to be decisively different from Aristotle's? Does Alexander at least give his fellow Peripatetic the credit he deserves?

1 Major contributions are those of Gatzemeier (1970), Furley (1985), Repici (1988), Berryman (1996), Desclos, and Fortenbaugh (2011).

2 I consider Jaulin (2011) most enlightening in this respect.

2 Strato on Time

Strato's observations recorded in Frs. 31–37 Sharples can be summarised as follows:[3]

1. Time is the number not only of *change*[4] but also of *rest* (Strato mentions sitting, sleeping and doing nothing as examples).[5]
2. Eternal things are not "in" time.[6]
3. Time is the quantitative aspect of actions (πράξεις), such as going abroad, sailing, going on a military campaign, or waging war; it also applies to actions that count as rest (sitting, sleeping, not doing anything).[7]
4. Aristotle said that time is the number of change. But there appears to be a mismatch between time and number in various ways:
   a. time and change are continuous quantities, but number is discrete quantity (cf. *Cat.* 6, 4b22), and the continuous is not countable. So how can time be *number*, i.e. the number of change?[8]
   b. if the continuous is countable because change has different parts which are ordered as earlier and later, all continuous magnitudes can be numbered, including time itself, so there will be time of time, and of all other quantities that come to be continuously.[9]
   c. Time continuously comes to be and passes away, whereas number does not (even if the things counted do).[10]

3 In Fr. 31 Sharples = Simplicius' *Corollarium de tempore* item [4] comes first, followed by [1] to [3] in that order, as will be clear from the footnotes. I discuss item [4] last for reasons of my own exposition.

4 In principle I translate κίνησις with "change" to allow for all kinds of change, although in many cases locomotion is the type of change under discussion. I have retained "movement" and "motion" in the translations I have used.

5 Fr. 31, 27–29 = Simplicius, *In Phys.* 789.15–18, with the examples at 789.35–37. Fr. 33 Sharples = Stobaeus, *Eclogae* I 8 40b, 102.17–103.8 Wachsmuth (1884) lists only this piece of information to the name of Strato. On the doxographical tradition for which Stobaeus is a source, see now Mansfeld, and Runia (2020), volume V, part 1, 605–634 on Aëtius *Placita* sections 21–22 concerning time, with translation of the reconstructed text of Aëtius ibid. part 4, 2083–2084.

6 Fr. 31, 40–43 Sharples = Simplicius, *In Phys.* 789.25–28.

7 Fr. 31, 50–74 Sharples = Simplicius, *In Phys.* 789.33–790.15.

8 Fr. 31, 9–12 Sharples = Simplicius, *In Phys.* 789.2–4.

9 Fr. 31, 12–18 Sharples = Simplicius, *In Phys.* 789.4–9. It is not clear exactly why it would be problematic if time measures time: long stretches of time (months) can be measured by shorter times (days), as well as by movements. See further below for a discussion of counting and measuring in this context.

10 Fr. 31, 18–24 = Simplicius, *In Phys.* 789.9–11.

d. All parts of number have to exist simultaneously (without all three units present, three does not exist), whereas it is impossible for all parts of time to exist simultaneously. If they did, then (i) prior and posterior time would exist simultaneously, (ii) the now and the unit would be the same, and (iii) if time is number, time would consist of nows like number consists of units.[11]

If we consider these statements separately from the projects of the later sources that transmitted them to us, we can discern a clear effort to interpret Aristotle by means of Aristotle. Item [1] draws attention to the fact that despite the overwhelming concern in Aristotle to determine the precise relation between time and *change*, Aristotle also explains that the *rest* of things is equally in time—at least the temporary lack of change (ἀκινησία) of things that have the capacity to move.[12] Both change and rest are in time in the sense that the number, or quantity, that belongs to them is measured by means of time. This is how time measures the being of what it belongs to *per se*: these physical entities are either in motion or in rest.[13] This definition of rest excludes the lack of change of something that cannot move at all, and is thus in an eternal state of *akinēsia* (the first unmoved mover being the most obvious example).

The exclusion of eternal rest matches perfectly with mention of item [2], which reminds us that Aristotle states that eternal things are not "in" time in the sense that they cannot be numbered by time.[14] For Aristotle eternal motion and rest are not "in" time because the duration of eternal motion and rest cannot be measured exhaustively by any measure since there is always more of them. This is an application of Aristotle's general discussion of the infinite's inexhaustibility, or there being always more beyond what has been taken, which plays a crucial role in Aristotle's famous replies to Zeno's paradoxes of motion.[15] Hence, the restriction of being "in" time to the duration of finite changes and rests in the universe makes us aware that all finite beings in the physical universe, which are necessarily changing or resting, can all be measured by time.

Nevertheless, the topic needs further clarification because it is immediately evident that sempiternal motion itself is not "in" time, in the sense of being

---

11 Fr. 31, 24–26 = Simplicius, *In Phys.* 789.11–15. I take all of (i)–(iii) to be consequences from applying the characteristics of number to time. These consequences are not acceptable because they make no sense of time, and contradict Aristotle: *ad* (i) *Phys.* IV 10, 218a25–30; *ad* (ii) *Phys.* IV 11, 219b33–220a26; *ad* (iii) *Phys.* IV 10, 218a8, 18–19.

12 Cf. Aristotle, *Phys.* IV 12, 221b7–23 with *Phys.* III 2, 202a3–5.

13 Cf. Aristotle, *Phys.* IV 12, 221b25–28; IV 13, 222b16–17; cf. IV 14, 223a30–31.

14 Aristotle, *Phys.* IV 12, 221b3–7.

15 On the infinite see Aristotle, *Phys.* III 4–8, esp. III 6, 206a27–b24; on Zeno's paradoxes *Phys.* VI 9–10.

measured by time. It gives rise to the misconception already mentioned by Aristotle, viz. that the motion of the heavens as such would *be* the time everything else is "in".[16] There is a further difficulty because Aristotle also pointed out that the circular motion of the heavenly sphere is used as the unit of measurement for other movements and changes.[17] By allowing the statement that the sempiternal motion of the heavens is "swiftest", he allows a time-related term to be used of an eternal motion. We shall see below how Alexander interprets this source of misunderstanding.

As to item [3], we should observe that action (πρᾶξις) is mentioned in relation to time in the discussion of quantity in *Categories* 6,[18] where action seems to figure on a par with change. So by introducing action in his discussion of time, Strato is not providing a rival definition of time, nor innovating or criticizing as Simplicius suggests, but he is simply taking account of an example from an obvious parallel text in the Aristotelian corpus.[19]

Although *Categories* 4, 2a2 briefly mentions the category of "when" (ποτέ)—with "yesterday" (χθές) and "last year" (πέρυσι) as examples—a more elaborate discussion of this category has not been preserved in the text of the *Categories* as we have it, nor do the ancient commentators reflect any knowledge of such a discussion. A commentary on the category of "when" will therefore have to rely on references to time in *Cat.* 4,[20] and supplement everything else from *Phys.* IV 10–14 and other texts in the corpus. We have evidence that Andronicus proposed to establish "time" and "place" as categories proper, with "when" and "where" as subordinate designations of items in those categories.[21] It is unclear, though, what this would have meant for the treatment of time and place as *per se* quantities in *Cat.* 6. Andronicus' proposal was rejected by Boethus. We shall see that Alexander follows Boethus in this regard, but develops his own solutions to the discrepancies he notes between Aristotle's texts.

16 Aristotle, *Phys.* IV 10, 218a33–b9.

17 See *Metaph.* X 1, 1053a10–12 (below T8 for further discussion). Contrast Boethus *In Cat.* Waitz (1844–1846), vol. 1, 21.8–20 = Sharples (2010), 172, text D: "time will be the number of the movement which is in a circle both as numbering it and as being numbered by it, but of the remaining movements only as numbering but not also as being numbered."

18 *Cat.* 6, 5b2–7.

19 For the importance of the *Categories* for philosophy in the early empire see Griffin (2015). In his commentary on the relevant passage Simplicius, *In Cat.* 6, 5a15–b10, 135.30–141.9 does not discuss πρᾶξις; elsewhere we find that each πρᾶξις is "in" time, cf. Simplicius, *In Cat.* 341.11–12, 348.10–12, 21–23.

20 *Cat.* 6, 4b24, 5a6–8, 5a26–30, 5b3–7, 6a22–23, 6a28–29.

21 For Andronicus and Boethus see Chiaradonna, and Rashed (2020), 227–234.

In *Cat.* 6 action is mentioned as an accidental quantity, whereas Aristotle designates time, along with number, speech, line, surface, body, and place, as quantities properly speaking.[22] Accidental quantities are designated by quantitative terms like "much" (πολύ) and "long" (μακρός) only by reference to one of the *per se* quantities. E.g., we say there is "much" white because the *surface* to which it belongs is large; in the same way, an action as well as a change are called "long" because they take a lot of *time*. As Aristotle explains, when we render the "size" (πόση τις) of an action we determine it by means of time as, e.g., "a year long". Clearly actions are measured by time just like changes.[23] In this context it is important that time is regarded as a *per se* quantity, because accidental quantities like action owe their quantitive attributes to it.

It is noteworthy that in terms of being a quantity, in *Cat.* 6 time is called a *per se* quantity, but action and change are quantities merely derivatively, or *per accidens*. This seems odd given the priority that change seems to have over time in *Phys.* IV 10–14. For in *Phys.* IV 10–14 our awareness of time passing, and the existence of time, rather depend on the existence of change (and rest), although from other perspectives they seem to have a reciprocal relationship.[24] For commentators there is every reason to specify exactly what type of relation holds between time and change since Aristotle calls time an affection (πάθος) or state (ἕξις) of change.[25]

In *Metaphysics* V 13 we find a related, but by no means identical classification. Here Aristotle distinguishes two types of *per se* quantity and two types of *per accidens* quantity, but now he assigns time to the *accidental* quantities. In 1020a14–32 we learn that

per se 1 line etc. are *per se* quantities because "quantity" is part of their definition,[26] and so are

per se 2 their *per se* properties and states (πάθη καὶ ἕξεις): much/little, long/short, broad/narrow, deep/shallow etc.[27] At 1020a26–32 Aristotle distinguishes two types of accidental quantities:

22 *Cat.* 6, 5a38 κυρίως ποσά; cf. 5b8–9 κυρίως καὶ καθ' αὑτὰ ποσά.

23 This is not surprising because in *EE* 1220b26–27 πρᾶξις is called a kind of change: ἡ μὲν γὰρ κίνησις συνεχές, ἡ δὲ πρᾶξις κίνησις. In ethics πρᾶξις refers to a specifically human action based on reasoning; Strato's examples are clearly all human actions in this sense. For an elaborate study on the significance of time in ethical praxis see Vigo (1996). In Aristotle's biology and *Parva naturalia* πρᾶξις is a more general term closely related to ἐνέργεια, see e.g. Corcilius (2019) and De Haas (Forthcoming).

24 Dependence of time on change: see e.g. *Phys.* IV 11, 219a9–14, 219b2–3; reciprocity: see e.g. *Phys.* III 1, 200b20–21; IV 11, 218b29–219b2; IV 12, 220b14–32 etc.

25 Aristotle, *Phys.* IV 14, 223a18–19.

26 These are the *per se* quantities of *Cat.* 6, with the exclusion of time as we shall see.

27 For the distinction between kinds of belonging *per se* cf. *APo.* I 4: a property belongs *per*

per accidens 1 the musical and the white are accidental quantities because that to which they (primarily) belong as accidents is a quantity.[28] Their case is different from

per accidens 2 both change and time which are here called accidental quantities, as well as accidental continua, because the things they are properties of (πάθη) are divisible as continua.

In addition, in this context Aristotle makes clear that time and change are not on a par: change is a continuous quantity because the distance covered is a quantity; time is a continuous quantity because change is. This framework more closely reflects *Phys.* IV 11, 219a10–30 where we find that change must be continuous because the distance covered in locomotion is (locomotion being considered as the primary species of change); in its turn, time must be continuous because it is an affection of change. But whereas in *Phys.* IV 11 Aristotle *derives* the continuity of time and change from the distance covered, in *Metaph.* V 13 he uses it to *explain* their status as accidental quantities—presumably because the quantity and continuity of the distance covered has explanatory priority to both, although in different ways. This leaves us with a discrepancy between *Cat.* 6 and *Metaph.* V 13 with respect to time: is it a (continuous) quantity *per se* or *per accidens*, and in what sense?

The classification of time is a larger issue which starts with the distinction that is basic to both *Cat.* 6 and *Metaph.* V 13, viz. that between discrete and continuous quantity. This is the issue addressed in Strato's arguments under item [4]. As we have seen, Strato notices that Aristotle's repeated stipulation that time is the *number* of change might suggest that time is a discrete quantity, from which a series of unwelcome consequences would follow. At the same time Aristotle makes abundantly clear that time and change are to be regarded as continuous quantities, and they are both assumed to be continuous quantities in *Cat.* 6 and *Metaph.* V 13—despite the discrepancy discussed above.

Again, in *Metaph.* X 1, 1152b20–27 Aristotle claims that a measure (μέτρον) is that by which quantity *qua* quantity is known. The indivisible unit of number

---

se either when the property is part of the definition of what it belongs to, or when that to which the property belongs is part of the property's definition. This seems to match the distinction between *per se* 1 and *per se* 2 quantities in *Metaph.* V 13.

28 White as a colour exists in body, cf. *Cat.* 1a2, or rather in a surface, *Cat.* 5b2–3, *Metaph.* 1022a16–17, 1070b20, which are both *per se* quantities. Musical is commonly said to belong to human being (e.g. GC 319b24–32 and *passim*), or to an individual human being like Socrates (e.g. *Metaph.* 1007b14 and *passim*), which could both count as bodies. However, it should be noted that a change from musical to non-musical is regarded as a change in the soul (which is not a quantity), *APo.* 70b9, GC 334a11–12, given that μουσική is a kind of ἐπιστήμη.

(the 1) which is used to count numbers (regarded as sets of ones) constitutes the basic notion of "one". Measures for continuous quantities (the examples are length, breadth, depth, weight and speed) derive their unity from this primary "one", i.e. they are regarded as one and indivisible (see further below T7–8 with discussion). In this context "measure" occurs as the more general term covering both the unit of counting discrete quantities, and the unit of measuring continuous quantities.[29]

To make matters worse, in *Metaph.* V 13, 1020a8–9 Aristotle clearly stipulates that "a quantity is a plurality (πλῆθος) if it is countable (ἀριθμητόν), and a magnitude (μέγεθος) if it is measurable (μετρητόν)". However, judging from the fluid vocabulary of counting and measuring in relation to time in the whole of *Phys.* IV 10–14 it seems we would have to conclude that time is both a plurality and a magnitude. Strato's [4a] and [4b] show that he is well aware of the counting/measuring issue, although the remaining fragments do not reveal how he proposed to deal with them.

Aristotle himself had already tried to clarify his use of "number" in "time is the number of change" by stipulating that time is not numberable or numbered number by means of which we count (ἀριθμὸς ἀριθμητόν ᾧ ἀριθμοῦμεν)—which would be the familiar discrete number of *Cat.* 6—but the numbered number (ἀριθμὸς ἀριθμούμενον), or "the way in which change has number" (ᾗ ἀριθμὸν ἔχει ἡ κίνησις).[30] However, both ancient and modern commentators provide different interpretations of this distinction. All this goes to show that interpreters have to deal with at least two major classification issues regarding time, viz. is time continuous or discrete, and is it a *per se* or *per accidens* quantity? These problems will become more apparent as commentators are trying to combine all relevant texts from the Aristotelian corpus, after Strato's example.[31]

29 For the intricacies of the argument in *Metaph.* X 1–2 and related texts, see Castelli (2018), 20–24 with the ensuing detailed commentary. Like Aristotle, Alexander wishes to steer clear from any suggestion that the number one, or any measure, has the status of a metaphysical principle of the Pythagorean or Platonic kind.

30 *Phys.* IV 11, 219b2–9.

31 Alexander, *In Metaph.* V 13 notices two more discrepancies between this chapter and *Cat.* 6. *In Metaph.* 396.6–8 Alexander asks whether number consists of parts that have no position (like the accidental quantities time [!], speech, and change in *Cat.* 6) or parts that have position after all (in V 13); Alexander, *In Metaph.* 397.31–34 concerns the question whether "large" and "small" (as opposed to "larger" and "smaller") are relatives (in *Cat.* 6) or exist in themselves (in V 13). There is no need to pursue these issues here.

3 Alexander and Strato's Observations on Time

Here we shall be concerned only with Alexander's response to Strato's observations, and to the problems of classification indicated above. For this purpose I shall take my cue from Alexander's treatise *On Time*, which discusses three of the four observations that Simplicius ascribes to Strato (95.16–40), although Alexander does not mention Strato by name.[32] In addition I shall use what remains of Alexander's commentary on *Phys.* IV 10–14,[33] along with his commentary on *Metaph.* V 13 to see how he dealt with the issues of classification we encountered above.

Alexander *On Time* is a short work best compared with the short treatises on assorted topics we are familiar with from the *Mantissa* and *Quaestiones*.[34] It has a clear structure:[35] Alexander first frames the issue of time along the lines of existing views of time some of which had already been discussed by Aristotle (92.31–93.35).[36] He then presents Aristotle's theory of time as he understands it (93.35–95.15), followed by a discussion of questions and specifications some of which echo Strato's concerns (95.16–97.5).

In order to be able to investigate whether and how Alexander's discussion of Strato's observations and the classification problems help him to formulate

---

32 In his remaining work Alexander only refers twice to Strato by name: *In Top.* 340.3, *In Sens.* 126.20.

33 See Rashed (2011), 56–81 (discussion); 252–293 (fragments). Within the confines of this paper I cannot do justice to his reconstruction of the wider aims of Alexander's concept of time. The Greek fragments are preserved in the later Greek commentaries on *Phys.* IV 10–14 by Themistius (2003), Philoponus (2011), and most of all Simplicius (1992), not least in his *Corollarium de tempore*, translated in Simplicius (1992). This state of transmission renders it impossible to verify Alexander's claim at the start of *On Time* (93.3–4) that perhaps (*forsitan*) he may have written something in the short treatise he did not mention in his commentaries.

34 Alexander, *On Time* (or as one manuscript has it: *On Motion and Time*) was first translated in English by Sharples (1982), which is the translation I shall use here, with modifications. The Greek original is lost. References are to the Latin translation by Gerard of Cremona (1114–1187), published in Théry (1926) 92–97. Gerard of Cremona translated a later Arabic version of the same text, edited by Badawi (1971) 19–24. Sharples cooperated with Fritz Zimmerman for a comparison of the Arabic and Latin texts; where indicated I follow the Arabic text as interpreted by Zimmerman. I also note my debt to Alessio Santoro's insightful comments on the basis of the Arabic text for which I am grateful.

35 From a distance, the structure rougly resembles that of Aristotle's discussion of time: discussion of untenable rival views (IV 10), development of his own view (IV 11–13), reply to residual questions (IV 14). But Alexander reshuffles arguments from all of Aristotle's chapters into an order of his own.

36 Aristotle, *Phys.* IV 10, 217b32–218a30.

his view of time, let us first outline Alexander's view of time as it emerges from *On Time*. Interestingly, he sets up his topic as "a difficulty to dissolve in what is said by those who think [that time] is neither" (i) the heavenly sphere, nor (ii) its movement, nor (iii) movement as such, nor (iv) the number of the movement of the heavenly sphere. Aristotle had already objected to (i) to (iii), but we shall see that Alexander will eventually support the fourth option once it has been properly specified. Hence a difficulty arises if the fourth option is denied, too. The fourth option itself is immediately pitted against a further claim about time:

> T1 Some however think that time is not any of these things, nor yet the number of the movement of the heavenly sphere; they think it is a single thing that encounters things. For they say that if we were to think of the sphere standing still, its standing still would be in time; and since its standing still would be in time, its movement would be in time.[37] But what they say is not true; for if we were to think of the sphere standing still, time would be done away with. (Alexander, *Temp.* 93.22–28, transl. Sharples modified)[38]

Themistius ascribes a similar view to Galen, viz. that time is defined by itself, separately from change, which is a view that may well have been inspired by Plato's *Timaeus* in which time figures as the image of eternity, without any intrinsic relation to change.[39] Alexander, on the other hand, wishes to preserve some kind of relation between time and change, esp. the movement of the heavenly sphere.[40] It is interesting to see that the concept that standing

---

37 At this point Sharples adds: ⟨and if its movement is in time, the attendant number of its movement is in time too⟩. But the argument is not about time being in time, but about time being unrelated to motion because rest, too, is in time.

38 Alexander, *Temp.* 93,22–28: *Quidam autem putant quod tempus non est omnia illa, neque numerus motus orbis; estimant quod ipsum est unum occurrens rebus. Dicunt enim quod si estimaremus orbem stantem, esset stacio ejus in tempore, et cum stacio ejus esset in tempore, tunc motus esset in tempore. Verum quod ipsi dicunt non est verum; quoniam si estimaremus orbem stantem, destrueretur tempus.*

39 Themistius, *In Phys.* 149.4–20; cf. 144.23–145.2: Galen is supposed to have claimed that change rather makes us think of unchanging things, such as the poles of the cosmos and the centre of the earth. This association is explicitly rejected in Alexander, *Temp.* 96.17–20.—On Galen's view of time, and on the possibility (raised by Zimmerman) that *On Time* was originally part of a polemical treatise by Alexander against Galen **On Motion and Time* referred to in Arabic sources, see Chiaradonna (2009).

40 Alexander, *Temp.* 93.28–32 also tries to undermine the argument by suggesting that thought experiments have no relevance to gaining knowledge of the actual nature of time.

still is also in time (Strato item [1]) was apparently used to separate time from change altogether, against Aristotle. This in itself should be sufficient reason for Alexander to be interested in the topic Strato drew attention to.[41]

Needless to say, Alexander's bold claim in T1 that time would be done away with if the sphere were to stand still, requires more elucidation, which he provides in his second section devoted to the Aristotelian notion of time. A series of texts in *On Time* paints the following picture of time:

> **T2** Time is number which is numbered by something else, namely by movement. [...] and time is number only because of the first and last which are determined in the movement; for it is by number that one speaks of first and last, since the unit is what is first in number, and the last is what exists after the first. [...] So, since movement exists, time exists through its quantity [*sc.* of movement]; so that time is the number of movement in this way. (Alexander, *Temp.* 94.3–4; 9–11; 15–16, transl. Sharples modified)[42]

> **T3** Although time is a single connected continuum, I say that it is number only because it is many in thought. For times only exist in potentiality and thought, not by being in actuality. [...] Time therefore is one in actuality, although it is divided potentially. (Alexander, *Temp.* 94.22–25; 41–42)[43]

> **T4** And movements differ as individuals and in form; but time is single and undivided.[44] (Alexander, *Temp.* 95.29–31)[45]

According to Alexander time is an undivided continuum that is one in actuality; time as such is the quantitative aspect of movement as such. Time is only numbered number in so far as it marks the first and last which exist in move-

41 Perhaps similar opposition to Aristotle already prompted Strato's remarks.

42 Alexander, *Temp.* 94.3–4; 9–11; 15–16: *Tempus autem est quod numeratur per aliud, scilicet per motum* [...] *non sit tempus numerus, nisi propter primum et postremum determinata in motu, quoniam numero dicit primum et postremum. Unum enim est primum in numero, postremum existens post primum* [...] *Cum ergo est motus, est tempus per quantitatem ejus, quia est numerus ejus in hoc modo.*

43 Alexander, *Temp.* 94,22–25.41–42: *Et non dico quod tempus est numerus quamvis sit unum conjunctum continuum nisi quoniam estimacione est multa; non sunt enim tempora nisi potencia et estimacione* [...] *tempus igitur est unum actu, quamvis dividatur potencia.*

44 Arab. "not different", "undifferentiated" (Santoro).

45 Alexander, *Temp.* 95.29–31: *Et dico quod motus diversi fiunt per individua et formam; sed tempus est unum non divisum.*

ments, and in that respect time can be numbered, by movement. The numbers used to do the numbering exist in rational souls; they are, as we shall see, the units of measurement, different as such from time as numbered number. Since individual movements are different, so are the corresponding individual finite times of movements (and rests). But marking the first and last in movements, and therefore in time, is an act of thought that does not affect the single continuum that is time. Against views of the Galenic type, Alexander maintains that ultimately this single continuum does not exist in its own right, but as the quantitative aspect of the eternal movement of the outer sphere, which is the principle of all other movements in the universe. After all, Aristotle did claim that time is "something" (219a9), or "an affection or state" of movement (223a21–22), in a sense that Alexander will try to define more clearly.

This concept of time, as we shall see, is built upon a series of passages in Aristotle that are all related to Strato's observations directly or indirectly. Hence, I propose, working through Strato's (and other commentators') observations is constitutive of Alexander's interpretation of Aristotle's theory of time.[46]

Strato's observations occur in *On Time* in the following way:

(i) the difference in genus between time and movement (95.16–21 = Strato [4]);

(ii) how time numbers change, and change numbers time (96.26–38 = Strato [4a–b]).

(iii) the fact that time is also the number of rest (95.21–26; 96.10–20 = Strato [1]);

(iv) the fact that eternal things are not in time (96.4–10 = Strato [3]).[47]

Alexander's response to (i) the perceived difference in genus between time as number (discrete) and movement (continuous) reads as follows:

> **T5** But if some, considering these things, have felt doubt, asking how time numbers movement, since it is not in the same genus as it, we say that ⟨time numbers movement even though movement is not in the same genus⟩, just as the cubit numbers the wood, even though the wood is not in the same genus. In this way time numbers every movement, of growth

46 From my arguments here it will be clear that I believe that Alexander is closer to Aristotle's intentions than e.g. Sharples (1982), 68–72, Coope (2005), 170–171, and Rashed (2011) seem to think. However, within the confines of this paper I cannot do justice to this complex issue.

47 He does not mention the fact that human action is measured by time (Strato [2]). The fragments of Alexander, *In Phys.* IV 10–14 do not discuss this point either.

> and of ageing and of change; for time itself numbers all these movements, but not qua growth and ageing and change. (Alexander, *Temp.* 95.16–21)[48]

Alexander takes the doubt which people have felt to be an issue concerning numbering, as if a number should be of the same kind as each thing it numbers. Alexander addresses the problem by pointing out, first, that a unit of numbering (e.g. the cubit) need not be in the same genus as what it measures (e.g. the piece of wood).[49] This reply turns on the same fluidity of the unit/measure vocabulary that caused the problem in the first place: a cubit taken as an indivisible unit of measuring can be regarded as a discrete quantity, a "one", whereas the piece of wood is a continuous quantity. This is compatible with the fact that the cubit is itself a particular length (made of wood or any other material) used to measure the length of the piece of wood: in that sense unit/measure and what is counted/measured are of the same kind.

At the same time Alexander alludes to Aristotle's discussion of the problem which kind of movement it is that time measures.[50] Aristotle explained that one and the same time numbers different kinds of movement *qua* movement, not *qua* the particular kind of movement each happens to be, e.g. growth or ageing or change. In that context Aristotle emphasises that one and the same time is the number of movement as such (ἁπλῶς), regardless of whether one is dealing with a particular kind of movement like generation, corruption, growth or alteration, or of movements that are simultaneous and equal, or not.

> **T6** For a time that is both equal and simultaneous is one and the same time, and even those that are not simultaneous are one in kind; for if there were dogs, and horses, and seven of each, it would be the same number. So, too, movements that have simultaneous limits have the same time, yet the one may in fact be fast and the other not, and one may be locomotion and the other alteration; still the time of the two changes is the same if it is both equal and simultaneous; and for this reason, while the movements are different and separate, the time is everywhere the same, because the *number* of equal and simultaneous movements is

48 Alexander, *Temp.* 95.16–21: *Quod si accidit ambiguitas aliquibus speculatoribus et dixerunt quomodo numerat tempus motum, cum non sit de genere ejus, dicemus sicut numerat cubitus lignum quamvis non sit lignum de genere ejus; sic dico quod tempus numerat omnem motum augmenti et vetustatis et mutacionis; ipsum enim illos motus omnes numerat, non per hoc quod ipsi sunt augmentum et vetustas et mutacio.*

49 Arab. *ḫašba* denotes "a piece of wood", "an object made out of wood" rather than the material in general (Santoro).

50 Cf. Aristotle, *Phys.* IV 14, 223a29–b12.

> everywhere one and the same. (Aristotle, *Phys.* IV 14, 223b3–12, transl. Hardie/Gaye in Barnes *RevOT*)

As the number 7 is the same, regardless of the kind of things it is used to count, dogs or horses, so the time is the same in all cases.[51] This text provides Alexander with a model for his own notion of a single, continuous and sempiternal time that only becomes differentiated by the different movements we are numbering by means of our rational soul (see T2–4 above).[52]

Two further texts from Aristotle may serve to explain the required concepts of numbering and measurement in more detail, and to shed light on the second of Strato's observations (ii). In *Metaph.* XIV Aristotle explains how "one" means "a measure", and how in each case of measuring the measure has a distinct nature of its own adapted to the purpose:

> T7 In every case it is some underlying thing with a distinct nature of its own (τι ἕτερον), e.g. in the scale a quarter-tone, in magnitude a finger or a foot or something of the sort, in rhythms a beat or a syllable; and similarly in weight, it is a definite weight; and in the same way in all cases, in qualities a quality, in quantities a quantity (and the measure is indivisible, in the former case in kind, and in the latter to the sense); which implies that the one is not, in any instance, in itself a substance. [...] The measure must always be something predicable of all alike, e.g. if the things are horses, the measure is horse, and if they are men, man. If they are a man, a horse, and a god, the measure is perhaps living being, and the number of them will be a number of living beings. (*Metaph.* XIV 1, 1087b34–1088a4; a8–11; transl. Ross in Barnes *RevOT*).

Because the measure must always be predicable of all things measured by it, the measures differ in each case, while the number that counts how often the measure is applied remains the same. In *Metaph.* X Aristotle clarifies how the measure is to be conceived as indivisible, even if it is itself something divisible.

> **T8** Now where it is thought impossible to take away or to add, there the measure is exact. Hence that of number is most exact; for we posit the unit as absolutely indivisible; and in all other cases we imitate this sort of measure. For in the case of a furlong or a talent or of anything large any

---

51 Cf. Aristotle, *Phys.* IV 12, 220b10–14; IV 14, 224a2–15 emphasizing both the sameness of number and (accordingly) time, over and against the variety of things counted.

52 See also *Phys.* IV 10, 218b13.

> addition or subtraction might more easily escape our notice than in the case of something smaller; so that the first thing from which, as far as our perception goes, nothing can be subtracted, all men make the measure, whether of liquids or of solids, whether of weight or of size; and they think they know the quantity when they know it by means of this measure.
>
> And they know movement too by the simple movement and the quickest; for this occupies least time. And therefore in astronomy a "one" of this sort is the starting-point and measure (for they assume the movement of the heavens to be uniform and the quickest, and judge the others by reference to it), and in music the quarter-tone (because it is the least interval) and in speech the letter. (*Metaph.* X 1, 1052b32–1053a13; transl. Ross in Barnes *RevOT*)

These passages illustrate that the confusion about the categorisation of time may well have its roots in the realm of measurement. If measures are supposed to be predicable of everything measured, they seem to belong to the same genus. The measure which is used in measuring or counting different types of entities has to be adapted to the purpose: quantities are measured by some definite quantity, a group of horses counted by the unit horse. In this sense the measure "has a distinct nature of its own" in each case. What is needed for the solution is the insight that measure and measured are in the same genus in all of these cases: a cubit is a definite length used as a unit of measurement for the length of the wood, which is the quantitative aspect of the wood. It is predicable of the wood as a length of a length, or a quantity of a quantity. In the same way a definite length of time can serve as the unit of measurement of the quantitative aspect of change. For the same reason movement can measure time, while each has a nature of its own.

Moreover, in so far as quantities or substances are used as a unit of measurement, they are considered to be as indivisible as the absolute unit of numbering, the number one. Hence the use of the term "number" for time in general is apt, even though time as such is a continuum. A cubit itself is a continuum, but as a unit of measurement it is considered indivisible. In this sense time is the number for counting movements in general, whereas in counting particular types of motion a suitable measure of time has to be chosen as the unit of measuring. Aristotle suggested that for this reason astronomers use the circular movement of the heavens (the length of a day), which they consider uniform and quickest, as a the unit measure for the movements of the heavenly bodies.[53]

---

53 Cf. Aristotle, *Phys.* IV 14, 223b12–23.

Alexander starts his report on Aristotle's theory of time with a reference to this passage in which, he says, Aristotle "maintains that [time] is said dialectically to be the number of the sphere that moves from east to west".[54]

Alexander all too readily adopts this measure as the natural unit for measuring time in the universe:

> **T9** Aristotle believed that time numbers movement qua movement, not as having any particular property; time does not number the movement of growth qua growth, or change qua change, but rather qua movement. So it numbers every one of the species of movement just as it numbers movement itself with reference to its intervals; for it numbers them with reference to that which they have in common, and not with reference to what is specific to each movement.
>
> Conversely, movement numbers time,[55] since a day is nothing but the number of the movement of the heavenly sphere for twenty-four hours; then, by means of those hours reduplicated, it measures other longer periods of time too. (Alexander, *Temp.* 96.27–38)[56]

These texts show why the definitions of time (i) to (iii) which Alexander listed in the first section of his treatise all fail. Time is not identical with the heavenly sphere itself, nor with its movement, nor with movement in general, and not even with (iv) the number of the movement of the heavenly sphere, at least not without further specification. To believe otherwise is to misunderstand not only the nature of time, but also the nature of number and the unit of measurement.

Nevertheless, Alexander believes that if we were to think of the heavenly sphere standing still, time would be done away with (T1). Without the primary

---

54 Alexander, *Temp.* 93.36–38: *Estimatur autem quod dicitur sermone dyalectico, numerus motus orbis orientalis.* Cf. Aristotle, *Phys.* IV 14, 223b22–224a2. Aristotle here refers to the common association of time and moving in a circle (223b21–22.29 δοκεῖ, 24 τὸ εἰωθὸς λέγεσθαι). On Alexander's broad conception of Aristotelian dialectic see De Haas (2021).

55 Cf. Alexander, *Temp.* 95.5–6. Cf. Aristotle, *Phys.* IV 12, 220b14–32.

56 Alexander, *Temp.* 96.27–38: *et Philosophus quidem estimavit quod tempus numerat motum per hoc quod ipse est motus, et non numerat ipsum propter hoc aliquid quod ei approprietur, cum tempus numeret motum augmenti, non per hoc quod ipsum sit augmentum, et numeret mutacionem non per hoc quod est mutacio, immo per hoc quod est motus. Ipsum igitur numerat omnem motum cum intencione, quia numerat omnem speciem ex speciebus motus, sicut ipsum numerat ex spaciis; ipsum enim enumerat eas ex re communi que est eis non per rem qua motus approprietur. Et dico iterum quod motus numerat tempus, quoniam dies non est nisi numerus motus orbis viginti quatuor horarum; deinde numerat per istas horas secundum duplicitatem et reliquum tempus multum.*

movement in Alexander's universe, which is the movement of the heavenly sphere, from which all movement in the universe is transmitted to the lower regions of the universe, time will not exist.[57] Alexander builds on Aristotle when he develops the view that the unwavering rational desire of the outer sphere towards the First Cause is the cause of its eternal motion. It is in this remarkable sense that Alexander can embrace the notion that if there were no soul, viz. the soul of the outer sphere, time would not exist.

The famous problem raised by Aristotle, whether time would exist if there were no soul to count time, thus changes its usual appearance. It should be noted that Aristotle does not *claim* that time does not exist if there is no soul to do the counting (although this is by far the most widespread interpretation of his doctrine of time). For he adds: "*except that* (my italics) there might be that thing (whatever it is) by being which time is [what it is]—for example, if it is possible for movement to exist without soul".[58] On this interpretation Aristotle most clearly invites Alexander's answer: no, movement cannot exist without the soul of the outer sphere, so time cannot exist if this soul does not exist.

What about the soul that does the numbering? Alexander is clear that the number that numbers something else is in the soul, whereas time is number which is numbered by something else, namely by movement (T2). This formulation neatly separates the nature and existence of time as such from the nature and existence of mathematical number in souls. Time can very well exist if there is no soul that wields units of measurement and numbers.[59]

Despite the fact that Alexander's Peripatetic account of time does not mention rest, and opponents used the duration of rest to raise difficulties for Aristotle, he does not spend much time on the issue that time is also the number of rest (iii): we use the same adjectives like "by day" and "by night" for rest and for change, so this should be obvious.[60] Like Aristotle, he needs both change and rest in order to support the claim that time measures the *being* of all things in the physical universe, which are all either in motion or rest, hence "in" time.

The fact that eternal movement is not in time (iv) is only touched upon briefly:

---

57 For the importance of the circular movement of the outer sphere and the divine power of movement which it imparts on the lower realm of the universe see e.g. Alexander, *On the universe*, and *Quaest.* II 13, with Genequand (2001), Fazzo (2002), Fazzo (2008a), Bodnár (2014).

58 Cf. Aristotle, Phys. IV 14, 223a21–29. A full discussion of this complicated issue is beyond the scope of this paper.

59 Cf. Aristotle, *Phys.* IV 14, 223b33–224a2.

60 When Aristotle *Phys.* IV 12, 221b7–8 makes this point he says "time is the *measure* of rest", whereas Alexander sticks to "number".

> **T10** I say that all things are in time because they move and rest, and time is the number of movement and of rest, and what neither moves nor rests is not in time. Moreover, the sempiternal things do not apply to any thing as a result of time, since they do not grow old as a result of time, as they are not in it or subject to it, and it is only the things which fall under time that grow old. (Alexander, *Temp.* 96.4–10)[61]

By way of argument Alexander merely repeats Aristotle's indication (*sēmeion*) for this fact: sempiternal things are not in time or subject to it, because they do not grow old because of time.[62] But the issue requires more of our attention when we compare two further statements of Alexander's:

> **T11** I say that time is the number of the movement of the heavenly sphere, apart from other movements, only because there is no movement swifter than this, and a thing is only numbered and measured and reckoned by what is less than itself. For a plurality of things is numbered by the unit, and long intervals are measured by the cubit, and measures of capacity are reckoned by the twelve-bushel unit or by what is less than the thing measured itself. And I say that movement neither is nor will be done away with. (Alexander, *Temp.* 94.16–22, transl. Sharples modified)[63]

> **T12** I say that the movements of the other spheres differ in speed, but the movement of the sphere that moves from east to west, without doubt, is the swiftest of them, and is in accordance with order. Therefore the number of this, without the number of the other movements of the rest of the spheres, is time; since a thing is only measured by what is less than it, as we said previously. (Alexander, *Temp.* 95.6–11)[64]

---

61 Alexander, *Temp.* 96.4–10: *Et dico quod res omnes sunt in tempore quoniam moventur et quiescunt et tempus est numerus motus et quietis et quod non movetur neque quiescit, non est in tempore. Et dico cum illo quod res sempiterne non occurrunt ex tempore alicui rei, quoniam ipse non veterascunt ex tempore, quia non sunt in eo neque manentes sub eo, et non veterascunt nisi res cadentes sub tempore.*

62 Cf. Aristotle, *Phys.* IV 12, 220a30–221b7; Aristotle associates time as the measure of change with decay and forgetfulness, rather than learning, rejuvenation and beauty. This passage immediately precedes Aristotle's claim that time is also the measure of rest discussed above p. 155.

63 Alexander, *Temp.* 94.16–22: *Et dico quod tempus non fit numerus motus orbis absque aliis motuum, nisi quoniam non est motus velocior eo, et non numeratur aliquid et cubitatur aliquid et mensuratur aliquid, nisi cum eo quod est minus illo. Numerantur ergo res multe cum uno et meciuntur spacia longinqua cum cubito et mensurantur mensure cum caficio aut cum eo quod est minus ipso, et dico quod motus non tollitur vel cessat, neque tolletur.*

64 Alexander, *Temp.* 95.6–11: *Et dico quod motus reliquorum orbium sunt diversi in velocitate,*

It would seem that according to Alexander time *does* measure eternal movement "but only because that movement is the swiftest",[65] and everything is measured by something smaller than itself. This seems to suggest that Alexander is willing to entertain the idea that eternal motion can be measured by a limited measure after all. Does this mean that time is *identified* as the number of the movement of the outer sphere? I believe not. In line with Aristotle, Alexander points out that astronomers (and ordinary people) use one rotation of the outer sphere as a measure of time. On the other hand, the term "swiftest" as such refers to time as part of a comparison between the speeds of single rotations of the different heavenly spheres, and is itself applied to the movement of the outer sphere.[66] One rotation of the outer sphere is 24 hours long, and therefore this movement (which is a part of the sempiternal movement of the outer sphere) is a limited movement which can serve as the measure of longer times (see T9). This is the point of the limiting clause "but only because that movement is the swiftest". After Aristotle, Alexander allows the comparison between limited single rotations which fits the general theory of measurement (time measures movements, and movements measure time). The *sempiternal* movement of the outer sphere is still not "in" time because the total number of its rotations cannot be counted in this way: they are infinite in number. So when Alexander states "Time is the number of the movement of the heavenly sphere, only because there is no movement swifter than this" (T11), and "the number of this [movement of the outer sphere] is time" (T12) this should still not be taken to mean that the single time Alexander believes to exist is *defined* as the number of the movement of the heavenly sphere—which cannot be exhaustively numbered. It is merely the preferred unit of time as measure of all movement in the universe, for which the movement of the outer sphere is the principle.[67]

Last but not least we have to see how Alexander deals with the classification issues of time. The first classification problem (is time continuous or discrete?) has been solved with reference to the theory of counting and measuring which Aristotle also referred to in *Phys.* IV 10–14. Time is a continuous quantity because change is, but chosen durations of time can be counted when

sed motus orbis orientalis, absque dubio, est velocior eis et est secundum ordinem. Numerus ergo ejus sine numero aliorum motuum reliquorum orbium est tempus, cum res non mensuretur, nisi per id quod est minus sicut diximus ante hoc.

65 Alexander repeats this point several times: *Temp.* 93.16–22, 94.16–18 (T11), 95.7–8 (T12), 95.26–29.

66 Cf. Aristotle, *Phys.* IV 10, 218b13 ff.

67 Cf. Alexander, *In Metaph.* V 13, 396.14–18.

regarded as indivisible measures of change.[68] The second classification problem (is time a *per se* or *per accidens* quantity?) is a thornier issue for Alexander. For he has to stay clear from associating time with a *per se* quantity in the sense of Galen, and he has to find room for the dependence of the continuity of time on change, and that of change on the distance moved, all within in the framework of distinctions we found in *Metaph.* V 13.

In his commentary on *Metaph.* V 13 Alexander addresses the discrepancy concerning the classifications of time in the *Categories* and the *Metaphysics*. Then he suggests different ways to understand the nature of Aristotle's subdivision of accidental quantities.

> **T13** It should also be pointed out that although in the *Categories* Aristotle said that time is a *per se* quantity, he says here that it is a quantity *per accidens*, and prior to it, of course, change, for time [is an accidental quantity] because of change.
>
> We must ask too why he does not locate change and time among those *per se* quantities that he called "affections or states" of quantities [our *per se* 2], but puts them with those quantities that are such accidentally [our *per accidens* 2]; for if change is an affection of something continuous (for it cannot take place except in another), the affections too would belong with things that are quantities in this way. (Alexander, *In Metaph.* V 13, 398.27–34; transl. Dooley (1993) modified, my brackets)[69]

Although at first sight the suggestion to regard change and time as *per se* 2 affections of quantities makes the best sense, categorising change and time in this way would change but not resolve the discrepancy between *Cat.* 6 and *Metaph.* V 13—for in *Cat.* 6 time is a *per se* quantity, but change *per accidens*, whereas they are lumped together as *per accidens* quantities in *Metaph.* V 13. That is probably why Alexander tries to develop a reason why Aristotle made this division in *Metaph.* V 13 in this way, but the density of his text makes it hard to understand.[70] Unfortunately, after the next paragraph the commentary on *Metaph.* V 13 ends abruptly.

---

68 Note that Damascius (Fr. 37 Sharples) transcends the discussion by claiming that time is *both* continuous *and* discrete.

69 Alexander, *In Metaph* 398.27–34: σημειωτέον δὲ καὶ τοῦτο, ὅτι τὸν χρόνον ἐν Κατηγορίαις καθ' αὑτὸ ποσὸν εἰπὼν ἐνταῦθα καὶ αὐτὸν κατὰ συμβεβηκός φησιν εἶναι ποσόν, καὶ πρὸ αὐτοῦ γε τὴν κίνησιν· διὰ γὰρ ταύτην καὶ τὸν χρόνον. ζητητέον δὲ καὶ τοῦτο, διὰ τί τὴν κίνησιν καὶ τὸν χρόνον οὐκ ἐν τοῖς καθ' αὑτὰ ποσοῖς ἔθηκεν, ἃ πάθη ἢ ἕξεις εἶπεν εἶναι τῶν ποσῶν, ἀλλ' ἐν τοῖς κατὰ συμβεβηκός· εἰ γὰρ πάθος ἐστὶν ἡ κίνησις συνεχοῦς τινος (οὐ γὰρ οἷόν τε ἄλλως ἢ ἐπ' ἄλλου κίνησιν γίνεσθαι), εἴη ἂν ἐν τοῖς οὕτω ποσοῖς καὶ τὰ πάθη.

70 See Dooley (1992), notes 311–314 for a slightly different reading.

> **T14** Well, he might be calling affections which are quantities in themselves [i.e. change and time] accidental quantities, but surely not in the same way as those accidents that are not affections of quantities, nor affections of quantities while being themselves quantities.
>
> Well, perhaps those [other affections] are quantities in themselves because they, as well as the [magnitudes] of which they are affections, are permanent, but neither of these affections, [change or time], is permanent; and also because the former affections are in the nature of [their subjects] and, if their subjects exist, they follow them everywhere, whereas change is no part of magnitude, but takes place in a magnitude. (Alexander, *In Metaph.* V 13, 398.34–39; transl. Dooley (1993) modified, my brackets)[71]

First he asks why Aristotle might have categorised time and change which are in fact *per se* 2 quantities as *per accidens* quantities. Well, Alexander suggests, Aristotle saw that they are not affections of their subjects in the same way as other *per accidens* quantities. What is their specific difference? The usual *per se* 2 properties are permanent, and the magnitudes they belong to also exist permanently, viz. they exist as wholes. Both change and time do not exist as wholes, but are different at any given moment. In that case the classification of change and time as a special type of accidental quantity highlights their peculiar fleeting nature. A further explanation focuses on the fact that the other *per accidens* quantities belong in the nature of their subjects, with the result that they exist whenever their subjects do. Change, on the other hand, is not part of what it is to be the magnitude it belongs to (the distance covered), but merely *supervenes on* such magnitude, if movement happens to occur at all. We have already seen that time only numbers change when a particular change has been defined by marking off an earlier and later that belong to change. From this perspective, time and change constitute a separate class of accidental quantities because their relation to their respective subjects is different from that of the other *per se* 2 quantities. The loosening of the bond between time and change which results from this discussion can be regarded as support for Alexander's view of time as a single undivided continuum in actuality whose pairings with the befores and afters in change are the result of incidental activities of an external mind applying its numbers.

71 Alexander, *in Metaph* 398.34–39: ἢ ἃ καθ᾽ αὑτὰ μέν εἰσι ποσά, κατὰ συμβεβηκὸς λέγει, οὐ μὴν ὁμοίως τοῖς ἃ μή ἐστιν ὡς ποσῶν καὶ καθὸ ποσὰ πάθη. ἢ ἐκεῖνα μὲν ποσὰ καθ᾽ αὑτά, ὅτι ὑπομένει ὡς καὶ τὰ ὧν ἐστι πάθη, τούτων δὲ οὐδέτερον ὑπομένει. καὶ ὅτι τὰ μὲν ἐν τῇ φύσει ἐστὶν αὐτῶν, καὶ ὄντων ἕπεται καὶ αὐτὰ πάντως, ἡ δὲ κίνησις οὐ τοῦ μεγέθους τί ἐστιν, ἀλλ᾽ ἐπὶ μεγέθους γίνεται.

4 Conclusion

I have discussed Strato's remarks on time as independently as possible from the late sources that use him for their own agenda. It turns out that each of them can be interpreted as pointing the attentive reader of Aristotle to issues that need further clarification if Aristotle's treatment of time across different works (*Categories*, *Physics*, and *Metaphysics*) has to be shown coherent. We have seen that Alexander incorporates most of Strato's remarks in his short treatise *On Time* (albeit without mentioning Strato by name). I have argued that a precise formulation of Alexander's own theory of time requires clear decisions and/or further development of most of Strato's points in the light of the earlier Peripatetic and anti-Peripatetic traditions. We do not know whether Alexander knew he was relying on points Strato made, or was merely building on the specifications the Peripatetic tradition required of him. I hope to have shown that Alexander saw how important these demands for clarification were for the correct understanding of the notion of time in a Peripatetic framework.[72]

72 I am grateful to Gweltaz Guyomarc'h and Alessio Santoro for organising a wonderful conference in Lyon, and to the participants and Alessio Santoro in particular for their detailed criticism and advice concerning my paper. Of course all remaining mistakes are my own.

CHAPTER 8

What Is the Purpose of *De Coloribus*?

Richard Schorlemmer

Hans Gottschalk, who conducted a thorough investigation into the authorship of *De coloribus* and attributed it to an unknown Peripatetic active during Theophrastus' lifetime, criticised its failure to develop a comprehensive theory of colours or to explain colour phenomena beyond the merely conditional level.[1] Similarly, in the Loeb edition, W.S. Hett remarks disparagingly that the work "is of more value as a collection of observed facts than for any theory of the origin and development of colour in physical life".[2] These criticisms are justified, not only in pointing to the work's lack of stylistic and conceptual refinement,[3] but also in highlighting how the numerous observations and examples of colour changes seem to challenge, or at least detract from, a unified theory implied by the title Περὶ χρωμάτων and the work's concluding sentence: "From what has been set forth in this treatise one may best understand the theory of colours."[4]

In the most recent edition, however, Maria Ferrini, while building upon the criticisms made by earlier commentators, praises the author's modernity for recognising that every colour perception is mediated, such that no colour can be observed in its pure hue—except in the cases of fire, water and air, whose natural colours the author believes to be evident. Ferrini further argues that the primary purpose of the treatise is to explain how and why identical objects take on a different appearance under different conditions, i.e. how colours are

1 Gottschalk (1964), 77: "When our author wants to establish a general law he does so by enumerating instances in which it can be seen at work, without further explanation. Often too his attempts to explain a phenomenon amount to no more than an enumeration of the conditions which must be fulfilled if it is to occur; no clear distinction is made between nearer and more remote causes or the way in which different factors contribute to the end-effect. [...] What matters to him is that the true colour of the object and the optical factors which modify it before it is perceived, are equally important in determining what is actually seen."

2 Hett (1936), 3. This view is echoed in the more recent translation and commentary by Georg Wöhrle (1999), 32: "Vor dem Hintergrund des als originell Aristotelisch Faßbaren bleibt auch heute [...] der Eindruck einer nicht in allen Teilen konzisen und durchdachten Abhandlung."

3 Ferrini (1999), 44, states that the style of the work often causes translations to resemble paraphrases.

4 Ps.-Aristotle, *Col.* 799b19 ff.: τὴν μὲν οὖν περὶ τὰ χρώματα θεωρίαν μάλιστ' ἄν τις ἐκ τῶν εἰρημένων δύναιτο συνιδεῖν. Unless otherwise indicated, the translations are taken from Loveday and Forster with minor modifications.

perceived differently from their original hues over a distance, and how natural colours correlate with the elements of their material composition.[5] On this view, the author does not present a comprehensive theory of colours, but merely addresses colour changes and the impediments to the clear perception of colours.

Building on Gottschalk's and Ferrini's interpretations, I argue that the author proposes a method to establish the true colour of an object, especially in the two cases Ferrini mentions: first, since colours that are perceived over a distance may not appear as they are, the colour an object really has (to which I will refer as "original colour") can be established via his method. Second, since natural objects—in particular plants, fruits as well as hair and skin of animals—may change their colour over time, his method allows for a definition of a true, mature colour of the object (to which I will refer as "natural colour").[6] The objective of the treatise is, therefore, not to start with a natural or original colour and to outline the factors that influenced its altered perception, but to equip the reader with the scientific tools necessary to analyse colour changes in order to identify an object's true colour and explain deviations from it.

This interpretation leads to three conclusions. First, the well-known relativistic passage which asserts that colours are never seen in their pure form is not something that the author endorses, but rather an issue he seeks to address. If the opposite were true, the author would face the epistemological problem of having to explain how any original or natural colour could ever be discerned, given that all colours are always mixed. Second, while colours correlate to their material, they remain, in Aristotelian terms, accidents, as colours and their associated materials are not always congruent. Third, the multitude of examples serves to illustrate the method across various instances of colour perception and to guide the reader in its practical application. If this interpretation

5 Cf. Ferrini (1999), 32–46, esp. 37–39. See also Ferrini (2020), 19: "In this way, everything can have a permanent colour and a transitory colour, an original colour and a colour that is formed over time, again starting from the original, linked to the element of which a substance is predominantly made up. The only distinction that the author invites us to make is between the colour based on observation, or presupposition, and the colour that occurs in successive phases of growth (plants and animals reach their own colour as they mature), or in dyeing processes, or following mechanical phenomena, for example, the rubbing of stones and metals, physical (and chemical) actions due to light and heat. We can, therefore, observe initial colours, intermediate colours, final colours [...]. There is a colour naturally linked to things, as they are mainly formed by an element, but we do not always see it; therefore analysis and interpretation focus on the visible and very mutable colour, on the colour that appears at the beginning of a process, for example, green in plants."

6 In the fourth chapter, Ps.-Aristotle, *Col.* 794a15–794b12, the author also mentions artificial colour changes like dyeing as a third kind of colour change.

is correct, the extensive use of examples in *De coloribus* appears more relevant and the underlying theory more coherent.

The first part of this paper outlines what colours are and introduces the method of how colour mixtures can be analysed. The second part contextualises the remark that no colour appears as it is and discusses its implications for the mixture of colours, while the third part is concerned with the application of the method to identify natural colours in plants and animals.

1

At the outset of the treatise, the author distinguishes between simple primary and secondary colours. Simple colours—white, yellow and black—correlate to or are concomitant with (συνακολουθεῖ) the elements.[7] Air, water and earth are naturally white (καθ' ἑαυτὰ τῇ φύσει λευκά), fire and the sun are yellow,[8] whereas black (τὸ μέλαν χρῶμα) arises when the elements change into one another.[9] This transformation is understood both as a process, since moisture

7 The meaning of συνακολουθεῖ is discussed below. I thank Katerina Ierodiakonou for sending me her new translation for the Loeb volume, in which she suggests translating συνακολουθεῖ as "concomitant". See also Hahm (1978), 84. I also benefitted from reading her forthcoming article on the nature of mixed colours.

8 Cf. Ps.-Aristotle, *Col.* 791a2 ff.

9 The much-discussed passage, Ps.-Aristotle, *Col.* 791a12–b6, distinguishes between three types of black appearances (μέλαν, instead of τὸ μέλαν χρῶμα). Following Loveday and Forster, I omit μή in τὸ μὴ ὁρώμενον (791a14) and interpret the first type at 791a13–15 as the colour black, of which some form of black light is reflected. Thus, the phrase τι φῶς μέλαν must be retained, as it refers to light that reflects a black colour. The second and third type describe perceptions of darkness (σκότος), which, as privations of light, are not directly seen (τὸ γὰρ μὴ ὁρώμενον, 791a16), but still appear black. The second type refers to the appearance of black darkness, such as at the entrance of a cave, which reflects no light at all, while its surroundings remain visible (791a15–17). The third type refers to the appearance of shadowy areas, which reflect some light and colour but are overlaid by a dark-black effect (791a17–b6). The term "shadowy-black" (τὸ μέλαν καὶ σκιερόν, 792a9), then, describes a black object that reflects little light.

Since the commentator Sophonias also reads τὸ μὴ ὁρώμενον at 791a14, the added μή must have already been present in the textual tradition. This addition may have arisen from an *aberratio oculorum* to 791a16 or as a conjecture by an Aristotelian interpreter who assumed that the first kind of blackness also referred to darkness. Although the Latin translations support the omission of μή, they are too unreliable for reconstructing the text.

The primary reason for omitting μή is that the three types follow the structure outlined in the introduction: the author begins by explaining how black arises as a colour, before differentiating it from darkness, leading to the classification of the three kinds of black and dark appearances. Subsequently, the author describes light, in contrast to darkness, before providing a more detailed account of how the colour black (τὸ μέλαν χρῶμα) is generated

darkens and eventually turns black through either concoction or cold,[10] and as a subsequent state, since naturally black objects, such as metals, also exist. The author further notes that, although earth is naturally white, it often appears dyed, because it readily mixes with other substances, such as water. Its true natural colour, however, is visible in burnt white ash, when all moisture has evaporated.[11] Consequently, earth is excluded from the author's investigation into natural colours, as it is deemed to be an unreliable foundation for such an enquiry.[12] Significantly, this clarification defines natural colours as those displayed by an object in its pure state and any deviation from this natural colour can be explained by external factors.

Secondary colours are created by combining the primary colours in varying proportions and intensities (τῷ μᾶλλον καὶ ἧττον).[13] For example, crimson (φοινικοῦν) can result from the mixture of black with the yellow of sunlight or fire which is seen in phenomena such as the glowing red appearance of burning coal and, presumably, objects in a dark room taking on a crimson hue when illuminated by firelight.[14] Thus, the mixture of colours from different objects may lead to a similar hue, as is further illustrated by two examples involving wine-coloured purple (οἰνωπόν / ἁλουργόν):

> For example, a wine-like colour arises 1) when airy sun rays mix with pure and shiny black, 2) as with grapes that appear wine-coloured during the ripening process. When they darken, the crimson turns into purple.[15]

These examples are consistent with other passages in the treatise: the first example is similar to weak sunlight interacting with shadowy white air, which creates a shiny purple,[16] as well as to black plumage that appears purple when

---

(791b17–792a2). Retaining μή makes the distinctions between the three types more problematic. For further discussion and alternative interpretations, see Gottschalk (1964), 61 ff., and Ferrini (1999), 122–125.

10 Also, hair and feathers turn black when enough moisture is concocted during their production. Cf. Ps.-Aristotle, *Col.* 793a21–25 and 797b1–7.

11 Cf. Ps.-Aristotle, *Col.* 791a4 ff.

12 Earth is not mentioned among the elements from which reliable conclusions can be drawn in Ps.-Aristotle, *Col.* 792b19–23.

13 Cf. Ps.-Aristotle, *Col.* 791a1–12 and 792a4–6.

14 Cf. Ps.-Aristotle, *Col.* 792a9–12. The author does not give an example of a shadowy-black that is mixed with light, but only of black things that mix with light.

15 Ps.-Aristotle, *Col.* 792b7–11: […] οἷον ὅτι τὸ οἰνωπὸν χρῶμα γίνεται, ὅταν ἀκράτῳ τῷ μέλανι καὶ στίλβοντι κραθῶσιν αὐγαὶ ἠεροειδεῖς, ὥσπερ καὶ αἱ τῶν βοτρύων ῥᾶγες· καὶ γὰρ τούτων οἰνωπὸν φαίνεται τὸ χρῶμα ἐν τῷ πεπαίνεσθαι· μελαινομένων γὰρ τὸ φοινικοῦν εἰς τὸ ἁλουργὲς μεταβάλλει.

16 Cf. the explanation of sunrise and sunset in Ps.-Aristotle, *Col.* 792a15–20, which will be dis-

light reflects from it in a certain way (which will be discussed shortly). In these cases, the rays fall upon a darkened or black simple colour, producing a mixed purplish colour. The second example is elaborated in the fifth chapter: as they ripen, grapes turn from the secondary colour crimson into a wine-like colour, as their moisture is gradually concocted.[17]

While the same colour can thus arise from different underlying colours, the author also discusses cases where the same object can exhibit different colours, focusing particularly on the recurring example of plumage:

> The same thing [as waves appearing purple when light falls upon them in a certain way] may also be observed in birds' plumage, which get a purple (ἁλουργόν) colour if extended in a certain way against the light, but if the amount of light falling on them is diminished the result is the dark colour called dark-violet (ὄρφνιον),[18] whilst a great quantity of light blended with primary black (πρώτῳ μέλανι) gives crimson.[19]

The primary black (πρώτῳ μέλανι) refers to the inherent blackness of the plumage, which remains constant.[20] Yet, given that the colour of the plumage may change depending on the lighting conditions, an analysis of its appearance is necessary to determine that black is its original colour.

To confront these two problems—that the same object may appear to have different colours and that the perception of the same hue may arise from different underlying colours—the author develops a method of colour analysis that neither tries to analyse a colour mixture with a single set of rules nor to assume any colour as given from the outset. Accordingly, the author begins the introduction into his method of colour analysis with a caveat:

---

cussed below. τῷ μετρίῳ λευκῷ καὶ σκιερῷ κραθῶσιν ἀσθενεῖς αἱ τοῦ ἡλίου αὐγαί and ἀκράτῳ τῷ μέλανι καὶ στίλβοντι κραθῶσιν αὐγαὶ ἠεροειδεῖς differ only in whether they describe the white and black sky (τῷ μετρίῳ λευκῷ καὶ σκιερῷ) as mixing with the yellow rays or the white and yellow rays (αὐγαὶ ἠεροειδεῖς) as mixing with the black of the sky. Possibly, this refers to different configurations of the sky during sunrise and sunset.

17 Cf. Ps.-Aristotle, *Col.* 795b24–30. The identification of purple and wine-colour is certain in this instance.

18 Following Wöhrle's translation of ὄρφνιον.

19 Ps.-Aristotle, *Col.* 792a24–28: ὃ καὶ ἐπὶ τῶν πτερωμάτων θεωρεῖται γιγνόμενον· ἐντεινόμενα γάρ πως πρὸς τὸ φῶς ἁλουργὲς ἔχει τὸ χρῶμα. ἐλάττονος δὲ τοῦ φωτὸς προσβάλλοντος ζοφερόν, ὃ καλοῦσιν ὄρφνιον· πολὺ δὲ καὶ τῷ πρώτῳ μέλανι κραθὲν φοινικοῦν.

20 Cf. Gottschalk (1964), 65.

> One must proceed in this way regarding the mixing of colours with each other, starting from an observed colour as a base (ἐξ ὑποκειμένου τεθεωρημένου χρώματος) and deriving the mixture from it. However, one should not assume the same origin for all colours.[21]

In the case of the plumage, the already established observed base colour (ἐξ ὑποκειμένου τεθεωρημένου χρώματος) is black, while light drives the observed changes as agent. However, since the original colours of such mixtures may vary, and since not all colour changes can be explained uniformly, the author proposes an empirical two-step method to determine the original colours: 1) comparison (ὁμοιότης) and 2) verification (πίστις).

> After the manner indicated we must treat all differences of colours, getting comparisons (ὁμοιότητα) by moving [coloured objects], keeping our eye on actual phenomena, assimilating different cases of mixture on the strength of the particular known instances in which a given origin and blending produce a certain chromatic effect, and verifying our results (πίστιν προσφερομένους). But we must not proceed in this inquiry by blending pigments as painters do, but rather by comparing the rays reflected from the aforesaid known colours, this being the best way of investigating the true nature (κατὰ φύσιν) of colour-blends. We must take as verifications and grounds of comparison those cases in which the origin of colours will be manifest; these are especially the ones produced by the interaction of sunlight and firelight with air and water.[22]

The comparison to the painter is usually connected with Aristotle's remark that the colours of a rainbow cannot be produced artificially,[23] suggesting that the

21 Ps.-Aristotle, *Col.* 792a29–32: κατὰ γὰρ τὴν πρὸς ἄλληλα κρᾶσιν οὕτως ληπτέον, ἐξ ὑποκειμένου τεθεωρημένου χρώματος ποιοῦντας τὴν μίξιν, ἀλλὰ μὴ πάντων ὁμοίαν γένεσιν ποιοῦντας.

22 Ps.-Aristotle, *Col.* 792b11–23: κατὰ δὲ τὸν ὑποδεδειγμένον τρόπον θεωρητέον πάσας τὰς τῶν χρωμάτων διαφοράς, ἐκ κινήσεως τὴν ὁμοιότητα λαμβάνοντας κατ᾽ αὐτὸ τὸ φαινόμενον, τὴν ἐν ἑκάστῳ μίξιν ὁμοιοῦντας καὶ ἐπὶ τῶν κατὰ μέρος ἐν γενέσει τινὶ καὶ κράσει ποιούντων φαντασίαν, καὶ πίστιν προσφερομένους. δεῖ δὲ καὶ πάντων τούτων ποιεῖσθαι τὴν θεωρίαν μὴ καθάπερ οἱ ζωγράφοι τὰ χρώματα ταῦτα κεραννύντας, ἀλλ᾽ ἀπὸ τῶν εἰρημένων τὰς ἀνακλωμένας αὐγὰς πρὸς ἀλλήλας συμβάλλοντας· μάλιστα γὰρ δύναιτ᾽ ἄν τις κατὰ φύσιν θεωρῆσαι τὰς τῶν χρωμάτων κράσεις. τὰς δὲ πίστεις καὶ τὰ ὅμοια δεῖ ἐν οἷς ἡ γένεσις ἔσται φανερὰ τῶν χρωμάτων. ταῦτα δὲ μάλιστά ἐστι τό τ᾽ ἀπὸ τοῦ ἡλίου φῶς καὶ τὸ ἀπὸ τοῦ πυρὸς καὶ ὁ ἀὴρ καὶ τὸ ὕδωρ. Following Gottschalk's (1964), 66, translation of the last sentences and retaining the difficult ἐκ κινήσεως.

23 Cf. Aristotle, *Meteor.* 372a5–11.

passage differentiates between an artificial and a natural way of mixing colours. However, given that the author draws this comparison in the context of his method, he rather denies that any recreation of a colour by mixing portions of other colours is sufficient to explain a certain hue. Instead, the true nature of a blend to be analysed by the proposed method is found on light rays. The case of the plumage was already an example of its application: the observation in differing lighting conditions revealed that black is the original colour from which the deviations can be explained. Thus, by comparing the changes in colour on the reflected light several times their underlying mechanisms may be discovered which in turn allows for an estimation of the colours in the blend. Verification, while more challenging, refers to an retrograde explanatory rule governing how the perceived mixture was produced by the colours that were mixed together. This is easiest when an original colour is simple as these colours can be identified with the highest degree of probability. In every case, both steps in the method are essential. Comparing straightforward cases of colour production establishes certainty with which subsequent observations can be verified.

2

Yet, the problems the author defines seem to challenge the notion that any colour could be identified with certainty, especially since the author posits that both air and light possess colour themselves and that colour perception depends on light rays. In a notable passage the author even argues that these rays not only carry colours of objects to the eyes and affect them with their own colour, but that the light rays also acquire the hue of each surface they fall upon, as well as the one of the air. Thus, these rays may accumulate a composite of various colours. If this assertion were true, his proposed method could never yield certain results. Since this seems at odds with many examples given in the treatise, a closer examination of the passage is warranted:

> We see no colour in absolute purity: it is always blent, if not with another colour, then with rays of light or with shadows, and so it assumes a tint other than its own. That is why objects assume different tints when seen in shade and in light and sunshine, and according as the rays of light are strong or weak, and the objects themselves slope this way or that, and under other differential conditions. Again, they vary when seen by firelight or moon-light or torchlight, because the colours of those lights differ somewhat. They vary also in consequence of mixture with other colours,

> for they are coloured by passing through one another (δι᾽ ἀλλήλων γὰρ φερόμενα). For if light falls on a given object and is coloured by it crimson or herb-green, and then the light reflected from that object falls on another colour, it is again modified by this second colour, and so it gets a new chromatic blend. This happening to it continuously, though imperceptibly (οὐκ αἰσθητῶς), light when it reaches the eye may be a blend of many colours, though the sensation produced is not of a blend but of some colour predominant in the blend (ἐκ πολλῶν μὲν κεκραμένον χρωμάτων, ἑνὸς δέ τινος τῶν μάλιστα ἐπικρατούντων ποιοῦν τὴν αἴσθησιν). This is why objects under water tend to have the colour of water, and why reflections in mirrors resemble the colour of the mirrors, and we must suppose that the same thing happens in the case of air. Thus all hues represent a threefold mixture of light, a translucent medium (e.g. water or air), and underlying colours from which the light is reflected.[24]

The mechanics underlying colour mixture are clear only in outline: a colour mixture is carried by light rays, resulting from reflections from objects, as well as the medium's and the light's inherent colour. However, what is perceived is not a mixture of colours, but a predominant colour in the mixture. In order to better understand this concept, several questions must be addressed. First, how much influence does the author attribute to the medium as an impeding factor at a distance? Second, what is the nature of light, and how does it transport colour mixtures? Third, how does this predominant colour within a mixture correspond to the proposed method?

Starting with the first question, the author concludes by analogy that air, like water or a tinted mirror, must exert some influence on colour perception,

---

24 Ps.-Aristotle, *Col.* 793b12–794a4: τῶν δὲ χρωμάτων οὐδὲν ὁρῶμεν εἰλικρινὲς οἷόν ἐστιν, ἀλλὰ πάντα κεκραμένα ἐν ἑτέροις· καὶ γὰρ ἂν μηδενὶ τῶν ἄλλων, ταῖς γε τοῦ φωτὸς αὐγαῖς καὶ ταῖς σκιαῖς κεραννύμενα ἀλλοῖα, καὶ οὐχ οἷά ἐστι, φαίνεται. διὸ καὶ τὰ ἐν σκιᾷ θεωρούμενα καὶ ἐν φωτὶ καὶ ἡλίῳ, καὶ σκληρᾷ αὐγῇ ἢ μαλακῇ, καὶ κατὰ τὰς ἐγκλίσεις οὕτως ἢ οὕτως ἔχοντα, καὶ κατὰ τὰς ἄλλας διαφοράς, ἀλλοῖα φαίνεται. καὶ ταῖς πρὸς τῷ πυρὶ καὶ τῇ σελήνῃ, καὶ ταῖς τῶν λύχνων αὐγαῖς, διὰ τὸ καὶ τὸ φῶς ἑκάστου τούτων ἀλλοιοτέραν ἔχειν χρόαν. καὶ τῇ πρὸς ἄλληλα δὲ μίξει τῶν χρωμάτων· δι᾽ ἀλλήλων γὰρ φερόμενα χρώζεται. τὸ γὰρ φῶς ὅταν προσπεσὸν ὑπό τινων χρωσθῇ, καὶ γένηται φοινικιοῦν ἢ ποῶδες, καὶ τὸ ἀνακλασθὲν προσπέσῃ πρὸς ἕτερόν τι χρῶμα, πάλιν ὑπ᾽ ἐκείνου κεραννύμενον ἄλλην τινὰ λαμβάνει τοῦ χρώματος κρᾶσιν. καὶ τοῦτο πάσχον συνεχῶς μὲν οὐκ αἰσθητῶς δὲ ἐνίοτε παραγίνεται πρὸς τὰς ὄψεις ἐκ πολλῶν μὲν κεκραμένον χρωμάτων, ἑνὸς δέ τινος τῶν μάλιστα ἐπικρατούντων ποιοῦν τὴν αἴσθησιν. διὸ καὶ ⟨τὰ⟩ καθ᾽ ὕδατος ὑδατοειδῆ μᾶλλον φαίνεται, καὶ τὰ ἐν τοῖς κατόπτροις ὁμοίας ἔχοντα χρόας ταῖς τῶν κατόπτρων. ὃ καὶ περὶ τὸν ἀέρα οἰητέον συμβαίνειν. ὥστε ἐκ τριῶν εἶναι τὰς χρόας ἁπάσας μεμιγμένας, τοῦ φωτός, καὶ δι᾽ ὧν φαίνεται τὸ φῶς, οἷον τοῦ τε ὕδατος καὶ τοῦ ἀέρος, καὶ τρίτου τῶν ὑποκειμένων χρωμάτων, ἀφ᾽ ὧν ἀνακλᾶσθαι συμβαίνει τὸ φῶς.

though not necessarily in a significant or perceptible way. A few lines after the passage above, the author employs a concept of density to account for colour differences in the air:

> Air seen close at hand appears to have no colour, for it is so rare that it yields and gives passage to the denser rays of light, which thus shine through it (διαφαινομένων δι᾽ αὐτοῦ); but when seen in a deep mass it looks practically dark blue. This again is the result of its rarity, for where light fails the air lets darkness through and looks dark blue. When densified, air is, like water, the whitest of things.[25]

For air to appear in its natural white colour, it must be sufficiently dense. Otherwise, at the extremes, it either allows light to pass through unimpeded, rendering it colourless (οὐδὲν χρῶμα), or it mingles with the surrounding darkness—likely that of the cosmos—and takes on the dark-blue appearance of the sky. Since the absence of colour in rarefied air is due to the fact that it lets light pass through, colourlessness means that the air does not impart any colour to the light. Thus, when light rays acquire an object's colour by reflection and transport it to the perceiver with no interference from the air, the medium does not alter a colour perception in every instance.

This leads to the second question: what is the nature of light and how does it transport colour mixtures? Since light travels in rays, falls on objects, changes its density and pushes itself through air or water, it must be material.[26] However, the author also defines light as the colour of fire or the sun, suggesting that light is the yellow colour of another body, and adds the following qualification:

> Light is clearly the colour of fire; for it is never found with any other hue than this, and it alone is visible in its own right whilst all other things are rendered visible by it. But there is this point to be considered, that some things, though they are not in their nature fire nor any species of fire, yet seem to produce light. So we cannot say that the colour of fire is identical with light, and yet light is the colour of other things besides fire, but we

25 Ps.-Aristotle, *Col.* 794a8–15: ὁ δ᾽ ἀὴρ ἐγγύθεν μὲν θεωρούμενος οὐδὲν ἔχειν φαίνεται χρῶμα (διὰ γὰρ τὴν ἀραιότητα ὑπὸ τῶν αὐγῶν κρατεῖται, χωριζόμενος ὑπ᾽ αὐτῶν πυκνοτέρων οὐσῶν καὶ διαφαινομένων δι᾽ αὐτοῦ), ἐν βάθει δὲ θεωρουμένου, ἐγγυτάτω φαίνεται τῷ χρώματι κυανοειδὴς διὰ τὴν ἀραιότητα. ᾗ γὰρ λείπει τὸ φῶς, ταύτῃ σκότῳ διειλημμένος φαίνεται κυανοειδής. ἐπιπυκνωθεὶς δέ, καθάπερ καὶ τὸ ὕδωρ, πάντων λευκότατόν ἐστιν.

26 Cf. Gottschalk (1964), 76.

> can say that this colour is to be found in other things besides fire, and yet light is the colour of fire. Anyhow, it is only by aid of light that fire is rendered visible, just as all other objects are made visible by the appearance of their body.[27]

Since the author holds that things other than fire might also produce light,[28] it seems that all bright objects emit or produce light in the sense of producing brightness (τὸ λαμπρόν or τὸ στίλβον), a characteristic of light.[29] It is reasonable to infer that non-fiery objects emit light that reflects their material properties, which may simply possess brightness without being yellow. In any case, the author does not clearly distinguish between light as a transporting material of colours, light as brightness and light as the colour yellow, and the precise relationship between the colour, the materiality and brightness of light remains ambiguous.

This ambiguity extends to how light interacts with the colours of other objects. When light shines through the air (διαφαινομένων δι' αὐτοῦ), the new colour cannot be a result of a superposition of two simple colours, as defined by Aristotle in *De sensu*.[30] Instead, the light's inherent colour mixes with the air and surfaces it encounters along the way via material processes such as collision and reflection, as was previously mentioned in *Col.* 793b12–794a4, and is reiterated in *Col.* 792a15–20:

---

27 Ps.-Aristotle, *Col.* 791b6–17: τὸ δὲ φῶς ὅτι πυρός ἐστι χρῶμα, δῆλον ἐκ τοῦ μηδεμίαν ἄλλην ἢ ταύτην ἔχον εὑρίσκεσθαι χρόαν, καὶ διὰ τὸ μόνον τοῦτο δι' ἑαυτοῦ ὁρατὸν γίνεσθαι, τὰ δ' ἄλλα διὰ τούτου. ἐπισκεπτέον δὲ τοῦτο. ἔνια γὰρ οὐκ ὄντα πῦρ οὐδὲ πυρὸς εἴδη τὴν φύσιν φῶς ποιεῖν φαίνεται. εἰ μὴ ἄρα τὸ μὲν τοῦ πυρὸς χρῶμα φῶς ἐστίν, οὐ μέντοι καὶ τὸ φῶς πυρός ἐστι χρῶμα μόνου, ἀλλ' ἐνδέχεται μὴ μόνῳ μὲν ὑπάρχειν τῷ πυρὶ τὴν χρόαν ταύτην, εἶναι μέντοι χρῶμα τὸ φῶς αὐτοῦ. οὐδενὶ γοῦν ἄλλῳ τὴν ὅρασιν αὐτοῦ συμβαίνει γίνεσθαι πλὴν τῷ φωτί, καθάπερ καὶ τὴν τῶν ἄλλων σωμάτων ἁπάντων τῇ τοῦ σώματος φαντασίᾳ. Prantl suggests reading χρώματος for the transmitted σώματος in the last sentence.

28 According to Aristotle, these things may be fungi, horns, heads, scales and eyes of fish, but none of them are seen in their own colour (cf. Aristotle, *An.* 419a1–6). In the debate among the Peripatetics about whether light was fire or not, the author apparently chose the latter side, cf. Wöhrle (1999), 68.

29 Ps.-Aristotle, *Col.* 793a10–16: "Difference of hue may also depend on the brightness and lustre or dimness and dullness of the blend. Lustre is simply continuity and density of light; e.g. we have a glistening gold colour when the yellow colour of sunlight is highly concentrated and therefore lustrous. That explains why pigeons' necks and drops of falling water look lustrous when light is reflected from them."

30 Cf. Aristotle, *Sens.* 440a6–15.

> But a vivid bright violet is obtained from a blend of feeble sunlight with a thin dusky white. That is why the air sometimes is purple at sunrise and sunset, for then the air is especially dusky and the impinging (προσβάλλουσιν) rays feeble.[31]

Thus, the light rays impinge or fall upon on the air, acquire its colour and carry the mixture to the perceiver. It remains to be determined, however, whether the air and light changes its material constitution to become purple as a result of the mixture or undergoes an accidental change. In the former case, light would act as an agent of colour change, causing both the air and itself to take on a new colour. In the latter case, the rays of the sun would function solely as a transport medium, leaving the air's colour unchanged.

The former interpretation is suggested by the many examples of material processes underlying colour changes—plants and water become darker when the moisture is concocted, earth turns white when the dark moisture is burnt out completely and so on—, as well as the idea that all light rays either affect their own colouration by passing through each other (δι' ἀλλήλων γὰρ φερόμενα, 793b22 ff.) or change the colour of the objects they fall upon:

> For all water in process of time first turns yellow-green on blending with the rays of the sun; it then gradually turns black, and this further mixture of black and yellow-green produces herb-green.[32]

If this interpretation were taken at face value, every object—not just the light rays—would undergo constant colour changes which can be understood in a strong and a weak sense. On the strong view, every light ray would colour everything it touches and would be coloured by it,[33] but this view must be

31 Ps.-Aristotle, *Col.* 792a15–20: τὸ δ' ἁλουργὲς εὐανθὲς μὲν γίνεται καὶ λαμπρόν, ὅταν τῷ μετρίῳ λευκῷ καὶ σκιερῷ κραθῶσιν ἀσθενεῖς αἱ τοῦ ἡλίου αὐγαί. διὸ καὶ περὶ ἀνατολὰς καὶ δύσεις ὁ ἀὴρ πορφυροειδὴς ἔστιν ὅτε φαίνεται, περὶ ἀνατολὴν καὶ δύσιν ὄντος τοῦ ἡλίου· ἀσθενεῖς γὰρ οὖσαι τότε μάλιστα πρὸς σκιερὸν ὄντα τὸν ἀέρα προσβάλλουσιν.

32 Ps.-Aristotle, *Col.* 794b26–29: τὰ γὰρ ὕδατα πάντα χρονιζόμενα κατ' ἀρχὰς μὲν γίνεται χλωρά, κεραννύμενα ταῖς τοῦ ἡλίου αὐγαῖς, κατὰ μικρὸν δὲ μελαινόμενα, πάλιν μιγνύμενα τῷ χλωρῷ, γίνεται ποώδη.

33 This may have been the position of Strato, who was also proposed as a candidate for the authorship of *De coloribus*. In the doxographical statement 64 Sharples in Desclos & Fortenbaugh (2011) = Aëtius, *Placita* 4.13.3 (only Stobaeus), Strato holds that colours tint the intervening air: "Strato (says that) that colours travel from bodies and give their colour to the intermediate air." (Στράτων χρώματά φησιν ἀπὸ τῶν σωμάτων φέρεσθαι συγχρῴζοντ' αὐτοῖς τὸν μεταξὺ ἀέρα, transl. Mansfeld and Runia). If φέρεσθαι is read in the passive voice, light might pick the colours up and carry them through the air which becomes tinted,

objected to on a general level. If the author's assumption is that underlying colours are identifiable after a careful analysis, extending this strong material explanation to the colouration of light introduces unsolvable problems: on the one hand, there would be a cloud of indistinguishable colours between the perceiver and the perceived due to the constant mixture, while on the other, light, as something material, would acquire materials different from its own at the first instance of mixture. If material light and air combined to create the colour purple, the purple perceived by the eye would not originate from the light ray itself but from a light/air mixture. If this mixture then fell upon iron, the result would be a light/air/iron mixture and so forth.

On the weak view, light rays would primarily pass through the air and themselves without significant alterations of any colour and interact with the colours of every object they fell upon only imperceptibly or in certain instances. However, since light receives a colour from an object through reflection at least to a noticeable extent, this also seems unlikely.

To better understand the colour change light produces in water, it is necessary to distinguish the transient effect of light and its impact on plants. The claim that the sun's yellow light turns water yellow-green seems unequivocal, but it is made in the fifth chapter on plants, which introduces the sun's heat as an agent in growth and maturation. The author argues in this chapter that due to the sun's influence, the white shoots of plants become yellow-green as they emerge from the ground and that the sun's heat concocts the moisture further which produces their colouration.[34] Prantl has therefore proposed that the change in the water's colour to yellow-green is more plausibly attributed to the growth of lichen rather than any alteration undergone by the water itself,[35] which is supported by the immediately following passage, according to which it is not the water that changes colour, but rather the colours that appear on

but this seems difficult to reconcile with *De coloribus*. However, as Pavel Gregorić pointed out at the conference in Lyon, Strato also assumed both that light is material and that the sun illuminates planets and stars at night by reflection (Fr. 43 Sharples). Cf. the helpful discussions of authorship in Wöhrle (1999), 42–51, and Wöhrle (2002).

If we possessed more than the few fragments of Aristarchus, Strato's student, he could possibly have been another candidate for authorship of *De coloribus*. According to Aëtius, *Placita* 1.15.5 (only Stobaeus): "Aristarchus of Samos the astronomer, disciple of Strato, (says that) colour is light falling on what it falls upon." (Ἀρίσταρχος Σάμιος μαθηματικός, ἀκουστὴς Στράτωνος, φῶς εἶναι τὸ χρῶμα τοῖς ὑποκειμένοις ἐπιπῖπτον.) On Aristarchus, cf. Diels (1893), 119.

34 Cf. Gottschalk (1964), 71ff.

35 Cf. Prantl (1849), 128ff. See also Ferrini (1999), 208ff. Gottschalk (1964), 73, notes that the author "confused the changes caused by the sun's heat with the more transient effects brought about by its light shining on coloured surfaces."

the walls of water reservoirs as the water evaporates.[36] It seems likely, then, that the author does not clearly differentiate between light as the sun's heat—presumably produced from its fiery material nature—and its yellow hue, which are both seen as contributing factors to plant colouration, and that it is at least imprecise to attribute the change of colour to the water instead of the plants, which are seen through the water.

Further, the author distinguishes the concoction produced by the sun from the concoction generated by animals' innate heat. In animals, the moisture which colours the hair, for instance, remains concealed from the sun beneath the skin. Consequently, since the colour of hair is not exposed to yellow light during concoction, it cannot turn purple or green.[37] Yet, since the hair outside the body is not affected by the sunlight, the author must claim that sunlight affects only the maturation process with its heat, while already produced colours remain constant. As a result, the transport of a colour impression from an object must be different from the role of sunlight in the colouration of plants.

If light rays, then, neither colour nor are coloured by the objects they interact with materially, how do they receive and mix the colours they transport? For Aristotle, there is a relationship between material elements and accidents such as colours, but this connection does not suffice to explain colours in every instance. In *De generatione et corruptione*, Aristotle contends that qualities correspond to the mixture of the underlying simple bodies, as substances in which water preponderates may more readily yield to the touch, but these qualities cannot be fully explained through mere ratios of the underlying materials.[38] A similar perspective appears to be implied when the author of *De coloribus* states that colours correspond to or are concomitant with (συνακολουθεῖ) the elements. Although συνακολουθέω denotes a necessary relationship in Aristotelian logic, the author must have a looser correlation in mind. As is illustrated in the case of hair, its natural colour corresponds to the colour of the skin:

> So [the colour of hair] corresponds (συνακολουθεῖ) in all cases the colours of the hide and skin; for when men are reddish in colouring, their hair too is of a pale red; when they are black, it is black; and if white leprosy has broken out over some part of the body, the hair on that portion is

---

36 Cf. Ps.-Aristotle, *Col.* 794b29–795a4.

37 Cf. Ps.-Aristotle, *Col.* 799b4–6. Exceptions include features like plumage, which undergoes concoction at different stages both inside and outside the body. Cf. Ps.-Aristotle, *Col.* 799b9–16.

38 Cf. Aristotle, *GC*, 329b25–330a24.

> also always white, like the marking on dappled animals. Thus all hair and feathering follows (συνακολουθεῖ) the colour of the skin, both regional hair and hair which is spread over the whole body.[39]

This passage establishes a general rule, but subsequent passages focus on instances where the colour of the hair differs from that of the skin. For example, the author explains why long hair is darker at points that are nearer to the body than at the ends, or why a teenager's first beard may initially appear red before darkening later in life.[40] Thus, the author recognises that objects generally possess a characteristic colour, but also acknowledges deviations from the corresponding or concomitant colours. Therefore, in accordance with Aristotle's position, the ontological status of colours must be accidental, but, in a development of Aristotle's view, material light acquires these colours through its physical reflection from objects. A colour mixture must consequently be understood as an accidental change with a possible, but not necessary underlying change in materials.

As for the third question, the concept of a predominant colour within a blend (ἐκ πολλῶν μὲν κεκραμένον χρωμάτων, ἑνὸς δέ τινος τῶν μάλιστα ἐπικρατούντων ποιοῦν τὴν αἴσθησιν) might refer to the Anaxagorean argument that an object composed of various materials is defined by the material that predominates within it, although this idea has already become a topos by the time the treatise was written.[41] In any case, this argument is adapted to the discussion of colours in *De coloribus*. As in Anaxagoras, where an object remains fundamentally the same even when its composition changes slightly—e.g. adding or subtracting a small amount of sugar to cake dough does not create a different object after baking—the author of *De coloribus* allows for the possibility that reflections from multiple differently coloured surfaces may alter the mixture

---

39 Ps.-Aristotle, *Col.* 797b11–19: διὸ καὶ πάντα τοῖς δέρμασι καὶ τῷ χρώματι συνακολουθεῖ, ἐπεὶ καὶ τῶν ἀνθρώπων τῶν ἐμπύρρων καὶ τὰ τριχώματα γίνεται λευκόπυρρα, τῶν δὲ μελάνων μέλανα· κἂν κατὰ μέρος τι τοῦ σώματος ἐξανθήσῃ λεύκη, καὶ τὰς τρίχας ἴσχουσιν ἅπαντες λευκὰς κατὰ τὸν τόπον τοῦτον, καθάπερ καὶ τὰ ποικίλα τῶν ζῴων. οὕτως ἅπαντα τὰ τριχώματα καὶ τὰ πτερώματα τοῖς δέρμασι συνακολουθεῖ, καὶ τὰ κατὰ μέρος καὶ τὰ κατὰ σῶμα ὅλον.

40 Cf. Ps.-Aristotle, *Col.* 797b26–29 and 798a3–8.

41 Anaxag. B12, transl. Curd (adapted): "each one is and was most manifestly that of which there is the most in it." (ὅτῳ πλεῖστα ἔνι, ταῦτα ἐνδηλότατα ἓν ἕκαστόν ἐστι καὶ ἦν.) See also Simplicius' introduction to Anaxag. B1: "all things are in everything but each is characterized by what predominates." (πάντων μὲν ἐν πᾶσιν ὄντων, ἑκάστου δὲ κατὰ τὸ ἐπικρατοῦν ἐν αὐτῷ χαρακτηριζομένου.) Cf. Gottschalk (1964), 68. A similarly phrased argument appears in the Derveni papyrus (XIX, 1–4 *KPT*), but Betegh (2004), 303–305, highlights the differences with regard to Anaxagoras' ontology. On Anaxagoras, cf. Curd (2007), 178–191, esp. 188.

only imperceptibly (οὐκ αἰσθητῶς, 793b27). Thus, the predominant colour in the blend may accurately represent the colour of the most recent object the light was reflected from, although it is mixed with the colours of light, other surfaces and the air.[42] The author, then, proposes his method to distinguish whether a perceived colour was perceptibly altered through the mixture with other colours: comparing the predominant colour to different situations with varying influences of light, media and reflections, so as to identify the original colour, and to verify it with reference to rules of colour change.

3

The method of comparison and verification can also be applied to determine the natural colours of natural objects, in particular of plants and hair. Since deviations from their natural colours are explained in the many examples in the treatise, the author must assume that natural colours are displayed only at the stage of most complete maturation. Epistemologically, this suggests that colours may contribute to the definition of a natural being, without being a necessary part of it. This is evident from the author's discussion of rabbit fur: naturally, rabbits are brown, but well-fed individuals develop black fur,[43] while malnourished ones turn white.[44] This allows for phenomenological variation within the species, but presumes a correct, natural colour.

Deviations from the natural colour can be analysed in a similar way to the changes in colour appearances at a distance, although natural processes generate natural colours in the opposite direction. Whereas light becomes increasingly mixed as it travels, and thereby loses its natural colour, objects like fruits, plants and hair develop their colouration as they mature. According to the author, fruits generally transition from herb-green to their natural colours through concoction.[45] This transformation is compared to the prepar-

42 Similarly Gottschalk (1964), 72: "The phenomena produced in this way come to an end as soon as the light changes, e.g. if a black feather appears red when the sun shines on it at a certain angle (792a27), it will no longer do so when the sun goes behind a cloud."

43 Cf. Ps.-Aristotle, *Col.* 797b33–35.

44 Cf. Ps.-Aristotle, *Col.* 798a25–b2.

45 Ps.-Aristotle, *Col.* 795a30–b5: "All fruits, as soon as they begin to ripen, change from herb-green to their natural colour (εἰς τὸ κατὰ φύσιν χρῶμα). They become white and black and grey and yellow and blackish and dusky and crimson and wine-coloured and saffron—in fact, assume practically every variety of colour. Since most hues are the result of the blending of several colours, the hues of plants must certainly also be due to the same blends; for the moisture percolating through the plants washes and carries along with it all the

ation of purple dye, with the mixing of multiple colours over time producing the final hue:

> The same thing happens in the process of dyeing with purple; when, after breaking up the shell and washing all the moisture from it, they pour it into earthenware vessels and boil it, at first no definite colour is noticeable in the dye, because, as the liquid boils more and more and the colours still remaining in the vessels mix together, each of the hues gradually undergoes a great variety of alterations; for black and white and brown and hazy shades appear, and finally the dye all turns purple, when the colours are sufficiently boiled up together; so as a result of the blending no other colour is separately noticeable.[46]

On this basis, the author studies variations and incomplete colourations in an attempt to uncover general rules. For instance, in the case of the iris and similar flowers, the petals fail to achieve their natural colour completely because the moisture is insufficiently concocted or does not adequately accumulate everywhere.[47] Thus, the natural colour manifests itself only where perfectly concocted moisture percolates through the entire petal.[48]

However, the author does not discuss how the stages of colour development and a mature state can be discerned, but he does provide many examples. Dates

---

ingredients on which their colours depend." Since the author also distinguishes between single-coloured and multi-coloured plants, but states that single-coloured fruits initially appear herb-green before they mature into their natural colour, single-coloured fruits cannot display just one colour. Uni-coloured plants are those in which the colour of the fruit and the flower are the same. In multi-coloured plants, the colours of the flower and the fruit differ. Cf. Ps.-Aristotle, *Col.* 796b2–21.

46 Ps.-Aristotle, *Col.* 795b11–22: καὶ γὰρ ταύτην ὅταν κόψαντες ἅπασαν ἐξ αὐτῆς τὴν ὑγρασίαν ἐκκλύσωσι, καὶ ταύτην ἐγχέαντες ἕψωσιν ἐν τοῖς χύτροις, τὸ μὲν πρῶτον οὐδὲν ὅλως ἐν τῇ βαφῇ τῶν χρωμάτων φανερόν ἐστι διὰ τὸ κατὰ μικρὸν ἕκαστον αὐτῶν τοῦ ὑγροῦ συνεψομένου μᾶλλον καὶ τῶν ἔτι ὑπαρχόντων ἐν αὐτοῖς χρωμάτων μιγνυμένων ἀλλήλοις πολλὰς καὶ ποικίλας λαμβάνειν διαφοράς· καὶ γὰρ μέλαν καὶ λευκὸν καὶ ὄρφνιον καὶ ἀεροειδὲς καὶ τότε ἅπαν ⟨ἁλουργὸν⟩ γίνεται συνεψηθέντων, ὥστε διὰ τὴν κρᾶσιν μηκέτι καθ' αὑτὸ μηδὲν τῶν ἄλλων χρωμάτων φανερὸν εἶναι.

47 Cf. Ps.-Aristotle, *Col.* 796a31–797a2.

48 Although the author consistently describes the colouration process in these material terms, the role of other materials within the body of plants remains unclear. It is reasonable to assume that different parts of the plant influence the distribution, affect the moisture and thereby determine the specific natural colour of each plant, too—similar to the production of flavour in Theophrastus' *De causis plantarum* 6.6.1–6. However, the author attributes natural colours to the interplay of heat and moisture alone, which suggests that colour is also an accident of the unqualified material moisture.

and grapes transition from wine-coloured to red to wine-coloured again, before becoming dark-blue (κυανοειδεῖς), whereas naturally black or dark fruits (μελάνων καρπῶν) are initially herb-green, then shift through hues of red (ἐπιφοινικίζουσι καὶ γίνονται πυρροί), before also turning dark-blue.[49] Thus, the stages of maturation vary among fruit species, but on the basis of comparisons, the author formulates the general rule that red colours precede the dark ones. This conclusion is verified (δηλοῖ δέ, 796a11) by observing the ground, which changes from herb-green to red to black and dark-blue in shaded, watery areas.[50]

Although this explanation of colour change is odd and, as Prantl and Wöhrle note,[51] apparently reflects the author's ignorance of Aristotle's and Theophrastus' references to the smoke-coloured grapes (κάπνειος ἄμπελος), the concept of black fruits as well as the many examples in the fifth and sixth chapters illustrate that the author has a conception of what the natural colours of certain fruits are and how the mechanisms of colour change can be established, since verification phrases like δηλοῖ or φανερόν ἐστιν appear repeatedly.

As the author thus establishes general rules, the examples provided serve as guides to differentiate between the stages of concoction and colour development and are meant to be practical reference points against which other colour changes can be compared. For instance, to determine the natural colour of a blackberry, one could consult *De coloribus*, review its observations on black fruit and identify the blackberry's stage of maturation and its natural colour. This enables us not only to draw comparisons with other fruits and helps us to classify colour changes in fruits that are otherwise unclear, but also provides a basis for identifying counter-examples, which might require a correction of the author's inductive reasoning in the form of the rejection of certain rules or by introducing *ad hoc* adjustments. As in modern medicine, where patients with the same condition may benefit from different treatments, the author acknowledges the need for the latter, when a rule is not universally applicable. For instance, while well-nourished animals with naturally brown hair usually develop black hair, some animals—such as certain dogs and horses—turn white instead under the same conditions.[52]

---

49 Cf. Ps.-Aristotle, *Col.* 795b22–796a5. The sap of black fruit is wine-coloured, which is presumably the reason why they seem to change colour from red to dark-blue, without becoming wine-coloured in between.

50 Cf. Ps.-Aristotle, *Col.* 796a11–14. This verification was used before as it refers to the stagnant water in reservoirs eventually turning black.

51 Cf. Prantl (1849), 85, and Wöhrle (1999), 91.

52 Cf. Ps.-Aristotle, *Col.* 798b6–13.

4

What clearly matters to the author is to find a way to establish original colours, which might not be the ones that we perceive, and natural colours, which might not yet or not anymore be the ones we see. Yet, the author firmly believes that original and natural colours can be identified by his empirical method of comparison and verification. Thus, the treatise does not aim to present a comprehensive theory of colours, but, judging from the theoretical fragments it contains, rather presupposes one, while its purpose is to serve as a practical guide and toolkit for identifying original and natural colours. In order to achieve this goal, the treatise is packed with examples, and thereby advances the empirical study of colouration through inductive reasoning or the falsification of established rules.

However, the criticisms levelled at this treatise with regard to its underdeveloped theoretical framework are certainly valid. Notably, the theoretical implications of simple and secondary colours, as well as single- and multicoloured plants, have minimal explanatory value and appear merely as appendages. Furthermore, ontological questions about the mixing of colours on light or the significance of a natural colour appear to be of little interest to the author, although they directly pertain to his method and strengthen his observations. Gottschalk suggested that the work's fragmented theoretical foundation and frequent inconsistencies may stem from its basis in lecture notes, possibly taken by a student.[53] If the treatise originated from a lecture series, albeit one not fully transcribed or understood, mainly the empirical part would have survived while the whole lecture series would have integrated these examples into a broader and more coherent theoretical framework of the Peripatetic teacher—whoever that may have been.

---

53 Gottschalk (1964), 82.

CHAPTER 9

Eternalism in Aristotle and in the Post-Aristotelian Tradition

Andrea Falcon

1 Eternalism Defined

"Eternalism" is not a word that is found outside philosophy. In fact, it is one of the many "isms" coined by historians of philosophy to capture a salient feature of a certain style of thinking, or else an important theoretical commitment that resurfaces in various philosophical positions. My first and most urgent task consists in giving the reader at least an idea of what I mean by "eternalism." By so doing, I will also circumscribe my topic in the pages to follow.

"Eternalism", as I understand and use the term in the rest of this essay, entails a commitment to the view that the world has always existed and will always exist in the same way as it does now. In other words, whoever holds the view that the world order (*kosmos*), as we know it, is not subject to generation or destruction but is everlasting is to be regarded as an eternalist. Eternalism so understood is primarily a cosmological doctrine. It can (but need not) imply a commitment to the fixity of the biological species. But as a rule, the biological version of eternalism is parasitic on the cosmological doctrine. So, in the rest of this essay, I will concentrate on eternalism as a cosmological doctrine. The reason for my narrow focus is that it is important to be as clear as possible about the cosmological version of eternalism before turning to its implications for how to think about animals and plants.

While eternalism as defined above may (but need not) depend on a certain metaphysical commitment (more on this front shortly), it is a *physical* doctrine through and through. As such, it falls squarely within the boundaries of physics (*alias* natural philosophy). It must not be confused with a metaphysical stance that emerges for the first time with Parmenides. I have especially in mind fragment 8, where we are told that what is (*to on*) "never was and never will be since it is now all together one".[1] For Parmenides, what is (*to on*) exists outside of time *in a timeless present*. Melissus adopted a creative version of this view

1 DK 28 B 8, 4–5.

when he said, right at the outset of his lost work, "it *always* was whatever it was and *always* will be."[2] This quote makes it clear that Melissus conceived of the Eleatic one as something that is temporally unlimited in the past as well as in the future. By his lights, this thing (whatever it is) exists forever in time rather than outside time as argued by Parmenides. I call this position Eleatic eternalism.[3] Eleatic eternalism is a metaphysical rather than physical doctrine even after the modification introduced by Melissus. Recall what Aristotle tells us in *Physics* I about the Eleatic challenge: Parmenides and Melissus denied the existence of change by positing a single thing that is not subject to change. By doing that, they ended up denying that there is a physical world. Aristotle does not consider Parmenides and Melissus to be failed physicists; rather, he equates them to opponents who were narrowly focused on winning an argument or proving a point with no interest in tracking the truth about the physical world.[4] We may find Aristotle's description harsh, but his point is clear: Parmenides and Melissus removed themselves altogether from the ancient tradition of natural investigation (the *peri physēos* tradition).[5] While their challenge is not without philosophical interest for a natural philosopher, it does not pertain to physics.[6] Accordingly, Eleatic eternalism remains beyond the scope of this essay.

2 Varieties of Eternalism

Eternalism, even when it is defined and circumscribed in the way I have done above, is at most an umbrella term used to cover a set of different views. Claiming that the world is everlasting does not by itself secure that eternalism is a homogenous position.[7] Quite the opposite: I can think of at least three distinct views falling under the label "eternalism." I added in round brackets names next to each position to give the reader a better sense of what I have in mind.

---

2 DK 30 B 1. I am adopting the translation offered by Benjamin Harrison (and defended in Harrison (2019), 56–59).

3 Can we use the label "eternalism" for Parmenides as well? The answer depends on whether we make eternalism contingent upon accepting that eternity must unfold in time. If we accept that eternity can be understood as duration in a timeless present, then we are entitled to use this label in connection not only with Melissus but also with Parmenides. In this case, however, we will have two distinct versions of Eleatic eternalism.

4 Aristotle, *Physics* I 1, 185a8.

5 Aristotle, *Physics* I 1, 184b25–185a1.

6 Aristotle, *Physics* I 1, 185a17–20.

7 I made this point in Falcon (2022), 7–22. What I say here is an attempt to elaborate further on the results reached in that earlier essay.

1. The world is everlasting, out of necessity, both in the past and in the future (Aristotle).
2. The world is everlasting, out of necessity, only in the future (Plato, on a literal reading of the *Timaeus*).[8]
3. The world is everlasting (Critolaus, Ps.-Ocellus).

What I have just outlined is the result we obtain when we distribute two key ingredients relevant to the eternalist position: *directionality* and *necessity*. I will show how this list is generated based on the extant textual evidence. But before I try to fulfil this promise, I would like to stress that there is little or even no awareness of the complexity of this conceptual landscape. In fact, scholars tend to speak of the eternity of the world without further elaborating on what exactly they mean when they ascribe such a position to an ancient philosophical authority. By so doing they are simply following our ancient witnesses, who are less than helpful on this delicate point. Take our main ancient source of information: Philo of Alexandria and his *De aeternitate mundi* (*On the Eternity of the World*).[9] Philo does not clearly distinguish between positions 1 and 3. This is hardly an accident; on the contrary, this lamentable feature of his overall discussion can be traced back to his source (or sources). More on this front in due course. For the time being, let me try to elaborate on the first two positions outlined above by looking at what Aristotle and Plato have to say on the topic of the eternity of the world.

3 Aristotle

Aristotle has come down to us as the champion of eternalism.[10] He is not simply committed to the view that the world does not admit of generation and destruction. His view is that our world has no beginning and no end in time *out of necessity*.

8 On the non-literal reading of the *Timaeus*, Plato ends up being committed to the view that the world is everlasting, out of necessity, both in the past and in the future. On the non-literal reading, his position ends up being close to, but not identical with, the one defended by Aristotle. In the rest of this essay, I will focus on the literal reading because it has given rise to a clearly distinct position from the one endorsed by Aristotle.

9 I take the *De aeternitate mundi* to be a genuine work by Philo of Alexandria (1st century CE).

10 Aristotle has contributed greatly to transmitting the image of himself as the eternalist *par excellence*. Aristotle argues that all his predecessors (*apantes*) maintained that the world has come to be (*De caelo* I 10, 279b12). This claim is contingent on reading Plato's *Timaeus* literally as a creation story. Those who opts for a non-literal reading of this story are free to consider Plato the first eternalist.

Let us take a close look at the textual evidence for this view. The most obvious starting point is *Physics* VIII, where Aristotle is programmatically concerned with the question whether change (or motion, *kinēsis*) is eternal.[11] The enquiry is conducted at the most general level in accordance with Aristotle's requirements of general physics. In other words, Aristotle is not asking whether the circular motion of the heavens is eternal. Nor is he asking whether the cycle of generation and corruption in the sublunary world is eternal. Rather, he is asking whether change *in general* is eternal. The first step in the enquiry consists in establishing *that* this is in fact the case. This step constitutes the *hoti*-stage of Aristotle's investigation. Part of Aristotle's strategy, at this early stage of his argument, consists in exploiting a well-known structural feature of time understood as the number of change. As soon as we take any point of time (a "now"), we thereby establish that is a before and an after that point of time. The same is true for change. As a result, there is no beginning and no end to either time or change. In *Physics* VIII 1, Aristotle establishes that there was never a time when there was no change and there will never be a time when there will be no change.[12] This is enough to establish that the physical world understood as the realm of change has no beginning and no end. This is nothing more (but also nothing less) than an *interim* result. At the same time, it is enough to secure the first conceptual ingredient for the version of eternalism Aristotle endorses, namely bidirectional everlastingness.

But this is far from being the end of the story. As soon as a positive answer to the question whether there is eternal change is in place, Aristotle is obligated, *by his own conception of the scientific enterprise*, to look for an explanation of this fact. The rest of *Physics* VIII is a long, and at times tortuous, search for a scientific explanation of this fact. This is the *dioti*-stage of Aristotle's investigation. There is no need to recall how Aristotle's complicated argument unfolds in the rest of *Physics* VIII.[13] What matters here is that Aristotle *reasons to* the existence of a first principle that is also the ultimate (moving or efficient) cause of change and is not itself subject to change. This principle explains how a continuous and inexhaustible chain of change is possible while it is itself outside that chain of change. More to the point: to perform its causal role, this principle must be without magnitude, parts, and must also be indivisible. This entirely negative description is enough to establish that this principle is not itself part of the physical world. It is best described as a *metaphysical* principle.

---

11 I am relying on the results reached in Falcon (2015), 263–283.

12 Aristotle, *Phys.* VIII 1, 252b5–6.

13 I refer the reader to Falcon (2015), 263–283.

This metaphysical principle is enough to ensure that the world order, as we know it, cannot be otherwise. This is the second conceptual ingredient for the version of eternalism that Aristotle endorses. Aristotle's considered position can be appreciated by turning, briefly, to his *De motu animalium* (*On the Motion of Animals*). In the first part of this short but remarkable treatise, Aristotle is concerned with two distinct, but also closely related, *aporiai*. His discussion of these *aporiai* go a long way toward establishing eternalism as is understood by Aristotle.

The first *aporia* is prompted by the observation that every motion requires a point of rest. Celestial motion is no exception to the rule. The pressing question for Aristotle is where exactly that point of rest is located. To be sure, it cannot be located inside the world. This claim is illustrated with the help of an image: a giant, Atlas, who could set the heavens in motion by supporting himself against the earth. But it does not take long to see that it would be physically impossible for Atlas to support itself on the earth: the earth is too small to perform the function of a point of support. The earth would be displaced from its natural place as soon as Atlas tried to set the heavens in motion.

The discussion of the first *aporia* goes some way toward explaining (by exclusion) why the relevant point of rest must be located *outside* the physical world. At the same time, it insinuates the thought that it would be possible, at least in principle, to displace the earth from the centre of the world if we could generate enough force to do it. More to the point: had we a sufficient force, we could not only displace the earth but also stop the heavens from rotating around the earth. When we reflect on this point, we see that Aristotle is not content with holding the view that the earth occupies the centre of the world because there is no force available that can displace it. Nor is he content with saying that the heavens rotate around the earth because there is no force that could stop them. Rather, he is committed to a stronger, more interesting, and at the same time more controversial, thesis: the arrangement that has the earth at the centre of the world and the heavens rotating is a necessary one. It is Aristotle's commitment to this stronger view this generates the second *aporia*, which can be introduced with the help of the following question: How in the world is it possible to establish the desired conclusion, namely that out of necessity the earth is forever at the centre of the cosmos, and that out of necessity the heavens rotate forever around it? Aristotle stops short of answering this question.[14]

---

14 Aristotle is content with quoting Homer in *De motu* 4:

> But you could not drag from the sky down to earth
> Zeus, the highest counsellor, not even if you worked very hard
> Take hold, all you gods and all goddesses. (*Iliad* 8, 20–22).

Zeus stands, of course, for the relevant metaphysical principle.

However, it is pretty clear that his answer should invoke the existence of a principle that serves as a point of support by being outside the physical world is the solution to the *aporia*. This is nothing else than the principle that Aristotle discovered toward the end of *Physics* VIII.

When we reflect on the second *aporia* and its proposed solution, it becomes clear that the necessity we want, indeed need, to ascribe to Aristotle when we ascribe him the view that the world is bidirectionally everlasting out of necessity is metaphysical rather than physical. We want and need to ascribe him a metaphysical rather than physical necessity because the ultimate source of the eternity of the world is a controlling principle that is envisioned by Aristotle as being removed from the physical world.[15]

Aristotle has a great deal more to say on the topic of the eternity of the world in *De caelo* I 11–12. A full discussion of the complex strategy he adopts in this stretch of text goes emphatically beyond the scope of this essay. Here suffice it to say this much: Aristotle does not deem sufficient to say that something is not subject to generation (*ageñeton*) and is not subject to destruction (*aphtharton*). As a matter of fact, the terms *ageñeton* and *aphtharton*, like their contraries *geñeton* and *phtharton*, can be used in more than one way. Consequently, Aristotle first engages in a conceptual analysis of what we mean when we use those words (*De caelo* I 11), then argues for the view that it is impossible for something that exists forever to be subject to destruction (*De caelo* I 12). By his lights, if *X* is eternal (*X* exists forever), there cannot be a time at which *X* does not exist, and this means that *X* is not subject to destruction without qualification.[16] The arguments linking eternity to necessity are notoriously difficult and open to more than one interpretation. But one feature of those arguments is immediately obvious even when one looks at them from the outside, so to speak: there is no appeal to the way the world is structured and how this structure depends on an external (*alias* metaphysical) principle in this stretch of text. Rather, Aristotle insists on the conceptual truth that eternity entails necessary existence. Put differently, Aristotle tries to establish that what exists eternally does so of necessity on purely conceptual grounds. Whether or not he is able

---

15 This crucial aspect of Aristotle's thought is missed in the most recent analysis of *De motu animalium* 3 and 4 (Morison (2020), 273–298). For the correct interpretation of this passage and its philosophical implications, I refer the reader to the interpretation offered by Klaus Corcilius in Primavesi-Corcilius (2018), 87–95.

16 Aristotle, *De caelo* I 12, 281b25. By adding "without qualification" (*haplôs*) Aristotle wants to concentrate on what is not subject to destruction in the strictest sense of the word. We in fact use the word "indestructible" in a loose sense also for things that are difficult to destroy (*De caelo* I 11, 281a1).

to achieve this remarkable feat is still debated.[17] What matters for my present purposes is only this: Aristotle's arguments in this stretch of text are not only of a whole different type from those offered in *Physics* VIII (and *De motu*), but they also rely on altogether different conception of necessity. I suggest speaking of conceptual necessity for the arguments offered in *De caelo* I 12 so as to distinguish it from the metaphysical necessity we encountered in *Physics* VIII (and *De motu*).[18]

To sum up the results reached so far: when Aristotle is committed to the eternity of the world, his claim entails a non-expendable reference to necessity. Moreover, the necessity in question is either metaphysical or conceptual. To be sure, the source of this necessity is never meant to be merely physical.[19] This is a remarkably strong conclusion. It is also a fitting conclusion for a philosophically ambitious thinker like Aristotle.

4 Plato

The likely account (*eikōs logos* or *eikōs mythos*) offered in Plato's *Timaeus* is open to a literal as well as a non-literal reading. Both readings can be traced back to the Old Academy. Xenocrates, the head of the Academy after Plato and Speusippus, reportedly defended a non-literal reading of Plato's *Timaeus* by arguing that the creation by the divine craftsman is only for the sake of instruction (*didaskalias charin*), just like mathematicians draw diagrams to facilitate their students' understanding.[20] By contrast, Aristotle took the likely story of Plato's *Timaeus* literally. He criticised Plato for holding the view that the world is generated but is not subject to destruction. He describes this position as an impossible one.[21] The impossibility Aristotle has in mind is physical: everything

---

17 For a start, the reader may want to read Judson (1983), 217–255, which a critical review of Waterlow (1982b). The disagreement between the two scholars can be put in the following terms: while Sarah Waterlow thinks that Aristotle has a good argument for the necessary existence of the world, Lindsay Judson thinks that Aristotle uses a bad argument in the service of a deeply felt commitment to the view that the world is eternal out of necessity. See also Mignucci (1990), 323–334, who sides with Lindsay Judson.

18 I would like to thank Pavel Gregorić for pushing me to trying to spell out this difference. I speak of conceptual rather than logical necessity because it seems to me that something is logically necessary if it cannot be otherwise given the rules of logic. But this seems too strong a characterization for what Aristotle is trying to achieve in *De caelo* I 12.

19 *Pace* Morison (2020), 281–286.

20 Aristotle, *De caelo* I 10, 279b32–280a2. Aristotle names no name, but Simplicius tells us that his immediate critical target was Xenocrates (Simplicius, *In De caelo* 303.33).

21 Aristotle, *De caelo* I 10, 279b17–19.

that comes into existence in the natural world goes out of existence sooner or later.[22] As a result, positing that the world comes into existence but does not go out of existence has the unwelcome consequence that there is a natural (physical) object that deviates from the way in which all the other things behave in the natural (physical) world.

At least two observations are in order in connection with Aristotle's criticism. To begin with, it is far from obvious why Aristotle finds it so compelling to saddle Plato with a view he considers untenable. Why not go for the non-literal reading, if adopting this reading saves Plato from holding an implausible thesis? It is not easy to answer this question. My educated guess is that in the absence of clear instructions from Plato on how the reader is expected to receive the likely account of the *Timaeus*, Aristotle felt entitled to take this account literally. To be sure, Aristotle also criticises the non-literal reading.[23] But his criticism is only directed at the analogy with the way in which mathematicians use diagrams. It may work against Xenocrates, if the latter made his non-literal reading contingent upon accepting this analogy, but it cannot rule out other ways to make the non-literal reading, if not compelling, at least plausible. Let us move to the second observation, which is immediately relevant to my present purposes. On any literal reading of Plato's *Timaeus* (*pace* Aristotle), the world is *in principle* liable to destruction. Like everything that is composite, the world too can be dissolved into its constituents.[24] Consider the following passage in which the divine demiurge addresses the lesser gods involved in the creation of the world:

> Oh Gods, works divine whose maker and father I am, whatever has come to be by my hands cannot be undone but by consent. *Now, while it is true that anything that is bound is liable to being undone, still, only one who is evil would consent to the undoing of what has been well fitted together and is in fine condition.* (Transl. D.J. Zeyl, italics are mine)[25]

22 Aristotle, *De caelo* I 10, 279b20–21.

23 Aristotle, *De caelo* I 10, 279b32–280a10.

24 The thesis that all that consists of parts can be taken apart resurfaces elsewhere in Plato. For example, think of the arguments for the immortality of the soul in the *Phaedo*. One of those arguments depends on the thought that the soul does not have parts. But if the soul does not have parts, it cannot be destroyed. By contrast, the body has parts, so it can be destroyed (*Phaedo* 78b–80c).

25 Plato *Timaeus* 41a–b: θεοὶ θεῶν, ὧν ἐγὼ δημιουργὸς πατήρ τε ἔργων, δι᾽ ἐμοῦ γενόμενα ἄλυτα ἐμοῦ γε μὴ ἐθέλοντος. τὸ μὲν οὖν δὴ δεθὲν πᾶν λυτόν, τό γε μὴν καλῶς ἁρμοσθὲν καὶ ἔχον εὖ λύειν ἐθέλειν κακοῦ.

The main idea conveyed in this passage is that the world, like everything else that is composed of more than one thing, would dissolve sooner or later if it were not for its creator. Here is how this idea was received in the doxographical tradition that goes back to Aëtius:

> Plato [says that] the world is perishable as far as nature is concerned; for [he says that] it is sensible because it is corporeal but [says that] it will never be destroyed through the providence and supervision of god. (Aëtius, *Placita* II 4,9)[26]

But why should the divine demiurge, after having created the world, also secure its eternal existence? An answer to this question can be found in the way Timaeus, the spokesperson for Plato, answers the question that is also the starting point and key assumption of the whole likely account. I give both the question and the answer:

> Why did he who frame this whole universe of becoming frame it? Let us state the reason why: *he was good, and one who is good can never become jealous of anything*. And so being free of jealousy, he wanted everything to become as much like himself as was possible. (Transl. D.J. Zeyl, italics are mine)[27]

Timaeus' answer contains not only an explanation for the creation of the best possible world but also for its everlastingness once it has been created. His answer has to do with the goodness of the divine craftsman. But what does it mean for the divine demiurge to be good? By my lights, the goodness in question need not be understood in moral terms. It can, indeed should, be understood in purely epistemological terms. The divine demiurge is a good craftsman because he is an expert who applies the rules of his craftsmanship, which are rules of practical rationality, in the most perfect and beautiful way.[28] Such an ideal agent cannot be a capricious god like the Homeric and Hesiodic gods. While the latter can do or refrain from doing something, and if they do something, they can do it either well or not so well, the divine craftsman *qua*

26 I adapted the translation offered by Jaap Mansfeld and David Runia in Mansfeld-Runia (2020).

27 Plato, *Timaeus* 29e: δι᾽ ἥντινα αἰτίαν γένεσιν καὶ τὸ πᾶν τόδε ὁ συνιστὰς συνέστησεν. ἀγαθὸς ἦν, **ἀγαθῷ δὲ οὐδεὶς περὶ οὐδενὸς οὐδέποτε ἐγγίγνεται φθόνος**· τούτου δ᾽ ἐκτὸς ὢν πάντα ὅτι μάλιστα ἐβουλήθη γενέσθαι παραπλήσια ἑαυτῷ.

28 The gendered language is required since the divine craftsman is envisioned as a father.

craftsman has no choice but to exercise its craft proficiently so as to produce the best possible result. More to the point: he has no choice but to take care of the world once it is created.

A fuller discussion of the line of interpretation that makes the divine craftsman an ideal rational agent who applies the rule of his craft goes beyond the scope of this essay.[29] What matters here is only this: the eternalist position emerging from a literal reading of the *Timaeus* is not so distant from the one adopted by Aristotle. To be sure, the literal interpretation argues that the world is everlasting in one direction only (future). For Aristotle, by contrast, its everlastingness is unlimited in both directions (past and future). In both cases, however, the everlastingness is anchored to a principle that is external to the physical world. In both cases, the world is not contingently everlasting. There remain important differences in how they conceive of this external principle. Aristotle will never accept to ground necessity on goodness and divine craftsmanship. Among other things, the very idea of a divine craftsman would violate the demanding explanation of motion that Aristotle establishes in *Physics* VIII (as well as elsewhere).[30]

5 Critolaus

Critolaus of Phaselis was the champion of the Peripatetic tradition in the late Hellenistic period. As head of the Peripatos, he took part in the diplomatic mission of philosophers to Rome in 155 BCE. In natural philosophy, Critolaus endorsed the Aristotelian thesis that the world is everlasting. The remaining evidence is scarce, but it suggests that the activity of defending this thesis (as well as others) was very important, if not even essential, to Critolaus' identity as a Peripatetic philosopher.[31]

It is, to say the least, unsurprising that a member of the school founded by Aristotle argued for the eternity of the world. After all, we know that Theo-

29 For the identification of the divine demiurge with the rules of craftsmanship as instantiated in a craft, see Johansen (2004), 83–86. Recent (helpful) attempts explore Plato's idea that a craft is a complete and perfect area of expertise can be found in a series of articles by Tamer Nawar: Nawar (2017), 1–20; Nawar (2018), 359–391 and Nawar (2021), 39–61. Whoever has mastered the rules of craftsmanship does not make mistakes (as Trasimachus reminds us in *Resp.* I).

30 For an introduction to this large question, I refer the reader to Menn (2012), 422–464.

31 On this point, see Hahm (2007), 47–101. For a discussion of Critolaus' arguments on the eternity of the world as preserved by Philo, I refer the reader to Sharples (2008c), 55–73.

phrastus defended eternalism.[32] What is surprising, however, is that Critolaus defended a *weaker* version of eternalism. This version is identical with the third position highlighted above. Here are four arguments as they are preserved by Philo of Alexandria. While the first two are explicitly ascribed to Critolaus, the last two are not. But I follow Robert Sharples in considering them part of his carefully orchestrated defence of the view that the natural world is eternal.[33]

> Critolaus, one of those who cultivated the Muses, a lover of the Peripatetic philosophy, supporting the doctrine of the eternity of the world used arguments like these: if the world was generated, it is necessary that the earth too was generated; and if the earth is subject to generation, certainly the human race [will be too]. But the human being is not subject to generation since the race has existed from eternity, as will be shown. So the world too is eternal. (Transl. R.W. Sharples, with modifications)[34]
>
> Critolaus also used an argument like the following one: What is the cause of its own health does not suffer disease; what is the cause of its own wakefulness, too, does not sleep. But what is the cause of its own existence is eternal. But the world is the cause of its own existence, if it is the cause of the existence of all other things. So the world is eternal. (Transl. R.W. Sharples)[35]
>
> It is worth considering this point too: everything that is generated is completely imperfect and only as time progresses does it grow into complete perfection. So if the world were generated, then it too would be—if I can borrow the terms applied to the stages of human life—a mere infant, and then progressing through the revolutions of the years and long stretches of time was at long last, and with great difficulty, brought to perfection. [But we do not see evidence that the world passes through the above stages]. (Translation and amplification are mine)[36]
>
> In addition to all this [Critolaus] says that apart from external causes of death to living beings, there are three causes to which they are subject: disease, old age and defect. He says that the world order cannot be affected by any one of them [an argument follows to show that the world

32 See Philo, *De aet. mundi* 117–149. For a discussion of Theophrastus' arguments and their fortune in antiquity, see Sedley (1998b), 131–142.

33 Sharples (2010), 175–179.

34 Philo, *De aet. mundi* XIV 55 (= Fr. 13 W).

35 Philo, *De aet. mundi* XIV 70 (= Fr. 12 W).

36 Philo, *De aet. mundi* XIV 71.

is not subject to disease, old age, and is self-sufficient so it does not lack anything] (Translation and amplification are mine).[37]

We lack a context for the above arguments, so our ability to evaluate them is seriously limited. The overall impression is that they are not based on any of the theoretical commitments we have discussed so far in connection with Aristotle. This may well be an inevitable feature of the dialectical context in which those arguments were originally offered. All dialectical arguments must be tailored to our interlocutors, so they must start from premises that they are willing to grant us. In this case, we cannot rule out that at least his second argument is an *ad hominem* argument against the Stoic orthodoxy.[38] In addition, we cannot rule out that Critolaus was self-consciously trying to reduce the distance that separated the Peripatetics from their philosophical rivals.[39] Last but not least, Critolaus was presumably updating Aristotle's philosophy to make it acceptable to his audience. Still, this dialectical engagement ends up producing a *new* result: Critolaus is only able to defend a weak version of eternalism, namely the view that the world order has always existed and will always exist, *but not out of necessity*. At most, his arguments succeeded in proving that the world is contingently eternal: it is a contingent fact that the world exists, and it is a contingent fact that it exists forever.

Based on what we read in his *extant* works, we can safely say that Aristotle would have found this conclusion deeply unsatisfying. The qualification "extant" is important. Aristotle defended the eternity of the world also in his lost dialogue *On Philosophy*. Philo of Alexandria is our only source of information for three arguments Aristotle may have employed in this lost work. They can be compressed as follows:[40]

---

37 Philo, *De aet. mundi* XIV 74.

38 Hahm (2007), 83–87 and Sharples (2010), 178–179. I speak of Stoic *orthodoxy* because there is evidence that eternalism made inroads into the Stoic tradition. Panaetius of Rhodes, the most important Stoic philosopher of the second century BCE, adopted a version of this theory (Philo, *De aet. mundi* 76–77). Before Panaetius, Boethus of Sidon and his teacher Diogenes of Babylon had already given up on the Stoic thesis of the universal conflagration (Philo, *De aet. mundi* 76–77 [= *SVF* III, *Diogenes Babylonius* 27 and *Boethus Sidonius* 7]). The extant information is too scanty to speculate on the relation between the version of eternalism defended by Critolaus and the view adopted by the Stoic philosophers who abandoned the doctrine of universal conflagration. But, at least in my view, it is likely that the doctrinal shift within the Stoic tradition is the outcome of dynamics that are all internal to the Stoic tradition.

39 This is the main message I drive home from reading Hahm (2007), 49–101.

40 My compressed presentation of the three arguments is an adaptation of what we read in Sharples (2008c), 61. Philo does not explicitly ascribe the argument to Aristotle, so we can-

> The world cannot be destroyed by anything outside it or by anything inside it; hence, the world is not subject to destruction.[41]
>
> Destruction entails return of parts to their natural places, but the parts of the world are already in their nature places; hence, the world is not subject to destruction.[42]
>
> God has no motive to destroy the world, whether permanently or in order to replace it with another [better] world; hence, the world is not subject to destruction.[43]

In the best-case scenario, the above arguments preserved by Philo go back to the lost work *On Philosophy* via an unknown source. In this scenario, they are at most *vestiges* of Aristotle's original arguments. And yet, one thing is clear: the arguments transmitted by Philo have very little in common with those we can read in Aristotle's extant work. By contrast, they are closer in style and spirit to the arguments that Philo ascribes to Critolaus. In light of this affinity, it is tempting to consider Critolaus an important link in the chain of transmission of these arguments from Aristotle to us. It is quite possible that Critolaus collected these arguments and supplemented them with his own as part of his effort to defend eternalism.[44] But it is remarkable that none of them depend on starting points that are specific to Aristotle's natural philosophy. Furthermore, none of them, taken in isolation from the others, or together with them, can secure the conclusion that the world is not subject to generation or destruction out of necessity. They all fall short of securing the stronger version of eternalism Aristotle defended in his extant works.

I complained that scholars failed to distinguish the version of eternalism Aristotle endorses in his extant writings from the weaker version that the doxographical tradition ascribes to him. I am now able to add a reason for their failure: they all relied, uncritically, on what they found in Philo, who is our most important source of information for eternalism in the ancient world. At this point, however, it should be abundantly clear that what we read in Philo is a selection of arguments that have little or nothing to do with what we read in Aristotle's extant writings on natural philosophy.[45]

---

not rule that they do not go back to him. For a full defense of the Aristotelian provenance of these arguments, I refer the reader to Effe (1970), 7–17.

41 Philo, *De aet. mundi* V 20–24 (= Fr. 19a Ross).

42 Philo, *De aet. mundi* VI 28–34 (= Fr. 19b Ross).

43 Philo, *De aet. mundi* VIII 39–44 (= Fr. 19c Ross).

44 I am following Sharples (2010), 178. I agree with him that Critolaus cannot be the immediate source of Philo, who must rely on a source of information that is not known to us.

45 For more on this point, with a special concentration on the lost work *On Philosophy*, I refer the reader to Falcon (2020), 93–108.

6 Ps.-Ocellus

With this conclusion in place, we can now turn to a little known and little studied work: the treatise *On the Nature of the Universe* attributed to Ocellus of Lucania. We cannot date this work with precision, but both Harder, its modern editor, and Theiler, who reviewed his critical edition for *Gnomon*, dated it to the second half of the second century BCE.[46] If they are right, this work is the earliest extant evidence of the reception of Critolaus' arguments for the eternity of the world outside the Peripatetic tradition narrowly understood. More directly, the work that goes under the name of Ocellus of Lucania contains indisputable evidence that its author appropriated ideas and arguments not only from Aristotle but also the Peripatetic tradition, and in particular Critolaus, for the Pythagorean tradition.

At least the following argument goes back to Critolaus:

> Fire, which imparts the heat to another thing, is self-hot, honey which is sweet to the taste is self-sweet. The principles of demonstrations, which conclude to things unapparent, are self-evident. Therefore, the cause of the perfection of other things is itself perfect. What preserves and renders permanent other things must itself be preserved and permanent. What harmonizes must itself be self-harmonic. Now, as the world is the cause of the existence, preservation, and perfection of other things, the world must itself be perpetual and perfect; and because its duration is everlasting, it becomes the cause of the permanence of all other things. (Transl. K.S. Guthrie)[47]

In addition to the appropriation of arguments for the eternity of the world that go back to the Peripatetic tradition, the author speaking for Ocellus makes use of a physical theory that can be traced back to the second book of Aristotle's *De generatione et corruptione*. He does not acknowledge the original source of his theory, nor does he invoke the authority of Aristotle. Instead, he presents the original material as ideas and arguments put forward by Ocellus. And yet, his reliance on Aristotle is a fact whose significance can hardly be overestimated. The use of ideas and doctrines that can be traced back to Aristotle's extant writ-

46 Harder (1926 [1966²]), 149–150. Cf. Theiler (1926), 585–597, who reviewed the first edition in *Gnomon*.

47 Ocellus Lucanus 13,5–17 (= §9). We know that this argument goes back to Critolaus because Philo (*On the Eternity of the World* 70) credits Critolaus with a version of it. Compare this argument with the second argument I reported above.

ings on natural philosophy are extremely rare in the Hellenistic period to the point that some scholars have concluded that these writings were not known and did not circulate in this period.[48]

But what use does the unknown author who speaks for Ocellus make of the Aristotelian doctrines that he appropriates for Pythagorean tradition? To answer this question, we need to look at the contents of his arguments. We can divide this work into four more or less self-sufficient pieces.[49] This does not mean that our text is just a compilation of materials of various provenance. On the contrary, there is a single overall argument running through the first three pieces and the main thesis that the author defends is in place only at the end of the third piece. It is telling that the third piece ends with a recapitulation of the main results reached up to that point:

> So much about the whole and the universe, the generation and corruption of natures generated in it, of how they subsist, and forever, one part of the universe consisting of a nature which is perpetually moved, and another part being passive and of how the former rules and how the latter is ever ruled.[50] (Transl. K.S. Guthrie, with modifications)

Clearly, the universe is envisioned as a causal arrangement in which one part—the region above the moon—*always* acts on the other. Moreover, this other part—the region below the moon—is *always* acted upon such that generation and corruption go on forever without interruption. The following aspect of the explanatory strategy adopted in this work is especially interesting: no cause from outside the world is invoked to explain the existence of this specific causal arrangement. It really looks as though our author, presenting himself as Ocellus, regards the world as a causally closed system that has within itself the resources to keep itself in existence forever. In other words, the world is conceived as a self-maintaining system.

There are two possible ways of understanding how the unknown author speaking for Ocellus employs Aristotle in his own work. The first is that he uses Aristotle, and only Aristotle, *selectively*. In this scenario, the main textual

---

48 *In primis* F.H. Sandbach, who discusses Ps.-Ocellus in Sandbach (1985), 63–64 because this text poses an obvious challenge to his thesis that Aristotle's writings were not known in the Hellenistic time. Unsurprisingly, Sandbach argues that Ocellus had only indirect access to Aristotle's *GC*. But it is hard to resist the impression that this is a foregone conclusion for Sandback given the thesis that he defends in his book.

49 Piece 1: §§2–17; Piece 2: §§18–37; Piece 3: §§38–43; Piece 4: 44–57.

50 Ocellus Lucanus 21,15–18 = §43.

basis for the picture of the physical world as a self-maintaining causal system is Aristotle's *De generatione et corruptione* (hereafter GC). In addition to relying on (a selection from) GC II 2–3, the author appropriates for Ocellus the thesis that the sun is the *proximate* efficient cause invoked to explain the continuity of generation and corruption.[51] This is a thesis that Aristotle defends in *Gen. Corr.* II 9–10. However, we cannot rule out an alternative scenario in which our unknown author does not use Aristotle but rather ideas and arguments that go back to Critolaus. There is in fact some evidence that may support this second scenario. For one, Critolaus appears to have adopted the view that the physical world is the result of the causal action of the imperishable celestial world (a world made of an intelligent, and indeed rational, element, which he may have called *aithēr*) upon the sublunary world. He seems to have invoked this element to explain not only the imperishability of the heavens but also the Aristotelian view that world is a causally closed system in which the heavens act upon the sublunary world, but they cannot be acted upon by it. If this element is imperishable and intelligent, it is also divine. It may be not just *a* god but rather *the* god listed as one of the two principles (the other is matter) in the doxographical tradition preserved by Epiphanius. All of this would have made the unmoved mover expendable. As a result, the gap between Stoic and Aristotelian physics would be reduced considerably.[52] This hypothesis works well with the suggestion that Critolaus is concerned with updating the Peripatetic philosophical outlook so as to make it acceptable to the philosophical discourse of his own time. Promoting eternalism without relying on a metaphysical principle that is outside of the physical world would be part of this strategy.

In connection with the last remark, it is worth stressing a further possible point of contact with what we know about Critolaus. The latter seems to have adopted the view that the heavens are made of *aithēr*. Our unknown author who presents himself as Ocellus has very little to say on the region above the moon. But toward the end of his second piece, he sums up his position as follows:

> Thus, what generates in something else is in the region above the moon while what generates in itself is in the region below the moon; what con-

51 Ocellus Lucanus 20,1–7 (= § 37 *partim*).

52 For this reconstruction of Critolaus' physical theory, I am relying on Sharples (2002a), 1–40 (especially 14) and Hahm (2007), 82–87. The two key texts for this reconstruction are found in doxographical tradition. They are Epiphanius (*apud* Diels (1879), 592.9–20) and Aëtius (*Placita* II 7,21).

> sists of both of them, *the ever-running divine part* and the ever-mutable generated part, is the cosmos. (Ocellus Lucanus 20,7–11 = § 37 *partim*)[53]

It is tempting to read in the description of the celestial part of the cosmos a reference to the etymology of *aithēr* as ever running (*aithēr* = *aei thein*).[54] This is certainly how Sextus Empiricus (or rather his doxographical source) seems to have understood our text:

> Ocellus of Lucania and Aristotle [hold that all things have come to be] from five: for, along with the four elements, they also adopted the fifth body that moves in a circle, and from which they say that the heavenly objects are derived. (Sextus, *Adversus Mathematicos* x 316, transl. R.G. Bury)

But even if we adopt the second scenario and argue that the conception of the physical world our unknown author appropriates for the Pythagorean tradition ultimately derives from Critolaus, we still have to say that the doctrine of the mutual transformation of earth, water, air, and fire—offered as the explanation for how generation and corruption takes place in the sublunary world—derives from a direct and vigorous engagement with the text of Aristotle's work.[55]

Ps.-Ocellus is an important witness for the transmission of ideas and material that go back not only to the Peripatetic tradition but also to Aristotle himself. Ps.-Ocellus confirms that eternalism, *in its weak version*, was popular and widespread in the late Hellenistic and early post-Hellenistic times. His work is also evidence that Critolaus was an important source for eternalism in the post-Aristotelian tradition. Critolaus may have selected and made available a few arguments that enjoyed great success toward the end of the Hellenistic times beyond the narrow boundaries of the Peripatetic school.

---

53 Guthrie translates as follows: "Different, therefore, is the generator, which is superlunary, while that which is generated is sublunary; and that which consists of both of these, namely, an ever-running body, and an every-mutable generated nature, is the world itself." This conception of the world occurs also in Ps.-Philolaus, *De anima*. In addition to the eternity of the world, both texts are committed to the claim that the world is a self-maintaining system due to its division into a superlunary and a sublunary part.

54 For this etymology, see Plato, *Cratylus* 410b6–7. Cf. Aristotle, *De caelo* 270b20–23. For more on this etymology, as well as its theoretical implications, see Falcon (2005), 113–121.

55 *Pace* Sandback 1985: 63–64, who claims without a single thread of evidence that Ps.-Ocellus relied on a compendium of Aristotelian physics.

7 Ps.-Aristotle, *De Mundo*

The treatise transmitted under the title *Peri kosmou* (Latin *De mundo*) is a protreptic piece meant to promote the study of philosophy over and above the sciences. The work is addressed to Alexander, who is to be identified with Alexander of Macedonia (Alexander the Great). Since Aristotle was the teacher of the young Alexander, the whole work is meant to present itself as an authentic work by Aristotle. Very few today would accept Aristotle's authorship.[56] There is in fact overwhelming consensus that this is a work written by a later author defending what was perceived to be an Aristotelian philosophical outlook.[57] The date of the work is still debated. Pavel Gregorić has offered new arguments to date this work back to the first century BCE. If he is right, this work would be significantly older than most scholars have thought.[58]

Unsurprisingly, eternalism is part and parcel of the philosophical outlook promoted in this work. But given that there is a weaker as well as a stronger version of the eternity of the world, we want to know which one of these two versions is endorsed by the unknown author of *De mundo*. An initial answer can be found toward the end of the treatise, where the author defends the view that the world order consists in the cosmic balance of its constitutive ingredients understood as opposite things. This balance is also invoked as an explanation for the eternal preservation (*sōteria*) of the world:[59]

> And yet, someone indeed wondered how in the world the cosmos consisting of opposite principles—I mean dry and wet, cold and warm—has not long ago been destroyed and perished. [...] But perhaps nature longs for opposites and creates what is consonant (*to symphōnon*) out of opposite things, and not out of similar things [...]. So, in this way, a single harmony (*harmonia*) has arranged the composition out of wholes—I mean heaven and earth and the entire cosmos—by means of a mixture of the most opposite principles. (*De mundo* 5, 396a33–b25, transl. Johan C. Thom, slightly altered)

---

56 Reale and Bos stand out of the most notable defenders of the minority view. See Reale and Bos (1995). Their arguments for Aristotle's authorship are restated by Abraham Bos in his BMCR review of Gregorić and Karamanolis (2020).

57 For a discussion of the majority view and a review of most obvious reasons why this is not likely to be an authentic work by Aristotle, I refer the reader to Gregorić and Karamanolis (2020), 6–8.

58 Gregorić (2025). See also his piece in this volume. Selene Brumana has recently argued for dating the work to the late Hellenistic era (in Brumana (2023), 300–307).

59 For more on this front, see also the essay by Pavel Gregorić in this volume (Sections IV).

We do not know if the author of *De mundo* has a theoretical or a real target in mind in the above passage. There is no shortage of potential targets given that the perishability of the world was defended, with different arguments, by Epicureans and Stoics alike.[60] In the absence of any clear indication on the part of the author, it is safer to take the question that concerns him to be a theoretical one: What licenses the view that the world order is eternal if it consists of a complex structure that emerges out of the combination of its constitutive ingredients? Why not think, instead, that the world order at some point will collapse into its constitutive ingredients? His answer is ingenious: consonance (*symphōnia*) and harmony (*harmonia*) are not only outstanding features of the way things are arranged in the world; they are also what secures the everlastingness of the whole arrangement. To clarify: according to our unknown author, the right kind of ingredients are present in the physical world, they are present in the right proportion so as to be sure that none of them wins over the other, and they are also well organised. It does not take long to see that any defence of the eternity of the world along these lines makes the everlastingness of the world a property that is contingent on the way its constituents are arranged or mixed. But a property of this kind is a physical property that supervenes on the properties of the original ingredients and the special or even unique way in which they are arranged. It is because those ingredients are well arranged that the whole is indestructible.[61] It does not take long to see that any defence of the eternity of the world along these lines ends up making the eternity of the world rest on physical rather than metaphysical principles. In this case, the eternity of the world would be a merely contingent (as opposed to necessary) truth. Such a defence would therefore fall short of the stringent requirements we have seen at work in Aristotle's demanding conception of the eternity of the world.

If this were the last word by the unknown author of the *De mundo* on this front, we would have to conclude that he aligned himself with post-Aristotelian attempts to defend a weaker version of the eternity of the world. He would be in broad agreement with the strategies adopted by Critolaus and Ps.-Ocellus. But this is emphatically not his last word on this front. While he may have deemed his defence sufficient to answer the question raised at the outset of the last passage, his reflection on the causal origin of the physical world does not stop here.

60 For the Stoics the world order is perishable because it undergoes periodic conflagrations. Like all living beings that are part of the world, also the world as a whole must undergo the natural processes of growth and decline. For more on this front, see Gregorić (2025).

61 For a fuller discussion of the defence of the eternity of the world attempted in this passage, see Gregorić (2020), 149–175.

On the contrary, our author goes on with an elaborate and ingenious series of analogies that are meant to make the view that the world order depends on a divine principle plausible.[62] This view is captured by the following passage:

> The most divine characteristic [of the supreme god] would be this: to produce all kinds of forms with ease and a simple movement, just as indeed the engineers do, producing by means of the single release mechanism of an engine of war many different activities. In the same way puppeteers, by pulling a single string make the neck and hand and shoulder and eye and sometimes all the parts of the animal move with a rhythmical movement. So also the divine nature, by a simple movement of the first region gives his power to the next thing and from these again to those further dawn until it permeates the whole (*De mundo* 6, 398b10–19, transl. Johan C. Thom, slightly altered)

This is a remarkable passage for several reasons. For one, the author recalls the image of the puppeteers that we find in Aristotle's *De motu* 6. If this work is to be dated back to the first century BCE, then this would be the oldest evidence for the appropriation and use of Aristotle's *De motu* in the indirect tradition. For another, the author endorses the idea that the supreme god imparts motion to the first heaven and from there indirectly to the world as a whole. The author of this work glosses over many details, so we do not know if he adopts the view that god is an unmoved mover. It has been argued that the text is self-consciously underwritten as the most technical doctrine would not be appropriate for a protreptic work like this one.[63] While this seems right, it is hard to avoid reading the doctrine of the unmoved mover into the divine nature described in this passage. But this also means that the world order and its eternity end up being anchored on a metaphysical principle. But this, in turn, means that the author endorsed a stronger version of the eternity of the world which is line with the view we can extract from Aristotle's extant writings.[64] This is a conclusion that is worth underscoring. Unlike Ps.-Ocellus, our author does not inhabit a liminal space between the Hellenistic and post-Hellenistic

62 For more on this front, I refer the reader to the essay by Pavel Gregorić in this volume (Section IV).

63 Gregorić and Karamanolis (2020), 7. See also the essay by Pavel Gregorić in this volume (Section I).

64 He may have accepted the stronger version as defended in *Physics* VII, but not the one that Aristotle defends in *De caelo* I 12. As Pavel Gregorić reminded me, there is not even a gesture in *De mundo* toward the claim that the world is necessarily eternal in the sense that it could not be otherwise than it is.

times: rather, he belongs squarely to the post-Hellenistic era. Quite tellingly, he chose to bypass the Hellenistic debates on and around the eternity of the world altogether.[65]

---

65 I am pleased to acknowledge the expert feedback I received from Pavel Gregorić on an earlier version of this essay.

CHAPTER 10

Physics and Eternity of the Cosmos in the Pseudo-Aristotelian *De Mundo*

Pavel Gregorić

1

It is not self-evident that a chapter on *De mundo* belongs to this volume, since we do not know for sure who composed it, when and why. There are very few scholars who believe that it was written by Aristotle himself, some time after his most famous pupil, Alexander of Macedon, rose to prominence as a military leader, that is between 336 and 323 BCE.[1] Doctrinal and linguistic evidence against such attribution is very strong, if not overwhelming. Most scholars would agree that *De mundo* was written by an anonymous author at some point between the third century BCE, when the Stoic school started to dominate the intellectual scene, and the second century AD, when the novelist and Platonic philosopher Apuleius of Madaura wrote his Latin adaptation of the same title. To be more fine-grained, some scholars put the date in the earlier half of that period, say around 200 BCE, whereas others push it to the turn of the eras, largely depending on their judgement about the parallels with Posidonius in the meteorological chapter of *De mundo*. The parallels are indeed striking, but they are not conclusive, because the order of authorship could go both ways. However, there are additional reasons to believe that *De mundo* was composed shortly after Posidonius, in the latter half of the first century BCE, when Aristotle again became an authority, someone whose doctrines required explication and defence against rival schools of thought.[2]

Although some scholars have argued that the treatise is an exercise in eclecticism,[3] the predominant view in recent scholarship seems to be that it was

1 Reale and Bos (1995). Overview of reasons against the Aristotelian authorship of *De mundo*, with some bibliography, can be found in Gregorić and Karamanolis, eds. (2020), 6–7.

2 Further arguments for this dating can be found in Gregorić (2025).

3 Zeller (1923), 670 characterises *De mundo* as "ein merkwürdiges Denkmal des Eklekticismus", repeated by Festugière (1949), 513: "un monument typique du dogmatisme éclectique". For the concept of eclecticism in general, and its meaning since Zeller in particular, see Donini (1988).

written in defence of Peripatetic philosophy, cosmology and theology.[4] More precisely, what we find in *De mundo* is a Peripatetic vision of philosophy as a grand project dedicated to understanding the cosmos, and a broadly Peripatetic picture of the structure of the cosmos and its relation to God, or so I shall argue in this paper. Assuming that the treatise was written at some point in the second half of the first century BCE, and in defence of Peripatetic philosophy, I believe that a chapter on this treatise very much belongs to this volume.

Interestingly, the author of *De mundo* thought that the best way to promote Peripatetic tenets was to pass the treatise as Aristotle's own. *De mundo* is directed to an "Alexander" (391a2), who is said to be "the best of leaders" (391b6). This Alexander is most probably meant to be identified with Aristotle's pupil, Alexander of Macedon, contributing to the impression that the treatise was composed by Aristotle himself. And what a captivating impression it is! One of the greatest philosophers in the world writes to one of the greatest military leaders of all times, and *De mundo* makes the reader privy to that exchange of titans. The impression is reinforced by the fact that the author espouses some *bona fide* Aristotelian doctrines, taking care not to quote from post-Aristotelian authors, to refer to events after 322 BCE, or to disclose its inauthenticity otherwise. This was not uncommon in the Hellenistic and early Imperial period. As Sir Ronald Syme wrote, "philosophical sects had fostered the production of *pseudepigrapha*. The purpose was to defend, expound (or even modify) the doctrines of a school or teacher".[5] Certainly the author of *De mundo* knew the doctrines of Aristotle well and probably wrote this little treatise in order to expound them—possibly aiming at a wider educated audience, which was rapidly growing at the time—and to defend it, primarily against the Stoics, but also against the Epicureans and the Platonists. In the course of doing so, the author of *De mundo* modified the Aristotelian doctrine, as we shall see. Apart from knowing his Aristotle well, the author was clearly conversant with the doctrines of the other philosophical schools, he held Plato in highest regard, and he was a master of Greek prose style, which all together made him confident that he was up to the task of writing a treatise that is supposed to pass as Aristotle's.

Whether the author was a card-carrying member of the Peripatetic school tasked with promotion of Aristotelian doctrines, a maverick champion of Peripatetic philosophy, or an immensely gifted forger intent on making a fortune by selling the manuscript of "Aristotle's letter to Alexander" to the highest bidder among ancient libraries or private collectors, it is difficult to say. Indeed,

4 See Federspiel (2018), Gregorić and Karamanolis, eds. (2020), Brumana (2023).

5 Syme (1972), 9.

it is unlikely that we will ever be able to identify the author with certainty. It is fortunate, however, that the authorship and date do not pose as interesting questions as the contents of *De mundo*.

A quick overview of the contents is in place here. Chapter 1 is an introduction with praise of philosophy as the only discipline which has exalted itself to the study of the cosmos. Chapter 2 starts with two definitions of the cosmos and gives an overview of its shape and structure, beginning with the heavens and stars made of aether. Under the lowest aethereal sphere, that of the moon, comes the sphere of fire, and then the sphere of air. Chapter 3 describes the lowest sphere of water and earth, explaining how the Ocean surrounds the continents, followed by a geography of the inhabited world. Chapter 4 is dedicated to meteorology. Assuming the doctrine of two exhalations, it overviews meteorological phenomena produced by wet exhalation (clouds, rain, snow, etc.) and then by dry exhalation (winds, thunders, lightnings etc.). The interaction of the elements under earth and water explains earthquakes, volcanoes, geysers, springs, etc. This chapter uses a lot of technical terms and definitions, as if compiled from some compendium of meteorology, possibly by Posidonius.[6]

Chapter 5 marks a difference in both style and content. It shows that the interaction of the elements, far from causing destruction of the cosmos, keeps it forever in existence, because the elements are well-balanced so that none prevails over the others. Chapters 6 and 7 are theological. Chapter 6 explains that God is the sustaining cause of the cosmos, but not by being inside the cosmos and governing it himself, directly; rather, God governs the cosmos indirectly, through a *power* that permeates the whole cosmos. The divine power causes celestial revolutions, and as it descends towards the centre of the cosmos, it weakens, so that order decreases with distance from the periphery. However, the divine power does reach the terrestrial region, ensuring regular cycles of generation and destruction. Most of Chapter 6 consists of a chain of analogies that are supposed to make the reader amenable to the idea that God can be causally active in the world without being in it, which was most probably directed against Stoic immanentism. The final chapter, Chapter 7, is an exercise in etymology of various names or attributes of Zeus, indicating the variety of effects that God produces in the world.

---

6 Posidonius' elementary treatise on meteorology is mentioned by Diogenes Laertius VII 138; cf. Frs. 14 and 15 (Edelstein-Kidd). The definition of the cosmos at 2, 391b9–10, the definition of rainbow at 4, 395a32–35, and the classification of earthquakes at 4, 395b36–396a14 might have been borrowed from there.

2

From all that has been said so far, it is hardly surprising that *De mundo* does not contain an enquiry into the principles of natural philosophy or a discussion of the concepts of nature, change, place, time, etc., familiar from Aristotle's *Physics*. More to the point, in *De mundo* we do not find Aristotle's standard philosophical repertoire of the four causes, form and matter, potentiality and actuality. In the cosmological part of *De mundo*, there are no multiple concentric spheres and counter-revolving spheres, probably because the author makes no note of the retrograde motion of planets. He only mentions the order of planets and their periods. The order largely corresponds to Aristotle's in the *Metaphysics*, except that Aristotle has Venus and Mercury in descending order, whereas *De mundo* has Mercury and Venus.[7] The periods of the planets, presented in Chapter 6, 399a6–12, correspond to those of Eudoxus.[8] In the theological parts, there is no concept of God as the activity of thinking of thinking, nor the concept of causing motion as an unmoved object of desire. A remark to the effect that God is "invisible to anything other than thought" (6, 399a33) is as close as we get to Aristotle's description of the prime mover in *Metaphysics* XII 7 and 10.

Nevertheless, what we do find in the text, stripped to the bare bone, is a recognisably Aristotelian picture of the cosmos. The universe is one and it is a genuine *cosmos*, i.e. an ordered whole marked by regularity and outstanding beauty. This ordered whole manifests a clear division into two domains, the supralunar domain made of aether, in which things are free from generation and destruction, and the sublunar domain made of the four elements, in which things are subject to generation and destruction, being exposed to environment replete with all sorts of phenomena and processes, some of them rather unpredictable and destructive, even catastrophic. However, all these phenomena and processes somehow contribute to the generation and nourishment of other things, so that the sublunary domain is stable in the long run. Thus, the ordered and dynamic whole is eternal, despite the impermanence of individual things and destructiveness of local events that we witness in the sublunary sphere.

Most importantly, this ordered and dynamic whole is sustained by something external to it but not removed from it. The author locates God at the

7 As Furley notes when the author of *De mundo* first introduces the planets in their order from the highest to the lowest one (2, 392a23–30): "This is the 'Pythagorean' order of planets, adopted by Aristotle, Eudoxus, Eratosthenes, and probably the early Stoics" (Furley 1955, 352 n. *a*).

8 See Simplicius, *In De caelo* 495.20–29 and Dicks (1970), 186.

periphery of the cosmos: "He has been allotted the highest and first place, and is therefore called supreme, established according to the poet 'on the highest peak' of the whole heaven" (6, 397b24–27). A little later, the author repeats that God is "based at the highest place" (6, 398b7). It is unclear whether God, according to our author, has a particular spatial location on the periphery of the cosmos, or if he is the sort of entity that cannot have spatial location except accidentally, like a soul or a law "being immovable in the souls of those who use it",[9] or indeed if he is the sort of entity that cannot have spatial location even accidentally, like an immaterial substance and pure actuality, as Aristotle argues in *Physics* VIII and *Metaphysics* XII. The author does not go to that level of detail. Nor does he go into any detail when it comes to the sort of causal efficacy that God exerts in the world. What the author says is that "his power, pervading through the whole cosmos, moves the sun and moon and causes the whole heaven to revolve and is the cause of preservation for the things on earth" (6, 398b8–10). However, it is unclear how exactly this power sets in motion the highest heaven of the fixed stars. Again, the author glosses over the details, probably unwilling to complicate matters and thus repel the wider audience with technical details that would call for further explanation.

This should suffice to convince the reader that *De mundo* indeed features a simplified, but recognisably Aristotelian cosmology. Let us now look at the author's account of the four sublunar spheres. In Chapter 2, we read:

> The substance of heaven and the stars we call "aether", not, as some would have it, because it "burns", being fiery—erring about its nature, which is far removed from fire—but because it "always runs", being carried around in a circle, as a different element from the four, pure and divine. (2, 392a5–9)[10]

First, this passage contains a piece of criticism of all those who think that stars are made of fire—which includes Anaxagoras and some Peripatetics such as Strato of Lampsacus and Xenarchus of Seleucia[11]—and also those who agree that stars are made of aether but conceive of aether as something akin to fire,

9 The analogy of God with the soul is proposed in Chapter 6, 399b12–19, whereas the law comes last, as the culmination of the series of analogies, at 400b11–24. For the structure of the analogies in Chapter 6, see Betegh and Gregorić (2014).

10 Οὐρανοῦ δὲ καὶ ἄστρων οὐσίαν μὲν αἰθέρα καλοῦμεν, οὐχ, ὥς τινες, διὰ τὸ πυρώδη οὖσαν αἴθεσθαι, πλημμελοῦντες περὶ τὴν πλεῖστον πυρὸς ἀπηλλαγμένην δύναμιν, ἀλλὰ διὰ τὸ ἀεὶ θεῖν κυκλοφορουμένην, στοιχεῖον οὖσαν ἕτερον τῶν τεττάρων, ἀκήρατόν τε καὶ θεῖον. All translations are by Johan Thom, often with some modification.

11 For Anaxagoras, see Aristotle, *De caelo* III 3, 302b4–5 and *Meteorologica* I 3, 339b16. For

which seems to include the Stoics, since the Stoics spoke of aethereal fire, or aether, in connection with the heavens and celestial bodies.[12]

Be that as it may, the passage formulates the flagship tenet of Aristotelian philosophy: aether is the element of which the heavens and the stars are made, it is a different element from the other four, it is pure—in the sense that it does not combine with the other four elements—it engages in unceasing circular motion, and it is divine, probably because it is not subject to generation and destruction or to any sort of external affection or disruption. As the author writes a little latter, it is "unchangeable, unalterable and unaffected" (ἄτρεπτον καὶ ἀνετεροίωτον καὶ ἀπαθῆ, 392a33). As such, aether "contains the divine bodies and the order of their motions" (392a30).[13]

After the aethereal domain follows the domain which is "throughout subject to change and alteration, and, in a word, destructible and perishable".[14] The top-most layer of that domain is made of fire. The author writes:

> The subtle and inflammable substance, which is ignited by the aethereal element because of its magnitude and its quick motion. In this so-called "fiery" and unordered element lights rush across and flames flash; and "beams" and "pits" and "comets", as they are called, are often fixed in place and then extinguished. (2, 392a34–b5)[15]

Here we learn about the element of fire, which constitutes the second nested sphere, right under and in contact with the sphere of aether. Fire is said to be made of fine particles and inflammable, actually ignited by aether, on account of "its magnitude and its quick motion". I suppose it is by some sort of friction that aether sets celestial fire ablaze, local manifestations of which are different phenomena in the high atmosphere. One thing to note here is that the element of fire is not described in terms of its elementary qualities or in terms which

---

Strato, see Fr. 42 (Sharples, in Desclos and Fortenbaugh [eds.] 2011), 104–106. For Xenarchus, see Falcon (2012), 37, 89, 163.

12 *SVF* I 102, II 504, 527, 571, 668, 684.

13 It should be noted, however, that Aristotle himself rarely uses the word αἰθήρ for the celestial element. Instead, he speaks of "the first body" or "the first element"; cf. Falcon (2005), 113–121.

14 This order of the explanation of the cosmos, from the top down to the centre, is in line with the order of study envisioned by Aristotle himself; see *Meteorologica* I 1.

15 Ταύτης δὲ αὐτῆς πρώτη μέν ἐστιν ἡ λεπτομερὴς καὶ φλογώδης οὐσία, ὑπὸ τῆς αἰθερίου πυρουμένη διὰ τὸ μέγεθος αὐτῆς καὶ τὴν ὀξύτητα τῆς κινήσεως· ἐν δὲ τῇ πυρώδει καὶ ἀτάκτῳ λεγομένῃ τά τε σέλα διᾴττει καὶ φλόγες ἀκοντίζονται καὶ δοκίδες τε καὶ βόθυνοι καὶ κομῆται λεγόμενοι στηρίζονται καὶ σβέννυνται πολλάκις.

would otherwise explain the behaviour of fire universally, that is, in the upper layers of the atmosphere as well as in fire-places on the ground.

Contained within the sphere of fire comes the sphere of air:

> Next to this, the air is spread out below, opaque and frosty in nature, but when it is illuminated and at the same time heated by the former element, it becomes bright and warm. In this ⟨element, i.e. the air⟩, which also has a capacity to be affected and undergoes alterations in all sorts of ways, clouds come together and rain showers fall down, also snow-flakes, frost and hail-stones, blasts of wind and whirlwind, further thunder and lightning and falling thunderbolts, and the collision of countless storm-clouds. (2, 392b5–13)[16]

So, the sphere of air is marked by transient phenomena such as clouds, rain, snow, winds, thunder and lightning. Again, the element of air is not described in terms of elementary qualities, but in terms of second-order properties of darkness and frostiness, which are replaced by the familiar properties of brightness and warmth only under the agency of fire in the next higher sphere.

It is interesting to note, as many interpreters of *De mundo* have done, that air is said to be dark and cold by nature, which is precisely what the Stoics claim.[17] According to Aristotle, by contrast, air is by nature hot. What is worse, if air were cold, the distribution of the four elementary qualities among the four elementary bodies would be disrupted and the transformation of elements would become impossible. That is to say, if air were cold by nature, and presumably such are also earth and water, then air could not be distinguished from earth and water by adding the quality of wetness or dryness (because earth is cold and dry, whereas water is cold and wet), and the Aristotelian scheme of pairs of elementary qualities for each element would be undermined. However, the

---

16 Ἑξῆς δὲ ταύτης ὁ ἀὴρ ὑποκέχυται, ζοφώδης ὢν καὶ παγετώδης τὴν φύσιν· ὑπὸ δὲ ἐκείνης λαμπόμενος ἅμα καὶ διακαιόμενος λαμπρός τε γίνεται καὶ ἀλεεινός. Ἐν δὲ τούτῳ, τῆς παθητῆς ὄντι καὶ αὐτῷ δυνάμεως καὶ παντοδαπῶς ἀλλοιουμένῳ, νέφη τε συνίσταται καὶ ὄμβροι καταράσσουσι, χιόνες τε καὶ πάχναι καὶ χάλαζαι πνοαί τε ἀνέμων καὶ τυφώνων, ἔτι τε βρονταὶ καὶ ἀστραπαὶ καὶ πτώσεις κεραυνῶν μυρίων τε γνόφων συμπληγάδες.

17 See *SVF* II 429–431 and Maguire (1939), 123–124. For possible reasons as to why the Stoics argued that air is cold and dark, see Plutarch's *De primo frigido* Chs. 9–10 (948d–949b). In Ch. 16 (952d) Plutarch suggests that Chrysippus maintained that the air is cold and *therefore* also dark. For the suggestion that the author of *De mundo* might be following Theophrastus' doctrine of the elements, of which we have rather patchy knowledge, see Moraux (1984), 14–15. Reale and Bos (1995), 130–131 go as far as to argue that Aristotle himself subscribed to the view that air is cold by nature.

author does not seem to operate with the concept of transformation of elements. He neither mentions it nor implies it anywhere in the text, although he does bring up the four elementary qualities later in Chapter 5. I will return to this point shortly.

In Chapter 3 the author proceeds to the spheres of water and earth, but he says nothing about their properties as elements. Rather, he focuses on second- or even third-order properties of being "filled with plants and animals", on their geographic distribution, and on the immobility of the earth in the middle of the cosmos. "All of earth," he says, "being firmly set and tightly compacted, exists in the depths right in the middle of the cosmos, immovable and unshakeable" (3, 392b32–34).

Here is how the author concludes the overview of the structure of the cosmos, before moving on to geography:

> These five elements, then, situated in five spherical regions, the smaller always being encompassed by the larger—I mean, earth within water, water withing air, air within fire, and fire within aether—make up the whole cosmos, and they make the whole upper part into a dwelling for the gods and the lower part into a dwelling for transient living beings. (3, 392b35–393a5)[18]

In Chapters 2 and 3, then, the focus is not on the elements as such, but on the constitution of the five nested spherical regions that make up the cosmos. The author describes the properties of the elements relevant for his description of the basic structure of the cosmos, not the properties of the elements which would explain the constitution and behaviour of things around us. In other words, the author's perspective in Chapters 2–4 when he speaks about the elements is *cosmological*, not one of elementary physics.

The main things we learn about the elements in Chapters 2–4 of *De mundo* is (i) their distribution in nested spheres; (ii) the fact that whatever is made of aether is ungenerated, immutable and divine, whereas things occurring in the four lower spheres are subject to generation, destruction and mutation; (iii) the fact that the elements constituting the nested spheres are mostly passive, receiving their properties from the agency of the higher element. With the exception of aether, which is not moved by any higher elemental body, but

18 Πέντε δὴ στοιχεῖα ταῦτα ἐν πέντε χώραις σφαιρικῶς ἐγκείμενα, περιεχομένης ἀεὶ τῆς ἐλάττονος τῇ μείζονι—λέγω δὲ γῆς μὲν ἐν ὕδατι, ὕδατος δὲ ἐν ἀέρι, ἀέρος δὲ ἐν πυρί, πυρὸς δὲ ἐν αἰθέρι—τὸν ὅλον κόσμον συνεστήσατο, καὶ τὸ μὲν ἄνω πᾶν θεῶν ἀπέδειξεν οἰκητήριον, τὸ κάτω δὲ ἐφημέρων ζῴων.

either on account of its nature (2, 392a5–9) or by the "single power which penetrates through all things" (5, 396b29 and 6, 397b23–25, 398b7–11), fire is set ablaze by aether, air is lit up and warmed by fire, seas are put into motion by winds, and land-masses are shaped by the flow of the Ocean. So, there is a downward causality in the cosmos, that is from the topmost periphery to the centre, which intercepts the possible objection that the cosmos, according to the Peripatetic conception, is episodic.[19] Although there is a clear division between the supralunar and the sublunar domain, the cosmos is unified by downward causation effected by the "single power that penetrates all things", which is said to originate from God.[20]

From the passages surveyed so far, we learn precious little about the essential properties of the elements, nothing about their combination in the sublunary world and still less about their transformation into one another. Indeed, given that air is said to be cold by nature, it is questionable whether the author can even accommodate the Aristotelian theory of elemental transformation.

3

The author mentions mixing of the elements for the first time in the conclusion of the meteorological Chapter 4:

> To summarize, since the elements are mixed with one another, it is reasonable that similar phenomena occur in the air, on earth, and in the sea, which bring about individual destructions and generations, but keep the whole indestructible and ungenerated. (396a27–32)[21]

A closer look at this passage shows that it is not about mixing of the elements in particular ratios, so as to produce inorganic stuffs or uniform parts of organisms. The passage is, rather, about the *interaction* of elements that bring about earthquakes, chasms, eruptions, as there are pockets of fire and air trapped inside the earth and under the sea.

---

19 Aristotle argues that the universe is episodic in Speusippus' philosophy: *Metaphysics* XII 10, 1075b37–1076a3 and XIV 3, 1090b19–20. See Theophrastus, *Metaphysics* 2, 4a9–17. The author of *De mundo* seems to me to be impervious to that criticism, adhering to Aristotle's account of the natural world as a top-down causal system.

20 More on this power in Section 4 below.

21 Ὡς δὲ τὸ πᾶν εἰπεῖν, τῶν στοιχείων ἐγκεκραμένων ἀλλήλοις ἐν ἀέρι τε καὶ γῇ καὶ θαλάσσῃ κατὰ τὸ εἰκὸς αἱ τῶν παθῶν ὁμοιότητες συνίστανται, τοῖς μὲν ἐπὶ μέρους φθορὰς καὶ γενέσεις φέρουσαι, τὸ δὲ σύμπαν ἀνώλεθρόν τε καὶ ἀγένητον φυλάττουσαι.

The talk about mixing of elements continues in Chapter 5. The chapter opens with the following statement:

> Some have indeed wondered how the cosmos, consisting of opposite principles, I mean dry and wet, cold and hot, has not long ago been destroyed and perished—just as one would wonder how a city endures, since it consists of opposite classes, I mean of poor and rich, young and old, weak and strong, bad and good. (5, 396a33–34)[22]

The opening statement is significant because the four Aristotelian elementary qualities are here mentioned for the first time together, and precisely in the capacity of "principles" (*archai*) which stand in mutual opposition. The transition from the "elements" (*stoicheia*), mentioned at the end of Chapter 4, to the "opposite qualitative principles" at the beginning of Chapter 5 is quite abrupt, but it can be safely assumed that the opposite qualitative principles are instantiated by the elements. However, the claim at 2, 392b5 that air is cold by nature should warn us against assuming that the author operates with Aristotle's theory of elements, according to which each element instantiates *two* qualitative principles. I will return to this issue presently.

The question that our author addresses in the last passage is the following: if the cosmos is constituted of opposite principles, how come that it does not fall apart? This question might be viewed as a challenge from the Epicureans, possibly also the Stoics, and it must be adequately answered by a Peripatetic who holds the cosmos to be eternal. So, how does the author address this question? Without going into the details of the first part of Chapter 5, the author's strategy is to adduce instances of marvellous and lasting things achieved by combining opposites. The first instance is from politics, where "civic concord" (πολιτικὴ ὁμόνοια) accomplishes a stable government out of different classes and walks of life. The second instance is from nature, where individuals of opposite sex secure the survival of the species, and the next three instances are from arts, where painting achieves wonderful things by mixing opposite colours, music by combining opposite tones, and grammar by "creating a mixture of letters with and without sound, that is vowels and consonants". *Krasis* and *mixis* are

22 Καίτοι γέ τις ἐθαύμασε πῶς ποτε, ἐκ τῶν ἐναντίων ἀρχῶν συνεστηκὼς ὁ κόσμος, λέγω δὲ ξηρῶν τε καὶ ὑγρῶν, ψυχρῶν τε καὶ θερμῶν, οὐ πάλαι διέφθαρται καὶ ἀπόλωλεν, ὡς κἂν εἰ πόλιν τινὲς θαυμάζοιεν, ὅπως διαμένει συνεστηκυῖα ἐκ τῶν ἐναντιωτάτων ἐθνῶν, πενήτων λέγω καὶ | πλουσίων, νέων γερόντων, ἀσθενῶν ἰσχυρῶν, πονηρῶν χρηστῶν. For an interpretation of Chapter 5 as a whole, see Gregorić (2020). The present paper builds on that piece, but also moves beyond it.

the operative words in most of these cases. These two words figure prominently in a passage which wraps up the author's idea. For convenience, I divide the passage in two sections:

> [i] In this way, then, a single harmony has arranged also the composition of the wholes, I mean heaven and earth and the entire cosmos, by means of the mixture of the truly opposite principles: dry mixed with wet, warm with cold, heavy with light, straight with curved; [ii] a single power pervading all things has set in order all the earth and sea, aether, sun, moon and the whole heaven, having fashioned the whole cosmos from the unmixed and diverse—from air and earth and fire and water—and by holding them with the single surface of a sphere, compelled the truly opposite natures in the cosmos to agree with one another, and from these brought about preservation for the whole. (5, 396b23–34)[23]

In section (i), we learn that the cosmos as a whole, and each one of its large regions, is an outcome of mixing the opposite principles. The list of the opposite principles is now expanded: apart from dry and wet, warm and cold, we now hear of heavy and light, and straight and curved.

Heavy and light are treated as opposites on an equal footing, which is true to Aristotle, except that he envisaged heavy and light to be second-order properties of elements, presupposing the theory of natural places. I am inclined to read the heavy-light opposition here with reference to the natural tendencies of the four elements, in which case this passage contains a trace of Aristotle's theory of natural places. As far as I am aware, this is the only trace of that theory in *De mundo*, apart from the fact that each nested spherical region is made of one element. In any case, against Epicurus and Strato of Lampsacus, who gave precedence to weight and made lightness a derivative property,[24] the author of *De mundo* seems to agree with Aristotle in taking heavy and light as qualities of the same order.

As for straight and curved, this is also true to Aristotle's division of simple local motion from *De caelo* I 2, and it should probably be read with reference to

---

23 Οὕτως οὖν καὶ τὴν τῶν ὅλων σύστασιν, οὐρανοῦ λέγω καὶ γῆς τοῦ τε σύμπαντος κόσμου, διὰ τῆς τῶν ἐναντιωτάτων κράσεως ἀρχῶν μία διεκόσμησεν ἁρμονία· ξηρὸν γὰρ ὑγρῷ, θερμὸν δὲ ψυχρῷ, βαρεῖ τε κοῦφον μιγέν, καὶ ὀρθὸν περιφερεῖ, γῆν τε πᾶσαν καὶ θάλασσαν αἰθέρα τε καὶ ἥλιον καὶ σελήνην καὶ τὸν ὅλον οὐρανὸν διεκόσμησε μία ἡ διὰ πάντων διήκουσα δύναμις, ἐκ τῶν ἀμίκτων καὶ ἑτεροίων, ἀέρος τε καὶ γῆς καὶ πυρὸς καὶ ὕδατος, τὸν σύμπαντα κόσμον δημιουργήσασα καὶ μιᾷ διαλαβοῦσα σφαίρας ἐπιφανείᾳ τάς τε ἐναντιωτάτας ἐν αὐτῷ φύσεις ἀλλήλαις ἀναγκάσασα ὁμολογῆσαι καὶ ἐκ τούτων μηχανησαμένη τῷ παντὶ σωτηρίαν.

24 See Simplicius *In De caelo* 269.4–7 and Strato Frs. 26A, 49, 50A, 50B, 50D (Sharples).

the natural motion of the elements—aether moving in circle, the other four elements moving up or down in straight line. As we know, Xenarchus challenged not only Aristotle's theory of aether, but also the straight-curved division of simple motions,[25] so this passage of *De mundo* looks like an affirmation of the original Aristotelian position against the unorthodox Peripatetics.

An important thing to observe about section (i) is the author's insistence on mixtures *within* pairs of opposites: dry is mixed with wet and warm is mixed with cold. How should we understand this? Let us suppose the author has elementary qualities in mind. First, it is not clear that Aristotle himself subscribes to the idea that *qualities* mix, though they do combine to constitute the elements.[26] Second, in Aristotle's theory qualities do not combine *within* the pairs of opposites, but *across* them. It is not dry that is mixed with wet, and warm that is mixed with cold; rather, dry and moist combine with warm and cold to generate the elements. So, for all the talk of the four Aristotelian elementary qualities and their "mixing", the author does not seem to operate with the idea that the elements are constituted by combinations of these four elementary qualities. What he seems to have in mind, rather, is that "dry, wet, warm and cold" pick out the elementary bodies according to the quality that they instantiate, i.e. earth which is dry, water which is wet, fire which is hot, and air which is cold. This is confirmed by the silent transition from the "truly opposite principles: dry, wet, warm and cold" in section (i) to "air, earth, fire and water" in section (ii). The latter are said to be "unmixed and diverse" (ἄμικτα καὶ ἑτεροία, 396b29), which suggests that they instantiate diverse and indeed opposite qualities, and that they preserve their identities through various interactions, without ever transforming into one another. I would conclude, therefore, that it is not elementary *qualities* that are basic, according to our author, but the elementary *bodies* which instantiate each one elementary quality. In other words, our author operates with a one quality-one element theory, which seems to be borrowed from the Stoics. As Diogenes Laertius writes, summarising the Stoic theory of elements: "Fire is the hot element, water the moist, air the cold, and earth the dry element."[27]

25 See Simplicius, *In De caelo* 20.10–36 and 21.32–22.17 (Texts G and H in Sharples 2010, 182 and 183).

26 Later ancient philosophers argued that it was qualities, not substances that mix. See, e.g., Sextus Empiricus *Pyrrhoniae Hypotyposes* III 5; Plotinus: *Enneas* II 7, 1; Galen, *Commentary on Hippocrates'* De Natura hominis, XV 32K; *De elementis ex Hippocrate* I 9, I 489K; Stobaeus, *Eclogae* I 17, 2.

27 Τὰ δὴ τέτταρα στοιχεῖα εἶναι ὁμοῦ τὴν ἄποιον οὐσίαν τὴν ὕλην· εἶναι δὲ τὸ μὲν πῦρ τὸ θερμόν, τὸ δ' ὕδωρ τὸ ὑγρόν, τόν τ' ἀέρα τὸ ψυχρόν, καὶ τὴν γῆν τὸ ξηρόν (DL VII 137 = *SVF* II 580; LS 47B). See also *SVF* II 416 and Hahm 1985, esp. 40–41.

It seems, therefore, that our author espouses the view that air is cold by nature as part and parcel of the Stoic theory of elements which he appropriated. However, he did not simply adopt the Stoic theory, but he modified it in conformity with Peripatetic views. First, he expanded the Stoic theory of the four elements by adding the fifth element, which is the Aristotelian trademark, and by appending the dynamic tendencies of heaviness and lightness, which seem to explain why the four elements go up and down "in straight line", whereas the fifth element's motion is "curved" (περιφερής 5, 396b25), that is, circular (κυκλοφορουμένη; 2, 392a8). Second, whereas the Stoics allow for the transmutation of the elements into one another, as well as for their ultimate dissolution into fire, our author makes no room for the transformation of elements into one another or, for that matter, for their mixing with one another in the strict sense that was important for the Stoics, but also for Aristotle. As opposed to mere combination, in which ingredients preserve their properties, Aristotle argued that in mixtures ingredients continue to exist only in potentiality, whereas in actuality they give rise to new substances with new properties. Given what our author says about the elements, he seems to admit only *synthesis*, and no *mixis*, strictly speaking, let alone the Stoic *krasis dia holou*.

According to the author of *De mundo*, then, the four sublunar elementary bodies, each instantiating one elementary quality, undergo neither transformation into one another nor genuine mixture, but preserve their identity through all their interactions and combinations that account for the generation and corruption of things in the sublunar world. The author does not venture a statement, nor even a faint gesture, as to whether there is a threshold to the divisibility of elementary bodies, and there is no good reason to ascribe the atomistic hypothesis to him.[28] Even so, his position—somewhat old-fashioned and reminiscent of Empedocles' theory of the four roots—does not naturally combine with the Peripatetic hallmark tenet of the eternity of the cosmos. Why does the author of *De mundo* adopt such a theory of elements, then? Several reasons might be at play. First, such a theory is simpler and hence more appropriate for a treatise of this kind. Second, the Stoic theory of the elements was the most popular one at the time and likely to have been regarded as "the best scientific theory" available. Third, and most important, such a theory creates an excellent backdrop for the central point of Chapter 5: a theory that emphasises the opposition among the elements and their resistance to mixture that makes the question of the eternity of the cosmos especially pressing.

28 In *De caelo* III 6, 305a1–3, Aristotle writes: "If division will come to an end, then the body at which it stops will either be indivisible (ἄτομον) or divisible, but only insofar as it happens not to be ever divided further, which is what Empedocles wants to say."

4

I propose to analyse the author's account of the eternity of the cosmos in two parts, one structural and the other theological. The structural part can be found in the following passage of Chapter 5:

> The cause of this preservation is the agreement of the elements, and the cause of this agreement is equal share and none of them being more powerful than the other; for the heavy things have an equal bearing to the light ones, and the warm things to their opposite, since nature teaches in these greater matters that equality somehow preserves concord, and that concord preserves the cosmos, the parent of all things and most beautiful of all. (5, 396b34–397a5)[29]

Let me unpack the causal story in this passage. The preservation (σωτηρία) of the whole cosmos is the result of the agreement (ὁμολογία) of elements, and this agreement of elements is in turn the result of their having an equal share (ἰσομοιρία). By "agreement of elements" the author probably means that the elements keep to their own spheres, without encroaching upon one another. That is to say, despite the higher sphere affecting the lower one, so that regular motions descend from the aethereal sphere down to the earth, none of the elements dominates over the others. On the other hand, to have an "equal share" clearly does not mean to be of equal mass or volume: the four elements are arranged in concentric spheres of significantly different sizes, so their volumes must be different. Aristotle explains this in *Meteorology* I 3 by arguing that a small volume of earth transforms into a larger volume of water, a volume of water transforms into a still larger volume of air, and a volume of air transforms into an even larger volume of fire—and the other way around. "Any small quantity of water of given volume," Aristotle writes, "must necessarily bear the same proportion to the air which is formed from it, as the total aggregate of air bears to the total aggregate of water".[30] To have an "equal share", then, is best understood as having a total volume proportional to the total volume of the other elements with respect to their transformative potential. Or rather, since the author of *De mundo* does not accommodate elemental transformation, the

29 Αἰτία δὲ ταύτης μὲν ἡ τῶν στοιχείων ὁμολογία, τῆς δὲ ὁμολογίας ἡ ἰσομοιρία καὶ τὸ μηδὲν αὐτῶν πλέον ἕτερον ἑτέρου δύνασθαι· τὴν γὰρ ἴσην ἀντίστασιν ἔχει τὰ βαρέα πρὸς τὰ κοῦφα καὶ τὰ θερμὰ πρὸς θάτερα, τῆς φύσεως ἐπὶ τῶν μειζόνων διδασκούσης ὅτι τὸ ἴσον σωστικόν πώς ἐστιν ὁμονοίας, ἡ δὲ ὁμόνοια τοῦ πάντων γενετῆρος καὶ περικαλλεστάτου κόσμου.

30 Aristotle, *Meteorologica* I 3, 340a11–13.

total amount of each element must be such that there is neither too little of it, so as to be depleted by other elements, nor too much of it, so as to overwhelm them. In the simplest possible terms, to have an "equal share" is to have equal power.

This idea is closely related to the theory of elemental proportions in Plato's *Timaeus*: "For these reasons and out of elements of this kind, four in number, the body of the universe was created, being brought into concord through proportion (δι' ἀναλογίας ὁμολογῆσαν); and from these it derived friendship, so that coming to unity with itself it became indissoluble by any force save the will of him who joined it" (32b6–c4). The same idea is found in Timaeus Locrus, where it is said that "since the elements are all equal in power, their proportions are in equilibrium (ἐν ἰσονομίᾳ)", and in Philo's *De aeternitate mundi*, where it is stated that "the self-governing equilibrium (ἡ αὐτοκρατὴς ἰσονομία)" maintains itself forever and that the "equality of strength (τὸ ἰσοκρατές)" in the elemental transformations is the cause of "unswerving stability and unshaken permanence".[31] To say that all elements have equal power so that none overwhelms the others is to make a strong anti-Stoic point, since the Stoics argued that one element, namely fire, overwhelms the others in the universal conflagration. And even in the post-conflagration (cosmic) phase, according to the Stoics, fire seems to be the most fundamental and also the most active element, which implies that it does overwhelm the other elements. Indeed, Chrysippus distinguishes three senses of the word "element", the first and the strictest one of which is reserved for fire only, "because out of it the remaining elements are composed by alteration and into it they get their resolution".[32]

Anyway, the idea of an "equal share" is accompanied by the remark that in important matters nature teaches that equality (τὸ ἴσον) preserves concord (ὁμόνοια), and concord preserves the cosmos—"the parent of all things and most beautiful of all" (397a2–5). The reference to equality, pure and simple, seems to state the highest abstract underlying principle which resonated with the Greeks; one immediately thinks of apothegms such as "Equality makes no war", attributed to Solon, or "Equality is good in everything", ascribed to Democritus.[33]

Let me now summarise the causal story regarding the elements, as I understand it. First, there is the abstract principle of equality—*to ison*—which underlies the distribution of the elements with opposite properties in the cos-

31 Timaeus Locrus 217.13 (Thesleff); Philo, *De aeterinate mundi* 34.3 and 35.8–9 (Cumont). See also Brumana (2023), 278–280.

32 *SVF* II 413 = LS 47A.

33 Plutarch, *Solon* 14, 4 and Democritus B102 (Diels-Kranz).

mos. Second, there is an application of the abstract principle which is called *isomoiria*: the elements with opposite properties are distributed so that their powers are balanced. Their total volume is such that, when arranged in successive concentric regions within a single sphere, none of them depletes or overwhelms the others, given their various interactions. Third, the effect of the applied principle is *homologia tōn stoicheiōn*: the interactions among the elements—notably, the transmission of celestial motions and their effects on the Earth—are such that the elements, despite their opposite qualities, do not harm one another, but agree with one another. Fourth, the global result of the *homologia* is the eternal preservation—*sōteria*—of the cosmos: given that the elements cooperate, thanks to their well-proportioned volumes and suitable spherical arrangement, the everlasting persistence of the cosmos is secured.

What we find here is the idea that the elements in the cosmos are supplied in the right amounts and arranged in such a way that they can interact without encroaching upon one another, which makes the cosmos structurally predisposed, or "fine-tuned", for eternal existence. This is the structural part of the author's explanation of eternity of the cosmos. Thus far, no necessity is involved: it is a contingent fact that the cosmos is structured so that the elements are forever harmonised.

If the author had nothing else to say on the subject, his view of the cosmos would not be much different from what we find in the Hellenistic Peripatetic philosopher Critolaus of Phaselis (c. 200–c. 118 BCE) and in the work of Pseudo-Ocellus *De universi natura*, both discussed by Andrea Falcon in his contribution to this volume. On that view, the cosmos happens to be, by nature, a self-sustaining system. However, this is emphatically *not* the view that the author of *De mundo* endorses.

According to the author, the cosmos requires a "sustaining cause" (συνεκτικὴ αἰτία, 397b9), which is "the most important thing regarding the cosmos" (τὸ τοῦ κόσμου κυριώτατον, 397b12),[34] and that is God. What is crucial to understand, however, is that God does not sustain the cosmos by himself being in it and making things happen.

> For God is really the preserver of all things and the begetter of everything whichever way it is brought about in this cosmos, without, however, taking upon himself the toil of a creature that labours itself, but by making

34 The translations "that which is most important *in* the cosmos" (Thom), "that which is supreme *in* the cosmos" (Furley) and "the most important thing *in* the universe" (Forster) suggest that God is inside the cosmos, which is not implicit in the Greek original. Strohm got this right with "das Wichtige des Weltalls".

> use of an untiring power by means of which he masters even things that seem to be far away. (6, 397b20–23)[35]

Unlike the Stoic God, who permeates the entire cosmos and thus exercises his causal efficacy, our author insists that God operates only indirectly, by way of his "untiring power" (δύναμις ἄτρυτος, 397b23) which is said to pervade all things. Indeed, God's power "has set in order all the earth and sea, aether, sun, moon and the whole heaven, having fashioned the whole cosmos from the unmixed and diverse" (396b25–29), implying that the structural features of the cosmos have a divine origin—not in time, of course, but in causal ancestry.

Although the power pervading all things goes back to God, the author observes that its grip wanes with distance from its source, which explains why things around us, in the central regions of the cosmos farthest away from the source, are marked by disorder and instability.

> The body closest to him benefits most from his power, and then the body next to it, and so on in sequence until the places where we are. That is why the earth and the things on the earth, being at the greatest distance from the assistance of God, seem to be weak, incongruous and full of much disturbance; but nevertheless, insofar as the divine naturally reaches everything, even the things around us occur in the same way as the things above us, sharing more or less in God's assistance according to whether they are closer or further from him. (6, 397b27–398a1)[36]

Although its causal efficacy decreases with distance from the source, the divine power is said to be "untiring" (or "unabating", "limitless", ἄτρυτος) because it reaches all the way down to the most remote nooks of the cosmos, "being the cause of preservation for all things, even those furthest separated" (398a2–4). More importantly, it is "untiring" because its source cannot be depleted. Given that it comes from God, we can be confident that this power has always operated in the cosmos and will keep operating steadily through all times.

---

35 Σωτὴρ μὲν γὰρ ὄντως ἁπάντων ἐστὶ καὶ γενέτωρ τῶν ὁπωσδήποτε κατὰ τόνδε τὸν κόσμον συντελουμένων ὁ θεός, οὐ μὴν αὐτουργοῦ καὶ ἐπιπόνου ζῴου κάματον ὑπομένων, ἀλλὰ δυνάμει χρώμενος ἀτρύτῳ, δι' ἧς καὶ τῶν πόρρω δοκούντων εἶναι περιγίνεται.

36 Μάλιστα δέ πως αὐτοῦ τῆς δυνάμεως ἀπολαύει τὸ πλησίον αὐτοῦ σῶμα, καὶ ἔπειτα τὸ μετ' ἐκεῖνο, καὶ ἐφεξῆς οὕτως ἄχρι τῶν καθ' ἡμᾶς τόπων. Διὸ γῆ τε καὶ τὰ ἐπὶ γῆς ἔοικεν, ἐν ἀποστάσει πλείστῃ τῆς ἐκ θεοῦ ὄντα ὠφελείας, ἀσθενῆ καὶ ἀκατάλληλα εἶναι καὶ πολλῆς μεστὰ ταραχῆς· οὐ μὴν ἀλλὰ καθ' ὅσον ἐπὶ πᾶν διικνεῖσθαι πέφυκε τὸ θεῖον, καὶ ἐπὶ τὰ καθ' ἡμᾶς ὁμοίως συμβαίνει τά τε ὑπὲρ ἡμᾶς, κατὰ τὸ ἔγγιόν τε καὶ πορρωτέρω θεοῦ εἶναι μᾶλλόν τε καὶ ἧττον ὠφελείας μεταλαμβάνοντα.

Yet another reason why the divine power is described as "untiring" might be that it "*compels* the truly opposite elements in the cosmos to agree with one another and thus brings about preservation for the whole" (τάς τε ἐναντιωτάτας ἐν αὐτῷ φύσεις ἀλλήλαις **ἀναγκάσασα** ὁμολογῆσαι καὶ ἐκ τούτων μηχανησαμένη τῷ παντὶ σωτηρίαν, 396b32–34). That is to say, the divine power *enforces* the structure of the cosmos and the processes therein that makes it predisposed for eternal existence, thus introducing a sort of necessity into the author's account of the eternity of the cosmos. With God and his divine power pervading the whole cosmos, the author of *De mundo* makes it no longer a contingent fact that the cosmos eternal, but something *exacted* by God's power. Thus, the second, theological part of the author's account of eternity of the cosmos supplies necessity where the first, structural part of the account yielded only contingency.

To sum up, according to *De mundo*, the cosmos is composed of the four elements that do not transform into one another, nor genuinely mix, but preserve their identity through all their interactions and combinations that account for the generation and corruption of things in the sublunar world. However, the elements are distributed in the cosmos and perpetually moved in such a way that they cannot but cooperate harmoniously through circular motions of the heavens and regular cycles of generation and destruction in the sublunar world, so that no element can ever overcome the others and thus compromise the structure of the cosmos and disrupt its regular motions. The cause of the structure and motions in the cosmos is the "untiring power" that originates from God. It is said to "compel" the elements to such structure and behaviour, thus adding a sort of necessity to the account of eternity of the cosmos. At the same time, God is piously saved from the trouble of interacting with the cosmos, as demanded by his supreme position and dignity.[37]

---

37 There is an interesting exception to this in the text of *De mundo*. The episode with a stream of lava splitting to save young men carrying their old parents on their shoulders (6, 400a34–b6) illustrates how the divine (τὸ δαιμόνιον) honours pious humans. This seems like an *ad hoc* response to the charge levelled against the Peripatetics that they do not admit divine providence below the sphere of the moon, where it matters. For instance, Atticus argued that Aristotle was as bad as Epicurus on this score: "What difference is there as far as we are concerned whether you locate the divine outside the cosmos and allow us no contact with it, or whether you imprison the gods within the cosmos but keep them away from what happens on earth?" (Atticus Fr. 3 [Des Places]; translated by Boys-Stones).

5

On the picture presented in *De mundo*, God has nothing to do with governing the behaviour of each thing in the cosmos, as opposed to what the Stoics think, let alone with the creation of the cosmos, as opposed to what some Platonists and later Christians think. That does not mean, however, that divine causation can be dispensed with altogether, as opposed to what the Epicureans think and what some distinguished Peripatetics like Strato and Critolaus might have thought.[38] On the contrary, in our author's view, God is needed to ensure that the cosmos lasts through all ages as a dynamic whole marked by circular motion and large-scale regular changes, which I take to be a genuine Aristotelian position.

True, on Aristotle's account this is accomplished by an unmoved mover that brings about the motion of the first heaven as an object of its thought and desire, whereas the author of *De mundo* accomplishes this by the divine power which permeates the whole cosmos. This detail notwithstanding, it seems that the author of *De mundo* defends a broadly Aristotelian position on God and its relation to the cosmos, not only against rival philosophical schools of his time, but also against unorthodox Peripatetic thinkers such as Critolaus.

It has to be said, however, that we have no wider context of Critolaus' arguments for the eternity of the cosmos, so we cannot be sure that they present a complete story. Perhaps Critolaus too supplemented his arguments with reference to God and did so in a way similar to what we find in *De mundo*. There are two fragments of Critolaus, listed under the heading "On Rhetoric" in Wehrli's collection, which bear a strong parallel with *De mundo*. I quote only the first and longer fragment, preserved in Plutarch's *Praecepta gerendae rei publicae* (811c10–d7):

> Others, among whom is Critolaus the Peripatetic, think that Pericles' [tactic] is more solemn and dignified: judging that the Salaminia and the Paralos, the Athenian galleys, were not launched for every task but [only] for the necessary and major activities, he employed himself for the most important and greatest [tasks], like the King of the cosmos: "For god takes hold of what is greater, but the small matters he lets go and leaves to chance" according to Euripides. (Fr. 37a W)[39]

---

38 For Strato, see Frs. 18 and 19A (Sharples); see also Frs. 19B, 19C, 20 and 21 (Sharples).

39 Ἕτεροι δὲ σεμνότερον οἴονται καὶ μεγαλοπρεπέστερον εἶναι τὸ τοῦ Περικλέους, ὧν καὶ Κριτόλαός ἐστιν ὁ Περιπατητικός, ἀξιοῦντος, ὥσπερ ἡ Σαλαμινία ναῦς Ἀθήνησι καὶ ἡ Πάραλος οὐκ ἐπὶ πᾶν ἔργον ἀλλ' ἐπὶ τὰς ἀναγκαίας καὶ μεγάλας κατεσπῶντο πράξεις, οὕτως ἑαυτῷ πρὸς τὰ κυριώτατα καὶ μέγιστα χρῆσθαι, ὡς ὁ τοῦ κόσμου βασιλεύς· "τῶν ἄγαν γὰρ ἅπτεται θεός, τὰ μικρὰ δ' εἰς τύχην ἀνεὶς ἐᾷ" κατὰ τὸν Εὐριπίδην. Translated by Sharples (2010, 201), with modifications.

We do not know the context in which Critolaus introduced Pericles' tactic to show up in the assembly only when matters of utmost importance were discussed, in analogy with (i) employment of two largest Athenian battleships only in most important naval missions, and (ii) God who is concerned only with important matters, relegating all other things to chance.[40] Whereas the first analogy definitely goes back to Critolaus, as confirmed by fragment 37b, the second analogy may be Plutarch's addition as much as Critolaus' original contribution.

Regardless of its original context and source, the passage from Plutarch's *Praecepta* does capture the idea that God's dignity requires that he attends only to most important matters. We will probably never know if Critolaus himself entertained that idea or if he used it to argue that God's causal efficiency or providence reaches only as far as the heavenly bodies, which was the view traditionally associated with the Peripatetics.[41] What we do know, however, is that the idea captured in Plutarch's *Praecepta* is central to the author of *De mundo*. He develops this idea by placing God at "the highest and first seat" (ἡ ἀνωτάτη καὶ πρώτη ἕδρα, 397b25) and letting his "untiring power" do all God's work by a chain of motions that descend from the first heaven down to every corner of the earth. This ensures, as we have seen, that the cosmos retains its structure, circular motions and periodic changes, despite local irregularities, through all ages.[42]

The lines from Euripides, quoted also in Plutarch's *De cohibenda ira* 464a5–7, belong to an unknown work (Fr. 974 Nauck).

40 Observe the similarity between the "King of the cosmos" (ὁ τοῦ κόσμου βασιλεύς) here and the Great King of Persia (ὁ μέγας βασιλεύς) in the first and most elaborate of a series of analogies in *De mundo* 6. The author compares God to the Great King, who governs the whole empire by means of an elaborate infrastructure and hierarchy of offices, never leaving his palace and remaining invisible to everyone save for the few most trusted subjects. More on this analogy in Betegh and Gregorić (2020).

41 See Sharples (2002a).

42 An earlier version of this text was read at the conference *Peripatetic Physics after Theophrastus* in Lyon in June 2024 and then at Andrea Falcon's seminar in Tübingen in July 2024. I thank the organisers for the opportunity to present my views and to the audiences at both events for their questions and suggestions. Special gratitude goes to Andrea for allowing me to read a draft of his contribution to this volume and for his comments on a draft of my contribution. I have learnt a great deal from his publications as well as from several exchanges we have had on the subject. My thanks are also due to Selene Brumana and Gweltaz Guyomarc'h for their comments on different versions of the paper. Research for this paper was funded through the project "Kvalitete, duša i um u aristotelovskoj filozofiji" (KvaDAr) at the Institute of Philosophy in Zagreb, reviewed by the Ministry of Science and Education of the Republic of Croatia and financed through the National Recovery and Resilience Plan 2021–2026 of the European Union—NextGenerationEU.

CHAPTER 11

Andronicus of Rhodes on Aristotelian Physics

Myrto Hatzimichali

1 Introduction

Andronicus of Rhodes is best known to most scholars for his role in the fate of Aristotle's books, and it would probably be fair to say that this was his main contribution to the history of ancient philosophy. The evidence for his involvement with Aristotle's corpus is not very extensive, so it is out of proportion with the modern ink that has been spilt on the matter. Nevertheless, the evidence does seem to show that Andronicus' contribution is unlikely to have been a full critical edition of the entire corpus, but rather an influential organisation of a canon with the "esoteric" works clearly and separately marked out in a privileged way, and perhaps in something close to the order and shape in which they still survive.[1] Other than this evidence on canon-formation and organisation, Andronicus' name comes up several times in Simplicius' commentary on the *Categories*, a work to which he (like all the Peripatetics of his generation) must have devoted considerable attention, after placing it at the head of the privileged esoteric canon.[2] Outside the debate on the *Categories*, references to Andronicus are quite sparse, so the texts that will be quoted and analysed below represent pretty much all the instances where Andronicus' name appears in connection with some topic that Aristotle would have considered part of the enquiry into nature and thus under the purview of the *physikos*.

Before beginning the discussion proper of those instances, however, a brief mention should be made of the programmatic claim concerning where one should begin their study of Aristotle's work, which would be tantamount to the proper beginning of philosophy itself:

> **T1** Boethus of Sidon says that (the study of Aristotle's works should start) with physics, but Andronicus of Rhodes the Peripatetic, eleventh suc-

1 For discussion of Andronicus' "editorial" role see Barnes (1997); Hatzimichali (2013); (2016). More recently Rashed (2021) has cast doubt on Andronicus' impact on one important piece of evidence, namely the catalogue of Aristotle's works surviving in a letter from a Ptolemy to a Gallus, which is preserved by Arabic sources.

2 Sharples (2008b); Griffin (2015).

cessor of Aristotle's school said (that we should start) with logic, while among the Platonists some (said) from ethics, others from mathematics. (Elias, *In Cat.* 117.21–25)[3]

The very fact of raising the question in this manner already puts us in mind of possible Stoic influences, as the Stoics were the ones who elaborated the tripartition of philosophy into logic, physics and ethics.[4] But Andronicus' position doubtless is related to his wholehearted adoption of the view that logic constitutes an instrument (*organon*) rather than an actual part of philosophy, which we find a few lines later in Elias' commentary and which no Stoic would have endorsed. From Philoponus, who provides us with a parallel passage to T1 (Philoponus, *In Cat.* 5.15–20), we learn that Andronicus emphasised getting to grips with demonstration in the logical works (which would then enhance our understanding of all of Aristotle's other works), while Boethus advocated starting with physics on the grounds that its objects are more familiar to us, as indeed Aristotle would have it in the opening of his *Physics* (I 1 184a16–23).[5]

2 Dividing Aristotle's *Physics* into Two Sections

Our next testimony connecting Andronicus with Aristotelian physics comes from his influential work on Aristotle's books which constitutes his main claim to fame, as we saw above:

T2a It was remarked earlier (801.13–15) that they [sc. the Peripatetics] call the five books before this one the *Physics* and the next three *On Motion*; for this is the way that Andronicus arranges them in the third book of his *On the Books of Aristotle*, and Theophrastus too bears witness concerning the first group: when Eudemus wrote to him concerning one of the faulty manuscripts in respect to the fifth book, he said "As for what you asked me

3 Βόηθος γὰρ ὁ Σιδώνιος ἀπὸ τῆς φυσικῆς λέγει, Ἀνδρόνικος δὲ ὁ Ῥόδιος ὁ Περιπατητικὸς ὁ ἑνδέκατος διάδοχος τῆς Ἀριστοτέλους σχολῆς ἀπὸ τῆς λογικῆς ἔλεγε, τῶν δὲ Πλατωνικῶν οἱ μὲν ἀπὸ τῆς ἠθικῆς οἱ δὲ ἀπὸ τῶν μαθηματικῶν.

4 See Ierodiakonou (1993) for discussion and qualification of this view in connection with the Stoics.

5 For more parallel evidence and further discussion of Andronicus' views on the foregrounding of demonstration at the beginning of the *Organon* see Griffin (2015), 32–77, esp. 32–41. The material on Boethus is more suspect, see Chiaradonna (2020), 169. As Chiaradonna also shows, however, Boethus did draw a connection between the *Categories* (the first work on everyone's canon) and natural philosophy.

> in your letter to write out and send you from the *Physics*, either I misunderstand you or there is something very small in the middle of the passage 'Of the things that are motionless, this alone I call "at rest"; for rest is the contrary of motion, so that it would be a privation in that which can have it'" (= *Phys.* V 2 226b14–16). Thus, Theophrastus considered the fifth book to be "from the *Physics*". (Simplicius, *In Phys.* 923.7–16, transl. Konstan, modified)[6]

This extract is particularly valuable in giving the title of the work, which we can reconstruct as simply Ἀριστοτέλους βιβλία (*Aristotle's Books*) or Περὶ τῶν Ἀριστοτέλους βιβλίων (*On Aristotle's Books*), depending on whether we assume that it was compressed here by Simplicius. In this work, which comprised at least three volumes, Andronicus provided a catalogue (*Pinax*) and he must have also discussed the rationale behind the construction of his canon (including e.g. his reasons for putting the logical works first as we saw above), issues of authenticity (we know about his athetesis of the *De interpretatione* and his suspicions around the *Postpraedicamenta*),[7] and finally issues of internal structure within the different treatises. The evidence before us now need not be taken to suggest that Andronicus questioned the existence of the *Physics* as an 8-book treatise, which goes back to the time of Theophrastus and Eudemus;[8] rather our passage suggests that he took a view on its internal subdivision—at least this is Simplicius' impression and he had more information than we do. In fact, Simplicius' information is so abundant that he provides a number of versions of this report on the title and sub-titles of the *Physics* (although T2a is the only place where Andronicus' name appears): in his *prolegomena* (4.11–16; 6.4–10) he cites Adrastus for the report that some gave the title *On Principles* (Περὶ ἀρχῶν) and others the title *Lectures on Physics* (Φυσικὴ ἀκρόασις) to the entire 8-book treatise, whereas others gave the title *On Principles* to the first five books and the title *On Motion* to the last three. At the beginning of his commentary on book V (which is probably the intended cross-reference in T2a) he names "Aristotle and his associates" as the group who include book V among those belonging to the

6 εἴρηται δὲ καὶ πρότερον, ὅτι τὰ μὲν πέντε βιβλία τὰ πρὸ τούτου Φυσικὰ καλοῦσιν, τὰ δὲ ἐντεῦθεν τρία Περὶ κινήσεως· οὕτω γὰρ καὶ Ἀνδρόνικος ἐν τῷ τρίτῳ τῶν Ἀριστοτέλους βιβλίων διατάττεται, μαρτυροῦντος περὶ τῶν πρώτων καὶ Θεοφράστου γράψαντος Εὐδήμου περί τινος αὐτῷ τῶν ἡμαρτημένων ἀντιγράφων κατὰ τὸ πέμπτον βιβλίον· 'ὑπὲρ ὧν, φησίν, ἐπέστειλας κελεύων με γράψαι καὶ ἀποστεῖλαι ἐκ τῶν Φυσικῶν, ἤτοι ἐγὼ οὐ ξυνίημι ἢ μικρόν τι παντελῶς ἔχει τὸ ἀνὰ μέσον τοῦ "ὅπερ ἠρεμεῖν καλῶ τῶν ἀκινήτων μόνον· ἐναντίον γὰρ ἠρεμία κινήσει, ὥστε στέρησις ἂν εἴη τοῦ δεκτικοῦ,"' ὥστε καὶ τὸ πέμπτον βιβλίον 'ἐκ τῶν Φυσικῶν' ὁ Θεόφραστος νομίζει.

7 See Hatzimichali (2016), 91–92; Griffin (2015), 36–41.

8 This appears to be the view of Barnes, cf. Barnes (1997), 35–36.

On the First Principles of Physics (Περὶ ἀρχῶν φυσικά), while giving the title *On Motion* to the final three books (801.13–15). Thus, in Simplicius' pages, the first five books are variously reported as having the title *On First Principles, On the First Principles of Physics* and finally just *Physics* (in T2a).[9] None of this maps neatly onto the report from Andronicus, while not contradicting it flatly either.

It is true that the 8-book *Physics* as we know it is not characterised by strong thematic unity—one can see a connecting thread around the issue of change and its principles, which crystallises more around locomotion as the work progresses, but the different books or even sections of books can be read as individual essays. There is therefore no obvious place where one would divide the work, but if pressed, most people nowadays would opt for a division after the fourth book rather than the fifth. In antiquity this approach was represented by Porphyry in the pages of Simplicius (*In Phys.* 802.7–13) and also by Philoponus (*In Phys.* 2.16–21). But Simplicius in T2a (as at 801.13–15) attributes the 5 + 3 division to an early Peripatetic tradition, and offers further support a few lines later:[10]

> **T2b** That the three books are those *On Motion* and the five the *Physics*, Damas, who wrote the *Life of Eudemus*, also testifies, when he says "and of the three books[11] *On Motion* from Aristotle's treatise *On Nature*" (*In Phys.* 924.12–14, transl. Konstan)[12]

The first question that arises for current purposes from T2a and T2b is how much we should attribute to Andronicus (because the information in T2a is that Andronicus was in agreement with the subdivision of another group, referred to by Simplicius in the preceding lines as "the Peripatetics" in general). It is quite possible that Andronicus was the one who originally reported on the correspondence between Theophrastus and Eudemus in T2a, and perhaps also on the biography of Eudemus by Damas in T2b[13]—he may even have

9 See Golitsis (2008), 47–50. He also comments on the fact that Simplicius endorsed the 4 + 4 division in his *De caelo* commentary (226.19–21), but probably changed his mind after consulting Adrastus (most likely indirectly through Porphyry).

10 In the intervening lines Simplicius cites *De caelo* I 5 272a29–31 and I 7 275b21–23 as evidence that Aristotle himself refers to the last three books with the title *On motion*.

11 Reading τριῶν, emended from τρία, with Wehrli and Konstan. Diels proposes τὰ Περὶ κινήσεως τρία in his apparatus.

12 ὅτι δὲ τὰ τρία ἐστὶ τὰ Περὶ κινήσεως, καὶ τὰ πέντε Φυσικά, μαρτυρεῖ καὶ Δάμας ὁ τὸν βίον Εὐδήμου γράψας λέγων "καὶ τῶν ἐκ τῆς Περὶ φύσεως πραγματείας τῆς Ἀριστοτέλους τῶν Περὶ κινήσεως τριῶν."

13 Thus also Golitsis (2008), 74.

found the correspondence quoted in Damas' work. The issue that was raised between Theophrastus and Eudemus is not entirely transparent from T2a; it seems that Eudemus was concerned about a reading or a possible lacuna in his manuscript (apparently there was a whole group of manuscripts in Eudemus' possession that were thought to be faulty—ἡμαρτημένα), but Theophrastus responded that he could not identify any significant problem.[14] As Ross pointed out, this exchange between the two early Peripatetics on its own does not prove Andronicus' point: it just says that Theophrastus took our book v (the book from which the quoted extract comes) to belong to a Φυσικά, but he may have the complete 8-book Φυσικά in mind.[15] It is only when we combine this with Damas' report in T2b, indicating that three books of the *Treatise on Nature* (Περὶ φύσεως) were known as Περὶ κινήσεως, that we can understand how Andronicus would have concluded in favour of a 5+3 division (there are three books on motion, book v belongs to Φυσικά, therefore the division must come after book v). However, as Ross noted,[16] Eudemus' version of the *Physics* did not include book vii (cf. Simplicius, 1036.11–15),[17] so the three books on motion referred to by Damas could be books v, vi and viii. On this interpretation, favoured by Ross, Andronicus is simply the victim of a misunderstanding and, more relevantly for broader questions about his practice and rationale, he is not looking to compile Aristotelian *pragmateiai* based on internal criteria, but is rather reporting and following what he takes to be an authoritative tradition.

It is still worth asking, however, whether he might have had any independent reasons for favouring a subdivision after book v with the respective titles Φυσικά for books i–v and Περὶ κινήσεως for vi–viii. Among modern interpreters, Pierre Pellegrin is in a minority in speaking explicitly in favour of such a division, on the grounds that book v is closer to the early chapters of book iii because it discusses motion and change in general terms, while kinematics proper begins with book vi, resting on and presupposing equally all the analyses and specifications from books i–v. Moreover, according to Pellegrin, there is a stronger break between books v and vi, as Aristotle switches to a more

14 Here the precise significance of "something very small in the middle of the passage" (μικρόν τι παντελῶς ἔχει τὸ ἀνὰ μέσον) is particularly hard to decipher.

15 Ross (1936), 1.

16 Ross (1936), 3–4. This is questioned by Barnes (1997), 61.

17 διὸ καὶ παρέλκειν ἔδοξέ τισι τοῦτο τὸ βιβλίον ἐν τῇ πραγματείᾳ μαλθακωτέραις ἤ, ὥς φησιν Ἀλέξανδρος, λογικωτέραις χρώμενον ταῖς ἀποδείξεσι καὶ ὅ γε Εὔδημος μέχρι τοῦδε σχεδὸν τοῖς ὅλης τῆς πραγματείας κεφαλαίοις παρακολουθήσας τοῦτο παρελθὼν ὡς περιττὸν ἐπὶ τὰ ἐν τῷ τελευταίῳ βιβλίῳ μετῆλθεν.

markedly mathematical approach evidenced especially in its heavy use of diagrammatic letters and its axiomatic structure.[18] Could Andronicus have been motivated by similar considerations? It is not possible to say, and the little that we can glean from Simplicius' text and the "evidence" cited there is more about a tradition that Andronicus looks to have followed.

3 On the Internal *Dynamis* of That Which Undergoes Change

The central question about Andronicus' methodology and the principles guiding his approach to Aristotle's text surfaces again in the next set of evidence concerning the entity that undergoes change in *Physics* III 3, where we will wonder again whether he is being a pedant, defending at all costs what he takes to be an authoritative tradition, or whether his choices are motivated by independent reasoning and reflection on the philosophical issues at stake. The two passages quoted below as T3 and T4 come from Simplicius' discussion of *Physics* III 3; they are separated by ten pages in the commentary, and Simplicius' concern is quite different in the two instances, but Andronicus seems to be making essentially the same point. It is therefore possible that Simplicius was drawing on the same source for Andronicus in both cases.[19] As is well known, in *Physics* III 3 Aristotle is preoccupied with showing that in the cases of change that he has been discussing since the beginning of book III, the action of the agent and the passion / undergoing of the patient are in fact one single change, and this one change takes place in the patient. The vocabulary he uses is primarily that of "causing change" / "being changed" (κινητικόν / κινητόν, e.g. 202a14 and κινοῦν / κινούμενον, e.g. 202a21), but we do also find "acting" and "undergoing" (ποιητικόν / παθητικόν, e.g. 202a23; ποιοῦν / πάσχον, e.g. 202a27). Aristotle will argue that the action and the passion may be one, but they are not one in the account of their essence, just like the road from Thebes to Athens and from Athens to Thebes (202b12–14). Similarly, teaching and learning are one and the same change, taking place in the learner, but what it is to teach is not the same as what it is to learn. Aristotle here is not explicit about the sense in which these are one, so Simplicius supplies that they are one in substrate (τῷ ὑποκειμένῳ, 440.5, 445.5 and *passim*, cf. *GC* I 3 319b3–4).

The first passage involving Andronicus comes from Simplicius' comments on the beginning of the chapter where Aristotle first states his claim, and the

18 See Pellegrin (2000), 36–57, esp. 56–57.

19 The case for taking the two passages T3 and T4 together has also been made by Lautner (2002), whose main preoccupation is to show the consistency between the two passages.

key detail for our purposes is that the change is said to be an actuality of the patient through the agency of the agent: "again—a point which makes difficulty—it is manifest that the change is in that which is changeable. For it is the actuality of this, brought about by that which is productive of change. Yet the operation of that which is productive of change, also, is not other." (*Phys.* III 3 202a12–15, transl. Hussey).[20] Andronicus was in favour of a different text for this passage:

> **T3** It should be known that in this area most write this passage more clearly as follows: "for it is the actuality of this, brought about by the source of change"; but Andronicus writes as follows: "for it is the actuality of the changed and from it". He gives the interpretation that, even if the changer be external, the changed seems to be changed by itself, since it is brought to actuality through its inherent potentiality. (Simplicius, *In Phys.* 440.13–17, transl. Urmson, slightly modified)[21]

We could paraphrase Andronicus' alternative reading as "an actuality of the patient and through its own agency". Scholars since Moraux are in agreement that this must be a reading that he found in some manuscript rather than his own conjecture.[22] One of the surviving manuscripts of the *Physics*, the 10th-century MS J now in Vienna carries Andronicus' reading, but it is not considered to be a high-value manuscript, and the reading has been dismissed by Moraux as a corruption later in the tradition.[23] Andronicus, however, proceeded to defend this reading by offering some exegesis: on his interpretation, it is the inner potentiality of the patient that carries most of the weight in the process of change, such that all changes would approximate to something like self-motion ("the changed seems to be changed by itself", T3 above), which would amount to a subtle challenge against Aristotle's total rejection of self-motion.[24]

---

20 καὶ τὸ ἀπορούμενον δὲ φανερόν, ὅτι ἐστὶν ἡ κίνησις ἐν τῷ κινητῷ· ἐντελέχεια γάρ ἐστι τούτου [καὶ] ὑπὸ τοῦ κινητικοῦ. καὶ ἡ τοῦ κινητικοῦ δὲ ἐνέργεια οὐκ ἄλλη ἐστίν·

21 ἰστέον δὲ ὅτι ἐν τούτῳ τῷ χωρίῳ οἱ μὲν πολλοὶ σαφέστερον οὕτω γράφουσι ταύτην τὴν λέξιν· *ἐντελέχεια γάρ ἐστι τούτου ὑπὸ τοῦ κινητικοῦ*, ὁ δὲ Ἀνδρόνικος οὕτως· *ἐντελέχεια γάρ ἐστι τοῦ κινητοῦ καὶ ὑπὸ τούτου*. καὶ ἐξηγεῖται ὅτι κἂν ἔξωθεν ᾖ τὸ κινοῦν, ἐκ τῆς ἐνούσης δυνάμεως εἰς ἐνέργειαν ἀγόμενον, ὑφ' ἑαυτοῦ κινεῖσθαι δοκεῖ τὸ κινούμενον.

22 We know that Andronicus "received and made public" a set of ancient manuscripts of Aristotle's works from the grammarian Tyrannion (Plutarch, *Sulla* 16). See Moraux (1973), 113; Gottschalk (1987), 1112–1114; Barnes (1997), 30. Lautner (2002), 126 with n. 2 expresses doubts about this approach.

23 Moraux (1973), 113–114.

24 Thus Moraux (1973), 114.

This need not be shocking or surprising, as Peripatetics at this time were not looking for an orthodox vindication of Aristotle on all counts.

The second time that Simplicius turns to Andronicus in T4 below, he is discussing the conclusion of *Physics* III 3, where Aristotle offers a slightly modified definition of change: "qualitative change, for example, is the actuality of that which admits of qualitative change, qua admitting of qualitative change. More intelligibly, it is the actuality of the potential agent and the potential patient as such, both in general and again also in each particular case, e.g. building or healing. And an account will be given in the same way about each of the other kinds of change" (202b25–27).[25] The reference to the "potential agent" in this passage was found to be puzzling by Simplicius, and indeed Hussey judged the entire expression τοῦ δυνάμει ποιητικοῦ καὶ παθητικοῦ ("the potential agent and the potential patient") to be illogical, leading him to athetise the entire sentence.[26] Simplicius worries that the actualisation of the agent's potential is a separate change that should precede the change under consideration, for example a teacher must be in a state of actuality throughout the process of teaching and the realisation of his potential is not part of the teaching process. Eventually he tries to reassure himself by suggesting that Aristotle may be drawing a distinction between cases of natural and artificial change where movers are always moved on the one hand, and permanently actual unmoved movers on the other. He concludes with the solution that we find in the opening of T4:

> **T4** Therefore, through the whole time of initiating change both potentiality and actuality are in them (sc. in the things that initiate change naturally and artificially), and they are changed through their progress from being imperfect, but initiate change through the perfection of their condition and actuality. Nature, which is also so predisposed, gives substrates their disposition from within in relation to each form of change, as Andronicus also said. For even if water be heated by fire, still the natural constitution in the water first becomes warm and thus warms, or warms with itself, the substrate. (*In Phys.* 450.14–20, transl. Urmson, slightly modified)[27]

---

25 ἀλλοίωσις μὲν γὰρ ἡ τοῦ ἀλλοιωτοῦ, ᾗ ἀλλοιωτόν, ἐντελέχεια. ἔτι δὲ γνωριμώτερον, ἡ τοῦ δυνάμει ποιητικοῦ καὶ παθητικοῦ, ᾗ τοιοῦτον, ἁπλῶς τε καὶ πάλιν καθ᾽ ἕκαστον, ἢ οἰκοδόμησις ἢ ἰάτρευσις. Τὸν αὐτὸν δὲ λεχθήσεται τρόπον καὶ περὶ τῶν ἄλλων κινήσεων ἑκάστης.

26 Hussey (1983) ad loc.: "exclude ἔτι δὲ γνωριμώτερον … ἰατρεύσις. A bad attempt at a general explanation, disrupting the thought of the section, and in which τοῦ δυνάμει ποιητικοῦ καὶ παθητικοῦ is an illogical expression."

27 διὸ παρ᾽ ὅλον τὸν τοῦ κινεῖν χρόνον ἄμφω ἐστὶν ἐν αὐτοῖς τό τε δυνάμει καὶ τὸ ἐνεργείᾳ, καὶ κατὰ μὲν τὴν ἐξ ἀτελοῦς προκοπὴν κινοῦνται, κατὰ δὲ τὸ τέλειον καὶ τῆς ἕξεως καὶ τῆς ἐνεργείας κινοῦσιν. ἡ δὲ φύσις καὶ προδιατιθεμένη διατίθησι τὸ ὑποκείμενον ἔνδοθεν καθ᾽ ἕκαστον

A more satisfactory solution than Simplicius' appeal to unmoved movers is provided by Ursula Coope, who accounts for the potential agent as follows: "By 'potential agent' here, he must mean not the thing that has the potential to be an agent, but the agent, qua having the potential in virtue of which it is an agent, i.e. the agent, insofar as it contains the potential for a patient of a certain kind to be different."[28] She emphasises that the agent's action cannot itself be a change that the agent undergoes (even if a "reciprocal", but different, change is produced upon the agent)[29]—an expectation that is probably behind Simplicius' interpretation. Simplicius did not see that the agent can be viewed as having the potential for *something else* to be other than it is.

What is the relevance of Andronicus to this discussion? Simplicius for his purposes may have taken this as a good example of nature as a moved mover (consistent with Neoplatonic thought)[30] that does not have to be in permanent actuality but is subject to external stimulus. But from Andronicus' perspective, in the light of T3, we may have here some more material coming from his defence of the internal agency of the patient. The similarity of wording between T3 and T4, especially with the concessive clause, is striking, and it appears also in another passage given here as T5 that does not name Andronicus but could well go back to him as it makes the same point:

> T5 So the activity of both is one, being called more strictly an activity as from the changer, but in the changed it is an affection rather than an activity. But perhaps this is also called an activity in so far as, even though it is externally initiated, still, because of its internal nature, it is able to be changed according to its own suitability. (Simplicius, *In Phys.* 439.29–33, transl. Urmson)[31]

The point of this entire line of interpretation is to ascribe an increased level of causal responsibility for a change to the internal condition of the patient. This is formulated in terms of the inner potentiality of the *kinēton* in T3, but in T4 and T5 the reference is to the internal *nature* that is or becomes predisposed in a certain way, and thus accounts for the change. It is tempting

κινήσεως εἶδος, ὡς καὶ ὁ Ἀνδρόνικος ἔλεγε. κἂν γὰρ θερμαίνηται ὑπὸ πυρὸς τὸ ὕδωρ, ἀλλ᾽ ἡ ἐν τῷ ὕδατι φύσις πρώτη θερμὴ γενομένη, οὕτως θερμαίνει ἢ συνθερμαίνει τὸ ὑποκείμενον.

28 Coope (2005), 220, n. 35.

29 Coope (2005), 207.

30 Cf. Simplicius, *In Phys.* 312.18–314.15.

31 μία οὖν ἡ ἀμφοῖν ἐνέργεια, ὡς μὲν ἀπὸ τοῦ κινοῦντος ἐνέργεια λεγομένη κυριώτερον, ὡς δὲ ἐν τῷ κινουμένῳ πάθος μᾶλλον ἤπερ ἐνέργεια. ἐνέργεια δὲ ἴσως καὶ αὕτη λέγεται, καθόσον κἂν ἔξωθεν ἔχῃ τὴν ἀρχήν, ἀλλ᾽ ὑπὸ τῆς ἐν αὑτῷ φύσεως ἔχει τὸ κινεῖσθαι κατὰ τὴν οἰκείαν ἐπιτηδειότητα.

to understand this nature in terms of the Aristotelian "internal principle of change", but examples of a nature that is changed and then produces change are hard to fit into the book II framework, where nature is itself the principle.[32] Perhaps it is more helpful to think here of the modified "principle of motion/ change" (κινήσεως ἀρχή) that we find in book VIII, chapter 4 where Aristotle considers things that move because obstacles were removed and balls bouncing off walls, and concludes: "hence it is clear that none of these things moves itself. But they do have a source of motion, not of causing motion or of acting on something, but of being acted on" (VIII 4 255b30–31)[33]—this would suggest that the κινήσεως αρχή here is a passive principle.[34] Even so, Andronicus' interpretation that foregrounds the internal power of the moved carries the opposite emphasis to that which Aristotle places on the mover at *Physics* VIII 4, where every motion requires a mover that is at least in some sense distinct from the moved. Andronicus is not inconsistent with this doctrine thanks to the concessive clauses, but the difference of emphasis is marked, and it is thus worth asking what might have led him to this departure, which comes dangerously close to endorsing that most un-Aristotelian of ideas, namely self-motion.

One option is that it all comes from a desperate effort to defend the inferior (according to Simplicius and all modern interpreters) text at T3. This would make him again a stickler for textual tradition with little consideration for the doctrinal implications. But perhaps this would be too uncharitable: it is still possible that Andronicus was motivated by questions about how to reconcile the internal principle of book II with the conclusions about external mover(s) in book VIII.[35] In his effort to find a congenial compromise he may have been helped by Chrysippus' analysis of causal powers, whose language is strongly reminiscent of Andronicus' concessive clauses at least as far as we can trust Cicero's reporting in T6:

> **T6** "Hence," he [Chrysippus] says, "just as the person who pushed the cylinder gave it its beginning of motion but not its capacity for rolling, likewise, although the impression encountered will print and, as it were, emblazon its appearance on the mind, assent will be in our power. And

---

32 It is possible to read the "which is also so predisposed" (καὶ προδιατιθεμένη) in a concessive way, "even though it is also so predisposed", which would create a contrast between this nature undergoing change and playing a causal role in the change of the composite object. I owe this suggestion to Inna Kupreeva.

33 ὅτι μὲν τοίνυν οὐδὲν τούτων αὐτὸ κινεῖ ἑαυτό, δῆλον· ἀλλὰ κινήσεως ἀρχὴν ἔχει, οὐ τοῦ κινεῖν οὐδὲ τοῦ ποιεῖν, ἀλλὰ τοῦ πάσχειν.

34 Cf. also the reference to passive potentiality at *Metaph.* IX 1 1046a11–13.

35 On this issue, with particular reference to the four elements, see Bodnár (1997).

> assent, just as we said in the case of the cylinder, although prompted from outside, will thereafter move through its own force and nature." (Cicero, *De fato* 43 = SVF 2.974 = LS 62C, LS translation)[36]

Chrysippus must have found himself in a similar need for compromise between the forces of fate and what is up to us, and some version of his argumentation would have certainly been available to Andronicus. Such a careful distribution of causality is not entirely foreign to Aristotle either, so we should not be too quick to attribute excessive Stoic influences to Andronicus. In an Aristotelian context, this would mean considering cases where there is external efficient causation but also a particular internal nature or "special potentiality" that makes possible the activity of the external agent: the complex case of sexual reproduction is a case in point, where Aristotle is at pains to explain how the male is the principle of motion even though it is not in continuous contact with the offspring. He compares the formation of an embryo to the operation of automatic puppets (τὰ αὐτόματα τῶν θαυμάτων), which possess an internal capacity that ensures the transmission and continued efficacy of the external cause.[37] Perhaps Andronicus had something similar in mind, and understood it also in similar terms to Chrysippus' cylinder, when he spoke of a "predisposed" nature that "gives their disposition from within" to the subjects of change (T4 above).[38]

4 On the Nature of the Soul

The next set of evidence concerns Andronicus' views on the soul.[39] These are included under "physics" on the Aristotelian grounds that it is the business of

36 "ut igitur" inquit "qui protrusit cylindrum, dedit ei principium motionis, volubilitatem autem non dedit, *sic* visum obiectum imprimet illud quidem et quasi signabit in animo suam speciem, sed adsensio nostra erit in potestate, eaque, quem ad modum in cylindro dictum est, extrinsecus pulsa, quod reliquum est, suapte vi et natura movebitur."

37 *GA* II 1 734b9–18. See also Connell (2016), 159–160.

38 Lautner (2002), 130 also considers a Stoic influence on Andronicus, citing Simplicius, *In Cat.* 306.19–27.

39 Additionally to the evidence presented here, Sharples (2007b), 611–613 and (2010), 247–248 discusses some references to Andronicus in the pseudo-Galenic *De spermate*, which will not be taken into account here. This text survives as a Latin translation of what must have been a Greek original from Late Antiquity. It attributes to Andronicus some confusing points about the will, followed by dualistic-sounding remarks to the effect that body and soul are not the same nature and therefore the inherently good soul finds it hard to dominate the impulses and evils of the body.

the natural philosopher to study soul, with the caveat that *nous* may be outside his purview. In any case, in the first passage T7 where Andronicus is named, coming from Galen, Galen has been talking explicitly about the desiderative and spirited parts of the soul only.

> T7 ⟨As for Andronicus the Peripatetic,⟩[40] in general I praise him for having the courage to declare the being of the soul,[41] [speaking] like a free man and not veiling the matter in obscurities (for I find him to be like this also in many other matters); but in that he says that it is either a tempering (mixture) or the power which follows on the tempering (mixture), I criticize the addition of "the power". (Galen, *Quod animi mores* 4, 44.12–20 Müller, transl. Sharples)[42]

Throughout the work in question (whose full title is the thesis *That the Faculties of the Soul Follow the Mixtures of the Body*, Ὅτι ταῖς τοῦ σώματος κράσεσι αἱ τῆς ψυχῆς δυνάμεις ἕπονται) Galen refuses to commit himself to a particular view of the soul, but he lets it be known that he is more in favour than against the thesis proposed in the title. A few pages earlier, at 3, 37.5–24 Müller, he claims that the Aristotelian body-soul hylomorphism only makes sense if we understand "form" not as "shape" but as the *krasis* of the four qualities (hotness, coldness, dryness and wetness) or of the four primary bodies that are hot, cold, dry and wet. It is against this background that Andronicus is singled out for praise as having been outspoken and bold in providing a definition of the soul in terms of *krasis*: scholars have tried to identify one of the two options (the mixture or the power which follows on the mixture) as the one Andronicus actually favoured,[43] but I wonder whether there is some irony here in Galen's praise, which would mean that Andronicus' definition was in fact quite tentative and non-committal, deliberately trying to allow for either the tempering or the power that follows it to be identified with the soul. In the lines that follow after T7, Galen critically points out that if we identify the soul with the power that follows upon the *krasis*, it will fail to be a substance, contrary to Aristotle's

40 The sentence containing the crucial reference to Andronicus is absent in the Greek MSS, but it is found both in the Aldine edition and the Arabic version.

41 Omitting Müller's insertion ⟨κρᾶσιν ἢ δύναμιν εἶναι τοῦ σώματος⟩, with Bazou (2011) and Singer (2013).

42 ⟨… Ἀνδρόνικον δὲ τὸν Περιπατητικόν⟩ ὅτι μὲν ὅλως ἐτόλμησεν ἀποφήνασθαι τὴν οὐσίαν τῆς ψυχῆς ὡς ἐλεύθερος ἀνὴρ ἄνευ τοῦ περιπλέκειν ἀσαφῶς, ἐπαινῶ τε πάνυ καὶ ἀποδέχομαι τὴν προαίρεσιν τἀνδρός (εὑρίσκω γὰρ αὐτὸν καὶ κατ' ἄλλα πολλὰ τοιοῦτον)· ὅτι δ' ἤτοι κρᾶσιν εἶναί φησιν ἢ δύναμιν ἑπομένην τῇ κράσει, μέμφομαι τῇ προσθέσει τῆς δυνάμεως.

43 Cf. Sharples (2007b), 610; Gottschalk (1987), 1113.

claims in *De anima* II 1: the substance-qua-form (which Galen has argued that the Peripatetics must identify with the *krasis* of the body) is what *possesses* the capacity, it cannot be the capacity itself (*QAM* 4, 44.24–30 Müller).

Either way, this evidence allows us, in line with the interpretation advanced by Victor Caston, to situate Andronicus in the Peripatetic tradition that favours a supervenience theory of the soul (in the sense that the soul is dependent upon the tempering of bodily elements): this tradition includes Aristoxenus and Dicaearchus who, as far as we can tell, opted for an epiphenomenalist view that denied the soul any causal efficacy.[44] This seems to be the version that Galen would have preferred on Andronicus' behalf, reducing the soul to the bodily mixture without any power of its own. The view that includes this emergent power is in fact that of Alexander, as found in his *On the Soul* 25.2–3: "For the soul is the power and form that supervenes on the blend of bodies in a particular proportion, not the proportion or composition of the blend" (cf. 26.21–22).[45]

Pronouncements about the nature of the soul, already in Simmias' proposal in the *Phaedo* (86c–d), mention tempering (κρᾶσις) and attunement (ἁρμονία) side by side, and this is reasonable, as what these views are trying to convey is the sense of combination, composition and holding together of the different elements that make up the body. This is also the way in which Aristotle presents the view in *De anima* I 4—that *harmonia* is meant to be a *krasis* and *synthesis* of opposites (407b30–31). It is therefore not surprising that we find Andronicus speaking in favour of a version of the *harmonia* view in T8b. What is perhaps a bit more surprising is that we find him doing this in combination with Xenocrates' "soul as self-moving number" view:

> **T8a** Next [Aristotle] refutes Xenocrates' definition of the soul, not by contesting the term "number", as Andronicus and Porphyry say, but by really inquiring into Xenocrates' meaning, as is clear from the latter's books *On Nature* where he would seem to make the soul a unit-like number. (Themistius, *In DA* 31.1–3, transl. Todd)[46]

---

44 Caston (1997), 351–353.

45 Transl. Caston. Cf. Caston (2012), 114 n. 230: "Alexander's position can be seen as directly opposing Galen's criticism, by defending the second of Andronicus' two alternatives."

46 μετὰ δὲ ταῦτα διελέγχει τὸν Ξενοκράτους ὁρισμὸν τῆς ψυχῆς, οὐ πρὸς τοὔνομα τοῦ ἀριθμοῦ μαχόμενος, ὥς φησιν Ἀνδρόνικος καὶ Πορφύριος, ἀλλὰ καὶ πάνυ τὴν διάνοιαν ἐπιζητῶν τὴν Ξενοκράτους ὡς δῆλόν ἐστιν ἐκ τῶν Περὶ φύσεως αὐτῷ γεγραμμένων, ἐν οἷς ἂν δόξειε μοναδικὸν ποιεῖν ἀριθμὸν τὴν ψυχὴν ὁ Ξενοκράτης.

T8b Instead, we ought to contrast Andronicus' arguments with those of that [critic], inasmuch as they offer at once a clearer and more plausible reconstruction of Xenocrates' theory. [Andronicus] says: "They called the soul number because no animal was derived from an uncompounded body, but only where the primary elements were blended in accordance with specific ratios and numbers. So they made essentially the same claim as those who posit the soul as an attunement, except that they more clearly formulated their theory by adding in their definition that the soul was not every number but just the self-moving number, as if these others too [had defined it] as not every attunement but just the one that attunes itself. For this type of soul is the cause of this type of blend, and thus of the ratio and mixture of the primary elements." (Themistius, *In DA* 32.22–31, transl. Todd)[47]

A few remarks are in order concerning the context of T8: Themistius is reporting together from Porphyry and Andronicus (possibly having found the Andronicus material in Porphyry), on their response to Aristotle's criticism of the self-moving number view in DA I 4 408b30–409b18, where Aristotle does not mention Xenocrates by name but calls the view "by far the most unreasonable" (πολύ … ἀλογώτατον, 408b32). In T8a Andronicus and Porphyry seem to have accused Aristotle of simply focusing on the use of the word "number" without addressing the substance of the view—but Themistius objects that Aristotle was right in taking Xenocrates to task for having a unit-like number in mind.

In any case, the more pertinent material is in T8b, where there is some doubt about the identity of "that [critic]" (ἐκείνου): Sharples takes it to be Porphyry,[48] whose views were discussed in the intervening text between T8a and T8b, but Rashed takes it to be Aristotle:[49] the text itself is quite unhelpful here because it is not clear at all where the report from Porphyry ends. This does not make a great difference for our purposes, given that what looks like a direct quotation from Andronicus is more clearly signposted. In the middle of the passage we have the key claim that identifying the soul with number is pretty much the

47 ἀλλὰ χρὴ παραβάλλειν τὰ Ἀνδρονίκου πρὸς τὰ **ἐκείνου**, ὅσῳ καὶ σαφέστερα καὶ πιθανώτερα πρὸς σύστασιν τοῦ λόγου τοῦ Ξενοκράτους. "ἀριθμὸν γὰρ ἐκάλουν" φησὶ "τὴν ψυχήν, ὅτι μηδὲν ζῶον ἐξ ἁπλοῦ σώματος, ἀλλὰ κατά τινας λόγους καὶ ἀριθμοὺς κραθέντων τῶν πρώτων στοιχείων. σχεδὸν οὖν ταὐτὸν ἀπεφαίνοντο τοῖς ἁρμονίαν αὐτὴν τιθεμένοις, πλὴν ὅσῳ σαφέστερον οὗτοι τῇ προσθήκῃ τὸν λόγον ἐποίουν, οὐ πάντα ἀριθμόν, ἀλλὰ τὸν κινοῦντα ἑαυτὸν τὴν ψυχὴν ἀφοριζόμενοι, ὥσπερ ἂν εἰ καὶ ἐκεῖνοι μὴ πᾶσαν ἁρμονίαν, ἀλλὰ τὴν ἁρμόζουσαν ἑαυτήν· αὐτὴ γάρ ἐστιν ἡ ψυχὴ τῆς κράσεως ταύτης αἰτία καὶ τοῦ λόγου καὶ τῆς μίξεως τῶν πρώτων στοιχείων."

48 Sharples (2007b), 610; (2010), 239.

49 Rashed (2004), 46.

same as identifying it with attunement. Andronicus' grounds for making this connection and the rationale he attributes to Xenocrates are not paralleled in the few other sources who report on Xenocrates (Proclus, *In Tim.* II, 165.3–12 = Fr. 189 Isnardi Parente; Nemesius, *Nat. Hom.* 2. 29.19–30.4 = Fr. 190; Philoponus, *In DA* 165.17–28 = Fr. 197). There, Xenocrates is reported to have thought that the soul is number because the Forms are numbers, and the soul is full of Forms and has the ability to distinguish and count them on the basis of sameness and difference.

By contrast, Andronicus' grounds for assimilating soul-as-number and soul-as-attunement seem to come from the *De anima* itself. In *De anima* I 4 Aristotle says that *harmonia* is either a ratio / proportion (λόγος) or combination (σύνθεσις) of the mixed elements (λόγος τίς ἐστι τῶν μιχθέντων ἢ σύνθεσις).[50] Andronicus, I take it, opted for the former interpretation of *harmonia*, as a *logos*, taking it presumably to mean a numerical ratio, which allows him to speak of "ratios and numbers" together in T8b—this move from λόγος (ratio) to λόγος καὶ ἀριθμός (ratio and number) is his innovation. Now, in his *De anima* discussion Aristotle objects that we cannot identify the soul with such a *logos*, because there are many different ratios in the different bodily parts, another one for flesh, and another for bone (he has Empedocles 31 B 96 DK (= D 192 LM) in mind here),[51] which would mean that the living thing has many souls differing according to the different ratios or combinations in its different parts (408a9–23). It is possible that Andronicus sought to address this objection with his introduction of number alongside ratio, for which he drew favourably on Xenocrates: the soul is not just any number that we might identify in the bodily ratios, but a special, self-moving one. We have already seen above in the discussion of T3 and T4 that Andronicus was not radically opposed to the idea of self-motion. This would also address Aristotle's initial objection to the *harmonia* theory in general, namely that an attunement cannot initiate motion (407b34–408a1). According to Andronicus (T8b) the *harmonia* theorists could have avoided Aristotle's criticism if only they had spoken of a self-attuning *harmonia* with causal powers. It is not entirely clear what this would involve, but perhaps Andronicus is simply providing a hypothetical analogy.

50 We should note that, unlike Themistius / Andronicus, Aristotle speaks in terms of μίξις (blend), not κρᾶσις (tempering), to indicate that the ingredients in question are fully mixed. On this material see Betegh (2021), esp. 8–11.

51 He quotes this fragment in the next chapter, 1.5, 410a4–6: ἡ δὲ χθὼν ἐπίηρος ἐν εὐστέρνοις χοάνοισιν τὼ δύο τῶν ὀκτὼ μερέων λάχε νήστιδος αἴγλης τέσσαρα δ' Ἡφαίστοιο· τὰ δ' ὀστέα λευκὰ γένοντο. The role of Empedocles in Aristotle's discussion is analysed in detail by Betegh (2021).

The very important point of course is the causal power with which the soul is invested here: it is much stronger than what we found in Galen, and stronger than Alexander's view, because the soul is now said to be the cause of the tempering and not even an emergent power over and above the tempering. However, the question of consistency with Galen's report in T7 should not be a matter of great concern, because this is meant to be Xenocrates' view (wrongly maligned by Aristotle according to Andronicus), as well as what ought to be the *harmonia* theorists' view—not Andronicus' own.

Thus far we have Andronicus oscillating between an epiphenomenalist and an emergentist view of the soul in T7, while also treating Xenocrates' self-moving number very sympathetically in connection with his *krasis / harmonia* view in T8b. It should not go unnoticed that this is not the only instance where Andronicus is associated with Xenocrates: in Simplicius' commentary on the *Categories*, they are cited together in connection with the view that there is a higher division of the items in the categories into the ones that are καθ' αὑτό (*per se*) and those that are πρός τι (in relation to something), Simplicius, *In Cat.* 63.22–24. In the light of this, it is very much worth asking whether it is possible to come up with a holistic explanation for Andronicus' interest in Xenocrates: on that score, Marwan Rashed has argued that Andronicus drew ammunition from the Xenocratean distinctions in order to deny the status of substance to the soul (like Boethus) since it is *in* the body as a substrate, and thus not *per se* (as we saw, this was the implication that Galen drew from the second of the two options he attributed to Andronicus, on which the soul is a power that follows upon the tempering of the body). A definition of the soul like the one defended by Andronicus on Xenocrates' behalf, Rashed argues, would allow him to place the soul in the category of quantity or relative.[52] But I will not pursue the matter further here, because it takes us quite far into metaphysical / ontological questions and away from the current focus on natural philosophy.

On the immediate question of Andronicus' definition of the soul, our primary source remains Galen's report in T7 which, as suggested above, testifies to an oscillation between an epiphenomenalist and an emergentist position that would allow causal powers to the soul via the addition of the *dynamis*. From Rashed's argument and the parallel, stronger evidence on Boethus[53] we can conclude that Peripatetics in the late Hellenistic period did not feel particularly bound to Aristotle's own placement of the soul in the category of substance, presumably emboldened by the fact that even in the definitions of the soul given by Aristotle in *De anima* II 1, the soul is always (said) *of* the body.

52 Rashed (2004), 44–48.

53 See Auffret (2020).

5 Conclusions

We may now draw some tentative conclusions. We have seen that Andronicus was in favour of subdividing Aristotle's *Physics* into 5+3 books and it is clear from Simplicius that he did this in his major work on Aristotle's books. But what was the aim of such subdivisions? It could be an effort to orient readers as Aristotle's work is not very transparent structurally, but it could also be a comment on the history of the text, indicating that two initially separate sets of books had been brought together, probably by the time of Theophrastus and Eudemus.

Secondly, Andronicus developed a distinct view of change which placed more weight on the inner passive potentiality of the thing that undergoes change. He may have developed this in an effort to justify an alternative MS reading of Aristotle's claim that change is *in* the patient, but we have seen that he offered some exegesis of the passage. We have seen three passages that bear striking similarities and we cannot exclude that Simplicius is paraphrasing in different ways a point that Andronicus made only once. But in what sort of work did Andronicus make this point? The simplest option is to posit some exegetical work on the *Physics*. Similarly, the context of his discussion of Xenocrates' view on the soul is tracking the text of the *De anima* quite closely, inviting us to think about a further exegetical work on that treatise too, which would have also included his definition of the soul.

On the other hand (and now I am venturing into more speculative territory), it might be worth asking whether all this could have come from his interpretation of the *Categories*. On this scenario, he could have brought up the reading of the *Physics* passage under a discussion of the categories of acting and undergoing (ποιεῖν and πάσχειν),[54] and the material on the soul under the discussion of substance, form and differentia. A similar argument has been made by Chiaradonna and Rashed in the case of Boethus.[55] However, the evidence for Andronicus is even more sparse than that on Boethus, and the surviving Neoplatonic commentaries do not mention Andronicus or views such as those discussed above in the relevant sections, so an agnostic conclusion as to the ultimate Andronican source is inevitable.

54 The fact that a treatise Περὶ παθῶν was falsely attributed to Andronicus could suggest that he had broader interests in the notion of πάθος in the sense of "affective quality"; we know for example that he had views on the παθητικαὶ ποιότητες at *Cat.* 8.9a35–b10, cf. Simplicius, *In Cat.* 258.15–18 and is also cited by Aspasius on emotions (πάθη), *In EN* 44.20–28.

55 Chiaradonna and Rashed (2020), 7.

CHAPTER 12

Aspasius on Providence

David Lefebvre

It has often been noted that the idea of a Peripatetic theory of providence is, at the very least, odd—if not outright paradoxical.[1] Aristotle's theology (if we take this to refer to the teachings of Book XII of the *Metaphysics* on the prime mover), insofar as it implies a rejection of the anthropomorphic view of divine action and an artisanal model of causation, seems to make providence impossible—and yet, Aristotle's philosophy is not devoid of the foundations needed to build a *sui generis* conception of providence.[2] Nonetheless, it must be acknowledged that Aristotle deliberately avoided making explicit references to providence, thus distancing himself from Plato and Xenophon.[3] According to our sources, Alexander of Aphrodisias was the first to develop a systematic teaching on providence (and fate) within the Peripatetic school. However, he was not the first Peripatetic to attribute a theory of providence to Aristotle.[4] On this point as on others, Stoicism and Middle Platonism put pressure on the Peripatos to update its doctrines, albeit in a minimal—or, one might say, "superficial"—manner. That said, we possess very little information on this subject from the Hellenistic and imperial Peripatos. In this context, Aspasius, writing in the second century CE, represents an important case study. He is the earliest extant commentator whose work has come down to us: apart from doxographers and Adrastus,[5] no other Peripatetic text on providence from this

1 Moraux (1970), 41; Sharples (1999), 159, and (2010), 206.

2 From the recent contemporary literature, see Babut (1974, 2019), 138–152; Bodéüs (1992), 191–281; Broadie (1993), 396, and Horn (2016) from *Metaph.* XII 10, 1075a11–25; and Sharples (2010), 206 n. 28. Sharples (2002a), 4–12 offers a still valuable *status quaestionis* on Aristotle's theology.

3 On the providence of the god (or gods) in Plato, see *Timaeus* 30c and 44c, as well as *Laws* X, 899d4–905d6; and Xenophon, *Memorabilia* I 4, 11–19, and IV 3.

4 See Fazzo (1999), 10 and 24 n. 27 (who quotes our passage from Aspasius) and Sharples in Moraux (2001), 524.

5 See Petrucci (2015) and Fazio and Petrucci in this volume. We should consider the pseudo-Aristotelian treatise *De mundo* dated "between the second half of the third century BCE and the first half of the first century BCE, though a later date cannot be excluded" by Gregorić and Karamanolis (2020), 8. See below n. 18.

period is available to us.[6] Furthermore, Aspasius demonstrates an extensive and detailed knowledge of Aristotle's corpus, particularly the treatises that are most relevant to the problem of providence.[7] We also know that Alexander was familiar with Aspasius' commentaries.[8] While it may seem surprising that a discussion of providence survived in his sole extant work—his commentary on Aristotle's *Nicomachean Ethics* (NE)[9]—he briefly addresses the topic in his commentary on a passage from NE III 3, which deals with the objects of deliberation.[10]

This stretch of text is not an isolated instance, which can be explained by the fact that providence is a topic of concern to both ethics and physics. There are several passages in both of his ethical treatises in which Aristotle introduces themes closely tied to providence. Commentators often point to NE X 7, 1177b26–31 (on the divine nature of the human intellect) and 9, 1179a24–26 (on the existence, ὥσπερ δοκεῖ, of care, ἐπιμέλεια, on the part of the gods for human affairs), as well as EE VII 10, 1242a32–35, and 12, 1244b7–10 (on the *philia* between god and human being).[11] As Aspasius' extant commentary covers books I–IV and VII–VIII of the NE, nothing has been preserved of his commentary on these passages. In other sections where Aspasius might have detected allusions to providence, he remains notably circumspect: regarding NE I 9, where, in the context of the discussion of the claim that happiness is "a gift of the gods" (1099b11–18), Aristotle affirms that the study of this question would be "more appropriate to another kind of investigation" (b14), Aspasius merely notes that Aristotle refers to a "more physical" type of enquiry (25.33).[12]

6 See Dragona Monachou (1994) for the period before Alexander; Moraux (1970), 41–65; Sharples (2010), 196–210. On the period from 80 to 250 for the Platonic philosophers, see Boys-Stones (2018), 323–343.

7 See Becchi (1994), 5368–5369; Barnes (1999), 8–13, attributes to Aspasius a commentary on the *Categories*, *De interpretatione*, *Physics*, *De caelo*, *De sensu*, *Metaphysics* and possibly the *Topics*, as does Kupreeva (2016), 140–141. Moraux (1984) 240 does not believe that Aspasius commented on the entire *De caelo*. The passage to be examined indicates that Aspasius had a thorough familiarity with the *De caelo* and the *Metaphysics*.

8 Barnes (1999), 9–11.

9 Aspasius' *Commentary* is dated shortly after 131 by Barnes (1999), 3.

10 I follow Bywater (1894) for the division of the NE into chapters.

11 See Moraux (1970), 51; Thillet (2003), 18–26; Fazzo (1999), 26–31; and Bodéüs (1992), 229–242. In NE VIII 7, 1158b11–14, Aristotle examines friendships involving superiority without mentioning the relationship between gods and humans. However, Aspasius uses this example (178.30–179.1) to affirm that there is no friendship between god and human beings, because the distance is too great. The gods can be "favourable" to human beings without any *philia*.

12 I quote the Greek text of Aspasius' commentary from Heylbut's edition (1889). I use the translation by Konstan (2006) unless otherwise indicated.

Aspasius does not elaborate further, but this clarification is significant:[13] as Moraux aptly put it, Aristotle's metaphysical conception of god implies that providence belongs exclusively to physics, since god's relation to the world is that of the mover to the moved.[14] Raising the question of whether the gods might grant happiness as a gift is, therefore, an inappropriate way of analysing god's relationship to the world. Any sound examination of divine intervention in the world must necessarily fall within the scope of physics. In his commentary on *NE* VIII 1, 1155a16–22, Aspasius offers a more explicit treatment of the quasi-providential aspects of nature. Aristotle explains here that there is a form of friendship that exists "by nature" (φύσει), as seen in the reciprocal *philia* between parents and offspring in humans as well as most animals, and that a natural friendship similarly exists between individuals of the same species, including human beings. Aspasius emphasises the teleological role of nature in this context, stating: "Nature has necessarily organised things so that there is friendship towards the little ones not only in human beings but also animals" (ἡ φύσις πρὸς τὰ ἔγγονα φιλίαν μηχανᾶται ἀναγκαίως, οὐ μόνον ἐν ἀνθρώποις, ἀλλὰ καὶ ἐν τοῖς ἄλλοις ζῴοις, 159, 26–27); he adds that nature "necessarily made those who are of the same species feel friendship with each other" (καὶ τοῖς ὁμοεθνέσι δὲ ἀναγκαίως ἡ φύσις ἐνειργάσατο πρὸς ἀλλήλους φιλίαν, 159, 30–31). While the analogy between nature and a craftsman is not unusual in Aristotle's physics,[15] Aspasius' commentary is noteworthy for making explicit, in this ethical context, nature's role as an agent that acts through conditional necessity. In Aristotle's original text, nature merely appears in the dative, without a corresponding explication. While Aspasius does not talk explicitly about providence in these passages, we can see that where it does emerge, it is in a *physical* context.

The handful of lines from Aspasius' commentary on *NE* III 3 reflect this general outlook. Despite the context, they do not involve any reference to the ethical dimension of providence (the individual care exercised by the gods to reward the virtuous and punish the wicked). Instead, Aristotle's explanations of the objects of deliberation, which are more on the order of physical

13 This is discussed in *EE* VII 14. Commentators have thought that Aristotle only wanted to get rid of a question that he thought belonged to the poets. See Burnet (1900), 46 ad loc., according to whom the question belongs to a "theological" study, and Gauthier-Jolif (1970), 72–73.

14 Moraux (1970), 51–52, writes: "It was not by studying the relationship of God to man but that of God to the world that the Peripatetics built their theory of providence" (my translation).

15 *PA* I 5, 645a9, II 9, 654b32; *IA* 711a18; *GA* II 1, 731a24, etc.

or ontological considerations, prompt Aspasius to digress briefly into physics and theology. He appears to rely on two key texts to clarify the role of providence within a Peripatetic framework: *Metaphysics* XII 7 and the treatise *De caelo*. From Aspasius' perspective, however, this is hardly a digression, but rather a response to an exegetical difficulty in Aristotle's text. His commentary at this point reflects, if not a misunderstanding, at least an overinterpretation of a passage where Aristotle likely intended to maintain some level of ambiguity or openness. Aspasius gives a strong doctrinal interpretation of a statement by Aristotle that could have been understood in a weaker (i.e. doxographic) sense. We can reasonably infer that this bears witness to a conception of providence developed by Aspasius elsewhere or, at the very least, to a conception that was probably common *doxa* in the Hellenistic and imperial Peripatos before Alexander of Aphrodisias. Put simply, this conception holds that providence operates first in the heavenly regions of the cosmos and then extends into the sublunary world, albeit in a "diminished" way.[16]

In the following sections, I will: (1) place Aspasius' position in its Peripatetic context, (2) outline the difficulties in the Aristotelian passage from *NE* III 3 regarding the objects of deliberation, and (3) examine Aspasius' commentary.

1 The Peripatos on Providence before Alexander

Sharples has summed up the standard view that providence is confined to the heavenly regions of the cosmos with the slogan "No Sublunary Providence" (NSP). In Alexander's words, providence "stops at the moon"[17] and does not extend below it into that part of the cosmos composed of the four elements and subject to generation and corruption.[18] Sharples identifies this as the central tenet of the Peripatetic position on providence during the Hellenistic and imperial periods before Alexander. The origin of this claim (at least chronolo-

16 I borrow the expression from Kupreeva (2009), 146.

17 In the *De providentia* (59.6–12), Alexander attributes this claim to Aristotle, which is nowhere to be found in the Aristotelian corpus. Sharples (2002a), 30, and (2007a), 603.

18 See Sharples (1982); Sharples (2002a), 14–18, 22–30; Sharples (2003) and (2007a); Moraux (1949), 33; Festugière (1981), 221–263. Moraux (1984), 571 n. 33, sees in this claim a simplification due to the doxographic tradition. I leave aside here the *De mundo*, whose conception of providence does not fall under NSP and whose date is uncertain (n. 5 above). See Sharples (2002a), 15, 25–26, Betegh and Gregorić (2020), 207–208, and Gregorić in this volume.

gically) is thought to be Critolaus of Phaselis, the scholarch of the Lyceum, as reflected in fragment 15 W:[19]

> Aristotle, son of Nicomachus, was a Macedonian from Stagira according to some, but a Thracian in race according to others. He said that there are two principles, god and matter, and that the things above the moon are objects of divine providence, but the things below the moon exist without providence and are borne along in some irrational way as chance has it. He says that there are two world orders, that above and that below, and that which is above is imperishable, but that which is below is subject to passing away. And he says that the soul is the continuous activity of the body. […] Critolaus of Phaselis held the same opinions as Aristotle.[20]

The value of this testimony has been contested, as the opinions are not directly attributed to Critolaus.[21] This fragment forms part of a series in which Epiphanius, having summarised the doctrine he attributes to Aristotle, asserts that Theophrastus shared Aristotle's views and Praxiphanes shared those of Theophrastus, while Critolaus, in turn, aligned himself with Aristotle. Critolaus' position is characterised by its proximity to Stoicism (acknowledging two principles: god and matter) as well as by its resistance to Stoic cosmological monism and its conception of providence.[22] The fragment attributes a strongly dualistic cosmological position to the Peripatetic school: the part "above" the moon, which is incorruptible, is the object of "divine providence", while the part "below" the moon is corruptible and "ignored by providence" (ἀπρονόητα). Moreover, sublunary motions are described as irrational and random, which explains why they are not, in fact, "foreseeable" (taking ἀπρονόητα literally here).[23] Critolaus represents the most robust version of this cosmolo-

---

19 Wehrli (1969), 66.

20 Critolaus of Phaselis, Fr. 15 W (*ap*. Epiphanius, *Adv. Haereses* III, 31 and 35): Ἀριστοτέλης ὁ Νικομάχου κατὰ μέν τινας Μακεδὼν ἀπὸ Σταγείρων, ὡς δὲ ἔνιοι Θρᾷξ ἦν τὸ γένος. ἔλεγε δὲ δύο ἀρχὰς εἶναι, θεὸν καὶ ὕλην, καὶ τὰ μὲν ὑπεράνω τῆς σελήνης θείας προνοίας τυγχάνειν, τὰ δὲ κάτωθεν τῆς σελήνης ἀπρονόητα ὑπάρχειν καὶ φορᾷ τινι ἀλόγῳ φέρεσθαι ὡς ἔτυχεν. εἶναι δὲ λέγει δύο κόσμους, τὸν ἄνω καὶ τὸν κάτω, καὶ τὸν μὲν ἄνω ἄφθαρτον, τὸν δὲ κάτω φθαρτόν. καὶ τὴν ψυχὴν ἐνδελέχειαν σώματος λέγει. […] Κριτόλαος ὁ φασηλίτης τὰ αὐτὰ τῷ Ἀριστοτέλει ἐδόξασε. Translation Sharples (2010), 202–203, Text 22 O; with apparatus FHS&G Fr. 162, 326–327.

21 And, of course, due to the two other doctrines attributed to Critolaus: the two principles and the definition of the soul. We cannot go any further into these questions here. See Moraux (1970), 55; Sharples (1998), 101–105, and (2002a), 22–23; Hahm (2007), 85–89, against Wehrli's skepticism. The fragment is discussed in Kupreeva (2009), 142–150.

22 Diogenes Laertius VII 134, 1–8 (SVF I 85).

23 See Festugière (1981), 226, and, e.g., *MM* II 6, 1203a31.

gical dualism. However, alongside this cosmological fragment, there is another fragment, which suggests that, according to Critolaus, the action of providence might extend to nature and human beings considered in general.[24]

A similar view can be found in Diogenes Laertius' *Life of Aristotle*, albeit with a potentially less pronounced dualism.[25] In his account, earthly bodies are said to be "organised" according to their "sympathy" with celestial bodies. It should also be noted that, for Diogenes, the idea that the god exercises providence in Aristotle does not prevent him from being considered "incorporeal" and "immobile". Things are somewhat different again in Adrastus' fragment,[26] which emphasises that sublunary beings can only follow noble and divine beings "by accident" (τῶν δ' ἐνταῦθα κατὰ συμβεβηκὸς ἐκείνοις ἑπομένων, 149.14–15). In this view, earthly beings are therefore not the intentional end of the divine providence, but benefit from it "by accident".[27]

The NSP label thus conceals a range of positions that are challenging to reconstruct.[28] The relative caution of the Hellenistic and post-Hellenistic Peripatos in elaborating a doctrine of providence is somewhat surprising, given the potential of the materials found within Aristotle's corpus to address such questions.[29] This apparent minimalism did, however, allow the Peripatos to assert a distinctive position relative to other philosophical schools. This position can be situated in relation to four contemporary conceptions in the Hellenistic and

24 This is the hypothesis of Mayhew (Forthcoming), based on two texts by Plutarch, *Political Precepts*, 42.15, and *Pericles* 7.7 (Frs. 35a and 35b W; Texts 22 K and L in Sharples 2010, 201). Both texts use political analogies that can be given a theological meaning. The first (which quotes two lines from Euripides) can be understood within the framework of the NSP doctrine, but the second seems likely to imply the god's use of intermediaries. See Kupreeva (2016), 147–148, and also Donini (1974), 112–113, based on the notion of "theoretical justice" in Aspasius, 2.2–4.

25 Diogenes Laertius V 32, 3–6: τὸν δὲ θεὸν ἀσώματον ἀπέφαινε, καθὰ καὶ ὁ Πλάτων. διατείνειν δὲ αὐτοῦ τὴν πρόνοιαν μέχρι τῶν οὐρανίων καὶ εἶναι ἀκίνητον αὐτόν· τὰ δ' ἐπίγεια κατὰ τὴν πρὸς ταῦτα συμπάθειαν οἰκονομεῖσθαι. Things are slightly different in Hyppolytus, *Refutation of all Heresies*, VII 19.2, who explains that under the moon there is neither order nor providence, "nature alone suffices".

26 Adrastus, *ap*. Theon, *De utilitate mathematicae*, 149.4–15.

27 See the same reference to accident in Aëtius (II 3.4). On the meaning of this expression, see Fazio and Petrucci here p. 311–312.

28 See Sharples (2002a), 22–26.

29 In addition to the references cited above from the *Ethics* and *Metaphysics*, one might add *GC* II 10, 336b30–32; *DC* I 4, 271a33, and II 9, 291a18–26. In this passage, Aristotle explains that everything happens *as if* nature had foreseen what would have happened (ὥσπερ τὸ μέλλον ἔσεσθαι προνοούσης τῆς φύσεως, 291a24–25), if the heavenly bodies moved in a mass of air or fire: the sound produced would be transmitted to earth and produce violent destruction.

imperial periods before Alexander:[30] (i) Epicurean anti-providentialism, which flatly denies divine care or intervention in the world;[31] (ii) Middle Platonic providentialism, as seen in figures like Atticus, who argued that a transcendent god, creator of the cosmos according to the artificialist model of the *Timaeus*, demonstrates, through the world's soul, moral and retributive concern towards the world and human beings;[32] (iii) Stoic pan-providentialism, which assumes that *pronoia* penetrates and governs the entire cosmos, extending individual and benevolent care to (almost) all particular phenomena and human beings;[33] (iv) Hellenistic and post-Hellenistic Peripatetic providentialism, which maintains that an immobile and non-demiurgic god exercises a purely kinetic and non-intentional providence, confined to the heavenly regions within an eternal cosmos.[34] This providence is transmitted to the sublunary world through a series of intermediaries, such as the celestial spheres and the motions of the sun and moon. However, it is impossible to claim in this context that generation and corruption on earth are the *purpose* of the first unmoved mover, let alone that moral providence or divine justice exists in the sublunary world. Adrastus reflects this view with his assertion that natural realities "follow" an order, but only "by accident".

This background explains the criticism of Atticus, who equates Aristotle with Epicurus on the grounds that neither of them offers divine care for human affairs, divine justice, or fear of the gods among the wicked.[35] The doctrine Alex-

30 In his *Discourses* I 12, Epictetus distinguishes five positions on providence. Sharples (2003) identifies them.

31 Cic., *ND* I VIII, 18–20.

32 See Boys-Stones (2018), 324; Burns (2020), 26–30; Chiaradonna (2023), 29–41. A full account of the different nuances in Middle Platonic conceptions of providence—especially in Apuleius and Pseudo-Plutarch—goes beyond the scope of this discussion. On this see Sharples (2003). On Atticus, in particular, who was a proponent of a literal reading of the *Timaeus*, see Michalewski (2024), 46–62. It is perhaps worth outlining here how Atticus portrays Aristotle's cosmology (see Frs. 3.1, 7–11, and 8.12, 2–4 Des Places). His interpretation constitutes another version of the broader NSP doctrine. According to Atticus, what defines the Aristotelian cosmos is the absence of governance by a soul, as in Plato's system. Instead, celestial realities are governed by fate, earthly realities by nature and human affairs by reason, prudence and the soul. Atticus distinguishes between a primary (non-psychic) principle, which serves as a source of motion, and nature, which functions as a final (non-animate) cause. Thus, while celestial realities follow a certain order, no divine providence oversees human affairs.

33 SVF I 176 and II 528, 933 and 937. However, the scope of providence remains unclear (i.e. everything including the details or just the main things). On this, see Sharples (2003), 110–112; Algra (2014), 119–126.

34 See Critolaus' arguments on this point (Frs. 12 and 13 W) and Falcon in this volume.

35 Fr. 3.7–10 Des Places and Text 22 N in Sharples (2010), 202. See Michalewski (2017) and

ander presents in his treatise (lost in Greek) on *Providence* and in his *Quaestiones* (I 25, II 19 and 21) is a response to Atticus' criticisms and the limitations of the earlier Peripatetic views.[36] Alexander addresses these issues within a purely physical or cosmological framework embedded in a set of theoretical constraints that are distinctively Peripatetic: the heavenly regions do not require providence (as they are eternal and inherently well-ordered),[37] while the sublunary world cannot benefit from providence either primarily (as this would make it the purpose of what is superior to it) or accidentally (as this would no longer constitute providence in the true sense).[38] For Alexander, providence, via the regularity of celestial motion, especially that of the sun, relates only to the sublunary part of the cosmos, insofar as it ensures the reproduction of living species, following *Physics* II 2, 194b13, and does not concern individuals. In this sense, Alexander takes Aristotle to mean that "things here below are not alien to providence" (οὐκ ἀπρονόητα κατ᾽ αὐτὸν τὰ τῇδε).[39]

2 *Nicomachean Ethics* III 3

The starting point for Aspasius' development with regard to providence is a passage from *NE* III 3, in which Aristotle examines the types of objects about which there is deliberation (βουλευτόν) and distinguishes between various causes of natural phenomena. It is no surprise, then, that this section sparked discussion during Hellenistic period. The passage reads as follows:

> Well, no one deliberates about eternal things, as for example about the universe or about the fact that the diameter and side of a square are incommensurable. But for that matter neither does anyone deliberate about things which involve change, but which always occur in the same pattern, whether from necessity, or also by nature,[40] or through some

(2024), 58–59. As a rule, Atticus opposes any attempt to reconcile Plato and Aristotle. He therefore emphasises the doctrinal differences between them in every part of their philosophy, particularly ethics. It is therefore as a result of the supposed proximity between Aristotle's ethics and that of Epicurus that Atticus associates the two philosophers.

36 Sharples (1987), 1216, and (2007a), 602; Guyomarc'h (2015), 289–291.

37 *De Providentia* 59.13–63.1; *Quaest.* I 25.41.4–19 and II 19.63.10–15.

38 *De Providentia* 53.1–13 and 63; *Quaest.* II 21.68.19–69.31. On this dilemma, see Sharples (1982), 204–208, and Sharples (2002a), 30.

39 Alexander of Aphrodisias, *In Meteor.* 83.9–9. On the Alexandrian doctrine of providence in a broader context, see Rashed (2007), 294–304, and Rashed (2011), 142–161.

40 The καί (translated by "also") in 1112a24 is found only in K^b, not in Aspasius (71.23) or the *Anonymous Commentary*.

> other cause, e.g. turnings and risings of celestial bodies; nor about things that happen sometimes one way, sometimes another, like droughts and rainstorms; nor about things that happen from chance, like discovering a cache of treasure. But there is no deliberation, either, about all human affairs, as for example no Spartan deliberates about how Scythians might best manage themselves politically—for none of these things will come about through our agency. What we do deliberate about are the things that depend on us and are doable, and these are in fact what is left once we have been through the rest. For the causes of things seem to be nature, and necessity, and chance, and then, in addition to these, intelligence and everything that occurs through human agency [...].[41]

The end of the text (1112a31–32) shows that Aristotle proceeds by elimination, drawing from a list of four causes inherited from the tradition:[42] nature, necessity, chance and intellect (and everything that comes from human beings). While the standard lists of causes in Plato's and Aristotle's works typically distinguish between nature, art and chance,[43] the list given here—nature, necessity, chance and intellect (or the human being)—is less common.[44] Notably, Aristotle omits art, but includes intellect, which encompasses all practical and technical human activities. On this basis, he excludes four groups of entities from the scope of deliberation:

Group 1: Eternal beings, exemplified by the world and the incommensurability of the diagonal to the side of a square.

Group 2: Beings in motion that always occur in the same way. Aristotle gives three causes for this type of phenomena: εἴτ᾽ ἐξ ἀνάγκης εἴτε καὶ φύσει ἢ διά τινα αἰτίαν ἄλλην (1112a24–25). He also lists two examples: the solstices and the rising of the stars.

---

41 *NE* III 3, 1112a21–33 (Bywater): περὶ δὴ τῶν ἀιδίων οὐδεὶς βουλεύεται, οἷον περὶ τοῦ κόσμου ἢ τῆς διαμέτρου καὶ τῆς πλευρᾶς, ὅτι ἀσύμμετροι. ἀλλ᾽ οὐδὲ περὶ τῶν ἐν κινήσει, ἀεὶ δὲ κατὰ ταὐτὰ γινομένων, εἴτ᾽ ἐξ ἀνάγκης εἴτε καὶ φύσει ἢ διά τινα αἰτίαν ἄλλην, οἷον τροπῶν καὶ ἀνατολῶν. οὐδὲ περὶ τῶν ἄλλοτε ἄλλως, οἷον αὐχμῶν καὶ ὄμβρων. οὐδὲ περὶ τῶν ἀπὸ τύχης, οἷον θησαυροῦ εὑρέσεως. ἀλλ᾽ οὐδὲ περὶ τῶν ἀνθρωπίνων ἁπάντων, οἷον πῶς ἂν Σκύθαι ἄριστα πολιτεύοιντο οὐδεὶς Λακεδαιμονίων βουλεύεται. οὐ γὰρ γένοιτ᾽ ἂν τούτων οὐθὲν δι᾽ ἡμῶν. βουλευόμεθα δὲ περὶ τῶν ἐφ᾽ ἡμῖν καὶ πρακτῶν· ταῦτα δὲ καὶ ἔστι λοιπά. αἰτίαι γὰρ δοκοῦσιν εἶναι φύσις καὶ ἀνάγκη καὶ τύχη, ἔτι δὲ νοῦς καὶ πᾶν τὸ δι᾽ἀνθρώπου. Translation Rowe in Broadie and Rowe (2002) with minor changes.

42 See the δοκοῦσιν (1112a32).

43 Gauthier-Jolif (1970) 199. See Plato, *Laws* X 888e; Aristotle, *Metaph.* VII 7, 1070a6; *Phys.* II 6, 198a9–10; *APo.* II 11, 95a7, etc.

44 See *EE* II 6, 1223a11–13; *Protreptic* B 11 Düring and somewhat differently *Rhet.* I 10, 1368b32–37.

Group 3: Phenomena that occur irregularly, such as droughts and rains.[45]
Group 4: Events that happen by chance (ἀπὸ τύχης).

What remains is the domain governed by intellect and dependent on human beings, in which Aristotle locates the objects of deliberation. While Aristotle's reasoning is clear and straightforward, it is essential that we delve into the physics or ontology underlying this argument, as this is the aspect that is of particular interest to Aspasius. The relationship between the three causes mentioned at the end of the text (necessity, nature and chance, 1112a32) and the four excluded groups of beings is not entirely transparent. The four groups do not correspond directly to the three causes, with two of the three causes being assigned to group 2 (1112a24–25): the cause of the first group of entities is unspecified (but is undoubtedly necessity); the second group has at least two causes (necessity and nature), but another unnamed cause (which cannot be chance) is also mentioned; no explicit cause is assigned to the third group (presumably nature); and the fourth group encompasses everything related to chance.

Given Aspasius' focus, the discussion can be narrowed down to the three causes mentioned in the second group: "either by necessity, or also by nature, or by some other kind of cause" (1112a24–25). The three words are coordinated (εἴτ' ἐξ ἀνάγκης εἴτε καὶ φύσει ἢ διά τινα αἰτίαν ἄλλην, 1112a24–25), and Aristotle adds an unnamed "other cause". This lack of specificity raises interpretative issues. Notably, Aristotle's statement confirms that all three terms refer to *causes*. Aristotle later reiterates that nature and necessity are among the causes (1112a31–32). While Aristotle's definition of nature as a principle and cause is well-established,[46] things are different with necessity. The expression ἐξ ἀνάγκης generally denotes events that "always" occur (the two terms are convertible) and is often contrasted with ὡς ἐπὶ τὸ πολύ, which refers to how *natural* phenomena typically occur, rather than a causal explanation.[47] Now, in our text, necessity seems to be understood not as a *modality* according to which phenomena occur, but as a *cause* in its own right. It is placed alongside nature and framed as a distinct causal principle. The phrase concludes with the mention of "some other cause" (διά τινα αἰτίαν ἄλλην), reinforcing the notion that necessity

45 On the explanation of droughts (when dry exhalation becomes more important than wet exhalation) and rains, see *Meteor.* II 5, 360a33–b26, and 8, 366a33–b14.

46 *Phys.* II 1, 192b20–23.

47 *Phys.* II 5, 196b10–17; *APr.* I 13, 32b9–10; *Metaph.* V 30, 1025a15; GA IV 4, 770b10–11; GC II 9, 335a34–b4.

is indeed presented as a cause, although Aristotle does not provide a precise definition of it. As we will see, this point is of central importance in Aspasius' commentary.

Before addressing Aspasius, let us examine how commentators have interpreted this passage. Broadly speaking, two main approaches emerge:

(1) The standard interpretation: Modern commentators have generally understood Aristotle to be distinguishing two classes of exclusive phenomena. On the one hand, there are those moved in a necessary manner (i.e. celestial bodies). On the other, there are all *natural* sublunary phenomena, such as the motions of simple bodies and biological phenomena, which occur "most of the time".[48] This interpretation suggests that the four groups excluded from deliberation follow a top-down order, from the outermost to the innermost regions of the cosmos, and from the most necessary to the most random:

Group 1: Eternal and necessary phenomena, such as the world and the properties of mathematical entities.
Group 2.1: Beings always moved in the same way *by necessity*: celestial bodies.
Group 2.2: Beings always moved in the same way *by nature*: sublunary beings.
Group 3: Natural phenomena that occur irregularly, such as droughts and rains.
Group 4: Events that occur by chance.

This interpretation subdivides Group 2 by distinguishing between what belongs to necessity and what belongs to nature. However, it also raises several issues. The two examples provided at 1112a25–26 at the end of the description of Group 2 (the solstices and the rising of the stars) do not support this subdivision, since they pertain to astronomical phenomena, even though they are supposed to illustrate all the phenomena in this group. Moreover, as Stewart has noted, necessity and nature are not mutually exclusive in Aristotle's philosophical framework: celestial bodies, whose motions are necessary, also move "according to nature", and necessity is not absent from motions occurring according to nature.[49] Finally, it is false to claim that natural *sublunary* phenomena always occur in the same way, as is required for inclusion in Group 2. Natural phenomena display a regularity that prevents them from fitting neatly into either Group 2 or Group 3. As Aristotle famously states: "Everything that is

48 See Stewart (1892), 253–259; Burnet (1900), 126; Tricot (1959, 1987), 133 n. 2 ad loc. Frede (2020), 474, brings the three causes together by speaking of what is "naturally necessary".

49 See *APo.* II 1, 94b37–95a3, and Stewart's lengthy commentary cited in the previous note. This likely accounts for the καί in K^b at 1112a24: εἴτε καὶ φύσει.

by nature occurs either always or for the most part."[50] Yet phenomena occurring "for the most part" are not explicitly accounted for here.

Furthermore, this interpretation largely overlooks the third, unnamed cause mentioned in the passage. Commentators have mostly remained silent on this point. Within an Aristotelian context, this other cause could refer to the separate substance (the prime mover of the motion of the heavenly bodies)[51] or to the matter of celestial bodies (the fifth element or aether). Both interpretations are supported by the astronomical examples at the end of the sentence, but it is unlikely that these two astronomical examples were given by Aristotle to illustrate *only* the last cause left anonymous by Aristotle.

As a result, an alternative reading of this passage emerges. Group 2 likely refers to phenomena that always occur in the same way, specifically all astronomical phenomena. These phenomena are described as occurring "either by necessity, or by nature, or by some other cause". The "third cause" could plausibly refer to either the prime mover or the material cause of celestial bodies—points on which Aristotle deliberately avoids taking a position in this ethical context. A significant implication of this interpretation is that natural sublunar beings are ultimately excluded from mention in any of the four groups.

(II) A contextual interpretation: In this ethical context, Aristotle may be deliberately refraining from becoming embroiled in a controversy over the precise physical causes of these phenomena. Whether these phenomena are explained by necessity, nature or some other cause is irrelevant for his purposes, as they are excluded from deliberation in any case. The essential point is that no deliberation is possible concerning the uniform motion of celestial bodies, which serves as the paradigmatic case of an eternally moving object. This perspective aligns with the interpretation offered in the *Anonymous Commentary*, which was written after Aspasius. The Anonymous suggests that Aristotle's primary aim is not to specify the exact causal hierarchy, but to underline the absence of deliberation regarding such phenomena in the context of ethical enquiry:[52]

50 *Phys.* II 8, 198b34–36.

51 This is the solution adopted by Thomas Aquinas, one of the few commentators, to my knowledge, to attempt to identify the unnamed third cause (*Commentary* on Book III, *Lectio* VII, § 461).

52 On the date of the anonymous commentator, see Eliasson (2013), 200, and Natali (2023), 1–4 and 18–20, for a comparison between him and Aspasius. Like Eliasson, I refer to the author of this commentator as "the Anonymous".

> (1) "About eternal things no one deliberates", e.g. about whether the world or god is or is not, no one in his right mind deliberates, because these things exist always and by necessity, nor [no one deliberates on the fact] that the diameter is incommensurable with the side of a square (for this too is necessary); (2) and [no one deliberates] either "about moving beings", for these things, without being eternal, always occur in the same way, and they too possess a necessary and definite generation and movement. But, since these things occur according to some, in one way, according to others, in another, and the object was not now to speak of this, he said "by nature or by some other cause." (3) But there is no deliberation either about things that do not always occur in the same way but most often, by nature, e.g. canities or teething; (4) [no one deliberates] "either about what happens by chance" and rarely, as about the discovery of a treasure (for it was not discovered by prior deliberation), "or about all human things".[53]

The Anonymous comments on each Aristotelian expression:

Group 1: Necessity is identified as the cause of eternal beings (the world and geometric properties). The Anonymous adds god to this list.

Group 2: He sets apart from the rest of the expression the phrase "either by nature or by some other cause." His description of the phenomena belonging to this second category is vague, but the entities in question are most likely celestial phenomena—beings that are all in motion yet whose motion exhibits a definite regularity. He explains that this open formulation reflects Aristotle's indifference, in this context, to giving the correct causality. Since the causes of these phenomena are debated and not the subject of enquiry at this point (οὐκ ἦν δὲ νῦν τὸ προκείμενον), there is no need to investigate what "by nature" means here, nor to identify the "other cause".

53 *Anonymous commentary on the* Nicomachean Ethics, CAG XX, 149.14–25: ⟨Περὶ δὲ τῶν ἀιδίων οὐδεὶς βουλεύεται,⟩ οἷον περὶ τοῦ κόσμον ἢ θεὸν εἶναι ἢ μὴ εἶναι οὐδεὶς νοῦν ἔχων βουλεύεται, ὅτι ἔστιν ἀεὶ καὶ ἐξ ἀνάγκης τὰ τοιαῦτα, οὐδὲ ὅτι ἀσύμμετρος ἡ διάμετρος τῇ πλευρᾷ (ἀναγκαῖον γὰρ καὶ τοῦτο), οὔτε ⟨περὶ τῶν ἐν κινήσει,⟩ ταῦτα γὰρ οὐκ ἀεὶ μέν ἐστιν ἀλλ' ὁμοίως γίνεται, καὶ ἀναγκαίαν καὶ αὐτὰ ὡρισμένην ἔχει τὴν γέννησίν τε καὶ κίνησιν. ἐπεὶ δέ τινες μὲν οὕτως, τινὲς δὲ ἐκείνως αὐτὰ λέγουσι γίνεσθαι, οὐκ ἦν δὲ νῦν τὸ προκείμενον περὶ τούτων λέγειν, εἶπε τὸ ⟨εἴτε φύσει ἢ δι' ἄλλην αἰτίαν.⟩ ἀλλ' οὐδὲ περὶ τῶν οὐχ ὁμοίως μὲν ἀεὶ γινομένων ἀλλ' ὡς ἐπὶ τὸ πλεῖστον, φύσει δὲ, οἷον τοῦ πολιοῦσθαι ἢ ὀδοντοφυεῖν, ⟨οὐδὲ περὶ τῶν ἀπὸ τύχης⟩ καὶ ἐπ' ἔλαττον, οἷον εὑρέσεως θησαυροῦ (οὐ γὰρ διὰ τὸ προβουλευθῆναι εὑρέθη) ⟨οὐδὲ περὶ τῶν ἀνθρωπικῶν πάντων.⟩ My translation. I have inserted the numbers.

Group 3: In Group 3 of *NE* III 3, the Anonymous places natural phenomena, things that occur not always in the same way, but most of the time and by nature; he gives the examples of the greying of hair and the growth of teeth.[54]

Group 4: Finally, he also recognises the group of phenomena that occur by chance.

A defining feature of the *Anonymous Commentary* is that Group 2 unifies necessary motions of the heavenly regions, while natural phenomena are assigned to Group 3, with both occurring "by nature". Furthermore, the Anonymous applies a sound exegetical principle: since the precise explanation of these phenomena falls outside the domain of ethics, there is no need to identify the unnamed third cause. Instead, he views Aristotle's discussion as a dialectical development, in which various opinions are enumerated without a definitive stance being taken.[55] That is why I call this interpretation contextual. Unlike the Anonymous, Aspasius takes the entire passage seriously, so to speak. I suggest that he has a rather systematic or unified conception of Aristotle's philosophy, a view likely reinforced by his comprehensive perspective on his entire philosophical system. He adopts a Type I interpretation, but one that diverges from the standard reading.

3 Aspasius on *Nicomachean Ethics* III 3

To fully grasp Aspasius' commentary, we must return to the Greek text. The passage is presented here as it appears in Heylbut's edition, with section letters added for clarity, followed by my translation (71.16–34):

> (71.16) τῶν δὴ πραγμάτων
>
> (71.17: a) τὰ μέν ἐστιν αἴδια καὶ ἀεὶ ὡσαύτως ἔχοντα, οἷον ὁ κόσμος· ἀγέννητος γὰρ καὶ αΐδιος, καὶ ἡ διάμετρος ἀσύμμετρος τῇ τοῦ τετραγώνου πλευρᾷ.

---

54 This distinction is found in Aspasius (see below p. 277) and has its origins in Aristotle (*EE* I 1, 1214a15–18; II 8, 1224b31–35).

55 This does not prevent the Anonymous from commenting a few lines below on the enumeration of the four causes (1112a32–33). He then specifies (although it may well be a gloss) that "according to these men, what is called fate would belong to nature because what belongs to fate is neither inevitable nor necessary" (εἴη δὲ ἂν καὶ ἡ εἱμαρμένη λεγομένη κατὰ τούσδε τοὺς ἄνδρας ὑπὸ τὴν φύσιν. οὐ γὰρ ἀπαράβατον τὸ εἱμαρμένον οὐδ' ἀναγκαῖον, 150.2–4). On this passage see Sharples (2007a), 596. As Sharples (2007a), 598, notes, Aspasius refers to fate and necessity further on, too. We return to this below p. 282.

περὶ δὴ τῶν τοιούτων οὐδεὶς βουλεύεται, οὐδὲ πῶς ἂν γένοιτο ἡ διάμετρος σύμμετρος, ἀλλ' εἴπερ ἄρα, ζητεῖ εἰ ἀσύμμετρός ἐστι, βουλεύεται δὲ οὐδαμῶς.

(71.20: b) ἄλλα δ' ἐστὶν, ἃ οὐκ ἀεὶ μέν ἐστι, γίνεται δὲ ἀεὶ ὡσαύτως, οἷον ἀνατολὴ ἄστρων καὶ δύσις καὶ τροπαί· τούτων γὰρ οὐδὲν μὲν ἀεί ἐστι, γίνεται δὲ ὡσαύτως· οὐδὲ περὶ τῶν τοιούτων βουλή ἐστιν. ἐξ ἀνάγκης δ' εἴτε φύσει ἢ διά τινα ἄλλην αἰτίαν τὰ τοιαῦτά φησι γίνεσθαι ὡσαύτως, ἀνάγκην λέγων οὐ τὴν βιαστικήν· οὐδὲν γὰρ τῶν ἐν τῷ οὐρανῷ βίᾳ γίνεται· ἀλλ' ἀνάγκην ὠνόμακε τὴν πρόνοιαν τοῦ ἀεὶ ὡσαύτως κινοῦντος διὰ τὸ πάντῃ πάντως κινεῖν καὶ μηδέποτε ἄλλως ἐνδέχεσθαι. ἀναγκαῖον γὰρ λέγεται τὸ μὴ ἐνδεχόμενον ἄλλως ἔχειν. τῇ τοιαύτῃ γοῦν ἀνάγκῃ αἱ κινήσεις τῶν ἄστρων καὶ αἱ ἀνατολαὶ δύσεις καὶ τροπαὶ ἢ φύσει ἢ κατὰ φύσιν· κατὰ φύσιν γὰρ κινεῖται τὰ ἐν τῷ οὐρανῷ ἢ καὶ ἀμφοτέροις· τῇ γὰρ φύσει τῶν φερομένων συμφωνοτάτη ἐστὶν ἡ τοῦ κινοῦντος πρόνοια.

(71.31: c) ἔνια δέ ἐστιν ἃ γίνεται μὲν κατὰ φύσιν, ἀτακτοτέραν δέ, οἷον αὐχμοὶ καὶ ὄμβροι. οὐδὲ περὶ τούτων βουλεύονται ἄνθρωποι.

(71.32: d) ἄλλα δέ ἐστιν ἃ ὡς ἔτυχε συμβαίνει, οἷον τὰ ἀπὸ τύχης· δῆλον δὴ ὅτι οὐδὲ περὶ τούτων εἰσὶ βουλαί· οὐδεὶς γὰρ βουλεύεται περὶ θησαυροῦ εὑρέσεως.

Among things,

(a) some are eternal and are always the same, like the world (indeed, it is ungenerated and eternal) and the diameter incommensurable with the side of the square; no one deliberates about this kind of thing, nor how the diameter would be commensurable, but if it is true that the geometer makes something, he investigates whether it is incommensurable, but he in no way deliberates.

(b) But there are other things which, without being eternal, always occur in the same way, such as the rising and setting of the stars and the solstices (indeed, none of these things are eternal, but they eternally occur in the same way); nor is there any deliberation about them. He says that it is "by necessity or by nature or by some other cause" that these kinds of things occur in the same way, by "necessity" he means not forcible necessity (indeed, none of the things in the heavens occur by force) but he has called "necessity" the providence of that which always moves in the same way by the fact that it moves absolutely, i.e. it is not possible for it ever to do anything else (for that is said to be necessary which cannot be otherwise). It is therefore by such a necessity that [occur] the motions of the stars (sunrises, sunsets, and solstices) either by nature or according to nature (for the things in the heavens move according to nature) or also by both[56] (for the providence of what moves is perfectly in harmony with the nature of the things transported).

(c) But there are certain things which, while occurring according to nature, occur according to a more disorderly nature, such as droughts and storms. Human beings do not deliberate on these things either.[57]

(d) But certain things happen as they happen, such as things that happen by chance. There is obviously no deliberation about this either, because nobody deliberates about the discovery of a treasure.

Aspasius gives a linear commentary on Aristotle's text, closely following its structure. It identifies and comments on the four *NE* III 3 groups. He adds parenthetical explanatory clauses with γὰρ, he clarifies Aristotle's statement on the eternity of the world (explaining that it is "ungenerated")[58] and he differentiates between deliberation and investigation in the case of the geometer. Most notably, Aspasius highlights the ontological framework underpinning Aristotle's argument. He elucidates the underlying reasoning by showing that Aristotle distinguishes between things based on two key variables: (1) whether they are eternal or not; (2) whether they always behave in the same way or not. The distinction between these two criteria lies in the fact that ungenerated beings (e.g. the world itself and mathematical entities) are eternal, whereas astronomical motions, while not identical and eternal in themselves, nonetheless always occur in the same way (e.g. the rising of the stars). Aspasius approaches the text in a different way than the Anonymous, interpreting it within what he perceives as its physical (or cosmological) and metaphysical framework, drawing on *De caelo*, the *Meteorologica* and the *Metaphysics*. As will be seen, Aspasius does not treat ethics in isolation from the rest of the corpus; instead, he draws on

56 Konstan has suggested a correction to Heylbut's text, taken up by Sharples (Konstan 2006, 72, 194 n. 153, and Sharples 2010, 200): ἢ φύσει is added in line 28 to justify the ἢ καὶ ἀμφοτέροις in line 30. The text then reads: κατὰ φύσιν ⟨ἢ φύσει⟩ γὰρ κινεῖται τὰ ἐν τῷ οὐρανῷ ἢ καὶ ἀμφοτέροις. This insertion is not required. As I read it, this proposition is a parenthetical clause that justifies the introduction of the κατὰ φύσιν just before it; and the phrase ἢ καὶ ἀμφοτέροις does not designate the pair "by nature" and "according to nature" but "by nature" (or "according to nature") and "by necessity". I will return to this point below.

57 Reading ἀτακτοτέραν (71.31) *pace* Konstan (xiii and n. 154).

58 "Ungenerated" is to be taken in its strict sense as defined in *DC* I 11, as that from which it is impossible for there to be generation (280b11–12; I 12, 282a27–28). See also *NE* VI 3, 1139b24 (eternal beings are ungenerated and incorruptible). Since the world is ungenerated, there can be no deliberation about it—and therefore no providence, if providence presupposes prior deliberation. The Aristotelian doctrine of the eternity of the world makes providence impossible on principle.

his expertise as a commentator on Aristotle's physics and metaphysics to elucidate this text. There is, for example, a close parallel between his commentary and *Metaphysics* VI 2, 1026b27–31:[59]

> Since, among things which are, some are always in the same state and are of necessity (nor necessity in the sense of compulsion but that which means the impossibility of being otherwise), and some are not of necessity nor always, but for the most part, this is the principle and this the cause of the existence of the accidental […].[60]

His most extensive commentary focuses on the second group of phenomena (70.23–31, my (b) above), corresponding to Aristotle's discussion of Group 2 in *NE* III 3. This analysis leads to a cosmological and theological excursus, in which Aspasius references providence twice. In examining this passage, I will distinguish three key points: (1) the interpretation of necessity as providence of the prime mover; (2) the distinction between "by nature" and "according to nature"; and (3) the meaning of the harmony between the prime mover and the nature of the celestial bodies moved.

3.1 *Necessity as Providence of the Prime Mover*

Lines 23–31 of Aspasius' commentary focus on the three causes mentioned in *NE* III 3, 1112a24–25: there is no deliberation about the motion of celestial bodies, for they occur "by necessity, by nature, or by some other cause". Aspasius begins by clarifying the meaning of necessity. Unlike the Anonymous, who associates necessity with the world or geometrical properties, Aspasius links it specifically to the motions of celestial bodies, which occur eternally in the same manner. This interpretation is grounded in Aristotle's *Nicomachean Ethics*, where everything that is "by necessity" is also "eternal".[61] Indeed, Aspasius had already noted earlier in his commentary that celestial bodies move "by necessity".[62] Yet Aspasius' interpretation here contains an important nuance:

---

59 *Metaph.* VI 2, 1026b27–31 (Ross): ἐπεὶ οὖν ἐστὶν ἐν τοῖς οὖσι τὰ μὲν ἀεὶ ὡσαύτως ἔχοντα καὶ ἐξ ἀνάγκης, οὐ τῆς κατὰ τὸ βίαιον λεγομένης ἀλλ' ἣν λέγομεν τῷ μὴ ἐνδέχεσθαι ἄλλως, τὰ δ' ἐξ ἀνάγκης μὲν οὐκ ἔστιν οὐδ' ἀεί, ὡς δ' ἐπὶ τὸ πολύ, αὕτη ἀρχὴ καὶ αὕτη αἰτία ἐστὶ τοῦ εἶναι τὸ συμβεβηκός. See also *Metaph.* X 8, 1064b32–36.

60 Translation Ross in Barnes (1984).

61 *NE* VI 3, 1139b23–25.

62 On a subject far removed from cosmology, Aspasius writes (26.6–11): "the things that result from nature and art and similarly also those that result from necessity, such as the things that are in the heavens (⟨τὰ⟩ κατὰ ἀνάγκην, οἷον τὰ ἐν τῷ οὐρανῷ), these things are all the most excellent they can be."

necessity is not only the modality of eternal beings' motions, but also an independent causal principle on a par with nature. As we saw above, Aristotle's wording permits this reading. Thus, Aspasius' interpretations can be explained as follows: Aspasius, building on Aristotle's formulation that equates necessity with a cause, and no doubt under pressure from Stoicism, which makes necessity one of Zeus' names, elevates necessity from a modality to an autonomous causal principle governing the necessary motions of celestial bodies.[63] As we will see, this interpretation also echoes Aristotle's description of the prime mover in *Metaphysics* XII 7, and we can assume that Aspasius relies on it. However, within the Aristotelian framework, it must be stressed that necessity *is not* a metaphysical principle. Strangely enough, Aspasius' recourse to providence serves to clarify how necessity operates as a causal principle.

Aspasius' explanation addresses an exegetical challenge. He takes Aristotle's threefold causal distinction in 1112a24–25 seriously and does not follow the Anonymous' deflationary approach, which treats Aristotle's formulation as noncommittal in an ethical context. Instead, Aspasius assumes that Aristotle's thought possesses a systematic unity, or, at the very least, he works with a unitary conception of the Aristotelian thought, interpreting passages in the *NE* through the lens of physics and metaphysics.[64] However, if, following Aspasius, necessity is to be understood as a distinct causal principle, its precise nature must be clarified, as, according to Aristotle, necessity is not the cause of the heavenly bodies' motions. In *De caelo*, Aristotle emphatically explains that the circular motion of the heavens and celestial bodies is "by nature" or "according to nature" and not "by force" (βίᾳ).[65] If Aristotle affirms in *NE* III 3 that celestial bodies move by necessity, then necessity cannot be a force that constrains them from outside. That is the reason why, as in the passage quoted above from *Metaphysics* VI 2, Aspasius specifies that this is not a coercive necessity (βιαστι-

63 SVF II 913 and 937; Cicero, *ND* I 39; see also Cicero, *Academica* I 6.29 and Kupreeva (2016) 141–142. On the polynomy of the Stoic god, see Diogenes Laertius VII, 135.9–136.1 (SVF I 580). Similarly, in chapter 7 of *De mundo* (401b8), necessity is one of the names of god, insofar as god is a cause that cannot be defeated (ἀνίκητον αἰτίαν ὄντα, 401b9).

64 Aspasius knows that the discussion of physical difficulties has no place in ethics. See 160.24–26, his commentary on *NE* VIII 1, 1155b8–10, on the difference between ethical and physical difficulties. But here we are dealing with a physical issue (the cause of celestial phenomena), not an ethical one. It is therefore legitimate to have recourse to texts from outside ethics.

65 See *DC* I 2, 268b15–16, 269a5–9, 269b1–2; II 1, 281a18–35; II 3, 286a12, 16–17; II 7, 289a16; II 12, 293a9.

κήν, 71.24):[66] celestial motion does not result from external force.[67] By pointing out the dual meaning of necessity, a distinction frequently made in Aristotle, Aspasius references one of the central claims of the *De caelo*. There, Aristotle argues that celestial bodies are moved neither by external compulsion nor by a soul (not even their own soul) exerting force upon them, but rather by their own nature—the nature of the aether.[68]

Now, instead of referring to it simply as necessity (a not forcible necessity), Aspasius calls it providence (*pronoia*), explicitly claiming that what Aristotle calls necessity is "the *pronoia* of that which always moves in the same way" (ἀνάγκην ὠνόμακε τὴν πρόνοιαν τοῦ ἀεὶ ὡσαύτως κινοῦντος, 71.25–26). Aspasius behaves like a true commentator;[69] he clarifies Aristotle's terminology, replacing necessity with what he considers a clearer term, providence, and then explaining this name: "Aristotle has called necessity the providence of what moves always in the same way, because it moves absolutely and can never do otherwise" (διὰ τὸ πάντῃ πάντως κινεῖν καὶ μηδέποτε ἄλλως ἐνδέχεσθαι, 71.26–27).[70] The defining characteristic of *pronoia* is that it cannot do anything other than move eternally in a certain way. So, providence, in this context, is best understood as the outcome of two intersecting interpretive moves: (i) necessity is not an external compulsion; admittedly, Aspasius retains an Aristotelian modal understanding of necessity, as that which "cannot be otherwise"; yet, following Aristotle's suggestion in *NE* III 3, he interprets necessity by giving it a strong causal status; thus, necessity is the cause of what cannot be otherwise. (ii) The point is that such a necessity taken as a moving cause does not correspond to any principle formally recognised within Aristotle's theory;

66 Aspasius defines force as a principle of external or unnatural motion (59.30–60.10) in his commentary on *NE* III 1, 1109b30–1111b3. The adjective βιαστικός is of Platonic origin. It occurs in Aristotle (*Top.* I 12, 105a18; *PA* II 9, 655a12, and III 1,662b3; *MA* 10, 703a22) and the Stoics (SVF II 976 = Aëtius I 27.3) in a definition attributed to Chrysippus: "they say that necessity is an immobile and constraining cause."

67 Aspasius does not distinguish here between all the meanings of the necessary. Given the context, he draws a contrast between "necessary" in the sense of external constraint and "necessary" as that which cannot be otherwise, without mentioning conditional necessity. For the latter sense, see, e.g., *PA* I 1, 639b21–24.

68 See the criticism of the explanation of the motion of celestial bodies by necessity in *DC* II 1, 284a15 (the heavens do not need a "forceful necessity"—βιαίας ἀνάγκης) and 284a18–35. The positions criticised are those of the ancient poets (Atlas), Empedocles, the atomists and Plato. On the double meaning of necessity (according to nature and against tendency), see *APo.* II 11, 94b37–95a3; *Metaph.* V 5, 1015a26–31 and 1015a34.

69 See also 53.11, 104.24, etc.

70 The phrase πάντῃ πάντως is an emphatic formula used by Plato, which was later adopted by Aristotle and Alexander.

this leads Aspasius to claim that what Aristotle calls here "necessity" actually is the providence of what moves always in the same way—that is, though Aspasius does not make it explicit, the providence of the first unmoved mover. Put briefly, the first unmoved mover causes the heavenly bodies to move necessarily, and Aspasius chooses to name this causal role its providence.

These two occurrences of *pronoia* (71.26 and 71.31) are the only ones in Aspasius' commentary. He does not use either the noun or verb elsewhere. The term *pronoia* is primarily associated with human foresight. As Mayhew correctly notes, it signifies "the ability to know the future and to what that knowledge makes possible: a long-range concern for and care of the objects of this knowledge".[71] This creates an apparent tension: if something is necessary and eternal (as the prime mover), how can it exercise *pronoia* over something equally necessary and eternal (as the world)? There can be no form of anteriority, foresight or deliberation on the part of the mover about what it moves. Similarly, if this providence is synonymous with necessity, the mover cannot possess the slightest dimension of contingency, unlike the notion of providence (if deliberation presupposes that the god is thinking about the best way to create the world). Since, for Aristotle, the causality of the prime mover is not artisanal causation, it must be some other form of providence. Rather than denoting intentional governance, *pronoia* expresses the absolute, unchanging causal role of the prime mover; what is "providential" is the unwavering necessity of the motion moved by the first unmoved mover.

The use of *pronoia* raises both a translation issue and a more fundamental issue regarding its application, as the usual notion of providence suggests more than what Aspasius intends—namely to confer necessity on the motion of the moved.[72] Instead of attempting to find a new, more appropriate translation of *pronoia*, it is better to refine our (traditional) understanding of providence, viewing it as the ultimate principle of a good for something, where the "good" in this case is the necessity or eternity of the motion of the mover. Boys-Stones explains that Platonists like Atticus criticise Aristotle not for the meaning he assigns to providence, but for the limited scope of its application: for the Peripatetics, providence is not care for human beings. That said, a broader concept of *pronoia* seems to have been shared by both the Middle Platonists and the Aristotelians. According to this general notion, providence is the impersonal

71 Mayhew (Forthcoming), n. 8, and Fazzo (1999), 26–31.

72 See Sharples (2002a), 30. It is preferable to retain the translation "providence" rather than adapting it to the context, as Longpré, for instance, does (1997), 63: she renders it in French as "caractère previsible", because she insists that the motion produced by the prime mover is predictable, since it is necessary.

principle of benefit.[73] It is in this sense that we must interpret the *pronoia* of the mover: it does not imply foresight or even attention on the part of the mover toward the moved. The use of the term *pronoia* therefore suggests that what the mover does (i.e. moving necessarily) provides non-intentional benefit to what is moved. In an Aristotelian context, it would be inconceivable to attribute intention or concern to the prime mover, but such an attribution is not directly implied by Aspasius. What he conveys is that the *pronoia* of the mover guarantees the necessity and eternity of the motion of the moved. This does not mean that the mover has an intention or purpose. By using the term "providence", Aspasius refers to what the mover does well in relation to the moved.

But we can go further. The term *pronoia* remains relevant because the mover has ontological rather than chronological priority over the moved. If the providence of the mover guarantees the eternity and necessity of the motions of celestial bodies, then what is called the "mover" is not the first moved (i.e. the first heavens), but the first unmoved mover. In Aristotle, the prime mover is, in fact, distinguished by its necessity, among other characteristics. Another passage from the *Metaphysics* might provide the basis for Aspasius' comment. In *Metaphysics* XII 7, while examining the relationship between the prime mover and the first moved, Aristotle states:

> Now if something is moved, it can also be otherwise than it is, so that the primary motion is indeed in actuality, in that it is moved; but in this way it can be otherwise—i.e. in respect of place—even though it cannot be otherwise in respect of substance. But since there is something which causes motion but is itself unmoved, and which exists in actuality, this thing cannot in any way be otherwise. For locomotion is the first of the kinds of change, and of this the first kind is locomotion in a circle; and this is the motion which this [the first mover] causes. It exists, then, of necessity; and inasmuch as it exists of necessity, it does so well, and in this way it is a principle. For the necessary is spoken of in this number of ways: that which is by force, because it is contrary to impulse; that without which things are not good; that which cannot be otherwise, but is necessary without qualification. On such a principle, then, depend the heavens and nature.[74]

---

73 Boys-Stones (2018), 324: "The question of whether god can be provident (insofar as that is not just a question about terminology) is quite distinct from the question of whether he is a 'personal' or 'impersonal' principle. To be 'provident', as far as this goes, may require no more than to be the ultimate source of *benefit*—as Aristotle's first unmoved mover is."

74 *Metaph.* XII 7, 1072b4–14: εἰ μὲν οὖν τι κινεῖται, ἐνδέχεται καὶ ἄλλως ἔχειν· ὥστ' ἡ φορὰ ἡ πρώτη

Aristotle defines the prime mover in terms of its necessity. As an immaterial, pure actuality, it cannot be otherwise than it is: it is absolutely necessary. The first moved differs from the prime mover in that, while it cannot be otherwise in substance (as it is ungenerated and incorruptible), it is still capable of being moved. From a spatial standpoint, the first moved possesses a potentiality that prevents it from being absolutely necessary. In contrast, the unmoved mover is absolutely necessary in both substance and place, as it is immaterial. The prime mover causes the first moved to move eternally in a circular motion, which aligns with Aspasius' use of "ὡσαύτως" (71.21, 22, 26).

Although Aspasius in this passage does not engage with the speculative depths of *Metaphysics* XII 7, his commentary on *NE* III 3 accords with the conception of the prime mover as necessity.[75] Both texts present the prime mover as necessary, or as necessity itself, eternally conferring on the first moved a circular motion. The first moved is not, however, necessary in the same way as the prime mover. This distinction implies that the prime mover has ontological priority, as pure actuality, and allows for a providential dimension in its relationship to the first moved.[76]

This passage from *Metaphysics* XII 7 also clarifies another aspect of Aspasius' commentary. As it will be seen, Aspasius emphasises the "harmony" between the *pronoia* of the prime mover and the "nature" of the celestial bodies (συμφωνοτάτη, 71.30–31). One could infer that this "harmony" points to the ontological proximity between the prime mover and the first moved: the prime mover has

---

καὶ ἐνεργείᾳ ἐστιν ᾗ κινεῖται· ταύτῃ δὲ ἐνδέχεται ἄλλως ἔχειν, κατὰ τόπον, καὶ εἰ μὴ κατ'οὐσίαν· ἐπεὶ δὲ ἔστι τι κινοῦν αὐτὸ ἀκίνητον ὄν, ἐνέργειά ὄν, τοῦτο οὐκ ἐνδέχεται ἄλλως ἔχειν οὐδαμῶς. φορὰ γὰρ ἡ πρώτη τῶν μεταβολῶν, ταύτης δὲ ἡ κύκλῳ· ταύτην δὲ τοῦτο κινεῖ. ἐξ ἀνάγκης ἄρα ἐστὶν ὄν· καὶ ᾗ ἀνάγκῃ, καλῶς, καὶ οὕτως ἀρχή. τὸ γὰρ ἀναγκαῖον τοσαυταχῶς, τὸ μὲν βίᾳ ὅτι παρὰ τὴν ὁρμήν, τὸ δὲ οὗ οὐκ ἄνευ τὸ εὖ, τὸ δὲ μὴ ἐνδεχόμενον ἄλλως ἀλλ' ἁπλῶς.—ἐκ τοιαύτης ἄρα ἀρχῆς ἤρτηται ὁ οὐρανὸς καὶ ἡ φύσις. Translation and Greek text by Judson (2019). I understand that the prime mover assumes the third sense of necessity (τὸ δὲ μὴ ἐνδεχόμενον ἄλλως ἀλλ' ἁπλῶς, 1072b13). See also Baghdassarian (2019), 260.

75 As said above, it is known from Alexander's references in his *Commentary* that Aspasius composed a commentary on Aristotle's *Metaphysics* (cf. 41.27, 59.6, 379.3). For pedagogical reasons, Aspasius does not employ technical ontological terminology. See Barnes (1999), 24–27, on Aspasius' work as a "beginner's" commentary that does not presuppose "background knowledge". Moreover, according to Aspasius (1.8), ethics is prior to first philosophy in the order of knowledge. Any explicit reference to first philosophy or theology would therefore be premature.

76 See also *Metaph.* IX 8, 1050b18–19. We cannot discuss here (*a fortiori* on the basis of this passage from Aspasius alone) the problem of how that which is in potential (as celestial bodies are) is also eternal and endowed with eternal motion. See, e.g., Frey (2015). In any case, Aspasius assumes that the nature of the prime mover alone does not allow it to move eternally and necessarily.

no potentiality or matter, which makes it absolutely necessary, while the first moved can be moved with a circular motion, but only possesses "local matter".[77] This point will be considered in more detail. Building on these similarities, it can be assumed that Aspasius retains an Aristotelian cosmological framework that accounts for the motion of celestial bodies through a twofold causal scheme: (i) the specific nature of the celestial bodies, i.e. the fifth element or aether (although neither these terms is used in this ethical context), and (ii) the prime mover. Against Xenarchus, but like Adrastus, Aspasius would therefore remain a proponent of the doctrine of the fifth element.[78]

Regarding the celestial bodies, Aspasius' commentary stays within the limits outlined in *NE* III 3, referring only to the rising and setting of the stars and the solstices. It can be assumed that this refers primarily to the motions of the sun and moon. Although Aspasius mentions "the things in the heavens" (71.25, 29), he does not specifically address the prime mover or the celestial spheres, suggesting that he views the providence of the prime mover as extending, through intermediaries, across the entire heavens, *reaching as far as the moon*.[79] However, he does not specify where providence ends, and further clarification may come later in the text.

Aspasius' strategy in this passage appears to involve focusing on the interpretation of "necessity", which plays a key role in the various doctrines underlying his commentary. (1) By distinguishing in a typically Aristotelian fashion between necessity and force, Aspasius aims to avoid attributing to Aristotle the idea that the motion of celestial bodies results from an external force, which would imply motion against nature. This distinction allows Aspasius to emphasise that what is "by necessity" is also "according to nature" (71.29–31), distancing his account from atomist positions, as well as from interpretations that align with Empedoclean or Platonic conceptions criticised by Aristotle in *De caelo* II 1. This accounts Aspasius' later insistence that what is "by necessity" is also "according to nature" (71.29–31). (2) Necessity also plays a role in Middle Platonism. According to the dualistic exposition of principles found in *Timaeus* 48a, it is one of the two principles (along with reason) responsible for the creation of the cosmos. Although this understanding of necessity differs from that of Aristotle, Aspasius' clarifications and his use of the term

77 *Metaph.* VIII 1, 1042b6, 4, 1044b8, and IX 8, 1050b20–24.

78 See Petrucci (2015), 196 n. 75. Barnes (1999), 8, emphasises the orthodoxy of Aspasius' Aristotelianism. See Falcon (2012), 35–36, on Xenarchus' critique and his own explanation of the motions of the heavens.

79 The description given by Boys-Stones (2018), 325–326, of a "providence exercised by proxy" from the prime mover to the sphere of the moon fits perfectly here.

pronoia may represent an attempt to distance himself from the Middle Platonic position.[80] (3) Moreover, by interpreting necessity as the providence of the prime mover, Aspasius comes close to a Stoic conception, in which necessity and providence are interchangeable names for Zeus.[81] However, necessity is seen here as a principle that sets heavenly bodies in motion according to their nature, meaning that the scope of Peripatetic providence is much narrower than that of Stoic providence. Nevertheless, Aspasius makes clear that providence is not exclusively Stoic, agreeing on this point with Middle Platonists such as Atticus.[82] (4) Finally, Aspasius emphasises that the motion of celestial bodies is not self-sufficient, but requires a cause. Indeed, Aspasius wants to show that the fact that the Aristotelian cosmos is eternal (he has just said that it is ungenerated, 71.17–18) does not mean that it does not need a first principle. For Aspasius, as for Aristotle, eternal motion still requires an absolutely necessary first principle.[83] Even in a non-creationist cosmology, providence guarantees the necessity and eternity of the motions of celestial bodies. To this extent, the supralunary world needs providence. The prime mover assumes the role of a demiurge, much like the demiurge in Plato's *Timaeus* (41b2–6), ensuring the eternity of the world.[84] According to Aspasius, there is, therefore, a Peripatetic providence that is not that of the craftsman who creates the world in time and takes care of his work, but rather that of the prime unmoved mover who guarantees the necessity and eternity of movement and, in this, ensures its good. This seems to be the strategy behind the interpretation that the necessity that causes the eternal motion of the stars is the providence of the prime mover.

The paradoxical aspect of Aspasius' passage lies in the introduction of providence as necessity in the context of a discussion on deliberation. While Plato's *Timaeus* maintains that providence includes deliberation, Aspasius clearly indicates that the providence he describes does not involve deliberation, as the world is eternal and ungenerated. It is noteworthy that Aspasius does not develop this aspect further, even though the absence of deliberation about the

---

80 Atticus refers to the necessity that would have "forced" the divine designs in Fr. 4. 16 Des Places. For the dualistic interpretation of the *Timaeus* principles in Atticus and Plutarch, see Michalewski (2024), 118–128.

81 See above n. 63.

82 On the "polemical resemantization" against an agonistic philosophical background, see Annas (1990), 81, and Bonazzi (2007), 126.

83 See Judson (2019), 227, and *Metaph.* IX 8, 1050b18–19.

84 On the contrary, Atticus argues that providence exists only for a created world. See Michalewski (2024), 47–48. On a non-literal reading of the *Timaeus*, such as Taurus' interpretation, to say that the world is generated is to say that it depends ontologically on a principle (Michalewski 2024, 70–71).

world suggests a different conception of providence than what might be expected from a god reasoning about what is best for the world.[85]

3.2 *"By Nature" and "According to Nature"*

The providence of the prime mover does not simply cause motion, but moves necessarily and ensures the necessity and eternity of the circular motion. This passage addresses the objection that the heavenly regions do not require *pronoia*. In line with the argument in *Metaphysics* XII 7, Aspasius demonstrates a lack of necessity in the motion of the sphere of the fixed stars while the prime mover is necessarily, the outmost sphere, as it is moving, can be otherwise than it is with respect to place; it therefore requires an absolutely necessary being.

The following sentence clarifies how this providence operates: τῇ τοιαύτῃ γοῦν ἀνάγκῃ αἱ κινήσεις τῶν ἄστρων καὶ αἱ ἀνατολαὶ καὶ δύσεις καὶ τροπαὶ ἢ φύσει ἢ κατὰ φύσιν (71.28–29). Aspasius coordinates the two types of causality that Aristotle distinguished in *NE* III 3: while Aristotle writes "necessity *or* nature" (1112a24–25), Aspasius shows that celestial bodies are moved by both. It is through the providence of the prime mover that the motions of the stars occur. Aspasius identifies three types of motions using examples: the rising, the setting and the solstices of the celestial bodies (such as the sun and moon). The conclusion of the sentence specifies that these motions occur "either by nature or according to nature" (ἢ φύσει ἢ κατὰ φύσιν, 71.29). This implies that what is "by nature" or "according to nature" is also by necessity. Thus, the causality of nature is coordinated with the absolute necessity of the prime mover. We have to understand that, in reality, there is only one cause, not two or three, as presented in Aristotle's passage. The three alternative causes in *NE* III 3 are replaced by the coordination of nature with necessity alone, i.e. with the providence of the prime mover.[86] Through this exegetical and philosophical move, Aspasius explicitly links together causes that Aristotle had distin-

---

85 Does the claim we find in *NE* III 3 to the effect that there is no deliberation about the world because it is eternal have an anti-Platonic meaning for Aristotle? The Anonymous is sensitive to this dimension, claiming that the gods do not deliberate about what they do because there is no deliberation of what there is science of, yet the gods have the science of what they do (*Anonymous commentary on the* Nicomachean Ethics, 150.8–9).

86 Sharples (2010), 200 n. 15, rightly states that "Aristotle in fact treats necessity, nature and some other cause as three co-ordinate possibilities. But the sequel shows that Aspasius treats the second and third as two alternative explanations of the first." It is questionable whether Aspasius identifies the unnamed third cause of *NE* III 3. In the following passage, he states that there is no cause other than necessity and nature. If there is a third cause, then it must be "according to nature", but Aspasius himself makes no distinction here between "by nature" and "according to nature."

guished, subordinating nature to necessity. Aspasius thus seems to erase or reduce the distinction between heavens and nature, as noted at the end of the passage from *Metaphysics* XII 7: "On such a principle, then, depend the heavens and nature" (1072b13–14).

Unlike modern commentators, Aspasius finds no room for biological phenomena in Aristotle's typology from *NE* III 3, i.e. for those things that Aristotle believes occur "most often": what occurs "by nature or according to nature" are the motions of the heavenly bodies, *not natural phenomena*. Aspasius' commentary is limited to the heavenly regions or sublunar meteorological phenomena. In the context of Aristotle's discussion on the objects of deliberation in *NE* III 3, however, Aspasius' exclusion of sublunary biological phenomena is problematic, as we do not deliberate about natural phenomena like the growth of teeth or hair.

Aspasius makes a seemingly puzzling distinction, at first glance, between "by nature" and "according to nature". We must now return to the tricky part of this digression. Heylbut's text reads as follows:

> τῇ τοιαύτῃ γοῦν ἀνάγκῃ αἱ κινήσεις τῶν ἄστρων καὶ αἱ ἀνατολαὶ δύσεις καὶ τροπαὶ ἢ φύσει ἢ κατὰ φύσιν· κατὰ φύσιν γὰρ κινεῖται τὰ ἐν τῷ οὐρανῷ ἢ καὶ ἀμφοτέροις· τῇ γὰρ φύσει τῶν φερομένων συμφωνοτάτη ἐστὶν ἡ τοῦ κινοῦντος πρόνοια. (71.28–31)

Aspasius argues that if the motions of the stars occur "by nature" or "according to nature," it is due to necessity as it has been defined—not as an external force, but as the necessity of what cannot be otherwise (τῇ τοιαύτῃ [...] ἀνάγκῃ). As we have seen, this implies that, for Aspasius, there are not two coordinated causes, but only one: necessity. But what exactly does the distinction introduced here by Aspasius between φύσει and κατὰ φύσιν signify? This legitimate question is further complicated by Konstan's emendation, which introduces an additional ἢ φύσει to account for ἢ καὶ ἀμφοτέροις. The resulting text and translation read as follows:

> τῇ τοιαύτῃ γοῦν ἀνάγκῃ αἱ κινήσεις τῶν ἄστρων καὶ αἱ ἀνατολαὶ καὶ δύσεις καὶ τροπαὶ ἢ φύσει ἢ κατὰ φύσιν· κατὰ φύσιν γὰρ ⟨ἢ φύσει⟩ κινεῖται τὰ ἐν τῷ οὐρανῷ ἢ καὶ ἀμφοτέροις· τῇ γὰρ φύσει τῶν φερομένων συμφωνοτάτη ἐστὶν ἡ τοῦ κινοῦντος πρόνοια. (71.28–31)

> By such a necessity, surely, occur the motions of the stars and their risings and settings and turnings, whether by nature or in accord with nature. For the things in the heavens move in accord with nature [or by nature] or

> even by both. For the providence of the one that moves them is entirely in harmony with the nature of the things that are borne along. (Transl. Konstan)

In the passage reproduced above, I have enclosed in square brackets the addition proposed by Konstan in line 29.[87] Since Aspasius distinguishes two modes of natural causality and concludes with ἢ καὶ ἀμφοτέροις, Konstan inferred that a missing ἢ φύσει should be restored in the second part of the sentence. The overall meaning would then be: "It is thus by such necessity that the motions of the stars [occur], either by nature or according to nature; celestial bodies move according to nature, by nature, or by both." If we adopt this correction, we might also modify ἀμφοτέροις to ἀμφοτέρως. However, this does not resolve the issue of the distinction between φύσει and κατὰ φύσιν. That said, this difficulty disappears if we arrange the passage as I proposed:

> τῇ τοιαύτῃ γοῦν ἀνάγκῃ αἱ κινήσεις τῶν ἄστρων καὶ αἱ ἀνατολαὶ δύσεις καὶ τροπαὶ ἢ φύσει ἢ κατὰ φύσιν (κατὰ φύσιν γὰρ κινεῖται τὰ ἐν τῷ οὐρανῷ) ἢ καὶ ἀμφοτέροις (τῇ γὰρ φύσει τῶν φερομένων συμφωνοτάτη ἐστὶν ἡ τοῦ κινοῦντος πρόνοια). (71.28–31)
>
> It is therefore by such a necessity that the motions of the stars (sunrises, sunsets, and solstices) [occur] either by nature or according to nature (for the things in the heavens move according to nature) or also by both (for the providence of what moves is perfectly in harmony with the nature of the things transported). (My translation)

When the passage is punctuated in this way, Aspasius' reasoning becomes clearer: the motions of celestial bodies occur by necessity, understood in this specific sense. Following Aristotle's text in *NE* III 3, Aspasius then adds "or by nature or according to nature". However, to justify this last addition—"or according to nature"—which does not appear in Aristotle's original text, he reiterates that celestial bodies move "according to nature" (a point already mentioned in 71.25). The key element of Aspasius' thesis follows: he asserts that the motion of celestial bodies takes place ἀμφοτέροις, meaning both under the effect of necessity and by nature. The subsequent explanatory sentence (71.30–31) leaves no doubt about the meaning of ἀμφοτέροις: to say that celestial bodies move "by nature", "according to nature", or by necessity (understood in this

87 There is nothing in Heylbut's apparatus.

way) amounts to the same thing. This is because necessity—identified with the providence of the prime mover—is "entirely in harmony with the nature of the bodies moved" (i.e. the nature of celestial bodies). Due to this harmony, there is ultimately no distinction between saying that celestial bodies move *by their nature* or by necessity. I will return to the significance of this "harmony" later.

For now, let us examine how Aspasius interprets the pair of terms "by nature" and "according to nature". In *Physics* II 1, Aristotle writes: "So we have explained what nature is and what is by nature and according to nature."[88] While Aristotle never explicitly clarified the distinction,[89] later commentators, such as Alexander, Themistius and Simplicius, showed increasing interest in it.[90] Aspasius himself provides a detailed discussion on the meaning of "by nature" in his commentary on *NE* II 1, 1103a18–26, where he identifies four distinct meanings:

> In order that it may become clear in what sense he says that they do not occur by nature, one must distinguish the several senses in which "by nature" is said.
>
> (i) For those things are said to exist by nature that are always co-present, for example everything heavy is said to be constituted by nature to be borne downwards and every light thing upwards.
>
> (ii) In another way, what is not co-present from the beginning, but after a certain time comes for the most part to be present, even though we do not busy ourselves about it whether through habit or through instruction, is said to be by nature, for example the natural growth of teeth or beards.
>
> (iii) In yet another way, that to which we are susceptible is said to be by nature. In this way even contraries are said to occur to the same object by nature, for example disease and health to a susceptible body.
>
> (iv) Further, that toward which what is susceptible is by nature more inclined and toward which it rather has impulses from its nature, is said to be by nature. Thus, health again is by nature, but disease becomes contrary to nature, since it is a kind of privation of health

88 *Phys.* II 1, 193a1–2: τί μὲν οὖν ἐστιν ἡ φύσις, εἴρηται, καὶ τὸ φύσει καὶ κατὰ φύσιν.

89 The two expressions are synonymous according to Bonitz's *Index aristotelicus*, 837A18. See, *contra*, Ross (1936), 501 and 524.

90 Following Aspasius, Alexander in his *Ethical Problems* 5 (125, 24–26) and 29 (160, 32–37), Themistius, *In Phys.* 37.7–10, and Simplicius, *In Phys.* 271.9–22, make similar distinctions. See Sambursky (1962), 93–98. I concentrate on these texts in Lefebvre (2026).

> and the body is constituted by its nature more in the direction of health than of disease. In the same way virtue too is more in accord with nature, and vice contrary to nature, and in general what is good is more in accord with nature, what is evil contrary to nature.[91]

Aspasius distinguishes four senses of "by nature". According to the third sense, what is by nature is the capacity to receive opposites, whereas according to the fourth sense, only what is "according to nature" aligns with a substance's "resources" or "natural tendencies" (ἀπὸ τῆς φύσεως ἀφορμάς, 38.17 and 29). From this perspective, health, virtue and goodness are more "according to nature" than disease, vice and evil, which are "against nature". Thus, while everything "according to nature" is also "by nature", not everything "by nature" is necessarily "according to nature". In our passage, Aspasius seems to suggest that when Aristotle states that celestial bodies move "by nature", we must understand this as necessarily "according to their nature"—and not contrary to it. As we saw earlier, and as Aspasius reiterates, celestial bodies move not "by force", but "according to nature". In other words, they move by nature, but also—since no external force constrains them—according to their nature.

Aspasius' innovation lies in how he coordinates nature with the necessity of the prime mover, redefining this necessity as the "providence of the mover". This implies that Aspasius wants to emphasise the providential dimension of nature—particularly the nature of celestial bodies, and perhaps even nature as a whole—through the necessity of the prime mover. More precisely, thanks to this necessity, nature is self-sufficient, at least in the sense that the nature of celestial bodies is endowed with necessary and eternal motion, requiring no external demiurgic intervention. The rest of the text may also suggest that nature as a whole is self-sufficient, insofar as it serves as a principle of motion.[92]

91 Transl. Konstan. Aspasius, 38.9–22: ἵνα δὲ δῆλον γένηται, πῶς λέγει μὴ παραγίνεσθαι φύσει, διαιρετέον ποσαχῶς τὸ φύσει. φύσει γὰρ λέγεται ὑπάρχειν ὅσα ἀεὶ συνυπάρχει, οἷον φύσει πᾶν τὸ βαρὺ κάτω πέφυκε φέρεσθαι καὶ τὸ κοῦφον ἄνω· καθ' ἕτερον δὲ τρόπον λέγεται φύσει καὶ ὃ μὴ ἐξ ἀρχῆς μὲν συνυπάρχει, ἀπὸ δέ τινος χρόνου παραγίνεται ὡς ἐπὶ τὸ πλεῖστον, μηδὲν ἡμῶν πολυπραγμονούντων μήτε δι' ἔθους μήτε διὰ διδασκαλίας, οἷον ὀδόντων φύσις καὶ γενείων. ἄλλως δὲ λέγεται φύσει, οὗ ἐπιδεκτικοί ἐσμεν· οὕτως καὶ τὰ ἐναντία τῷ αὐτῷ γίνεται φύσει, οἷον νόσος καὶ ὑγίεια τῷ δεκτικῷ σώματι. ἔτι δὲ λέγεται φύσει, πρὸς ὃ μᾶλλον τὸ δεκτικὸν πέφυκε καὶ πρὸς ὃ ἔχει ἀπὸ τῆς φύσεως ἀφορμὰς μᾶλλον. οὕτως δὲ καὶ ἡ ὑγίεια μὲν φύσει, ἡ δὲ νόσος παρὰ φύσιν γίνεται, στέρησίς τις οὖσα τῆς ὑγιείας καὶ διὰ τὸ τὴν φύσιν συστῆσαι τὸ σῶμα μᾶλλον πρὸς ὑγίειαν ἢ πρὸς νόσον. τὸν αὐτὸν δὲ τρόπον καὶ ἀρετὴ μᾶλλον κατὰ φύσιν, παρὰ φύσιν ἡ κακία, καὶ ὅλως τὸ μὲν εὖ κατὰ φύσιν, τὸ δὲ κακὸν παρὰ φύσιν. I add the subdivisions. The passage from *NE* II 1 was used by Alexander as part of an argument on providence (see Fr. 2 Grant in Thillet 2003, 58–59).

92 Aspasius draws on this reference to nature to highlight the distinctiveness of providence

3.3 *The Harmony between the Providence of the Prime Mover and the Nature of the Celestial Bodies*

Let us turn now to the last parenthesis in 71.30–31: "the *pronoia* of the mover is most in harmony [or perfectly in harmony] with the nature of the moved bodies" (τῇ γὰρ φύσει τῶν φερομένων συμφωνοτάτη ἐστὶν ἡ τοῦ κινοῦντος πρόνοια). Aspasius employs the superlative form of the adjective σύμφωνος to designate this harmony, which can be understood either absolutely or relatively. This is a very common term that appears in a variety of contexts, including music, arithmetic and astronomy, and implies that the two terms thus connected are not inherently in harmony or might not be.[93] The term is also frequently used in Aristotle's *Ethics* to discuss the relationship between desire and reason,[94] and it is found with the same meaning in Aspasius' commentary.[95] Since this word appears only four times in Aspasius, and the other three occurrences are all in an ethical context, it might be tempting to assume that the meaning here is the same. However, the relationship is not identical. In an ethical context, it is the appetitive or irascible part that is (or should be) in harmony with the rational part: achieving perfect harmony is regarded as the goal of educating the appetitive part, since, in the moderate person, desire is perfectly in harmony with reason.[96] In a cosmological context, by contrast, Aspasius states that provid-

---

in Aristotle. This may be a reference to the debate in the Atticus fragment between two conceptions of providence: a Peripatetic view, in which nature itself is providence as a principle of sufficient order, and Atticus' view, which is modeled on craft causality, in which providence involves deliberation and reasoning. See Atticus Fr. 3.81–85 Des Places, where Atticus reproaches Aristotle for having left human things governed "by a certain nature and not by the reason of the god" (φύσει τινὶ καὶ οὐ θεοῦ λογισμῷ διοικούμενα).

93 The word is found in the *De mundo* (396b7–11) in connection with the way in which opposites harmonise in the cosmos. See Gregorić in Gregorić and Karamanolis (2020), 154–155.

94 *EE* II 8, 1224a24–26 (Rowe): ἐν μὲν τοῖς ἀψύχοις ἁπλῆ ἡ ἀρχή, ἐν δὲ τοῖς ἐμψύχοις πλεονάζει, οὐ γὰρ ἀεὶ ἡ ὄρεξις καὶ ὁ λόγος συμφωνεῖ. "If, in inanimate beings, the principle is simple, on the other hand, in animate beings, it is superabundant, for desire and reason do not always agree." *NE* III 12, 1119b15–16 (Bywater): διὸ δεῖ τοῦ σώφρονος τὸ ἐπιθυμητικὸν συμφωνεῖν τῷ λόγῳ, σκοπὸς γὰρ ἀμφοῖν τὸ καλόν [...]. "This is why the appetitive part must agree with reason, because for both of them, the noble is the goal" (my translation).

95 Aspasius, 70.14: "the desire being in harmony with such reasoning" (σύμφωνον τῷ τοιούτῳ λογισμῷ ἔχων τὴν ὄρεξιν); 94.14–16: "The appetitive part of a moderate person is in harmony with reason and is not carried away by being tyrannized like that of the self-controlled person" (τοῦ δὴ σώφρονος τὸ ἐπιθυμητικὸν σύμφωνόν ἐστι τῷ λόγῳ καὶ οὐ καταδυναστευόμενον ἄγεται ὥσπερ τὸ τοῦ ἐγκρατοῦς); 119.19–21: "the high-tempered part of the soul must be educated so as to do nothing contrary to reason, but rather to sally forth along with reason, being in harmony with it" (χρὴ τὸ θυμοειδὲς μέρος τῆς ψυχῆς πεπαιδεῦσθαι ὡς μηδὲν παρὰ τὸν λόγον πράττειν ἀλλὰ συνεξορμᾶν τῷ λόγῳ, σύμφωνον αὐτῷ ὄντα).

96 In *NE* I 13, 1102b28, Aristotle uses ὁμοφωνεῖ, which belongs to the same family, and Aspasius uses the word again in his commentary (35.34).

ence is "most in harmony with the nature of the transported bodies." Therefore, there is no direct analogy between *pronoia* and reason, or between the nature of the moved bodies and the appetitive part. In an ethical context, the relationship flows from bottom to top, while, in a cosmological context, it flows from top to bottom. This is because in an ethical context it is assumed that a rapprochement between the two terms (the desiring part and reason) is both possible and desirable. There is no inherent agreement between them, but the goal is for the appetitive part to be brought into alignment with reason. In a cosmological context, however, the situation is different. Aspasius means to say that the mover is in harmony with the nature of the celestial bodies in the sense that they are doing the same thing. To say that there exists perfect harmony between the *pronoia* of the mover and the nature of the celestial bodies means that the mover does not impose any external constraint on them. It is in the nature of the celestial bodies to move as the *pronoia* of the prime mover directs them to. They are not moved by force, but by the providence of the mover without any external compulsion, because the mover's providence is in harmony with their nature. This harmony allows us to understand that necessity does not function as an external force, but as the actualisation of an inherent potentiality—that of local matter.

The suggestion could also be made that, in the relationship between the providence of the prime mover and the celestial spheres, the connection is downward in nature, implying that necessity gradually diminishes. The providence of the mover is "most in harmony" or "nearly perfectly in harmony" with the nature of the celestial spheres, but it is less in harmony with the nature of the bodies in the lower strata of the cosmos. This is why Aspasius does not claim that providence is *absolutely* in harmony with the bodies it moves. As we have seen, based on the passage from *Metaphysics* XII 7, the harmony is ontological: the necessity of the prime mover perfectly aligns with the nature of the celestial spheres, as the mover is absolutely necessary (it cannot be otherwise), while the celestial spheres can be otherwise depending on their location (they possess local matter). Thus, there is a very great ontological proximity between the two, but not an equality of ontological levels (the prime mover is immaterial, whereas the bodies moved are material). The use of the superlative συμφωνοτάτη could suggest that, according to Aspasius, there is a scale of harmony between the prime mover and the moved bodies. While Aspasius does not specify this in detail, it is almost certainly the *matter* of the celestial spheres that the *pronoia* of the prime mover is in nearly total harmony with. This is confirmed by the following sentence, if it is interpreted as building on what was just said:

> τῇ γὰρ φύσει τῶν φερομένων συμφωνοτάτη ἐστὶν ἡ τοῦ κινοῦντος πρόνοια, ἔνια δέ ἐστιν ἃ γίνεται μὲν κατὰ φύσιν, ἀτακτοτέραν δέ, οἷον αὐχμοὶ καὶ ὄμβροι. (71.30–32)
>
> For the providence of what moves is perfectly in harmony with the nature of the things transported, but there are certain things which, while occurring according to nature, occur according to a more disorderly nature, such as droughts and storms.

It is remarkable that Aspasius directly quotes, in the very middle of his commentary on the *Ethics*, a phrase from the opening of the *Meteorologica*.[97] This confirms his unitary conception of the Aristotelian corpus, in which the *Meteorologica* naturally finds its place within an ethical commentary. Moreover, this reference indicates that Aspasius situates his analysis at the boundary between the supralunary and sublunary parts of the cosmos. From this, we may infer that in the first part of the sentence, when Aspasius refers to the nature of moving bodies, he has in mind "the first element of bodies" or aether. In the final passage quoted (71.30–32), Aspasius shifts his focus from the primary spheres, which are made of a specific nature, to the irregular phenomena of the sublunary regions. These phenomena, while irregular, are initially described by Aspasius as occurring "according to nature", meaning "by necessity" and through the "providence of the mover". To clarify this point, he then provides two contrasting examples of irregular yet relatively common meteorological phenomena: droughts and storms. By doing so, he underscores the distinction between supralunary nature and the more disorderly nature of sublunary occurrences. This leads him to differentiate at least two types of nature: that of celestial bodies and that of sublunary meteorological phenomena. While all these movements ultimately stem from the necessity of the mover, this necessity is not equally in accord with all natures, since some align with it more closely than others. For this reason, the motions of celestial bodies occur by necessity, by nature and according to nature, whereas sublunary movements

---

97 *Meteor.* I 1, 338a25–b21 (Fobes): λοιπὸν δ' ἐστὶ μέρος τῆς μεθόδου ταύτης ἔτι θεωρητέον, ὃ πάντες οἱ πρότεροι μετεωρολογίαν ἐκάλουν· ταῦτα δ' ἐστὶν ὅσα συμβαίνει κατὰ φύσιν μέν, ἀτακτοτέραν μέντοι τῆς τοῦ πρώτου στοιχείου τῶν σωμάτων […]. "There remains for consideration a part of this inquiry which all our predecessors called meteorology. It is concerned with things that happen according to nature, though according to a more disorderly nature than that of the first of the elements of bodies." Translation Webster in Barnes (1984), with modifications. The same term ἀτακτοτέραν is used by Theophrastus in his *Metaphysics* (4a4).

occur according to nature, but not the same nature as that of celestial bodies. Rather, they follow a "more disorderly nature".[98]

The use of the superlative συμφωνοτάτη may also suggest that the *pronoia* of the prime mover does not terminate in the heavenly regions, but extends to quadri-elemental nature, albeit in a weakened form. This implies that there is no absolute break between the two natures Aspasius has just distinguished—that of celestial bodies and the "more disorderly" nature composed of the four elements. Aspasius thus avoids positing a radical discontinuity, even if an ontological dualism remains. Contrary to the strict dualistic interpretation of the NSP, the *pronoia* of the prime mover extends beyond the supralunary sphere, exerting an influence within the realm of corruptible nature and remaining the primary cause of motion.[99] However, necessity is progressively diluted due to the differences between these natures. If we acknowledge two levels in the causality of necessity—one in which the prime mover is "fully in accord with the nature of the bodies moved" and another where it is less so due to the "more disorderly nature" of the sublunary realm—Aspasius' passage reveals a two-tiered conception of providence. By tempering the necessity of the prime mover in the sublunary world, Aspasius avoids a Stoic determinism, as evidenced by his remark in the commentary on *NE* III 3, 1112b15–16: "The human being is always the principle of his own actions, and all things are not determined by necessity or fixed by fate."[100]

4 Conclusion

This extract from Aspasius' commentary, though brief and allusive, proves to be particularly valuable and significant. Indeed, it is the only extant Aristotelian text on providence from this period before Alexander, apart from a quotation from Adrastus. What distinguishes Aspasius' account is that (a) while he operates within the broader NSP framework, (b) he identifies necessity (also referred to as providence) as the sole first principle of motion and (c) he subordinates to this necessity/providence the motion "by nature" or "according

98 Like Aristotle, Aspasius uses the word φύσις to refer to the nature both of celestial bodies (in which case it means the aether) and of sublunary beings. In both cases, the bodies move "by" or "according to their nature". See Frey (2015), 99–125.

99 We might wonder whether the version of the NSP doctrine found in Aspasius is not closer to the one echoed by Diogenes Laertius when he says that earthly things are "organised" according to their *sympatheia* (κατὰ τὴν πρὸς ταῦτα συμπάθειαν, V 32.6).

100 74.11–12: ἄνθρωπος ἀεὶ ἀρχὴ τῶν ἰδίων πράξεων καὶ οὐκ ἐξ ἀνάγκης ὥρισται πάντα οὐδ' εἵμαρται. See Sharples (2007a), 599.

to nature" of celestial bodies, and possibly even that of natural beings in the sublunary realm. We can only speculate as to whether Aspasius expounded a fully developed doctrine of providence in his lost commentaries on the *Metaphysics* or *De caelo*. However, certain considerations in his commentary on this passage of *NE* III 3 suggest that he may be drawing upon a broader doctrinal framework. The cosmological micro-passage in *NE* III 3, with its reference to necessity taken as a cause, compels Aspasius to extend his analysis beyond the ethical domain. In this passage, Aristotle makes (or appears to make) necessity a cosmological principle, which leads Aspasius to reinterpret Aristotle's theory of principles: this necessity is, in fact, the providence of the first mover. The significance of Aspasius' interpretive move lies in its attempt to delineate the specificity of a Peripatetic model of providence—one that is not previsionist, demiurgic, intentional, creationis, or moral. Instead, Aspasius presents a conception of providence that is purely kinetic: the prime mover exercises providence insofar as, in virtue of its ontological priority, it serves as a causal principle for the first heaven and guarantees the eternity and necessity of its motion. Seen in this way, necessity itself is providence, since it is the source of benefit for what is moved. This designation is reinforced by Aristotle's own characterisation of the prime mover as "the good", in precisely the same passage where its necessity is emphasised.[101]

Aspasius further articulates the relationship between necessity and the nature of eternal bodies, specifically aether: moving eternally in a circle *by nature* and *according to their nature*, celestial bodies remain subject to the necessity of the prime mover. The reference to nature in his quotation from the *Meteorologica* suggests that the causal influence of necessity extends into the sublunary realm via the circular motions of the sun and moon. However, motions become increasingly disorderly, because the nature of what is moved is no longer the same: i.e. no longer aether, but the composite nature of the four elements. At this level, the *pronoia* of the prime mover is less in harmony with the nature of what is moved. The regularity of generation and corruption in living beings thus derives indirectly from necessity, offering a more nuanced view of Aristotelian cosmological dualism.[102]

What we find here is an instance of what Kupreeva has termed "diminished" providence: all natural motion, however contingent or disorderly, ultimately bears the imprint of the prime mover's necessity or providence. From this perspective, nature—whether supralunar or sublunar—exhibits a certain order,

101 See Broadie (1993) and Baghdassarian (2019), 259, with reference to *Metaph.* XIV 4, 1091b16–19.

102 See Aristotle's *GC* II 10, 336b27–337a17, and *GA* IV 10, 778a5–9.

since its principle of motion remains dependent on the prime mover's providence. A final distinctive feature of this Peripatetic conception of providence is that, in the absence of an intention on the part of the prime mover, the benefit that nature receives remains accidental. The fact that *pronoia* designates the necessity of the prime mover does not imply that the good received by the moved bodies (i.e. the eternity and necessity of their motion) is anything other than an *accidental* good. From the standpoint of both the Stoics and Alexander, this constitutes a fundamental weakness in the Peripatetic model.[103] It is likely that the use of the term *pronoia* to describe the necessity of the prime mover serves, at least in part, to mitigate this non-final and accidental character. That said, we should not expect too much from this brief passage in Aspasius. In just a few lines, he could hardly have set out his full conception of Aristotle's theology, and he remains silent on many critical issues, most notably the precise mode of causality exercised by the prime mover.[104]

103 For Alexander, there is no providence if its effect is accidental: *De providentia* 63.2; *Quaest.* II 21, 65.18–66.2.

104 I would like to thank Manuel Arbeláez, Fabienne Baghdassarian, István Bodnár, Pavel Gregorić, Gweltaz Guyomarc'h, Orna Harari, Myrto Hatzimichali, Thornton Lockwood, Alexandra Michalewski, Federico Maria Petrucci and Luca Torrente for their comments and suggestions. All remaining errors are, of course, my own.

CHAPTER 13

Providence and Method: Peripatetic Physics in Adrastus' *Commentary on the Timaeus*

Ludovica M. Fazio and Federico M. Petrucci

1 Introduction*

As is widely known, the post-Hellenistic Peripatos was a quite complex and prolific philosophical movement: several important figures who acknowledged Aristotle's authority and produced exegetical works on the Aristotelian corpus were active after Andronicus of Rhodes.[1] Their positions represented important reference points for rival schools (especially the Platonists), as well as providing the fundamental materials for Alexander's philosophy—a philosophy which constituted the apex of the movement and, roughly speaking, also coincided with the Peripatos' "swan song". Compared to this lively philosophical scenario, extant direct witnesses of post-Hellenistic Peripatetics before Alexander are, however, minimal: what we know of the philosophy of figures such as Herminus, Alexander of Aegae or Sosigenes is filtered through Alexander's mentions and criticisms, which are often not only biased, but also indirectly transmitted in turn. The most celebrated exception is probably that of Aspasius, whose *Commentary on the Nicomachean Ethics* has been extensively transmitted: reasonably enough, this work has received relatively widespread scholarly attention and might be used as a fundamental cornerstone to reconstruct some post-Hellenistic standards of the Peripatos.[2]

There is, however, another work of which substantial portions have reached us, albeit through complex and somewhat paradoxical mechanisms within the

* Although this paper derives from a close collaboration, Federico M. Petrucci's contribution mainly concerns sections 2–3, while sections 4–5 were mainly written by Ludovica M. Fazio.

1 Authoritative studies on the post-Hellenistic Peripatos before Alexander include Donini (1974), Moraux (1984), Gottschalk (1987), Rashed (2007), Sharples (2010), Kupreeva (2016), Falcon (2016) and Baltussen (2016b).

2 After Donini (1974), 63–125, and Moraux (1984), esp. 161–170, see Alberti and Sharples (1999) and Konstan (2006).

tradition, namely Adrastus' *Commentary on the Timaeus*. According to a classical reading, extensive sections of this commentary emerge from *ad litteram* quotations of Theon of Smyrna's *Expositio* and were translated by Calcidius in his own *Commentary on the Timaeus*.[3] Assuming that most of Theon's musical and astronomical discussions come from Adrastus, and that Calcidius used Adrastus as his sole technical source, it would seem that Adrastus' *Commentary* is the best-preserved post-Hellenistic Peripatetic work after that of Aspasius, and it certainly provides the most substantial testimony about post-Hellenistic Peripatetic discussions of physics. Given this background, it seems astonishing that scholars have devoted so little attention to Adrastus: after Moraux's classical discussions, the only significant contributions devoted to Adrastus are Richard Sorabji's 2007 chapter on Adrastus' discussion of Aristotle's astronomy and two papers by Federico Petrucci, concerning, respectively, the structure and text of Adrastus' *Commentary* and Adrastus' Peripatetic cosmology as it emerges from *Expositio* 178.3–186.16.[4]

At this point, a reader may expect us to claim that scholars should engage in an in-depth reconstruction of Adrastus' *Commentary* and perhaps encourage someone to edit it. If so, this chapter will be quite disappointing, since it has a seemingly more modest goal. In the second section, we will show that, even though it is true that Theon and Calcidius provide us with some kind of access to Adrastean materials, it is often virtually impossible to be certain of the extent to which they tampered with their source and, in some cases, even to determine whether Adrastus provides the basis for certain sections of their writings. The third section will make things even more puzzling, for we will argue that

3 This view, having its origins in Hiller (1871), and Switalski (1902), 58–91, was then strongly supported and made bolder as to the possibility of reconstructing extensive sections of Adrastus' text by Waszink (1964). Recently Bakhouche (2011), 36–37, has revived the view already expressed by Martin (1849), 74–79, that Adrastus would be the source of Theon, while Calcidius did not use Adrastus' commentary, but relied on Theon's *Expositio*. This hypothesis, however, was already refuted by Bergk (1850), 176; cf. also Switalsky (1902), 58–91, and Petrucci (2012b).

4 See Moraux (1984), 294–332; Sorabji (2007); Petrucci (2012a) and (2015). A rough idea of the scenario we are describing already emerges from skimming Sharples' 2010 sourcebook, in which, apart from very brief mentions of our author, just two texts ascribable to him are included as sources, namely *Exp*. 181.12–182.25 = 21M, which is part of the section also analysed in Sorabji (2007) and Petrucci (2015), 201.20–202.2 = 21N (it is also taken into account by Sorabji, but it is far from clear that this section goes back to Adrastus), and 149.4–150.4 = 22I (the last part of a text we will discuss, see T5 below), which, however, is not really analysed.

even those views which have been traditionally regarded as characteristic of Adrastus (especially in the field of harmonics) rest on quite fragile testimonies. However, taking these methodological caveats into account, we will also show that it is still possible to exploit extant evidence about Adrastus' exegetical method relating to his use of Peripatetic physics in his interpretation of the *Timaeus*, particularly with reference to his use of Peripatetic acoustics in his exegesis of Timaean harmonics. This provides a solid foundation to approach a much more substantial issue in section 4, where we will argue that, starting from Adrastus' astronomical exegesis of the *Timaeus*, it is possible to grasp certain key aspects of his theory of providence and its physical bases. This will lead us to suggest that the introduction of this theory in its philosophical and exegetical context has philosophical import: not only does it reveal Adrastus' peculiar conception of the relation between Plato(nism) and Aristotle concerning this important theme, but we will also be able to reassess Adrastus' role in the development of a post-Hellenistic Peripatetic theory of providence, a role that will eventually be seen to be much more important than usually believed.

2 The Hunt for Adrastus: Where and How to Chase Him

The overall conclusions reached in Federico Petrucci's 2012 paper on the possibility of reconstructing the structure of Adrastus' *Commentary*, that is, the topics it included and the order in which they were presented, were already relatively pessimistic: extant witnesses, as provided by Theon and Calcidius, diverge with regard to the composition of the text, as well as the selection and arrangement of passages, in a way that makes Adrastus' original position disappear in some cases. Still, a comparison of the sources may at least suggest what topics were probably included in the commentary: in this introductory section, we will offer a general outline of these topics, indicating their main sources, and emphasise those general points which will prove important for the following sections of this paper.

By bringing Theon's and Calcidius' texts together, the contents of Adrastus' *Commentary*—at least, those which have been transmitted in the sources—can be sketched out as the following table shows:

Topic Number	Adrastus' *Commentary*	Theon	Calcidius	Other Sources	Notes
			Harmonics		
1	Theory of proportions (related to the proportional arrangement of the world's body)		chs. VIII–XIX		No parallel to Calcidius; if this topic was in A., the scope of his commentary (at least, in its mathematical parts) would exceed the psychogony.
2	Philosophical discussion of the soul's composition		chs. XX–XXXI		It is likely that C. drew this from a different source.
3	On Plato's *tetraktys* and the lambdoic structure		chs. XXXII–XXXIX	Cf. Procl., *In Tim.* II, pp. 170.22 ff. and 177.25 ff. Diehl	See section 3 below.
4	Theory of *mesotēs*		chs. XL–XLIII		Theon has a different version; it is far from clear whether this section was in A. and in what form.
5	Further introductory notions of musical terminology	pp. 49.6–50.3	ch. XLIV, pp. 92.10–93.4		There is very good evidence that this material is from Adrastus.
		pp. 50.4–56.5	ch. XLIV, pp. 92.16–93.2	Porph., *In Ptol. Harm.* 7.24–8.5 and 96.2–6 Düring; cf. Procl., *In Tim.* II, pp. 169.27–170.1	

(*cont.*)

Topic Number	Adrastus' *Commentary*	Theon	Calcidius	Other Sources	Notes
6	Pythagoras' discoveries	pp. 56.9–57.10 (but only 56.9–57.1 is ascribed to Adrastus)	ch. XLV, pp. 93.16–94.16		See section 3 below.
7	Technical remarks on the harmonic system of the soul	pp. 61.20–63.24	ch. XLVI, pp. 95.1–96.16		Th.'s last section is probably heavily altered.
		pp. 63.25–65.9		Procl., *In Tim.* II, p. 170.5–21	
		pp. 65.10–66.11			
		p. 66.12–18	ch. XLVII, p. 97.11–12		
8	The *leimma* and the "numbers'" of the soul	Cf. pp. 68.12–69.3	chs. XLIX–L	Cf. Procl., *In Tim.* II, pp. 170.22 ff. and 177.25 ff.	C. is probably closer to A., but neither Th. nor C. really supports Proclus' testimony.
9	On ratios	pp. 73.16–74.14			
10	Proportions and means	pp. 106.12–107.14	Cf. ch. XXXII, pp. 81.26–82.3		
11	Philosophical exegesis		chs. LI–LV		It is likely that C. drew this from a different source.
			Astronomy		
12	Introductory remarks		chs. LVI–LVIII		It is likely that C. drew this from a different source.

(*cont.*)

Topic Number	Adrastus' *Commentary*	Theon	Calcidius	Other Sources	Notes
13	The shape of the Earth and of the world	pp. 120.1–129.4	chs. LIX–LXIV, pp. 106.17–111.20	Cf. Theon Alex., *In Ptol. Alm.*, pp. 394.13–398.5 and 399.9–400.15 Rome	The sources are very close to each other.
14	The celestial circles	pp. 129.10–133.25	chs. LXV–LXVIII, pp. 111.21–115.19		
15	The motions of the stars and of the planets	pp. 134.1–138.8	chs. LXIX–LXXI, pp. 116.1–119.10		
16	The order of the planets	pp. 138.9–140.3	chs. LXXII–LXXIII, pp. 119.11–120.13		Th. extensively modifies the source (esp. at pp. 140.3–142.6 and pp. 143.7–147.6).
		pp. 140.3–141.4			
		p. 142.7–15	ch. LXXIII, pp. 120.13–121.3		
		p. 143.1–6	ch. LXXIII, p. 121.3–6		
		Cf. pp. 143.7–147.6	ch. LXXIII, pp. 121.6–122.2; Cf. ch. XCV p. 148.6–9		
17	Introduction to the planetary motions	p. 147.7–14 pp. 147.19–150.12	chs. LXIV–LXXVII, pp. 122.3–125.4		Here too, Th. modifies the source.
		p. 150.12–18	chs. LXXVII–LXXVIII, p. 125.4–17		
		p. 151.8–19			
		pp. 152.8–153.15	ch. LXXVIII, pp. 125.17–127.15		
		pp. 153.16–154.23	ch. LXXIX, pp. 127.16–128.9		

(*cont.*)

Topic Number	Adrastus' *Commentary*	Theon	Calcidius	Other Sources	Notes
18	The eccentric	pp. 155.1–157.12 pp. 157.13–158.9	ch. LXXX, pp. 128.10–130.16		
19	The epicycle	p. 159.1–5	ch. LXXXI, p. 131.5–7		Here too, Th. modifies the source.
		pp. 160.22–162.18			
		pp. 163.1–165.11	chs. LXXXI–LXXXII, pp. 132.7–134.5		
		pp. 165.12–166.3	ch. LXXXII, p. 134.5–8		
20	The equivalence of epicycle and eccentric	pp. 166.14–171.2			C. just hints at this topic, while Th. modifies his source according to his own exegetical agenda.
		pp. 172.15–174.15	Cf. ch. LXXXIII, pp. 134.9–135.7		
		pp. 177.9–178.2			
21	Aristotle's model	pp. 178.3–179.6	Cf. ch. LXXXIV, pp. 135.8–136.6		Th. modifies his source.
		pp. 181.9–186.16		Cf. Theon Alex., *In Ptol. Alm.*, pp. 851, 11–856, 8	
		pp. 186.17–188,7	Cf. chs. CVIII–CXII, pp. 156.4–159.13, and chs. XCVIII–C		
		pp. 188.8–24			

(*cont.*)

Topic Number	Adrastus' *Commentary*	Theon	Calcidius	Other Sources	Notes
22	The motions of the superior planets	pp. 190.1–191.7	ch. LXXXV–LXXXVI, pp. 136.7–138.3		
		pp. 191.8–192.21	Cf. ch. LXXXVII, p. 138.4–6		
23	Eclipses	pp. 192.22–198.8	chs. LXXXVII–XCI, pp. 138.6–144.10		
24	Philosophical discussion		chs. CI–CVII		It is likely that C. drew this from a different source.
25	Technical clarifications		chs. CXIII–CXVIII		

There are cases in which the textual agreement between our sources is so close that one can conceive of the possibility of reconstructing Adrastus' text *ad litteram*. This emerges, for instance, in parts of "Topic 5", for which we have not only the agreement of Theon (*Exp.* 50.5–12) and Calcidius (*In Tim.* XLIV, 92.16–93.4), but also Porphyry's testimony (*In Harm.* 7.22–8.5). Nonetheless, these cases are much less frequent than those which confront the reader with extremely puzzling scenarios. Most clearly, not only is the presence of a third parallel (like Porphyry in this case) quite a rare event, but it also often happens that several sections in either Theon or Calcidius lack a parallel in the other source. In cases such as Topics 1 and 4 for harmonics and Topic 20 for astronomy, we cannot be certain that the source is Adrastus, or, even if we are, it remains unclear to what extent our witness tampered with Adrastus' text. Fortunately enough, in some of these cases one of our sources, namely Theon, is kinder to us, repeatedly mentioning Adrastus at the outset of certain discussions: this suggests that, even though doubts could be raised regarding the accuracy of the testimony, it is at least possible to state that this or that view can be traced back to Adrastus.

Things are made even more complicated by the fact that, as the table already suggests, Adrastus' overall interest in the *Timaeus* is mirrored in quite different ways in our more extensive witnesses, namely Theon and Calcidius, and in other witnesses, namely Porphyry and Proclus, who are at the same time (allegedly) more authoritative and less generous. As is widely known (and confirmed by the table), if we had to assess Adrastus' interests solely based on Neoplatonic testimony, we would be led to conclude that Adrastus' main interest was directed towards Timaean harmonics: passages from Proclus' *Commentary on the Timaeus* mention Adrastus as an authoritative exegetical source with regard to the structure of the harmonic series of the soul (Topic 3), some of its technical features (Topics 5, which is also paralleled in Porphyry, and 7), and the overall numerical values according to which the soul was harmonically arranged (Topic 8). We will revisit some of these topics in the next sections, but, at this stage, a more general point is worth mentioning: as far as we can infer from Theon and Calcidius, and contrary to what Neoplatonic testimonies would suggest, Adrastus was more interested in astronomy than in harmonics. Establishing to what extent this interest was developed is impossible, but for the sake of the present demonstration, it is enough to emphasise that it is legitimate to expect some kind of commitment on the part of Adrastus to emerge from the astronomical sections of our main sources. Moreover, if one had to speculate about the reasons for Adrastus' interest in astronomy, two straightforward—though still provisional—replies present themselves. First, while both harmonics and astronomy are involved in the exegesis of the *Tim-*

aeus,[5] from the point of view of a Peripatetic, the latter certainly looks for certain much more promising, since it is clear that Aristotle's engagement with harmonics is much less extensive and crucial than his treatment of astronomical matters:[6] astronomy may thus offer more profitable opportunities to bring together the *Timaeus* and Peripatetic philosophy. But it is even more important to consider that, as scholars have been emphasising, one of the most important Aristotelian texts in the post-Hellenistic age was the *De caelo*,[7] and more generally that the debate about the heavenly motions, their origin and their nature, was undoubtedly at the centre of the Peripatetic philosophical debate at the end of the Hellenistic age: dealing with astronomy thus enabled Adrastus to engage with this debate.

To sum up: even though in most cases it is impossible to reconstruct *ad litteram* Adrastus' text, and sometimes it is impossible even to be certain of the presence of Adrastus in our sources, there are specific cases that give us a glimpse of Adrastus' *Commentary*. Within this framework, if we have to search for Adrastus' views in the available materials, we are likely to find them *especially*, though not only, in passages devoted to astronomy.

3 Theon, Calcidius and Proclus: Tampering with Adrastus—and Getting Away with It

We will now expand on the conclusions reached in the previous section to emphasise to what extent and why some extant testimonies are unreliable, while others allow us to detect Adrastus' own commitments and determine the scope of his *Commentary*.

Let us begin with a telling test case. We have stressed that, according to authoritative Neoplatonic testimonies, Adrastus' main contribution to the ex-

5 It is not by chance that the practice of writing technical exegeses to the musical and astronomical sections of the *Timaeus* was widespread in the post-Hellenistic age, and perhaps already even in the Hellenistic age (see esp. Ferrari 2000, Petrucci 2016 and Petrucci 2020).

6 A (perhaps too) negative evaluation of Aristotle's competence in harmonics is provided by Barker (2007), 328–363. (Interestingly, the Peripatetic who dealt most extensively with harmonics, i.e. Aristoxenus, famously rejected a quantitative approach to the discipline.) Be that as it may, mentioning the *De caelo* and *Metaphysics* XII 8 is sufficient to emphasise to what extent astronomy (or, rather, "astrophysics", Judson 2015) was much more relevant as a discipline for Aristotle than harmonics.

7 See esp. Gottschalk (1987) and Rescigno (2004), 59–71. For a commented review of the works on celestial motions composed within the Peripatos before Alexander, see Sharples (2010), 180–195.

egesis of the *Timaeus* concerned harmonics. Indeed, the most characteristic view that Proclus ascribes to Adrastus is related to the *vexata quaestio* concerning the disposition of the numbers structuring the world soul in the *Timaeus* (35b4–36b5). From Crantor onwards, exegetes of the *Timaeus* have considered this issue a crucial *zētēma*,[8] and their solutions can be classified into two groups. Some readers construe the numbers (i.e. both the basic series 1, 2, 3, 4, 9, 8, 27 and the subsequent numbers deriving from the application of harmonic and arithmetic means) as arranged along a single straight line. Others, probably insisting on the opposition between even and odd, instead took the numbers to be arranged according to a lambdoic scheme—that is, along two branches that have their common origin in the number 1 and include, respectively, the series of even numbers and the series of odd numbers. Although the first and main supporter of this view was an important Academic, namely Crantor (T11a.70–74 Mette = Plut. *An. Procr.* 1022D10–1028A1), in his discussions Proclus instead considers Adrastus as its main supporter (*In Tim.* II, 170.22–171.4 and 187.17–188.8 Van Riel). Hence, the least one would expect is to find this view reflected, and emphatically so, in Theon and Calcidius. Interestingly, this is not the case. The clearest mention of the lambdoic scheme in our main witnesses comes from Calcidius, who hints at it very quickly, in a not entirely explicit way and without any emphasis:

> T1. Calc. *In Tim.* XXXIII 83.9–19
>
> Now, the apex of these numerical series is a unity which like the geometrical point lacks parts of any kind; the double of which is a line, length without width; the double of the line is a plane surface, or length with width; the double of that is the cube, a body divisible into length, width, and depth, 2×2×2, i.e., 8. Similarly, there is found on the other side an affinity between numbers in the ratio of the triple: the cube, in the sum of 27, is the first cubic power among odd numbers, for 3×3×3 produces the sum indicated. Hence, according to the ratio of a continuous proportion, i.e., a geometric ratio, 8:4 = 4:2 = 2:1, and conversely, 27:9 = 9:3 = 3:1.

The lambdoic scheme also appears in Theon's *Expositio* (94.10–96.9), where, however, it is introduced in the context of an arithmological section associating the Pythagorean *tetraktys* (1, 2, 3, 4) with sets of "items" whose articulation recalls that of the *tetraktys*; this textual nucleus probably does not derive from

8 See Petrucci (2019) for a survey of these Platonist readings.

Adrastus, and anyway the Platonic *tetraktys* plays no substantial role in it.[9] Even admitting that Theon tampered with his non-Adrastean source and supplemented it with Adrastean materials, we find no mention in the passage of Adrastus and no emphasis on his (alleged) commitment to this exegetical view. While other test cases could be invoked,[10] a relatively straightforward conclusion may already be drawn. Although we would not push the argument so far as to state that the views stressed by Proclus were not mentioned by Adrastus in his *Commentary* at all,[11] the testimonies of Theon and Calcidius at least show that these were not distinctive of Adrastus' exegesis and were not proposed with any particular commitment or emphasis. Given the importance of these topics in the commentary tradition, this suggests that Adrastus had a limited interest in technical issues related to harmonic theory: while discussing them, as required by the text he was commenting upon, the Peripatetic Adrastus must have seen no reason to commit himself to a specific technical view about the numbers of the harmonic system of the soul and their arrangement.

Despite this quite discouraging situation, however, there is at least one case indicating that Adrastus did, albeit perhaps implicitly, introduce some kind of

9 See Petrucci (2012b), 408–412.

10 For instance, one of the implications of ascribing to Adrastus the adoption of a lambdoic scheme including all numbers of the *divisio animae* is that the final values must be calculated in such a way as to allow the entire system to be represented. This has as a consequence that, in order to represent the series in integers, he must have used 384 as coefficient for all values starting from the first number of the series (i.e. 1), so that 1 coincides with 384 and 27 with 10368. One would expect this account to be mirrored in Theon and Calcidius, but once again this is not the case. The issue of the numbers according to which the soul is structured is addressed by Calcidius in his discussion of the divisibility of the tone (XLIX–L, 98.4–100.2): although the number 384 is briefly mentioned in this discussion (98.9–11), no preference for its adoption is spelled out. Theon's testimony is even more discouraging, since the system ascribable to Adrastus is very briefly referred to, but no mention of Adrastus is made and Theon even criticises the (allegedly) Adrastean view: at most, Adrastus might be included among "those" (ἔνιοι) who rather prefer to associate 1 with 384 instead of 192 (68.12–69.3), and by commenting on this view Theon does not side with this option nor emphasise it. Moreover, when Theon has to produce his model for the soul's structure (87.4–93.11), he explicitly adopts Thrasyllus' version, employing much lower values, and concludes the section by mentioning, as part of a different scheme and without citing his source, the possibility of considering 10368 as the highest value. In brief, while Theon and Calcidius neither confirm nor disprove the fact that Adrastus supported the adoption of 384 as the coefficient for the structure of the soul, they certainly do not present this option as distinctive of his exegesis.

11 One might object that these views may have been found in other Adrastean writings. Even in this case, however, given that they all concern harmonic issues related to the exegesis of the *Timaeus*, it is highly unreasonable for Adrastus not to have mentioned them also in the *Commentary*.

personal touch—and, more specifically, a Peripatetic touch—in his exegesis of Timaean harmonics, one related to the acoustic nature of sound.

Both Theon and Calcidius dwell on the issue of the discovery of the numerical values of consonances, albeit in quite different ways. What we find in Calcidius is just a quick hint at an experiment conducted by Pythagoras, through which he discovered these values (*In Tim.* XLV, 94.14–16). Theon pays much more attention to the empirical discovery of consonances, ascribing it to Pythagoras (*Exp.* 56.9–57.8). Subsequently, however, Theon gives an extensive report of a different story, which instead ascribes experiments to Lasus and Hippasus. But this cannot be drawn from Adrastus, since Theon concludes his discussion by saying: "This is what concerned the discovery of consonances: let's now get back to what Adrastus handed over" (61.18–19). Hence, Theon's strategy can be reasonably reconstructed as follows: he found in Adrastus the reference to Pythagoras' experiments, which he then wanted to supplement by adding materials on Hippasus and Lasus coming from a different line of transmission.

Now, the story of Pythagoras' empirical discovery of the numbers expressing the consonances was certainly not invented by Adrastus,[12] but what is remarkable about his approach is that the texts in which similar heurematologic stories occur are not exegeses of the *Timaeus* (with the notable exceptions of Theon and Calcidius, of course). Thus, it was not an obvious or necessary choice on the part of Adrastus to include this reference to the empirical discovery of consonances in his *Commentary*, meaning that his commitment to establishing a connection between the consonances' numbers and the physical nature of sound must have been substantial. This interest is confirmed later in the *Expositio*, when Theon explicitly draws from Adrastus a discussion of the relationship between numbers and pitch, and does so by appealing to an empirical experiment involving weights hanging on a string (65.10–66.11). It is thus worth trying to understand why Adrastus wanted to include this nucleus in his exegesis. Now, the story of the discovery of consonances and more generally the interest in the physical nature of sound are different from the previously discussed topics (i.e. those related to the numbers of the *divisio animae*), as they do not consider consonances exclusively in mathematical terms. On the contrary, they imply that the consonances' numbers are mirrored in phys-

12 It probably remounts to the Xenocrates (F6 Isnardi Parente) and various versions of it can be found in later sources: see, e.g., DL VIII 12 and Nicom. *Ench.* 245.18–248.26. References to the discoveries of Hippasus and Lasus can be found in a set of other sources, probably depending on Aristoxenus (Fr. 90). For more details and bibliographical references, see Petrucci (2012b), 359–368.

ical nature, since otherwise it would be impossible to discover them through experiments. In this case, however, it is easy to see why a Peripatetic could be interested in this point: while Plato and the musical texts inspired by him dismiss the reliability of hearing,[13] the idea that acoustics may be conceived of in quantitative terms and reflect the numerical nature of consonances can be already detected in Aristotle and is typical of some Peripatetic writings.[14] Thus, by including the story of Pythagoras' experiments in his *Commentary*, Adrastus could supplement Plato's (and the Platonists') account, which largely marginalised acoustics in favour of mathematical harmonics, by introducing elements taken from Peripatetic physics.

This test case is important in two ways. On the one hand, it allows us to outline Adrastus' method for making the *Timaeus* and Peripatetic physics interact: when he was confronted with cases that allow Peripatetic views to emerge, Adrastus introduced them to supplement Plato's account, even when they appeared at odds with Plato's most straightforward positions (or those standard from a Platonist standpoint). On the other hand, despite the challenges posed by the sources, it testifies more generally to the possibility of reconstructing a specific "Peripatetic" commitment ascribable to Adrastus—although in this case it does not seem to imply any polemic within the field of Peripatetic philosophy. As we will see in the next section, however, Adrastus' treatment of astronomy presents us with a much more interesting scenario.

4 Physics and Providence: Adrastus' Peripatetic Cosmology

Combining the methodological conclusions reached so far, one would now expect to find clearer and more relevant cases of Peripatetic supplements when astronomy comes to the fore. In this section we will thus analyse a famous Adrastean piece of astronomical exegesis, starting from some technical remarks, in order to identify an important theoretical nucleus that may be ascribed to Adrastus, namely his theory of providence.

---

13 This is done for important theoretical reasons: e.g. the fact that hearing is not reliable allows Platonists to stress that, even though we might think we perceive a perfect halftone, the mathematical calculation of the epogdoic interval (i.e. the tone) indicates that it is impossible to divide it into halves and that it rather corresponds to the Platonic *leimma*. See Petrucci (2020), 118–121.

14 This provides the theoretical basis for many of the problems included in *Problemata physica* XI, but it seems that Aristotle had already established a close connection between the quantitative and physical aspects of sound. See, e.g., *DA* II 8, 419b4–421a6; *DA* II 12, 424a21–24; *DA* III 3, 426a26–b7; *DS* VII 447a12–b21, 448a20–b2; *GA* V 7 786b7–788b2.

According to Theon's testimony (178.3–188.7),[15] Adrastus included in his *Commentary* his exegesis of Aristotle's planetary model as expounded in *Metaphysics* XII 8, but he also combined it with the *De caelo* (cf. esp. *Exp.* 178.3–17). Indeed, while most of the astronomical principles in Adrastus' model can be readily traced back to *Metaphysics* XII 8—for instance, the sequence of planets, references to Eudoxus and Callippus, the existence of solid spheres, the presence of "counteracting spheres" (regardless of the meaning of "counteracting"), the need for a "prime" ultimate cause, etc.—the reference to the *De caelo* is crucial to properly address the physical nature of the constituents of the astronomical model, as well as their relations. The text of the *De caelo* engages precisely with these issues, by enquiring into the nature and capacities of the elements—most notably, aether—while also providing a robust kinetic theory of the heavenly bodies. Thus, by incorporating *Metaphysics* XII 8 and the *De caelo*, Adrastus is able to account for planetary motions, without doing away with the spheres' order and regularity, as we will see later on.

For present purposes, it will suffice to highlight some aspects of Adrastus' reading that are relevant for the argumentative context.[16] In brief, Adrastus adapts Aristotle's astronomical model to a system of spheres as follows:

- Planetary motions depend on a system of aethereal spheres, as the planets are carried by them.
- The universe is enclosed by the external sphere of the fixed stars, which has the Earth as its centre and moves westwards.
- The motion of each planet is determined by:
  - A hollow aethereal sphere, which is concentric to the sphere of the fixed stars: this sphere acts as the deferent of another sphere, but is also identified with the one which, in Aristotle's model, determines the latitudinal motion of the planet.
  - This last sphere is solid and aethereal, and the planet is fixed on its surface. This sphere is located in the depth of the hollow sphere, in which it rotates around its centre. In sum, this sphere acts as an epicycle, possessing its own rotation while also being carried around by the deferent.

Of course, a rather desperate exegetical move is required to make this model appear consistent with that of Aristotle, since his model is strictly homocentric

15 Although the passage has no parallel in Calcidius, Theon's explicit attribution to Adrastus of the contents of these pages is confirmed by their focus on Aristotle, whose text is extensively paraphrased. A more detailed demonstration can be found in Petrucci (2012a), 22–23.

16 A discussion of the technical details of Adrastus' demonstration, along with bibliographical references concerning the technical background of the model, can be found in Petrucci (2015), while a different reading has been proposed by Sorabji (2007).

and was not meant to explain anomalistic motion, a phenomenon which was ignored at that stage of astronomical enquiry, while Adrastus' model substantially reproduces the epicycle model first developed in the Hellenistic age.[17] To achieve this goal, Adrastus places special emphasis on Aristotle's reference to the counteracting spheres, which he construes as if these spheres coincided with epicycles.

Accordingly, from the point of view of the overall structure of the heavenly spheres, the final result is a model which, at least to some extent, entails that some of the celestial motions, namely those of the epicycles, are not concentric to the universe. This is necessary in order to account for the planets' anomalistic motion. At the same time, with regard to the direction of the motions of the planetary spheres, namely the deferent (i.e. the hollow sphere) and the epicycle (i.e. the solid sphere), Adrastus explicitly states:

> T2. Theon, *Exp.* 182.4–9
> Moreover, let all the spheres move regularly (ὁμαλῶς) in the same direction with simple (ἁπλᾶς) motions from east to west, and only the one that determines the movement in latitude of the planet move in the opposite direction, or even in the same direction as long as it is left behind due to its slowness: in both ways, the phenomena will be preserved.

There is no ambiguity here about the fact that the epicycle moves westwards, that is, in the same direction as the external sphere of the fixed stars. If, however, the epicycle must be "counteracting", then it is reasonable that it is the case with respect to the sphere that is closer to it (i.e. the deferent). Indeed, some kind of eastward motion is required to account, even if in a very imprecise way, for the planetary motion. At the same time, it is remarkable that the passage suggests the possibility of explaining away the eastward motion of the deferent as a merely *apparent* motion: in that case, the actual motion of the deferent, too, would be westwards, but its slowness would make it appear to be moving eastwards (i.e. "counteracting"). However, the passage does not clarify whether Adrastus would be inclined to accept this explanation or would rather opt for the view of an eastward motion of the deferent—a view which would be much more usual from a technical perspective and much more consistent to the text he is commenting on, namely the *Timaeus* (where the planetary

17 At the same time, it can be noted that ascribing the epicycle model to authors preceding its discovery, and especially to Plato, was a typical move in the Platonic tradition. See esp. Petrucci (2016), 168–178.

motions, relying on the motion of the circle of the different, entail that the planetary orbits move eastwards).

Nonetheless, another section of Theon's *Expositio*, which contains an explicit mention of Adrastus, indicates that Adrastus endorsed the more peculiar reading ascribing a very slow westwards motion to the deferent:[18]

> T3. Theon, *Exp.* 147.8–19
>
> Regarding the planets, Adrastus states that some are always left behind, like the Sun and the Moon: in fact, these never move towards the preceding signs, but are always seen moving towards the following signs, and precisely for this reason, they do not produce either stations or retrogradations. Others, however, both advance and are left behind just like the others: for this reason, it is inevitable that they sometimes appear in station and retrogradation. According to Adrastus, 'being left behind' is the *apparent* retrogression of a planet towards the following signs and towards the east (ἔστι γὰρ ὑπόλειψις μὲν φαντασία πλάνητος ὡς εἰς τὰ ἑπόμενα τῶν ζῳδίων καὶ πρὸς ἀνατολὰς ἀπιόντος, ὥς φησιν ὁ Ἄδραστος), while according to Plato, it is not an appearance but a real movement towards the following signs of the planet, which, according to its own movement, retrogrades towards the east, for example, from Cancer to Leo (ὡς δὲ ὁ Πλάτων φησίν, οὐ φαντασία, ἀλλὰ τῷ ὄντι μετάβασις πλάνητος εἰς τὰ ἑπόμενα ζῴδια ἐπ' ἀνατολὰς ἀπιόντος κατὰ τὴν ἰδίαν κίνησιν, οἷον ἀπὸ Καρκίνου εἰς Λέοντα).

According to Adrastus, therefore, the eastward motion of the planets is merely apparent and is due to the peculiar slowness of their sideral periods, while their real motion is westward. Furthermore, although some ambiguity with regard to Theon's use of his source may persist,[19] one can hardly deny that Adrastus must have had quite a serious reason to commit himself to this account, given that, as already mentioned, it explicitly contradicts Plato's text (and, more generally, a shared and very basic view of astronomical motions).

18 The interpretation of this aspect provided here diverges from that of Petrucci (2015), but this has no impact on the conclusions reached there.

19 The opposition between Adrastus' and Plato's views may already be present in Adrastus' text (i.e. Adrastus would have contrasted his "Aristotelian" view with that of Plato) or it may result from Theon's reconstruction. Be that as it may, for the sake of the present argument it is enough to state that Adrastus' physical model had the deferent moving westwards, albeit very slowly.

This reason rests on a very specific conception of the notion of "regularity" that Adrastus has, a conception that emerges a few pages later in the same text, where Theon ascribes to him the following claim:

> T4. Theon, *Exp.* 151.20–152.8
> As Adrastus states, 'to move regularly' (ὁμαλῶς) means to traverse equal intervals in equal times, without ever increasing or decreasing one's speed from one moment to the next. 'To move in an orderly manner' (εὐτάκτως) is to never alternate between moments of standing still and retrograding, that is, to move in the same direction always in the same way. However, it seems to us that all the planets partake in an irregular movement (ἀνωμαλίας), and some even in a non-orderly one (ἀταξίας). What then is the cause of such an appearance (φαντασίας)? Primarily the fact that the planets, although they are in different spheres and different circles along which they move, seem to move through the zodiac, as previously mentioned. Moreover, as stated, it is only accidentally that the 7, while performing their own simple motion, describe multiple and different circles.

Adrastus' notions of regularity and order are thus quite specific: for the heavenly motions to be regular, it must be shown that they traverse equal intervals in equal times, while for the heavenly motions to be orderly, they must move "in the same direction always in the same way". Adrastus did not have the expertise to really deal with the first issue, and the demonstrations spread throughout the *Expositio*, which probably go back to Adrastus' *Commentary*, are simply based on geometric representations, without introducing observational data. The latter point, however, has a clear impact on the passages cited above, since it explains why Adrastus seems so eager to claim that the eastward motion of the planets is only apparent. If moving in an orderly way entails moving in the same direction, and if all the heavenly bodies are constituted of the same element (i.e. aether), then if the deferent were to move eastwards, this would introduce a kind of "disorderly" motion among the planetary spheres. Conversely, if, as Adrastus wants, all the spheres determining the motion of a planet move westwards, even though one of them (i.e. the deferent) does it so slowly that it seems to be moving eastwards, the conclusion is that the motion of a planet is determined by a series of motions that are orderly, both per se and collectively, in the specific sense that they all always move in the same direction.[20] If

20 One might wonder whether the appropriate subject of the orderly motion, as described in T4, should be the single sphere taken per se, the set of all spheres in the cosmos or

this demonstration is sound, Adrastus' model accounts for the planets' apparent anomalistic motions by reducing them to a series of orderly motions (i.e. in the same direction). At the same time, two facts remain: first, the trajectory of a planet's motion *can be reduced* to a series of regular and orderly motions, but, in itself, it appears irregular and disorderly; second, some of the spheres introduced to explain the complex planetary motions, namely the epicycles, cannot be homocentric to the universe. In brief, Adrastus' model seems designed in such a way as to sacrifice the homocentric structure of the heavens in favour of the regularity and orderly motions *of the spheres*, hence admitting some kind of irregularity for the planets' motions as such.

So far, so good. What we are confronted with is a technically rough epicyclic model characterised by a curious theory about the direction of the motions of the spheres. But why did Adrastus commit himself to this model? Technical reasons may be part of the story, but they do not explain the model's peculiarities, especially in relation to the direction of the motion of the deferent, which in standard epicyclic models is eastwards. In the rest of this section, we will argue that Adrastus had stronger and properly philosophical reasons for upholding his peculiar views.

The key passage comes again from the *Expositio*:

T5. Theon, *Exp.* 148.13–149.15

φησὶ δ' ὅτι ὁ μὲν πᾶς κόσμος τοιοῦτός τε καὶ ἐκ τοσούτων καὶ τοιούτων συνεστηκὼς οἵων καὶ ὅσων διειλόμεθα, φερόμενός τε φορὰν ἐγκύκλιον καὶ τοῦ σφαιρικοῦ σχήματος οἰκείαν ὑπὸ τοῦ πρώτου· ὅθεν καὶ κατεσκευάσθη τοῦ βελτίστου καὶ ἀρίστου χάριν. πρὸς δὲ τὴν χρόνου διαρίθμησιν καὶ τὴν τῶν περιγείων καὶ ἀπογείων μεταβολὴν ἐγένετο ἡ τῶν πλανωμένων φορὰ ποικίλη τις ἤδη συνεστηκυῖα, ὥστε ἀκολουθεῖν αὐτῇ τὸ ἐνταῦθα· ταῖς γὰρ τούτων τροπαῖς προσιόντων καὶ ἀπιόντων συμμεταβάλλει καὶ τἀνταῦθα παντοίως. τῶν μὲν γὰρ

the planets. The last option seems problematic. Even if the planet's final trajectory is considered a mere by-product (153.6–7), the complex articulation of its motion prevents us from classifying it as regular and orderly *stricto sensu* (i.e. as Adrastus does)—despite his claim that the disorder and irregularity of such motion is apparent (153.4). It remains the case that either (a) each sphere per se moves in an orderly manner or (b) all the spheres do so collectively. In the first case (a), adopting a multi-directional account of movement would pose no issue: as long as each sphere performs a uniform motion, whether eastwards or westwards, without changes in speed or direction, its motion would be regular and orderly. However, this would entail a significant problem, as cosmic spherical motion would not be orderly *as a whole*, but only in a fragmented way. This, along with certain physical considerations about aether and its natural motion, leads us to adopt (b) and regard the entire spherical system as orderly *stricto sensu*.

ἀπλανῶν ἁπλῆ καὶ μία φορὰ κύκλῳ, τεταγμένη τε καὶ ὁμαλή. τῶν δὲ ἄλλων πλανωμένων κυκλικὴ μέν, οὐ μὴν ἁπλῆ δοκεῖ καὶ μία, οὐδὲ ὁμαλὴ καὶ τεταγμένη. τῶν δ' ὑπὸ σελήνην καὶ περὶ ἡμᾶς καὶ μέχρις ἡμῶν πᾶσα μεταβολὴ καὶ κίνησις καί, καθάπερ φησίν, ἔνθα κότος τε φόνος τε καὶ ἄλλων ἔθνεα κηρῶν. καὶ γὰρ γένεσις καὶ φθορὰ περὶ πάντα τἀνταῦθα καὶ αὔξησις καὶ μείωσις ἀλλοίωσίς τε παντοία καὶ ἡ κατὰ τόπον ποικίλη φορά. τούτων δέ, φησίν, αἴτια τὰ πλανώμενα τῶν ἄστρων. ταῦτα δὲ λέγοι τις ἂν οὐχ ὡς τῶν τιμιωτέρων καὶ θείων καὶ ἀιδίων ἀγεννήτων τε καὶ ἀφθάρτων ἕνεκα τῶν ἐλαττόνων καὶ θνητῶν καὶ ἐπικήρων πεφυκότων, ἀλλ' ὡς ἐκείνων μὲν διὰ τὸ κάλλιστον καὶ ἄριστον καὶ μακαριώτατον ἀεὶ οὕτως ἐχόντων, τῶν δ' ἐνταῦθα κατὰ συμβεβηκὸς ἐκείνοις ἑπομένων.

He [i.e. Adrastus] then states that the entire cosmos has these qualities, it is composed of objects with qualities and quantities corresponding to those we have distinguished and moves with circular motion, i.e. the one which is proper of the spherical shape, due to what is prime. And it is precisely for this reason that it [the cosmos] was arranged for the sake of what is perfect and of the best. Moreover, the movement of the planets, though being a manifold composition, had generation for the calculation of time and the alternation of perigees and apogees so that the events here follow those up there: indeed, the things here change completely along with the revolutions of the stars that advance and retrograde. The motion in a circle of the fixed stars is one and simple, orderly, and regular, while that of the other heavenly bodies, the planets, is in a circle but does not seem one and simple, nor regular and orderly. To what lies below the Moon—near us and up to us—belong all change and movement, and, as someone says: 'here are hatred and death and here every other evil' (Emp. B121 DK). In fact, both generation and corruption concern all things here, as well as growth, decrease, any change, and the manifold local movement. But the cause of this, he states, are the wandering heavenly bodies. Someone might say that these characteristics of the lesser, mortal, and naturally perishable things are not caused by what is more worthy of honour, divine, eternal, ungenerated, and incorruptible, but that those realities are always the same because of what is most beautiful, perfect, and most blessed, while the things here follow them by accident.

Adrastus distinguishes three different kinds of motion, each characterised by a specific degree of regularity and order and belonging to distinct physical items: the motions of the stars, the motions of the planets, and the motions of the bodies below the Moon. The first two motions belong to the same realm, that is, the heavenly and aethereal one, but differ substantially from one another: while the motion of the stars is not only circular, but also simple, orderly, and regular, that of the planets is composed and neither regular nor orderly. So understood, a planet's motion is indeed neither regular nor orderly—provided that these notions have the meaning we identified in *Expositio* 151.20–152.8 quoted above. In that case, the composite motion Adrastus refers to cannot be that of individual planetary spheres, which is simple, meaning that it must be the motion of each planet as such, that is, the final trajectory of a planet's motion deriving from the combination of the motions of that planet's spheres. Accordingly, Adrastus implicitly distinguishes between heavenly bodies with a circular, regular and orderly motion, namely the stars and the planetary spheres, and those whose motion is composite, namely the planets as such.[21]

But the initial lines of the passage also indicate that there is a specific cause that determines the fact that all those heavenly bodies endowed with circular motion move in this way, and do so in a regular and orderly manner: it is τὸ πρῶτον. It is hard to understand what kind of object is being referred to here: the most suitable candidates would be the divine prime mover and the first heaven. Although the latter option cannot be completely ruled out, the former option seems to be much more plausible. First, as we have seen, Adrastus read *Metaphysics* XII, and hence he was fully aware of the central role that Aristotle ascribed to the prime mover—or at least he must have been aware of the difference that Aristotle established between the prime mover and the first heaven. Second, the first heaven itself has a circular motion and a spherical body, meaning that, if it were the prime cause, it would be the ultimate cause of its own motion—a position which is quite puzzling for any committed reader of *Metaphysics* XII. For these reasons, in the following pages we will consider "the prime" to be different from the first heaven, although opting for the other option would not undermine our argument.

Be that as it may, what matters is the kind of action that this entity exerts. Theon provides little to no information regarding the causal mechanism at stake here, but he does reveal something equally important: the outcome of this

21 The sentence ὁ μὲν πᾶς κόσμος τοιοῦτός τε καὶ ἐκ τοσούτων καὶ τοιούτων συνεστηκὼς οἵων καὶ ὅσων διειλόμεθα might sound ambiguous and one might suspect that Adrastus is also referring to sublunary bodies. This cannot be the case, however, for two reasons: first, there is no trace of a discussion of sublunary bodies in the previous pages of Theon's work; second (and decisively), the bodies at issue are then said to have circular motion and spherical shape, and this applies only to the heavenly bodies.

causation. At 148.15, he states that, thanks to the first cause, the cosmos moves with a circular motion (φερόμενός τε φορὰν ἐγκύκλιον). The question, then, is how we are to interpret this statement. Evidently, it would be odd, at the very least, to suggest that the prime cause makes the spheres (and the stars) move *in a circle*, since this would imply that they could potentially move in another way. But the very nature of (circular) motion should depend on the body that performs it and, since the bodies at issue here are aethereal, it would be logical to maintain that *if* they move, *then* they must do so in a circular manner. And this helps us make quite good sense of the aforementioned puzzle: the prime cause initiates motion, a motion which happens to be circular, simple, regular, and orderly in spheres due to the aethereal and spherical nature of the bodies involved.[22] In other words, the emphasis should be placed on the φερόμενος alone. Understood in this way, circular motion is caused by the prime cause only indirectly, that is since its proper role is to initiate the aethereal circular motion, preventing the cosmos from remaining at rest.

However, the overall outcome of heavenly motion depends on a broader set of conditions. Indeed, for any given heavenly motion, while the prime cause's action represents a necessary condition, the specific type of motion performed is conditioned by the shape and elementary constitution of the body (i.e. physical conditions), as well as by its position within the cosmic structure and its interaction with other heavenly bodies (i.e. configurational conditions). But what aspects of the resulting motion can really depend on the prime mover, provided that Adrastus is committed to the theory of aether?

First, let us recall that we are concerned here with the circular, simple, unique, orderly, and regular motion of the spheres and the fixed stars located on them (149.1–2)—consequently, there is no need to refer to configurational conditions, which are more relevant for explaining planetary motions. Accordingly, we can attempt to formulate a plausible answer:

22 This also depends on *Exp.* 178.3–10 (i.e. the first lines of Adrastus' exegesis of Aristotle's cosmology mentioned above), where Adrastus suggests that the stars and the planets are, in fact, fixed to the spheres and move because they are carried along by them. In that case, the direct object of the prime cause's causation would be the spheres, while stars and planets would be moved only as a consequence. Consistently, the view scholars have sometimes ascribed to Adrastus (see, e.g., Sorabji 2007, 587, followed by Sharples 2010, 195) based on *Exp.* 201.20–202.2, according to which the planets move "intentionally" thanks to a κίνησις προαιρετική, not only contradicts the account in 178.3–10, which is certainly Adrastean, but also belongs to a section of the *Expositio* which, upon closer inspection, derives from Dercyllides and not from Adrastus. For more detailed arguments, see Petrucci (2015), 166 n. 22.

- Circular motion (φορὰ κύκλῳ) derives from the spherical shape of the heavenly body and its aethereal constitution, features that also render it simple (ἁπλῆ)[23] and unique (μία).[24]
- In principle, the unidirectionality of this motion (T4) could likewise derive from aether, since, according to Adrastus, every sphere (both hollow and solid) naturally moves westward (T3). However, it is the prime cause that enables the spheres to move in accordance with their nature, thereby ensuring that their motion remains *always* unidirectional without ever stopping—traits which are equally essential to account for orderly motion (cf. ἀεὶ ὁμοίως, T4). Thus, the order of this motion (τεταγμένη) arises from both the prime cause and the aethereal constitution of the spheres.
- A major point of controversy concerns regularity (ὁμαλή), that is, the constant speed of the spheres' motion (T4). Given that spheres move regularly if taken per se, but they do not if taken collectively, and given that variations in speeds cannot be attributed to differences in magnitude nor to the propagation of movement (as the homocentric spheres are hollow—see also the following point), it is reasonable to conclude that the regularity of the spheres' motion does not depend on their aethereal constitution, but instead on the prime cause. This reasoning aligns with Adrastus' overall system and may also draw upon Aristotle's argument in *De caelo* II 6—a well-known chapter in post-Hellenistic debates[25]—stating that as long as aether moves according to its nature (i.e. in a circle), any acceleration or deceleration can only result from alterations in the mover. In other words, Adrastus may have relied on the unchangeability of the prime cause to account for the regularity of motion in heavenly spheres.
- Finally, it is reasonable that Adrastus had to explain the different speeds of planetary spheres. Extant witnesses do not clearly testify to Adrastus' treatment of this issue, for the astronomical sections of Theon's and Calcidius' writings that can be traced back to Adrastus' *Commentary* just refer to the *apparent* irregularities related to the planets' speeds. However, it is implausible that Adrastus did not notice at least that the first heaven moves much more quickly even if its circumference is larger than the deferents of the planets. It is no coincidence that Adrastus' spherical model also seems to imply that the deferent moves more slowly than the first heaven (*Exp.* 182.7–

23 Simplicity is directly linked to circularity since, according to Aristotle, the only simple motions are rectilinear and circular ones (cf. *DC* I 2, 268b17–20).

24 Cf. the principle stating that simple bodies perform simple, unique motions, while composite ones perform composite motions (*DC* III 4, 303b5–8).

25 See p. 326, n. 44.

8: ὑπολειπέσθω δὲ διὰ βραδυτῆτα). This could hardly depend on the aethereal composition of the spheres, since, in principle, spheres composed of the same body should move at the same speed, as the passage from the *De caelo* mentioned a few lines above also indicates. Hence, this must reflect an additional and crucial contribution of the first cause to the overall arrangement of the heavenly motions.

If this reconstruction is accurate, we can conclude that while physical conditions partially explain the kind of motion exhibited by the spheres, the prime cause must ultimately account for several fundamental aspects: the very presence of motion in the heavens, its appropriateness, its order, its regularity, and the different speeds of the spheres' motions—and it does so without undergoing any alteration.

In light of all this, a rich picture of τὸ πρῶτον emerges. On the one hand, it is a proper cause (ὑπὸ τοῦ πρώτου), at least in the sense that it produces an effect on what is subject to its action (in other words, αἰτία cannot refer here to an explanatory reason). More specifically, its causation results in the sempiternal *motion* (φερόμενος) of the heavenly objects, not only with respect to its order (τεταγμένη) and regularity (ὁμαλή), but also with respect to the complex articulation of the spheres' speeds. On the other hand, the circularity, direction, simplicity, and composite character of planetary trajectories do not depend on the prime, but rather on the elementary constitution and position of the objects upon which the cause acts. Within this framework, it is also remarkable that the prime's causation manifests in varying ways depending on different conditioning factors and that the highest level of perfection is displayed by aethereal spheres, since their circular motion is thorough. All this provides a firm foundation for the final and perhaps most important claim, namely that the causation at issue is not merely mechanical, but is devised for the sake of what is perfect and of the best (τοῦ βελτίστου καὶ ἀρίστου χάριν).[26] Even though it would be arbitrary to push this view to the point where it would ascribe to the prime the status of a craftsman-like divine cause, one can at least state that the kind of teleological orientation enacted by the cause through the production of the heavenly motions is not axiologically neutral, but aims at making the heavens *absolutely good*.

Within this framework, let us outline in what sense this divine organisation orients the heavens for the sake of the good. The heavens are composed

26 The reference to what is "perfect and best" may be a paraphrase of an important line in *Metaph.* XII 10, where Aristotle asks how the nature of the whole universe possesses the goodness and the best: Ἐπισκεπτέον δὲ καὶ ποτέρως ἔχει ἡ τοῦ ὅλου φύσις τὸ ἀγαθὸν καὶ τὸ ἄριστον (1075a11–12).

of spherical objects (i.e. aethereal spheres, stars, and planets), each of which has a certain shape and position. If the action of the prime cause is responsible for their motions, one can thus understand why the cause's action is beneficial: not only does it allow the heavenly bodies to move (for they could also rest), and to do so according to the motion properly belonging to their shape and elementary composition (i.e. circular motion), but it also guarantees their order and regularity, since they all move in the same direction and with regular speeds that allow the planets to "correctly", that is "regularly", reach their perigees and apogees.[27]

This account has two important implications. The first concerns the action of the cause, which is freed from any responsibility for the fact that some spheres are not homocentric to the universe[28] and, more generally, on the composite nature of the planetary motions: Adrastus seems to regard these as constitutive features of the universe's structure, while all positive aspects characterising the heavenly motions depend on the cause's action. In brief, it is legitimate to state that the cause's action is *providential*, albeit in a narrow and soft sense: although the providential cause exerts neither craftsmanship nor will, its action is oriented towards the good and ultimately beneficial, since it accounts for the orderly life of the cosmos (i.e. its motion). For this reason, it ensures the goodness of the heavens. The second (and related) implication is that this account leaves room for some irregularities, namely those emerging if one considers the trajectories of the planets as such: even though the motions of the spheres are individually and collectively perfect, the composition of the planetary motions is unavoidable and this makes their motion necessarily imperfect to that of the stars. However, this latter point does not jeopardise the value of the prime's action if we recall that, as mentioned above, the kind of motion that ensues depends on a variety of factors, specifically physical and configurational ones.

That being the case, we can return to the initial distinction between two kinds of heavenly motion, that of the stars and that of the planets, stressed by Adrastus in T5. Not only are we now in a position to clarify in what sense, and due to what causes, they have to be distinguished, but we can also identify them

27 One might speculate that this model also allowed Adrastus to credibly address and, at least in his own mind, solve the issue that led Aristotle to hypothesise the presence of several movers—that is, the need for a specific motion to be impressed on each sphere: by stressing the cause's beneficial action and its tendency towards what is best, it is possible to regard it as being able to determine the most suitable motions—and, more precisely, the most suitable speeds—of the spheres, case by case.

28 These are the "configurational features" already mentioned.

with two levels of realisation of the good causal action of the prime. On the one hand, at the level of individual aethereal spheres the conditioning in the outcome is minimal (since only physical factors are relevant in this context), allowing the prime's action to manifest in its highest form—that is simple, circular motion. On the other hand, where "disturbing" factors multiply, the prime's action, although equally imparted, manifests at a lower degree of perfection. This explains why planetary trajectories appear disorderly and irregular, as they result from a combination and stratification of different causes (setting aside the prime, consider the interaction between the hollow spheres and the epicycle-deferent mechanism). Overall, it is reasonable to assert that while the prime's action plays an essential role in causing celestial motions in the whole heavenly realm, it is not directly responsible for individual outcomes: its action is declined variably, for variable are the heavenly entities themselves. Thus, while the prime cause guarantees order and regularity in the cosmos, it should not be held accountable for the particular nature of the resulting motions. In other words, the cause's providence, although exerted on the whole heavenly realm, is less effective at the level of the planets for reasons which are independent of the prime.

Against this background, the last part of our passage, concerning the sublunary world, also becomes much more interesting. Adrastus[29] describes it negatively, as a dimension characterised by continuous change: it is clear that the motions of the sublunary region represent a third and much more imperfect dimension, namely one in which everything undergoes generation and corruption in relation to its specific spatial motion. It is not clear what objects Adrastus is referring to here, but the most straightforward way to read the passage is that it mainly concerns the elements: thanks in part to the strategic quotation of Empedocles,[30] it is easy to trace all the changes and motions of the

29 There is some ambiguity as to the identity of the supporter of this thesis: since the subject of the sentence is omitted, it might be the case that Theon is just reporting *ad litteram* Adrastus' text, in which case one cannot rule out that the implicit subject of φησίν is rather Plato. At the same time, it would be hard to construe the *Timaeus* as describing the world in such negative terms, and this makes it much more reasonable that φησίν is introduced again by Theon to emphasise that he is reporting Adrastus' theory—probably in order to stress his disagreement, since Theon is in general committed to the idea of a strong notion of cosmic harmony (Petrucci 2012b, 61–62). Even if this were not the case, given the consistency of the view with what precedes, we would anyway be faced with Adrastus' understanding of Plato's view on the sublunary world, which would be telling with regard to the Peripatetic theory at stake.

30 The choice of the quoted author could hardly be casual, as it may also suggest that Adrastus wants to challenge an "Empedocles-like" model of divine organisation that applies to sublunary bodies and not to the heavenly ones. Moreover, Empedocles' model

sublunary realm back to those of the elements, not least since they are the basic constituents of sublunary bodies (including in the *Timaeus*) and their manifold rectilinear motions are by nature opposed to the circular motions of the heavenly aethereal bodies. Moreover, in this case, too, Adrastus identifies the cause determining the sublunary dynamics, namely planetary motions.

The last view has quite a good Aristotelian pedigree,[31] and for this reason Adrastus does not feel the need to argue in its favour. However, T5 shows that Adrastus is aware that his account of heavenly motions might to some extent be at odds with it and that he faces a difficult objection, which is only partially made explicit in the passage. What he explicitly points out is that since planetary motions are divine and eternal (albeit less perfect than the stars), it is hard to explain how they could be the cause of motion for what is perishable and not divine.

The objection is dismissed with the claim that sublunary dynamics are caused by planetary motions only by accident (κατὰ συμβεβηκός). Now, by stating that sublunary dynamics are caused only accidentally, Adrastus is perhaps not suggesting that sublunary motions could also cease to occur at any time or stop depending on planetary motions. Rather, he seems to rely on a refined interpretation of the Aristotelian theory of "accident" (as defined in *Metaph.* V 30, 1025a14–30), according to which it means "lacking a *proper* cause"—that is, a cause specifically oriented towards that outcome. Such a reading is supported by Adrastus' consistent usage of the same phrase (κατὰ συμβεβηκός) in other contexts, where it has a further implication that is rather related to necessity. For instance, some lines after our passage it is said that the apparently complex motions of the planets supervene *by accident* following the combination of a set of regular motions (*Exp.* 150.12–18);[32] or, again, Adrastus repeatedly

---

implied a cosmological feature, namely the theory of cycles, which was later rehabilitated by the Stoics and which Adrastus would certainly reject. We are grateful to David Lefebvre for encouraging us to further consider this point.

31 See esp. *Meteor.* I, 338a–340b (esp. 340b4–10); I, 346b; II, 361a; *GC* II, 336a–b (with Rashed 2005, 172–174); *Metaph.* XII 5, 1071a13; cf. also *DC* II 3, 286b3–4.

32 Cf. also 152.6–8, 177.12, 179.2, 181.5, 185.16, 186.11. Although Calcidius simply paraphrases the quoted passages, we can be quite confident that the phrase κατὰ συμβεβηκός was used by Adrastus: first, it is more reasonable that it is introduced, albeit in a peculiar sense, by a Peripatetic; second, Theon repeatedly ascribes the views in question (and those spelled out in the passages listed in the following footnote) to Adrastus; finally, many of the quoted passages belong to the exegesis of *Metaphysics* XII we have discussed above, which is unquestionably by Adrastus. The fact that Adrastus typically uses the phrase in astronomical contexts and that Theon appropriates it might also explain why it occurs in the very last pages of the *Expositio*, too, which Adrastus is not the source for. But even on the hypothesis that the phrase rather belongs to Theon's technical vocabulary, our argument would

states that, at least in the case of the Sun, the model of the epicycle and that of the eccentric circle follow from one another *by accident*.[33] It is clear that, in all these cases, the relation established by the use of "by accident" points to some kind of necessity: the apparent motion of the planets necessarily follows from the combination of the motions of the planetary spheres, just as an eccentric circle is necessarily drawn when one develops the geometrical explanation of how the Sun moves according to the epicycle model. At the same time, in all these cases what determines something to be such and such by accident acts without intentionality or teleological orientation. Hence, the Adrastean usage of κατὰ συμβεβηκός aligns perfectly with the Aristotelian definition of accident as "lacking a proper cause" which was recalled above. This interpretation becomes even more convincing when we recognise that lacking a proper cause does not entail ipso facto the absence of any cause whatsoever, nor does it inform us about the modality of existence of the (accidental) effect as such (e.g. its recurrence, necessity, impossibility, etc.). Instead, a certain effect can be jointly accidental and necessary (as "always recurring"), in the Adrastean sense. For instance, the fact that the combination of the motions of the planetary spheres produces complex trajectories by accident has nothing to do with the nature, function, and orientation of the spheres' motions: it is just a necessary by-product of them. The same consideration applies to the case of the "derivation" of the eccentric model from the epicycle one: an eccentric is produced when one draws the Sun's trajectory according to the epicycle model, just because this is what results from the geometrical demonstration, but producing an eccentric circle is not the goal of the application of the epicycle model. In brief, Adrastus' use of *by accident* refers to a narrow application of the Aristotelian formula, which is shown by the following relation: X follows Y by accident when Y produces an effect on X necessarily, but independently from any intentionality or teleological orientation.

Accordingly, when Adrastus says that sublunary motions follow the planetary ones *by accident* he means that the planetary motions produce the sublunary ones necessarily, but independently of any intentionality or teleological orientation—that is, just by moving as they do. If so, the notion of causation by accident first solves the aforementioned puzzle: perishable and imperfect motions may be caused by imperishable and more perfect ones, because this happens without any orientation on the part of the planets (and hence any imperfection necessarily depends on the nature of the elements undergoing

not be affected: Theon would just be rephrasing, in his own words, the relation between planetary motions and sublunary ones as expressed by Adrastus.

33 Cf., e.g., *Exp.* 166.12, 168.13 and 22, 169.2, 170.4 and 19, 171.13, 183.14, 185.4, 188.15 and 22, 189.14.

sublunary motions). But there is more to this, since the introduction of this notion also has a remarkable payoff: the influence of the planetary motion is necessary, and this ensures that those dynamics that they determine in the sublunary world will not fail at any point. The latter aspect is particularly important, given that some of the effects of the planetary motions are "positive", especially as to the determination of natural rhythms (it is not by chance that, in his demonstrations of the Sun's motion, which heavily depend on Adrastus (see table, at p. 290), Theon repeatedly refers to the length of the seasons).

In light of all this, the sublunary realm represents the third and lowest level, not only from the point of view of permanence and the order of motions, but also from the point of view of its orientation towards the good (i.e. providence): while the motions of both the stars and the planets are arranged by the prime cause for the sake of what is best, each at its level, sublunary motions are determined by the planetary ones. This occurs necessarily and not for the sake of the good. Thus, there is no providence below the Moon, even though the providential order depending on the cause of the heavenly motions filters through this lowest part of the universe, as in a necessary chain of causations.

5 Adrastus' Physics and Its Philosophical Outlook in the Post-Hellenistic Peripatos

Let us now summarise the model that emerged in section 4. According to Adrastus, there is a sharp distinction between the heavenly region and the sublunary realm. This depends not only—and, indeed, not primarily—on the aethereal constitution of heavenly bodies, but on the fact that the motions of the heavenly bodies (including not only the stars and the planets, but also and above all the spheres determining their motions) are dictated by the prime cause, which organises them for the sake of what is best by ensuring their continuous appropriate motion, order and regularity (this claim, as we have seen, is supported by Adrastus' specific version of the epicycle model). This goal is achieved completely in the case of the stars, while in the case of the planets one must distinguish the movements of the individual spheres determining their motion and the overall trajectory each planet follows: while the former are simple, regular, and orderly, the latter are not. Accordingly, the prime cause's beneficial orientation successfully orients the heavens, although its goodness manifests itself to a lesser degree in planetary motions. Sublunary motions and objects, namely the four elements, are characterised by a much lower degree of order since they are non-circular and perishable. However, the divine heavenly order reaches them to some extent, since planetary motions determine sublunary ones in a necessary way and without a teleological orientation. Adrastus' model therefore envisages a multi-layered physical world, whose articulation coincides with that of a threefold providence: providence (at least in the sense of the realisation of a good arrangement and orientation determined by a divine cause) permeates the whole world to different degrees and in different ways, but the prime cause exerts a direct teleological arrangement only in the heavenly region, where its action manifests to two different degrees, while its beneficial ordering extends down into the sublunary world only through the necessary and non-teleologically oriented influence of planetary motions.

We have thus reached the first substantial conclusion, which is an outline of Adrastus' theory of providence. With this framework in place, we are now in a position to properly assess its theoretical import, as well as its historical relevance. A first remark concerns the very notion of "providence" in the Peripatos: despite the widespread criticism of Aristotelian theology in antiquity,[34] the voices that outright rejected *any* theory of providence whatsoever represented a minority in the Peripatos.[35] Rather, the Peripatos often sought to mediate between the absolute expunction of the divine and an excessive dependence on its ultimately arbitrary will.[36] However, we would like to point out that Adrastus' theory is remarkable within this nuanced history for three reasons: (i) the link it establishes between theology and physics, (ii) the teleological nature of the prime cause's action and (iii) the peculiar relation of this theory to the *Timaeus*.

34 A criticism famously championed by the Platonist Atticus, cf. Fr. 3.7–10 des Places = 22N Sharples.

35 In fact, even those thinkers who publicly rejected the theory of the unmoved mover felt the need to account somehow for the goodness of the cosmos and to search for a first principle of some kind (e.g. the first heaven or, in Strato's case, nature).

36 In the history of this mediation, two positions can be cited: the first asserts that providence stops at the Moon, while the second maintains that some providence reaches the sublunary world, but only indirectly. Evidence for the first view is found in various doxographical sources (Hippol. *Haeres.* 7.19.2–8 = 22I Sharples; Epiph. *De fid.* 9.35 and 39 = 22O Sharples), while the latter is supported by a report in Aëtius (2.3.4 = 22H Sharples), an excerpt from Aspasius (*In EN* 71.16–72.4 > Sharples 22J; for a discussion of this text, see chapter 12 by David Lefebvre in this volume) and the *De mundo*, whose philosophical affiliation is disputed. Furthermore, some philosophers, such as Critolaus, present contrasting evidence, making it difficult to classify him within one group or the other (Epiph. *De fid.* P. 39 = Fr. 15 W; Plut. *Praecepta* 15 = Fr. 35a W; Plut. *Pericles* 7.7 = Fr. 35b W). See Mayhew (Forthcoming) for a detailed analysis. On these Peripatetic views, see esp. Sharples 1983, 2003 and 2010, 196–210 (also including a collection of the relevant sources). On Alexander's theory, see also Donini (1996) and Orna Harari's chapter 14 in this volume.

Now, in very general terms, the association between a prime cause and celestial motion is not an outstanding original doctrine in Antiquity. Nonetheless, Adrastus' case is of particular significance insofar as he associates the prime cause's action with the *appropriate* motion of heavenly bodies, many features of which also depend on their physical features (i.e. shape and elementary constitution).[37] In other words, Adrastus provides a more precise theory of how the prime cause interacts with these physical features. This is particularly remarkable since, in this way, Adrastus restores credibility to the theory of aether, which had been famously rejected by Xenarchus of Seleucia in the first century BCE. Adrastus is thus forced to challenge this authoritative and relatively recent view, and he does so by refuting all of Xenarchus' major tenets regarding natural motion, thereby reaffirming the elementary theory of Aristotle's *De caelo*. To briefly illustrate this point, let us recall the key tenets of Xenarchus' theory of natural motion. In general, Xenarchus grounded his rejection of aether on three main arguments:[38]

(1) It is not true that for every simple body there is just one simple natural motion (e.g. downward rectilinear motion for earth, circular motion for aether), as water and air move both upward and downward relative to different elements.[39]

(2) Circular motion is not uniform, since the centre and the periphery of a sphere move at different speeds.[40]

(3) If a simple body moves by its natural motion until it reaches its natural place, then aethereal bodies should remain at rest, since the heavens constitute their natural place.[41]

Adrastus' theory of natural celestial motion can be mapped on these arguments. First and foremost, (1′) he re-establishes the one-to-one correspondence between simple bodies and simple motions, claiming that circular motion is

37 Moreover, the indirect doxographic reports on Peripatetic providence before Alexander mentioned before do not really explain the physical background of the views at issue.

38 Cf. Sorabji (2007), 591–594, and Falcon (2012), 25–32.

39 See Simplicius *In DC* 23.31–24.7 (= T5 Falcon). See also Falcon (2012), 29, 86–89.

40 See Simplicius, *In DC* 14.20–21 (= T2 Falcon), 24.20–27 (= T6 Falcon), 42.8–10. In addition to associating simplicity to uniformity, Xenarchus also argues that, in simple bodies, each part displays the same nature as the whole (cf. Falcon 2012, 29). For a discussion on the excerpts, see Falcon (2012), 63–71, 89–91, 96–99.

41 On Xenarchus' account of natural motion and place, see Simplicius *In DC* 21.33–22.17 and 20.10–36 (= T3 Falcon), 42.10–14. For a discussion on the excerpts, see Falcon (2012), 71–82, 96–99. In Xenarchus' positive doctrine, he resolves this difficulty by claiming that once the four elements reach their natural place, they either rest or move in a circle—there is thus no need for aether.

intrinsic to spherical aethereal bodies (148.16), while going even further by stating that all aethereal bodies move westwards, as their motion is orderly (T3, T4).[42] Then, (2′) he redefines the notion of regularity (T4) to account for the uniformity of celestial motions. Finally, (3′) while he may have acknowledged that the third critique had some validity, he rejected its conclusion, namely the abolishment of aether. Instead, he advanced a causal explanation for the motion of aethereal bodies by introducing the prime cause (ὑπὸ τοῦ πρώτου, 148.16), thereby also restoring the Aristotelian link between physics and theology (i).[43]

Beyond these dialectical considerations, there is also a historical reason why Adrastus is significant. While he may not have been the first to oppose Xenarchus—although this remains highly probable—the available evidence suggests that he appears to be the only one who challenged Xenarchus' theoretical framework in its entirety. For instance, a discussion arguing for the uniformity of heavenly motion (and thus opposing Xenarchus' argument 2) is attested already for Alexander of Aegae,[44] while Aspasius hinted at a connection between providence and celestial motion (thus opposing Xenarchus' argument 3 and perhaps aligning himself with Adrastus' reply 3′).[45] However, as far as we know, neither of them sought to provide a comprehensive account

42 In fact, Xenarchus' criticism could be framed in an even more compelling way: if upward and downward motion differ—and they are, in fact, assigned to different elements—why should westward and eastward motion not be distinguished in the same manner? Aristotle himself acknowledged a difference between the two (cf. *DC* II 5).

43 Now, it is important to note that we lack explicit evidence indicating that Xenarchus denied the existence of a prime cause altogether (although it seems probable and has been suggested, cf. Falcon 2012, 36). However, whether this was the case for Xenarchus is of only marginal relevance for our analysis, given Adrastus' own account of the prime. Even if Xenarchus did acknowledge a prime cause, it could not have played a kinetic role, since, according to Xenarchus, once fire reaches its natural place in the heavens (and becomes "properly fire"), it moves naturally in a circular fashion. See Simplicius *In DC* 42.10–14, 50.21–24, and a discussion of these texts in Falcon (2012), 96–106. The aim of this model was probably the resolution of the problem of the ὑπέκκαυμα as presented in *Meteor.* 339a15–19, 341b13–19. This account, evidently, leaves no room for a providential mover of any kind. By contrast, in Adrastus' view, a prime cause with no kinetic function would be a contradiction in terms. Thus, regardless of Xenarchus' position, Adrastus likely felt the need to reintroduce a genuine prime cause into his cosmological model.

44 See Simplicius *In DC* 430.29–32 and 430.32–431.11. We know that Alexander of Aegae, Aspasius and Herminus participated in the discussion of Aristotle's *DC* II 6 and that all of them defended the uniformity of celestial motion. Furthermore, their exegetical approach involved misinterpreting the Aristotelian passage, which only bears on the first heaven, in order to extend its argumentation to cover the entire heavenly realm.

45 See Aspasius *In EN* 71.25–31. Moreover, the case of Aspasius not only seems to diverge from that of Adrastus with regard to the contents of the theory, but it is also less detailed,

of celestial dynamics. Instead, they addressed specific criticisms in isolation—unlike Adrastus, whose approach was far more systematic with regard to cosmological and astronomical matters. Whether this makes him the only available witness to this kind of systematic approach, or whether he was in fact an isolated pioneer of it, is impossible to know and largely irrelevant for the assessment of the importance of his work.

Moreover, Adrastus' reintroduction of a prime cause is even more remarkable inasmuch as it is grounded in teleological considerations, which also characterise the theological model he supports (ii): the prime cause guarantees the sempiternal, regular and orderly motion of the heavens *for the sake of what is perfect and best* (148.17). While this claim is certainly significant in its own right, it assumes even greater importance given that subsequent Peripatetic debates on providence (in the second century AD) were indeed quite lively. Undertaking a thorough analysis of these developments and their connection to Adrastus' account lies beyond the scope of this chapter. However, it is worth mentioning the key themes of these debates. As *De intellectu* 113.6–15 attests, Alexander sought to distinguish Aristotle's theory of providence from that of Plato and the Stoics. For this reason, he criticises an anonymous Peripatetic predecessor—probably one of his teachers—for transforming the Aristotelian agent intellect into a "demiurge" that organises the sublunary realm in relation to heavenly motion and that exercises a form of "primary providence" in the world, similar to the Stoics.[46] Alexander thus testifies to the existence of an intense Peripatetic debate on the issue of divine providence, and, if our analysis is sound, Adrastus' criticism of Xenarchus' highly anti-teleological account and the alternative views he proposes must at least have established the framework for this debate.

We said that three reasons especially may be identified to account for Adrastus' importance. Alongside physical and teleological considerations (i–ii), there is a third reason why Adrastus' case is indeed exceptional and deserves particular attention, which becomes clear once Adrastus' theory of providence, the strategy he adopted to support it, and its location in the Peripatetic philosophical debate, are combined with what was established in the previous sections of this chapter. There, we tried to emphasise that, even though the assessment

depending on the nature of the texts in question. On Aspasius' theory, see chapter 12 by David Lefebvre in this volume.

46 Although several important studies have been written on the *De intellectu*, an in-depth analysis of these issues is currently lacking (highly synthetic readings of 113.6–15 can be found in Sharples 2004, 43 n. 112, and Sharples 2008a, 157). They are among the objects of Ludovica M. Fazio's ongoing research.

of extant sources is difficult and much less obvious than usually believed, it is possible to discover the method (or perhaps one of the methods) Adrastus exploited in his Peripatetic commentary on the *Timaeus*, which consists in supplementing Plato's account by introducing Peripatetic philosophical nuclei (iii). Now, the case that was briefly analysed in section 3, concerning the acoustic bases of the discovery of the numerical values of consonances, as intriguing as it may be, is not that interesting from a philosophical point of view. But now its broader importance has become clearer, since Adrastus seems to use the same method when a topic of greater significance is at issue, namely providence. Let us briefly consider an outline of the theory of providence that emerges—at least *prima facie*—from the *Timaeus*.[47] According to the *Timaeus*, Plato is committed to the following general tenets: god is good and causes the world to be good, and he is thus providential; this providence is mirrored in the order of the world, starting from its physical, and mainly astronomical, dimension. These fundamental tenets are then justified as follows. At least at a literal level, the perfect organisation of the world testifies to the action of a divine intellect, the demiurge: in the crucial and programmatic passage (30a–c)[48] in which Timaeus sketches out the demiurge's action in terms of the production of order from disorder, a key notion is introduced, namely providence (πρόνοια, 30c1). This also applies, more specifically, to Plato's astronomical account, since the reason why the demiurge produces the planets and provides them with specific motions is that they are meant to dictate time (38b4–e2), and time is produced by the demiurge because he is good and wants the world to be as perfect as possible—that is, to imitate the paradigm in the best possible way (37c6–d2). But the demiurge does much more than that: he also organises the sublunary realm and its dynamics, starting from those of the elements, which he directly produces according to geometric shapes, which in turn determine their kinetic and, in general, physical properties (53c4–56c7).

Let us now read Adrastus' theory within this framework. On the one hand, Plato's *Timaeus* certainly represented a reference point for anyone who wanted to contrast Xenarchus, not only because it was the text he himself commented on,[49] but also (and perhaps more importantly) for the theory of providence

47 We are not necessarily committing ourselves to this reading, which may be regarded as a sort of standard literalist one.

48 However, this reasoning applies to several passages in the first part of the tale and its proem, e.g. *Tim.* 28a6–b1 has been often regarded as a paradigmatic argument from design. See Hankinson (1998), 108–111; Sedley (2007), 105–107; Johansen (2014).

49 Even though we do not know the extent to which Xenarchus engaged with the *Timaeus*, it is certain that he commentated on at least some parts of it, as Proclus *In Tim.* 425.22–

contained in it. On the other hand, how Plato justifies his fundamental tenets, and more specifically the kind of causation he ascribes to god, is too much for Adrastus' prime cause (or that of any Peripatetic). Thus, while acknowledging the importance of claiming a theory of providence for the Peripatos, too, Adrastus' account radically diverges from Plato's on at least two major points: first, the nature of the divine cause's causation, which for Adrastus is limited to motion; second, the direct objects of the cause's providence, which for Adrastus are limited to the heavens. The reasons for these divergences do not emerge from the extant testimonies, so we can only speculate. It is reasonable to assume that Adrastus had some good exegetical reasons: for instance, his "moderate" account of the prime cause's providence fits quite well with Aristotle's image of the unmoved mover, as Adrastus' prime is responsible for the overall regularity and order of heavenly motion, but not for individual outcomes, just as the order of the army depends on its leader, while individual actions performed by the soldiers do not.[50] But there are also possible theoretical reasons to be considered. For instance, Adrastus might have thought that if the divine cause directly determines *all aspects* of the heavenly region, as well as being responsible for the organisation of the elementary dynamics of the sublunary region, then it fails miserably, since aspects of disorder remain in the heavenly region and are dominant in the sublunary realm (this must have been Adrastus' opinion, as T5 shows). However that may be, another point is important here: divergences notwithstanding, Adrastus seems to agree with Plato's fundamental tenets, namely that the prime cause has to be providential and hence effectively organises the world, starting from the heavens, for the sake of what is best. Interestingly, both divergences and points of contact are stressed in Adrastus' account of providence as emerged from T5: first, as we have seen, the fact that the divine cause is solely a cause of motion is related to the fact that the position and physical nature of the planets and their spheres do not depend on god, and this, in turn, ensures that planetary irregularities do not depend on god; second, the sublunary world is *not* characterised by order and regularity, as it should be if a divine cause organised it. Still, the overall system outlined by Adrastus is able (or at least he hopes it is able) to ensure that god beneficially organises the heavens in a specific way, and this, in turn, conditions the sublunary realm. And, of course, in support of his account, Adrastus proposes a complex physical model that he identified with that of Aristotle. In light of all this, a fuller account of Adrastus' attitude towards Plato emerges.

426.4 attests. Furthermore, to the best of our knowledge, the only two documented cases of *Timaeus* exegesis within the Peripatos are those of Xenarchus and Adrastus.

50 Cf. Aristotle, *Metaph.* XII 10, 1075a11–15.

Indeed, Adrastus must have been aware that his account contradicted Plato's in several regards. At the same time, his convergence with Plato's fundamental tenets cannot be a coincidence. If so, Adrastus' strategy cannot be reduced either to criticism of Plato or to a blind attempt to say that Plato and Aristotle were in agreement. Rather, his supplementing of the *Timaeus* actually amounts to an attempt to "bring aid" to Plato: Adrastus' theory of the prime cause is, after all, able to explain how it exerts providence and organises the world for the sake of what is best without putting its effectiveness at risk, while ensuring, to some extent, the regularity of the sublunary region. In other words, Adrastus admits that Plato's fundamental tenets are correct, but precisely because of this he cannot just agree with him: rather, Adrastus proposes (what he considers) an improvement on Plato's model, an improvement that saves Plato's major tenets, while avoiding the risks the Timaean model vainly takes.

At this point, however, we can go a step further, since we also know that Adrastus' agenda included quite strong views about how to reinvent Peripatetic physics by overcoming Xenarchus. All in all, therefore, acknowledging that Adrastus regards his model as the best Aristotelian one,[51] his final goal proves ambitious and radical. It is to show that, thanks to his brand new "Aristotelian" theory of providence, based on a specific physical model and ruling out concurrent Peripatetic theories, it is also possible to improve and save Plato's theory: it is not merely that Plato and Aristotle were in agreement, but that Plato should ultimately agree with Aristotle in order to see his agenda effectively realised.[52]

51 This is indeed quite reasonable, let alone for the fact that the overall physical model Adrastus proposes is meant to reproduce that of *Metaphysics* XII. For Adrastus' conception of Aristotle's authority, see Petrucci (2016).

52 If this reading is correct, one could even try to (speculatively) place Adrastus' strategy in the contemporary debate on Plato's account of providence. After all, the "literal" outline of the *Timaeus* we have provided above is also the core of the reading of the dialogue of an extremely influential Platonist active in the same decades as Adrastus, i.e. Plutarch: one can imagine that Adrastus' reading of the *Timaeus* was *also* aimed at drawing a contrast with Plutarch's interpretation by arguing that the only way in which Plato's theory of providence might be saved was not that proposed by Plutarch, but one rooted in the awareness that only a Peripatetic turn could have saved Plato from himself. This relies on chronological considerations: Adrastus' activity is usually dated to the first half of the second century AD (see, e.g., Sharples 2010, 289). However, considering that Theon's *floruit* was around 125 AD (Petrucci 2012b, 9–11), his activity must be placed sometime between the last decades of the first century AD and the first decades of the second. Plutarch's death is usually dated to around 120 AD.

CHAPTER 14

Alexander of Aphrodisias: Teleology without Design

Orna Harari

1 Introduction

Our understanding of Alexander of Aphrodisias' account of natural production is largely shaped by a fragment from his commentary on *Physics* II 3 found in Simplicius' commentary on this chapter—the text (henceforth, "the fragment") is quoted in full in the appendix—and a passage from his own commentary on Aristotle's *Metaphysics* I 9 (102.3–104.18 Hayduck) where he rejects the view that Platonic Forms or preconceived notions (νοήματα) serve as paradigms that guide natural production.[1] Interpreting this account from the viewpoint of his anti-Platonic polemic, studies of these passages highlight Alexander's contention that nature is a non-rational power (ἄλογος δύναμις), the analogy of its mode of production to marionettes (νευροσπαστούμενα) and his understanding of embryogenesis as an ordered transmission (διαδοχή) of changes, but they downplay his aim of offering an alternative teleological account of natural production.[2] This aim is most clearly expressed in section E of the fragment, where Alexander stresses that his account does not imply that natural processes occur by chance, arguing that the expression "for the sake of something" (ἕνεκά του) does not exclusively hold for processes caused by reason and choice, but for any well-ordered process that comes to be for a certain

1 The translation of this fragment and all other translations are mine. A brief formulation of this argument is also found in Alexander's commentary on *Metaphysics* V 2, 349.6–16. In the fragment from his commentary on Physics II 3 he counters the view that concepts are paradigms, whereas in his commentary on *Metaphysics* I 9 he opposes the view that Forms are paradigms. This difference has no significance for understanding Alexander's argument, since, in the latter discussion, he clarifies that it holds whether paradigms are Forms or concepts (103.1–2). The view that Alexander's argument in the commentary on *Physics* II.3 counters is propounded by Atticus, who understands paradigms as god's concepts, in analogy to a craftsman who first grasps what he plans to make and then makes, (Fr. 9 Des Places = Eusebius, *PE* XV 13, 1–6, 814d–816b).

2 See, e.g., Genequand (1984), 119; Accattino (1988), 82–87; Sharples (1994), 166–168; Kupreeva (2003), 328–330; Henry (2005), 7–11; Rashed (2007), 278–282.

purpose (ἄλλου χάριν), whether by choice and reason or without reason. The significance of this argument for understanding Alexander's account of natural production becomes apparent in a less-discussed fragment found in Simplicius' commentary on *Physics* II 8, where Alexander gives the following explanation of why early Greek philosophers and the Epicureans wrongly deny teleology:[3]

> T1: Those among the ancient natural philosophers and the Epicureans among later [philosophers] who say that material necessity is the cause of generation seem to be of this opinion. But as Alexander says, "This mistake arises from their assumption that all things that come to be for the sake of something come to be by choice and reasoning, and from seeing that things that come to be by nature [do not come to be] this way." But as we have already said above, he [scil. Alexander] says that this is not the case, when he firmly stated that nature makes for the sake of something but not by reason.[4] (Simplicius, *In Phys.* 372.9–15 Diels)

This passage calls into question the mechanical interpretation of Alexander's characterisation of nature as a non-rational power. It shows that, through this characterisation, Alexander rejects not only the Platonists' conception of teleological causation, but also the Epicureans' non-teleological understanding of natural processes, thereby indicating that Aristotle's discussion of non-rational capacities (δυνάμεις) in *Metaphysics* IX 5 is not the suitable context for understanding Alexander's account.[5] In this discussion, Aristotle distinguishes the rational capacity of human beings to produce opposite effects, e.g. the capacity of physicians to cause both health and disease, from inanimate things'

3 For a brief mention of this fragment, see Rashed (2007), 295 n. 798.

4 ταύτης δοκοῦσι τῆς δόξης τῶν μὲν ἀρχαίων φυσικῶν ὅσοι τὴν ὑλικὴν ἀνάγκην αἰτίαν εἶναι τῶν γινομένων φασί, τῶν δὲ ὑστέρων οἱ Ἐπικούρειοι. "ἡ δὲ πλάνη γέγονεν αὐτοῖς, ὥς φησιν Ἀλέξανδρος, ἀπὸ τοῦ ἡγεῖσθαι πάντα τὰ ἕνεκά του γινόμενα κατὰ προαίρεσιν γίνεσθαι καὶ λογισμόν, τὰ δὲ φύσει μὴ οὕτως ὁρᾶν γινόμενα. τὰ δὲ οὐχ οὕτως ἔχει, φησίν, ὡς ἤδη προειρήκαμεν, πάντως ὅτε ἔλεγεν, ὅτι ἡ φύσις ἕνεκά του μὲν ποιεῖ, οὐ κατὰ λόγους δέ." As the verb ἔλεγεν at line 15 indicates, the last sentence is not a part of the quotation from Alexander. The back reference is not to Aristotle, as Barie Fleet translates, but to section D in the fragment (311.10).

5 For this interpretation, see Henry (2005), 16–17. By offering an alternative to the Epicurean stance, Alexander addresses a criticism raised by his near contemporaries against the Peripatetics. For instance, Atticus says that he unhesitatingly puts these schools in the same basket, because they both deny that divine providence extends to human beings (Fr. 3 De Places). As I show later, Galen raises a similar charge regarding the Peripatetic attempts to explain teleology through mechanical analogies.

non-rational capacity to produce only one effect, e.g. fire's capacity to heat (*Metaphysics* IX 2, 1046b4–7). This non-rational capacity cannot serve Alexander in distinguishing his account of natural production from the Epicureans' view. It explains how inanimate things bring about effects by material necessity, while Alexander's understanding of nature as a non-rational power underlies his teleological account of natural processes.

In view of these considerations, in what follows I examine Alexander's account of natural production from the viewpoint of his disagreement with the Epicureans and address the question of how he secures the teleological character of natural processes. I focus on the three elements of his account that gave rise to the mechanical interpretation—the analogy to marionettes, the characterisation of nature as a non-rational power and the notion of a transmission of changes—and show that they underpin a strong conception of teleology that guarantees that natural processes are directive and come to be for the sake of their end. Regarding the analogy to marionettes, I show that rather than implying that Alexander models his account of natural processes on mechanical devices, it invokes his understanding of animal motion, according to which a power in the sinews that is subservient to the governing part of the practical faculty of the soul explains the directive character of this motion. So understood, this analogy reveals the dependence of nature on the divine body, thereby laying the groundwork for the discussion of the characterisation of nature as a non-rational power in the second section. Here, I discuss the divine origin of this power and clarify its contribution to teleological processes through Alexander's understanding of hypothetical necessity. I show that, in his view, nature as a non-rational power is an additional factor that overrides the indeterminacy of material and proximate causes, and thus secures the production of an end in processes whose course is hypothetically necessitated. Finally, I examine Alexander's view that the end is produced through an ordered transmission of changes, showing that he secures a strong conception of teleology by holding (1) that the efficient cause is causally dependent on the final causes and therefore produces for the sake of an end, and (2) that each stage in a natural process is the efficient cause of the subsequent stage. Moreover, this analysis of the fragment shows that Alexander not only offers an alternative to the Platonist and Epicurean conceptions of natural production, but also tackles difficulties that earlier Peripatetic thinkers, who appealed to a mechanical model in order to understand teleological processes, left unanswered.

2 Teleology and the Analogy to Marionettes

The analogy to marionettes appears twice in the fragment, but in each occurrence it illustrates a different point. Whereas in section C Alexander uses it to distinguish production by reason and choice from processes that bring about their effects through a successive transmission of changes, in section F it explains how nature inclines towards an end without choice. The latter use of this analogy is an expression of Alexander's view that nature produces for the sake of something, but it is nonetheless puzzling: how does a mechanical device whose parts successively move each other convey the directive character of the process?[6] Rather than calling for an answer, this question is an indication of the limitations of the mechanical interpretation of Alexander's account of natural production. It also points us to another context for understanding this analogy: the only other occurrence of the portmanteau νεῦρον and σπάω in the Alexandrian corpus found in the following passage from the *Mantissa*:

> T2: Of the impulsive and practical [faculty of the soul], one part is governing and the other is subservient. The governing part is properly called impulsive and desiderative, while the other [is called] *neurospastikon*, where the speech capacity is also present.[7] (Alexander, *Mantissa* 105.31–33 Bruns)

The fragment has been discussed in connection with this passage for textual reasons. It lends further support to Michael of Ephesus' version of a passage from Alexander's *De anima* where he draws the same distinction between these parts of the practical faculty by suggesting that the phrase "ὡς ὑπηρετικόν, ὃ καὶ

6 In raising this question, I do not assume that mechanical and teleological explanations are mutually exclusive. They are not. Sylvia Berryman has shown that analogies to mechanical devices served post-Aristotelian thinkers in resolving the tension between explanation based on material causes alone and teleological explanation in terms of intention and design. According to her analysis, they do so by suspending the question of the origin of suitable material structures that can underlie goal-directedness and by focusing on the material operation within the device (Berryman 2007, 366). Here, however, Alexander uses the analogy to illustrate the inner tendency of natural processes towards the end. The inner structure of mechanical devices cannot explain this feature. In what follows, I show that Alexander's analogy is not mechanical and that he focuses on the question that other thinkers, according to Berryman, suspend.

7 τούτου δὲ τοῦ ὁρμητικοῦ τε καὶ πρακτικοῦ τὸ μέν ἐστιν ὡς ἡγεμονικόν, τὸ δὲ ὡς ὑπηρετικόν, ἡγεμονικὸν μὲν τὸ ἰδίως ὁρμητικόν τε καὶ ὀρεκτικὸν καλούμενον, τὸ δὲ ἕτερον νευροσπαστικόν, ἐν ᾧ καὶ τὸ φωνητικόν.

νευροσπαστικὸν καλοῦμεν", which appears in Michael's text, should be inserted into the passage from *De anima*.[8] However, the connection between the fragment and the passage quoted above is more than terminological. This passage renders the use of the term νευροσπαστούμενα in section F of the fragment less puzzling, by showing that the image of being pulled by strings illustrates a directive activity that follows upon the motivational factors, impulse and desire.[9] It also facilitates a fuller understanding of Alexander's account of natural production. By highlighting the dependence of the mechanical motion of marionettes on a factor that acts by desire, it draws our attention to the fact that, in the fragment, Alexander elaborates upon only one aspect of his account, namely nature and its nonrational power, but does not specify what the analogies to the craftsman in section C and to the sower in section D represent.[10]

The importance of the latter analogies for understanding Alexander's account becomes clear from Simplicius' remark on Alexander's interpretation of Aristotle's maxim in *Physics* II 2, 194b13: "a man is begotten by a man and by the sun":

> T3: "And generally", Alexander says, "what comes to be by nature itself seems to be generated by something that is identical to it either in species or in genus, or in general by activity. Things that are generated by the sun are generated by an activity in this way; for they are heated and cooled by its activity" [...] And Alexander infers also from the things said here that, according to Aristotle, the generation of things in the sublunary world is linked to the divine body and is not detached from it.[11] (Simplicius, *In Phys.* 307.3–12 Diels)

By showing that, in Alexander's view, nature is linked to the divine body, this passage fills in the missing information from the fragment and also strengthens the analogy between Alexander's accounts of the practical faculty of the soul

8 On this passage, see Todd (1982), 48–49; Donini (1968), 318; Zingano (2008), 4–5; Fernandez and Mittelmann (2020), 278 n. 63.

9 This term is often used in the sense of being pulled by impulse in Marcus Aurelius' *Meditations*, e.g. 2.2.1, 3.16.1, 6.28.1, 6.16.1, 12.19.1.

10 See also Rashed (2007), 279–280 n. 753.

11 "καθόλου δέ, φησὶν ὁ Ἀλέξανδρος, τὸ γινόμενον φύσει δοκεῖ αὐτὸ ὑπὸ τοῦ αὐτοῦ γίνεσθαι ἢ εἴδει ἢ γένει ἢ ὅλως ὑπὸ ἐνεργείας. ὑπὸ ἡλίου γοῦν οὕτως τὰ γινόμενα γίνεται ὡς ὑπὸ ἐνεργείας· θερμαινόμενα γὰρ καὶ ψυχόμενα ὑπὸ τῆς ἐνεργείας τῆς ἐκείνου. καὶ δῆλον μὲν ὅτι καὶ τὰ κατὰ τέχνην γινόμενα, ὑπὸ ἐνεργείας γίνεται ἀλλ᾽ οὐ φυσικῆς" [...] σημειοῦται δὲ ὁ Ἀλέξανδρος καὶ ἐκ τῶν ἐνταῦθα λεγομένων, ὅτι κατὰ Ἀριστοτέλην συνῆπται τῷ θείῳ σώματι ἡ τῶν ἐνταῦθα γένεσις καὶ οὐκ ἀπήρτηται ἀπ᾽ αὐτοῦ.

and his account of natural production. It indicates that the craftsman and the sower represent the divine body, which, like the governing part of the practical faculty, is animate and moves by impulse and desire.[12] In *Quaestio* I 1, Alexander establishes the view that the first cause is unmoved on the basis of this understanding of the divine body. He infers that there is an unmoved eternal and simple substance that is the cause of the divine body's eternal and continuous motion from the assumptions that this body must be animate and that it is moved by the unmoved mover "by thinking it and by longing (τῷ ἔχειν ἔφεσιν) and having a desire (ὄρεξις) for becoming similar to it" (3.8–4.3 Bruns).[13] This understanding of the analogy to a craftsman and sower gains further support from the light it sheds on Alexander's choice of a biological process, embryogenesis, as an example of physical production in section D of the fragment. Having a share in the eternity of god is the explanation that Aristotle offers for reproduction in *De anima* II 4, 415a25–b7, in *Generation of Animals* II 1, 731b23–732a1, and elsewhere.[14] In his own *De anima*, Alexander formulates this explanation in the same terms that he uses in *Quaestio* I 1. He says that the faculty of reproduction is the cause of the generation of something similar through longing (ἐφέσει) for immortality to the extent that mortal beings can have a share in it (36.6–9 Bruns). A few pages earlier, he presents reproduction as a special case of this explanatory principle, saying that all things that are constituted by nature long (ἐφίεται) to become similar to god and do everything for the sake of this (32.14–15 Bruns). Similarly, in his *On the Principles of the Cosmos*, which is extant in Arabic translation, he says that natural things have a desire for the cause of their natural motion (section 5) and that all natural things desire the first cause (section 21).[15] Alexander's account of natural production rests on this explanation. In section A of the fragment, he states that Aristotle calls "paradigm" the form of the generated thing "because nature makes all the things that it makes by longing (ἐφιεμένη) for it" (310.27–28 Diels). He offers the same explanation in a similar passage in his commentary on *Metaphysics* V 2, where he says: "Everything that makes for the sake of something makes what it makes by a longing (ἐφέσει) for this thing" (349.12–13 Hayduck).[16]

12 As we see from *Quaestio* I 1, the impulse and desire of the divine body differs from the impulse and desire of the ruling part of the practical faculty of the soul with respect to their outcomes. The former's desire results in the divine body's own motion, whereas the latter affects the subservient part, i.e. the power in the sinews. I elaborate on this difference in section III below.

13 See also *Quaestio* II 18 and II 19.

14 GC II 10, 336b25–337a7, and *Pol.* I 1, 1252a26.

15 The references are to Genequand (2001).

16 For the extension of this principle to nature in general, see also Genequand (2001), 55.

Further examination of Alexander's account of the practical faculty of the soul shows that his understanding of the subservient part of this faculty corresponds to his understanding of nature as a non-rational power. In his *De anima*, he describes the practical faculty as follows:

> T4: Just as there is in the faculty of judgment a governing part and a subservient part, so in the practical faculty there is also a governing part, which we call impulsive and desiderative, and a certain part that resides in the sinews ⟨as subservient, which we call also *neurospastikon*⟩.[17] For in these too there is a power by which the body subserves the activities in respect of impulse. For since things done according to desire are done when a certain affection occurs, some of these affections slacken the body and produce an expansion in them [scil. the sinews] through a well-tempered heat, while others cool and contract [them] (for representations, perceptions as well as conceptions of objects that come to be by a certain similarity to them either slacken the body or contract and cause it to shiver, just as it does with these objects when they are present). When these contractions and expansions occur around the congenital pneuma and are transmitted from it to the sinews, they can perform the activities in respect of impulse, since they receive the principle of motion from desire.[18] (Alexander, *DA* 76.14–77.8 Bruns)

Regardless of whether the term "neurospastikon" in the phrase marked off by angle brackets is Alexander's, his description of the subservient part of the practical faculty corresponds to his understanding of the operation of nature. As in sections C and D of the fragment, Alexander explains animal motion here as the result of an ordered transmission of changes: heat and coldness bring about expansions and contractions in the congenital pneuma that are trans-

17 In Michael's text the phrase "as subservient, which we call also *neurospastikon*" is added here (*In anim. mot.* 115.5–6).

18 ὡς δὲ ἐν τῷ κριτικῷ τὸ μέν ἐστιν ἡγεμονικόν, τὸ δὲ ὑπηρετικόν, οὕτως ἐστὶ καὶ ἐν τῷ πρακτικῷ τὸ μὲν ἡγεμονικόν, ὃ ὁρμητικόν τε καὶ ὀρεκτικὸν καλοῦμεν, τὸ δέ τί ἐστιν ἐν τοῖς νεύροις. δύναμις γάρ ἐστι καὶ ἐν τούτοις, καθ' ἣν ὑπηρετεῖται τὸ σῶμα ταῖς καθ' ὁρμὴν ἐνεργείαις. ἐπεὶ γὰρ τὰ κατ' ὄρεξιν πραττόμενα πάθους τινὸς ἐγγενομένου πράττεται, ὧν παθῶν τὰ μὲν ἀνίησίν τε τὸ σῶμα καὶ αὔξησιν αὐτοῖς ἐμποιεῖ διὰ θερμότητα εὔκρατον, τὰ δὲ καταψύχει τε καὶ συστέλλει (αἱ γὰρ φαντασίαι καὶ αἰσθήσεις καὶ ἔννοιαι τῶν πραγμάτων κατὰ ὁμοίωσίν τινα αὐτῶν γινόμεναι ὡς ἐπὶ παροῦσιν ἐκείνοις ἢ ἀνιᾶσιν τὸ σῶμα ἢ συστέλλεσθαί τε καὶ φρίττειν ποιοῦσιν), ὧν γινομένων συστολῶν τε καὶ ἐκτάσεων περὶ τὸ σύμφυτον πνεῦμα καὶ ἀπὸ τούτου διαδιδομένων ἐπὶ τὰ νεῦρα, κινεῖσθαι ταῦτα δύναται τὰς καθ' ὁρμὴν ἐνεργείας τὴν ἀρχὴν τῆς κινήσεως παρὰ τῆς ὀρέξεως λαμβάνοντα.

mitted to the sinews, causing them to expand and contract. This explanation is based on Aristotle's *Movement of Animals* 7. There, too, animal motion is understood as being triggered by qualitative changes, specifically heat and cold—caused by perceptions, representations and thoughts—that slacken and contract the sinews (701b2–24). However, Alexander goes beyond Aristotle. He introduces a power in the sinews that is subservient to the governing part of the practical faculty, thereby explaining not just the motion of animals but its directive character—i.e. its conformity with the impulse and desire that arise in the governing part.[19] Much like nature, this power is non-rational and directive. It is a bodily power whose presence in the sinews explains how they move precisely in a way that is subservient to impulse through a succession of natural changes alone. Nature, as Alexander understands it in the fragment, plays the same explanatory role. It is non-rational (i.e. it brings about a succession of purely physical changes) and it is directive (i.e. it produces for the sake of something). Consequently, the similarity in the use of the portmanteau νεῦρον and σπάω in these two contexts, the fuller understanding that it facilitates, and its explanatory value cumulatively support my interpretation that Alexander models his account of natural production not on mechanical devices, but on his physiological understanding animal motion. The following discussion of Galen's engagement with earlier Peripatetic attempts to explain teleological processes reveals the theoretical advantage of this model over the mechanical model, thereby reinforcing this interpretation.

In *The Formation of the Embryo* 6, Galen explains why—considering he has shown that the formation of the embryo manifests the extreme wisdom and power of its maker—he hopes that the philosophers will show whether this maker is a certain god and whether its power is a soul:

> T5: For they will not admit that the substance of what is called "nature"—whether it is something incorporeal or corporeal—reaches the summit of wisdom, and they say that it cannot be believed at all that it operates in an artful manner in the case of the formation of the embryo. For hearing this from Epicurus and from those who hold that everything comes to be without providence, we are not convinced. Indeed, it is necessary that the formation of the embryo reaches its useful end either by a certain nonrational and non-artful motion or just as those who construct mechanical instruments who depart after they give them the principle

19 That Aristotle's model of animal motion does not require the assumption of power even in order to account for the movement of the limbs can be seen from Gregoric and Kuhar's (2014) explanation of why the muscles play no role in the model.

> of motion, while the constructions are moved in an artful manner not for a long time, so the gods too, having constructed the seeds of plants and animals, do nothing more than [give them] the propensity for so many transmissions of motion.[20] (Galen, *De foetuum formatione* 92.10–21 Nickel = IV 688 K)

In this passage, Galen questions the view that nature is the cause of embryogenesis in light of philosophers' understanding of its mode of production. He says that having rejected the assumption that nature forms the embryo in a wise and artful manner, they are bound to explain how this process comes to be in one of the two ways. In one, attributed to Epicurus, this process is caused by a nonrational and non-artful motion. In the other, which is most likely Peripatetic, it is caused by gods who, like operators of mechanical devices that move in an artful manner, merely initiate the process and give the instruments the propensity to continue it.[21] He dismisses the former view without argument and raises two queries about the latter view. First, he asks whether the moistness of the seed is strong enough to accomplish such an extended sequence of transmissions of motion, without failing before it reaches the end. He answers that this view is as implausible as the view that this process occurs by chance (92.22–94.2 Nickel = 689 K).[22] Next, he questions the plausibility of the view that the seed is constructed like mechanical devices (θαύματα) and acts without knowing what it does, on the grounds that the muscles perform motion that accords with wish (βούλησις) (96.12–13 Nickle = 692K).[23]

---

20 οὐ γὰρ δὴ τῆς ὀνομαζομένης φύσεως τὴν οὐσίαν, εἴτ᾽ ἀσώματός τίς ἐστιν, εἴτε σωματική, σοφίας εἰς ἄκρον ἥκειν ἐροῦσιν, ἥν γε μηδ᾽ ὅλως ἔχειν φασὶν, οὕτω τεχνικῶς ἐνεργῆσαι περὶ τὴν τῶν ἐμβρύων διάπλασιν οἷόν τε πεισθῆναι. τοῦτο γὰρ Ἐπικούρου τε καὶ τῶν ἄνευ προνοίας ἅπαντα γίγνεσθαι νομιζόντων ἀκούοντες οὐ πειθόμεθα. καὶ μὴν ἀναγκαῖόν ἐστιν, ἢ κατά τινα κίνησιν ἄλογόν τε καὶ ἄτεχνον εἰς χρηστὸν τέλος ἀφικνεῖσθαι τὴν διάπλασιν τῶν ἐμβρύων, ἤ, καθάπερ οἱ τὰ θαύματα κατασκευάζοντες, ἀρχὴν κινήσεως αὐτοῖς παρασχόντες, ἀπέρχονται μὲν αὐτοί, τὰ δὲ κατασκευασθέντα μέχρι τινὸς οὐ πολλοῦ χρόνου κινεῖται τεχνικῶς, οὕτω καὶ τὰ σπέρματα τῶν φυτῶν καὶ τῶν ζώων κατασκευάσαντας τοὺς θεοὺς εἰς τοσαύτην διαδοχὴν κινήσεων ἐπιτήδεια μηδὲν ἔτι πράττειν αὐτούς.

21 That this view is Peripatetic can be inferred from the fact that Galen attributes to Aristotle the view that the semen provides only the principle of motion and the analogy to mechanical devices (*On Semen* I 5.24, 82.32–84.4, De Lacy).

22 For a more detailed discussion of Galen's approach to mechanical understandings of embryogenesis, see Hankinson (2017), 259–263.

23 In *On the Utility of the Parts*, Galen endorses the idea that mechanical devices aptly describe physiological processes (*UP* XIV5: II 295.7–15 Hermelisch = IV 156–157 K). This objection may explain why he discards this view in *The Formation of the Embryo*.

These queries turn on two physiological phenomena that Aristotle explains by means of mechanical analogies: embryogenesis and animal motion. In the former case, this analogy explains how the parts of the embryo are formed by an external cause (*Generations of Animals* II 1, 734b10, II 6, 741b9). In the latter case, it explains how a small change around the heart causes great and varied changes in the whole body (*Movement of Animals* 7, 701b2–16). In these discussions, Aristotle uses the term αὐτόματα, which aptly illustrates how organisms move by themselves after an external factor initiates their motion. The thinkers that Galen criticises in the above passage use this analogy to explain teleology, thus exposing their position to the difficulties that he raises and that Aristotle's account does not aim to address. Alexander uses the marionette analogy precisely in these contexts, adapting Aristotle's mechanical explanations to the purpose for which these thinkers use it.[24] Unlike automatic devices, this analogy conveys the image of being pulled by strings, and hence can depict the directive character of teleological processes. Marionettes do not move by themselves, but need an operator who moves their parts according to his intention. Alexander's notion of power plays this role. In his account of animal motion, it explains how the body moves in accordance with the impulse that arises in the governing part of the practical faculty, thereby addressing Galen's concern over how bodily organs can perform directive activity without knowledge. In his account of natural production, it explains how nature as a non-rational power brings about teleological processes without reason and choice, thereby addressing Galen's contention that it is implausible that a complex process, such as the formation of the embryo, is accomplished through a mechanical transmission of motion. The following examination of Alexander's characterisation of nature as a non-rational power supports this interpretation. It shows that by means of the notion of power, he secures the teleological character of natural processes by adapting Aristotle's understanding of the relation between the divine body and the sublunary world to this purpose, just as he does in his account of animal motion.

3 Teleology and Nature as a Non-rational Power

In T4, Alexander explains the directive character of the expansions and contractions of the sinews through two factors. One is the power in them by means

24 In Pseudo-Aristotle's *On the Universe* 6, 398b17, operators of marionettes illustrate divine nature, but here too this model does not explain teleology but clarifies how god imparts his power to every region of the world.

of which they move the body in conformity with the impulse that arises in the governing part of the practical faculty; the other is that desire is the principle from which their motion originates. Power and the principle of motion also feature as explanatory factors in Alexander's description of the contribution of the divine body to the sublunary world in the following passage from his commentary on *Meteorology* I 2, but here they account for goal-directedness in a different way:[25]

> T6: Their efficient cause and that which serves as the principle from which their [*scil.* the elements'] motion comes is the eternal motion of the divine bodies and the power that comes to be in those bodies [*scil.* the elements] from these bodies [*scil.* the divine bodies] because of their proximity, which one can reasonably call nature; for this power is the cause of all the things that are believed to be generated by nature and according to nature.[26] (Alexander, *In Meteor.* 7.9–14 Hayduck)

This explanation of the effect of the divine body on the sublunary world differs from Alexander's explanation of the relation between the governing and the subservient parts of the practical faculty in two related respects: (1) while in both cases the principle of motion comes from an animate entity that acts by desire, in the case of the divine body it is its motion, rather than its desire, that provides the principle of motion; (2) while in the case of the subservient part of the soul power is an additional factor distinct from the sinews, power deriving from the gods is not added to nature, but identified with nature itself. The former point is an expression of the view that we encountered in *Quaestio* I 1, according to which the divine body accomplishes its desire to imitate the eternity of the unmoved mover through its own motion, and not by bringing about regular processes in the sublunary world. The latter point features in Alexander's other extensive discussion of the Platonists' view that nature produces in relation to a paradigm in his commentary on *Metaphysics* I 9. Here, as in section B of the fragment, he claims that nature is a non-rational power that causes an orderly process in which one change follows the other until this sequence of changes reaches its end (103.38–104.1 Hayduck), and he adds an aspect that in the fragment he left unspecified, saying that this power comes from the gods

---

25 See also *De mixtione* XI, 23.27–24.16 (Groisard) = 225.27–226.7 (Bruns).

26 ποιητικὸν δὲ αἴτιον αὐτῶν καὶ τὸ ὡς ὅθεν ἡ ἀρχὴ τῆς κινήσεως ἡ τῶν θείων σωμάτων ἀίδιος κίνησις καὶ ἡ δύναμις ἡ ἀπὸ τῆς γειτνιάσεως ἀπ᾽ ἐκείνων ἐπιγινομένη τοῖσδε τοῖς σώμασιν, ἣν καὶ φύσιν ἄν τις εὐλόγως ὀνομάζοι· αὕτη γὰρ ἡ δύναμις αἰτία πάντων τῶν φύσει τε καὶ κατὰ φύσιν γίνεσθαι πεπιστευμένων.

(104.8 Hayduck). Consistent with his denial of the Platonists' view, he stresses that the gods do not purposely bring about these orderly changes, arguing that nature is not a divine art in the sense that it is used by the gods as an art. Rather, he holds that owing to the gift that nature receives from the gods, it produces—without reason and with no reference to a paradigm—something that is similar to the cause that initiated the process, through well-ordered changes (104.3–14 Hayduck). The divine contribution to natural processes is discussed in detail in *Quaestio* II 3, which is devoted to the question of what the power that comes from the motion of the divine body to the sublunary world is. Here, Alexander responds to the objection that if this power gives the four elements their principle of motion and rest, it contributes nothing to them (48.8–12 Bruns). He considers two answers: the first is that this power ensouls natural bodies and accounts for their capacity to move to their natural places not only by inclination (ῥοπή), but also by animate (ψυχική) motion (49.1–4 Bruns). In the other answer, which reflects his own view, this power is nothing other than the nature and form of the elements. In this way, Alexander avoids the objection that the divine body contributes nothing to the sublunary world, by arguing that due to this power, the elemental qualities (hot, cold, moist, and dry) are present in unqualified prime matter and that it is the cause of their reciprocal generation from one another (49.28–50.17 Bruns).[27]

By contrast, the desire arising in the governing part of the practical faculty is accomplished not through its own motion, but through the changes in the subservient part. Here, material causes—heating and cooling and their transmission—explain why the sinews expand and contract, while the power in the sinews explains why they are expanded and contracted in the exact way required to move the animal towards or away from a certain object. Thus, in contrast to Alexander's claim that nature is a power from the gods, the power in the sinews is an additional factor that is irreducible to the material and mechanical effects of the heating and cooling brought about by the objects of desire grasped by the cognitive faculties. These differences, in turn, raise the question of what the contribution of the gods to nature is. In other words, what does it add to processes in the sublunary world that material causes alone cannot explain? Apparently, the forms and qualities of natural bodies can account for their orderly interactions, and natural causes alone can explain why regions

27 See also the fragment from Simplicius' commentary on *Physics* II 1, where Alexander says that nature is present principally and primarily in the simple bodies and not in organic bodies (*In Phys.* 268.18–22 Diels). In presenting this *Quaestio*, I follow Moraux's division in Moraux (1967), 161. For a discussion of this *Quaestio* see Donini (1996), 11–29; Rashed (2007), 288–290.

nearer to the celestial bodies are hot and dry (i.e. fire) or hot and moist (i.e. air), whereas remoter regions are cold and moist (i.e. water) or cold and dry (i.e. earth).

The following passage from *Quaestio* II 19 helps answer this question:

> T7: Rather, the whole cosmos does not need something that exercises providence over it neither for its being nor for its well-being, but the providence that comes to be in the cosmos comes to be since one part exercises providence, and another part is an object of providence. For insofar as its body is ungenerated and uncorrupted, being always ordered in conformity with its nature and soul and performing its motion by longing for imitating the first god, it needs nothing that exercises providence over it, since it has perfection in its proper nature with respect to being as well as to well-being. But insofar as [a part] of it is generated and corrupted, needing the assistance of the other [part] for its being as well as for the eternity with respect of species through well-ordered change, it is an object of providence, being governed (κυβερνώμενον) by the well-ordered motion of the divine part of the cosmos and its certain relations to it.[28] (Alexander, *Quaest.* II 19, 63.15–26 Bruns)

From this passage, we can see that in Alexander's account of natural production, providence plays the explanatory role that subservience plays in his account of animal motion. Like the latter, it accounts for a well-ordered motion that brings about an end, which is the preservation of species. The difference between these modes of explanation finds expression in Alexander's different understandings of the power that accounts for directive motion. In the case of animal motion, it is in the sinews. In the case of natural production, it is external: natural bodies have it owing to their proximity to the divine body. In section D of the fragment, Alexander expresses this idea when he says that, in the first stage of the embryogenetic process, nature and power are sown in the

28 ἡ πᾶς μὲν ὁ κόσμος οὐ δεῖταί τινος προνοήσαντος, οὔτε πρὸς τὸ εἶναι, οὔτε πρὸς τὸ εὖ εἶναι, ἡ δὲ γινομένη πρόνοια ἐν τῷ κόσμῳ γίνεται μέρους μέν τινος ἐν τῷ κόσμῳ προνοοῦντος, μέρους δέ τινος προνοουμένου. ὅσον μὲν γὰρ αὐτοῦ σῶμα ἀγένητόν τε καὶ ἄφθαρτόν ἐστι, καὶ ἀεὶ κατὰ τὴν αὐτοῦ φύσιν τε καὶ ψυχὴν τεταγμένον, καὶ τὴν αὐτοῦ κίνησιν κινούμενον ἐφέσει μιμήσεως τοῦ πρώτου θεοῦ, τοῦτ᾽ οὐδὲν δεῖται τοῦ προνοήσοντος, ἐν τῇ οἰκείᾳ φύσει τήν τε κατὰ τὸ εἶναι καὶ τὴν κατὰ τὸ εὖ εἶναι τελειότητα ἔχον, ὅσον δὲ αὐτοῦ γενητόν τε καὶ φθαρτὸν καὶ τῆς παρ᾽ ἄλλου βοηθείας δεόμενον πρός τε τὸ εἶναι καὶ πρὸς τὴν διὰ τῆς εὐτάκτου μεταβολῆς κατ᾽ εἶδος ἀιδιότητα, τοῦτ᾽ ἐστὶ τὸ προνοούμενον ὑπὸ τῆς τοῦ θείου μέρους τοῦ κόσμου κινήσεως εὐτάκτου καὶ ποιᾶς σχέσεως πρὸς αὐτὸ κυβερνώμενον.

semen, namely coming from outside. This understanding of nature is revealingly different from Aristotle's. According to Aristotle's *Physics* II 1, nature is an intrinsic principle of motion and rest with respect to place, growth and decease, as well as alteration (192b14–15). So understood, it alone can determine the regularity of natural processes and of the interactions between natural objects, while the unmoved mover and the motion of the heavenly body account for the continuity of motion and change in the sublunary world, but not for its order. In Alexander's view, however, the heavenly body accounts for order and regularity. Its eternal motion brings about well-ordered changes in the sublunary world and thus secures eternity in species. In this view, then, nature's divine power is the additional factor that explains the regularity of natural processes and their directive character.

Alexander's understanding of hypothetical necessity clarifies why the state of affairs in the sublunary world leaves room for this additional directive factor. In his commentary on *Physics* II 9, Simplicius raises the following question about Alexander's understanding of unconditional necessity and teleology:

> T8: I, for my part, am at a loss for how Alexander says that unconditional necessity abolishes that for the sake of which. For he says, "Unconditional necessity is found in those cases where the end follows by necessity from the things that precede it." However, in eternal things too the end follows by necessity from the things that precede it (for instance, the setting of the sun follows its rising, and from winter follows spring, and from it follows summer) and nonetheless each of them is for the sake of something.[29] (Simplicius, *In Phys.* 389.4–10 Diels)

It is indeed difficult to understand Alexander's view, but Simplicius' criticism is unwarranted. Here Alexander says that necessary processes lead to an end, and in the above passage from *Quaestio* II 19, he says that the motion of the divine body, which is, according to Aristotle, unconditionally necessary (GC 2 11, 338a4–b5), is performed by longing to imitate god. Thus, by maintaining that necessity excludes teleology, Alexander distinguishes two processes that lead to an end, regarding as teleological strictly speaking only those pro-

29 ἀπορῶ δὲ ἔγωγε, πῶς ὁ Ἀλέξανδρος τὸ ἁπλῶς ἀναγκαῖον ἀναιρετικὸν τοῦ ἕνεκά του λέγει εἶναι. "ἁπλῶς γὰρ τὸ ἀναγκαῖον, φησίν, ἐν οἷς ἐξ ἀνάγκης τοῖς πρὸ αὐτῶν τὸ τέλος ἀκολουθεῖ". καίτοι ἐν τοῖς ἀιδίοις καὶ ἐξ ἀνάγκης ἀκολουθεῖ τοῖς πρὸ αὐτοῦ τὸ τέλος (οἷον τῇ ἀνατολῇ τοῦ ἡλίου ἡ κατάδυσις αὐτοῦ, καὶ τῷ χειμῶνι τὸ ἔαρ καὶ τούτῳ τὸ θέρος) καὶ ὅμως ἕνεκά του τούτων ἕκαστον.

cesses whose course is determined *a fronte* by the end and not processes that bring about an end by necessity.[30] If this interpretation is correct, this passage implies that, in Alexander's view, teleological processes are distinct from necessary processes not just because they lead to an end, but because of the way they bring about that end. Whereas in teleological processes strictly speaking the preceding stages are necessitated by the end, in necessary processes the end is necessitated by the preceding stages. In this view, then, there is a room for teleology strictly speaking where *a tergo* causality does not suffice to bring about an end. Stated in Alexander's terms, the assistance of the gods or providence is needed to do so.

In *Quaestio* III 5, which shows that unconditional necessity is found in circular processes, Alexander clarifies why this assistance is needed in the sublunary world. Here, he addresses the question of whether unconditional necessity holds for any circular process, arguing that it does so to the extent that the circular motion of the divine body influences the process. In processes that are fully determined by this motion, unconditional necessity holds absolutely. For instance, if the winter solstice occurs, then the spring equinox does too; and if the spring equinox occurs then the summer solstice does too; and if the summer solstice occurs, then the autumn equinox does too; and then the winter solstice occurs again. Unconditional necessity holds for the change of the seasons in a less determinate way, because in this case not only the motion of the divine body, but matter, too, contributes to their coming-to-be. In the sublunary world, Alexander argues, the causal effect of the divine body is even less determinate, because the contribution of proximate productive causes is greater. Therefore, in this case, the motion of the divine body is the cause of the eternity of the species, and this process is circular and hence unconditionally necessary on the level of universals (e.g. a man begets a man), whereas for particular processes (e.g. Sophroniscus begets Socrates), hypothetical necessity holds: if the second then the first and not if the first then the second (89.2–24 Bruns). Of course, this does not mean that the motion of the divine body does not influence the sublunary world. As Alexander says in this *Quaestio*, species are not subject to generation (88.15–16 Bruns). Thus, securing the eternity of the species requires the contribution of external causes to override, for the most part, the indeterminacy of the proximate causes. This additional factor is the

30 To the best of my knowledge, the only other discussion of this fragment is found in Sharples (2012). Sharples does not notice that here Alexander does not merely hold that necessity excludes teleology, but also distinguishes two processes that lead to an end.

power deriving from the motion of the divine body that Alexander identifies with nature. It brings about well-ordered processes that are, strictly speaking, teleological: they lead to an end—the eternity of the species—that necessitates the preceding stages of the process and not the other way round.

4 Teleology and the Transmission of Changes

My discussion of Alexander's understanding of hypothetical necessity suggests that he distinguishes necessary processes that lead to an end from teleological processes in the strict sense, but it leaves two related questions open: (1) What is the characteristic of the latter processes? (2) How does he guarantee that natural processes have this characteristic? Within the framework of Alexander's account of natural production, these questions are particularly pressing. While he holds that the sublunary world is an object of providence, it is unclear how his notion of providence facilitates answers to these questions. According to *Quaestio* II 21, providence, though not accidental, is not essential. It is not exercised by an agent who sets the benefit of the object of concern as a goal and does the things through which this goal is attained for the sake of this benefit (66.21–25). This understanding of providence echoes Alexander's account of the divine body's contribution to the sublunary world. As we saw, in his view, the divine body secures the regularity of natural processes and the preservation of the species not because it moves for the sake of this end, but rather as a by-product of its circular motion whose end is imitating the unmoved mover in respect of eternity. Moreover, Alexander's view that unconditional necessity excludes teleology does not answer these questions. Even if hypothetical necessity is a distinctive characteristic of teleological processes, it implies that the preceding stages of natural processes are necessitated by their effect. It does not imply, however, that these stages come to be for the sake of bringing about this effect.[31]

Answering these questions is crucial for Alexander's attempt to distinguish his teleological account of natural processes from the Epicurean stance that denies not that natural things have certain useful and beneficial functions, but that they come to be for the sake of these function.[32] Indeed, from Simpli-

31 For a thorough justification of the view that hypothetical necessity is not a distinctive characteristic of teleological processes, see Charles (1988), 5–20.

32 See Lucretius' *On the Nature of Things*, 4.823–857, where he states that the eyes were not produced for the sake of seeing and that the knees and hips were not produced for the sake of taking long strides.

cius' commentary on the *Physics* we learn that Alexander was concerned about these questions and interpreted the following lemma from *Physics* II 8 in light of them:

> T9: Where there is an end, the prior and the subsequent stages are done for the sake of this end. Therefore, as each thing acts, so it is natural, and as it is natural, so it acts, if nothing prevents. But it acts for the sake of something and therefore it is naturally for the sake of something. For example, if a house had been one of the things that came to be by nature, it would have come to be in the same way as it comes to be now by art; and if things that come to be by nature came to be not only by nature but also by art, they would come to be in the same way as by nature.[33] (Aristotle, *Phys.* II 8, 199a9–15)

The distinction between acting (πράττεται) and being by nature (πέπφυκε) is not straightforwardly clear, but Aristotle's example suggests that the former refers to art and the latter to nature.[34] In this passage, he says that if a house had been generated by nature, it would have come to be in the same way as it comes to be by art, and the other way around: if natural things were also made by art, they would come to be in the same way as by nature (199a12–15). Themistius (*In Phys.* 60.23–29), Philoponus (*In Phys.* 308.21–309.8) and Simplicius (376.24–26) understand the passage this way,[35] whereas Alexander understands it in light of his concern over the teleological character of natural processes. He argues that here Aristotle shows that natural processes are not just for the sake of something but are naturally for the sake of something (375.8–9), and explains that Aristotle adds the phrase "it is naturally for the sake of something" at 199a10–11, "lest one would argue that in things that come to be by nature there is that for the sake of something, and yet nature does not make them for this

---

33 ἔτι ἐν ὅσοις τέλος ἔστι τι, τούτου ἕνεκα πράττεται τὸ πρότερον καὶ τὸ ἐφεξῆς. οὐκοῦν ὡς πράττεται, οὕτω πέφυκε, καὶ ὡς πέφυκεν, οὕτω πράττεται ἕκαστον, ἂν μή τι ἐμποδίζῃ. πράττεται δ' ἕνεκά του· καὶ πέφυκεν ἄρα ἕνεκά του. οἷον εἰ οἰκία τῶν φύσει γιγνομένων ἦν, οὕτως ἂν ἐγίγνετο ὡς νῦν ὑπὸ τῆς τέχνης· εἰ δὲ τὰ φύσει μὴ μόνον φύσει ἀλλὰ καὶ τέχνῃ γίγνοιτο, ὡσαύτως ἂν γίγνοιτο ᾗ πέφυκεν.

34 Charlton (1970), 123, offers two interpretations of this passage. In one, Aristotle draws an analogy between art production and natural production, while, in the other, he argues that the product of both natural and art production is for the sake of something.

35 Simplicius' view is slightly different. Reacting to Alexander's claim that Aristotle draws a certain distinction here, Simplicius argues that with the phrase "it is naturally for the sake of something" at 199a10–11, Aristotle distinguishes nature from art by stressing that the products of art come to be for the sake of something, but not according to nature (376.24–26).

sake" (375.5–11). The motivation behind this interpretation becomes clear later in Simplicius' commentary, when he quotes the continuation of this fragment:

> T10: But Alexander, as said above, says that "naturally" is presented in addition "lest one would argue that in things that come to be by nature there is that for the sake of something, and yet nature does not make them for this sake (as it was said about the teeth), that nature does not make them for this purpose, but the usefulness supervenes". Further, he says, "as Aristotle will mention as he proceeds, in things that come to be by chance there is that for the sake of which and for this purpose, but the preceding stages are not for the sake of the end. Thus, having shown that natural things are for the sake of something, now he adds that it is not for the sake of something as in things that come to be by chance, but they are naturally so". (Simplicius, *In Phys.* 376.11–17)

Here Alexander refers to the difficulty raised at the beginning of *Physics* II 8, that useful bodily parts, such as sharp front teeth and broad molars, can come to be by necessity and not teleologically, as Empedocles holds (198b23–27), and adds another difficulty that arises from Aristotle's own view. Considering that chance events lead to a certain end (e.g. a creditor who happens to collect a debt when he goes to the marketplace for another purpose), how can genuine teleological processes be distinguished from processes that lead to an end by chance? This understanding of chance events implies that not all natural processes that lead to an end are teleological, and hence it requires an account of teleological processes that shows that their end is not an incidental byproduct of material necessity, but is the goal for whose sake their stages come to be. This, then, is the answer to the first question I raised above: in genuine teleological processes, the preceding stages not only bring about an end, but come to be for the sake of bringing about this end. The following examination of the process of transmission of changes found in sections C and D of the fragment answers my second question and clarifies how Alexander guarantees that natural processes are teleological in this way.

The idea of a transmission of changes is prevalent in Alexander's writings. In his commentary on *Sense and Sensibilia* I 6, it accounts for the changes and affections that the parts of the medium successively undergo by sounds and odours until they reach the sense organs (128.20–129.2; 131.13–17). In *Quaestio* II 3, it describes the effects of the heat produced by the divine body on the elements (48.6–8), and in a fragment from his commentary on *Physics* VIII.10 it underlies Alexander's explanation of projectile motion, where the causes of motion and alteration are understood as factors that move and change their effects and also give them the capacity of moving and changing other things in the same way (1346.29–1347.37).[36] In the present fragment, the notion of transmission plays a similar role. As in the commentary on *Sense and Sensibilia*, here it describes a successive process and, as in *Quaestio* II 3 and in the account of projectile motion, a power is transmitted from one stage to the other. This similarity notwithstanding, in the fragment the idea of a transmission of changes accounts for a different process. Whereas, in the other contexts, it describes the transmission of the same change or affection from one stage to another (e.g. the transmission of heat from one body to another), in the fragment Alexander stresses that each stage not only moves its subsequent stage, but produces the determinate thing of which it is productive (311.1–6), adding that this transmission results in perfection (311.18–19). The example that Simplicius presents in his criticism of Alexander's account, which may go back to Alexander himself, provides a good illustration of this account: a grain produces a shoot, a shoot produces a stalk and a stalk produces an ear (312.24–26).[37] This understanding of transmission suits Alexander's purpose in the fragment. It explains generation rather than alteration and, as I show in what follows, can underwrite the distinction between teleological processes strictly speaking and necessary processes that lead to an end.

Alexander's view that, in natural processes, each stage not only moves but also produces the subsequent stage is stated more clearly in his commentary on *Metaphysics* V 2, where he argues that the external cause is primarily an efficient cause, whereas the intermediate stages leading from it to the effect are derivatively efficient causes:

> T11: He adds "first" because this is primarily the efficient cause; for that which is inherent in the generated thing and thus makes the subsequent things has its cause from the thing that generated it. And even the organs do not have their motion from themselves, but also in these the first cause is external to the generated thing.[38] (Alexander, *In Metaph.* 349.33–36)

We learn from Simplicius that Alexander offered the same interpretation of Aristotle's description of the efficient cause in his commentary on *Physics* II 3.

---

36 See also scholium 805 in Rashed (2011), 634–635. For a discussion of this fragment see Harari (2016), 516–518.

37 The attribution of this example to Alexander finds support in Simplicius' commentary on *Physics* II 7, where it appears in a testimony on Alexander (366.23–25).

38 τὸ δὲ πρῶτον πρόσκειται, ὅτι τοῦτο μάλιστα ποιητικὸν αἴτιον· τὸ γὰρ ἐνυπάρχον τῷ γιγνομένῳ καὶ οὕτως ποιοῦν τὰ ἐφεξῆς παρὰ τοῦ ποιήσαντος αὐτὸ τὴν αἰτίαν ἔχει· καὶ τὰ ὄργανα δὲ οὐκ ἐξ αὑτῶν τὸ κινεῖν ἔχει· τὸ δὲ πρῶτον καὶ τούτοις αἴτιον καὶ ἐκτὸς τοῦ γινομένου.

He reports that Alexander agrees with him that nature, as an inherent cause, is not strictly speaking an efficient cause, but rather a formal cause, because it is not the first among the efficient causes (315.13–15), and also that he concedes that the instruments are causes of motion, but are not efficient causes in the strict sense, because they move by being moved (315.17–18). While these reports leave the exact status of the intermediate stages unclear, the claims that nature is not the first among the efficient causes and that the instruments are not, strictly speaking, efficient causes imply that the intermediate stages are, in a way, efficient causes. In his discussion of the final cause, Simplicius more clearly attributes this view to Alexander, saying that Alexander understands Aristotle's list of the means leading to health—thinness, purging, drugs and surgical instruments—as efficient causes of health (316.8–9). Apart from supporting the conclusion that Alexander understands the intermediate stages as efficient causes in a derivative way, this report brings to light the significance of this understanding. It implies that efficient causes produce for the sake of an end, thus suggesting that by viewing the intermediate stages of natural processes as efficient causes, Alexander guarantees that the stages of these processes come to be for the sake of the end, as his conception of genuine teleology requires.

Alexander's discussion of the final cause in his commentary on *Metaphysics* I 3 lends support to this suggestion. Here, he explains Aristotle's claim that the final cause and the good are opposed to the efficient cause as follows (983a31–32):

> T12: He says that the final cause is opposed to the efficient cause insofar as it preexists whereas the former comes to be last, a point that he himself clarified by adding "for it is the end of all generation and motion." For if it [scil. the efficient cause] is the principle, whereas the former is the end, they are reasonably opposed. The end is the cause of the efficient cause; while it is last in generation, it is first in nature and in reason (λόγῳ), for due to it, we begin the action by first apprehending it in reason. But in things that come to be by nature too this is the aim, and the aim preexists. And, as I already said, when he opens the *Ethics* he clarified that the good is the cause as an end and that such a principle is the end of all generation and motion, saying "every art and method is for a certain good" [*EN* I 1, 1094a1–2]. And if all things long for their proper end as good, the end strictly speaking and that for which all things in motion long would be the good and the best of all beings.[39] (Alexander, *In Metaph.* 22.11–20)

39 ἀντικειμένην φησὶν αἰτίαν εἶναι τῇ ποιητικῇ τὴν τελικήν, καθὸ ἡ μὲν προϋπάρχει ἡ δὲ ἐσχάτη

From this passage, we see that Alexander does not regard the final cause and the efficient cause as two independent and complementary causal factors, but rather holds that the end is the cause of the efficient cause.[40] He does not elaborate on this matter, but it can be understood in light of his interpretation of Aristotle's discussion of the senses of the term "principle" (ἀρχή) in *Metaphysics* V 1. Here, he considers the possibility, which he later accepts, that Aristotle's description of the efficient cause, "that from which, not as inherent, a thing is first generated, and that from which motion and change naturally first results" (1013a7–8), does not refer to this cause alone, but that each of its parts refers to a different principle: the first (i.e. that from which, not as inherent, a thing first comes to be) to knowledge of the end (τέλος) and the aim (σκοπός), and the second (i.e. that from which motion and change naturally first results) to the efficient cause (346.7–17).[41] In interpreting Aristotle's description this way, Alexander assumes that not only the efficient cause, but also the final cause is a productive principle, the former being a principle of motion and the latter a principle of generation.[42] The difficulty in understanding this distinction aside, it helps clarify Alexander's claim in T12 that the end is the cause of the efficient cause. In describing the final cause as "knowledge of the end and the aim" (γνῶσις τοῦ τέλους καὶ σκοποῦ), it highlights the significance of Alexander's claim in T12 that we begin to act by first apprehending the end. Specifically, it shows that he bases his understanding of the relationship between these causes on his account of human actions, thereby clarifying that the end is the cause of the efficient cause as an object of desire that causes the bodily

γίγνεται. ὃ ἐδήλωσεν αὐτὸς προσθεὶς τὸ τέλος γὰρ γενέσεως καὶ κινήσεως πάσης τοῦτ᾽ ἐστίν· εἰ γὰρ ἐκεῖνο μὲν ἀρχὴ τοῦτο δὲ τέλος, εἰκότως ἀντίκειται. ἔστι δὲ τὸ τέλος αἴτιον τοῦ ποιητικοῦ, τῇ μὲν γενέσει ἔσχατον ὄν, τῇ δὲ φύσει καὶ τῷ λόγῳ πρῶτον· τοῦτο γὰρ τῷ λόγῳ προλαβόντες διὰ τοῦτο τῆς πράξεως ἀρχόμεθα. ἀλλὰ καὶ ἐν τοῖς φύσει γινομένοις σκοπὸς τοῦτο, ὁ δὲ σκοπὸς προϋπάρχει. ὅτι δὲ τὸ ἀγαθόν ἐστι τὸ ὡς τέλος αἴτιον καὶ ἡ τοιαύτη ἀρχὴ τέλος γενέσεως καὶ κινήσεως πάσης, καὶ τῶν Ἠθικῶν ἀρχόμενος, ὡς ἤδη προεῖπον, ἐδήλωσεν εἰπὼν "πᾶσα τέχνη καὶ πᾶσα μέθοδος ἀγαθοῦ τινός". εἰ δὲ πάντα τοῦ οἰκείου τέλους ὡς ἀγαθοῦ ἐφίεται, τὸ κυρίως τέλος καὶ οὗ πάντα ἐφίεται τὰ ἐν κινήσει εἴη ἂν τὸ ἀγαθὸν καὶ τὸ τῶν ὄντων ἄριστον.

40 This view has a basis in *Parts of Animals* I 1, where Aristotle says that the final cause is first and the efficient cause is second, arguing that in art and, more so, in nature, the good as an end determines the causes of things produced and why they should be produced in a certain way (639b12–21). Alexander's view is different from the modern understanding of efficient causality, in that he views it as dependent on final causality. For an elaborate justification of the view that efficient causes can operate without final causes see, Lewis (1988), esp. 61.

41 He accepts this interpretation in the commentary on *Metaphysics* V 2, which lists the senses of the term "cause" at 348.16–18.

42 See also 348.3–5. For a discussion of this distinction, see Bonelli (2014), 123–125.

changes that result in goal-directed motions and actions. The fact that Alexander presents this description of the final cause in this context and not in his interpretation of the passage where Aristotle presents choice and art as another sense of principle (1013a10–14) suggests that, in his view, this description does not exclusively hold for the ends of human actions. Indeed, in T12 he applies his understanding of the relationship between the end and the efficient cause also to natural processes. He understands the opening sentence of the *Nicomachean Ethics* as expressing his view, found in section A of the fragment and elsewhere, that all the things long for their end, and shows that natural processes correspond to human actions in having a pre-existent aim.

Section G of the fragment explains this correspondence. Here, Alexander presents a different interpretation of the contention that nature produces in relation to a paradigm from that found in section A of the fragment. Whereas there he contends that when Aristotle calls the form "paradigm" (194b26), he refers to the enmattered form of the generated thing (310.26–27), here he argues that the form of the producer, being the same in species or in genus as the form of the thing produced, is also the paradigm. Through this argument, he avoids the Platonists' anthropomorphic view, according to which nature first grasps the object of its production and then produces it (310.25–26), and still shows that the aim of natural processes pre-exists, not in the intellect or perception, but as the form of the efficient cause that brings about these processes. Thus, the model through which Alexander explains natural processes in the above passage is analogous to the model through which he explains human actions. Just as human beings are prompted to act by the end and thus produce a thing that is similar to a certain pre-existent form (e.g. the form of a house conceived of as a certain good), so are natural processes caused by the end that prompts the efficient cause to initiate the change resulting in the production of a thing similar to a pre-existent form (e.g. a man). This conclusion answers the second question I raised in this section. By showing that Alexander's claim in T12 that the end is the cause of the efficient cause holds for natural processes, it implies that they are to be understood as chains of efficient causes, which guarantees that they are teleological in the strict sense. Since efficient causes depend on the end and operate for the sake of bringing it about, processes that are initiated by efficient causes and proceed through derivative efficient causes are teleological in the sense presented in T10: they and each of their stages come to be for the sake of an end.[43]

43 The process that Alexander presents in section C is successive. Each stage produces the thing of which it is productive, and this thing, in turn, produces another thing until the form comes to be. This account prompted Simplicius' criticism that Alexander fails to spe-

Yet, this discussion leaves a crucial question open: what is the factor that plays the role of longing in natural processes? On the one hand, the view that nature produces by apprehending the end and longing for it as human beings do is incompatible with Alexander's criticism of the Platonists who explain teleology this way. Indeed, in T12 and in section G of the fragment, Alexander avoids this implication. He retains the distinction between natural processes and human actions and does not mention longing for the end as the factor that explains natural processes. On the other hand, in section A of the fragment and in the passages cited in section II above, he explicitly appeals to longing in explaining natural processes and states that all natural things long for the cause that initiated their natural motion. While Alexander does not address this question explicitly, T12 suggests an answer. Here he does not base his account on the Platonic model that he rejects in the fragment and elsewhere, according to which nature produces by setting a goal and acting for the sake of attaining it. Instead, he argues that all things long for their own proper end and for the best being (i.e. god), thereby understanding natural processes in the same way as he explains the motion of the divine body. According to this explanation, just as the divine body brings about regular processes in the sublunary world not by aiming to do so, but by longing to imitate the eternity of the unmoved mover through its regular motion, so natural things bring about directive processes not by aiming to bring about certain ends but by longing for their own proper good in imitation of the best being. This understanding of longing does not expose Alexander to the charge that he failed to offer a genuine alternative to the Platonist view that nature produces in relation to a paradigm. On his account of natural production, the production of something similar to a pre-existent entity that serves as a paradigm is not the aim of natural processes, but is rather a derivative outcome of natural things' longing for their own proper good. Moreover, since this longing is not a longing to attain or produce something, but a longing to become something, it admits a non-animistic construal as an inner tendency to perfection, analogous to the elements' inclination (ῥοπή) to move towards their natural places that Alexander explains in terms of perfection (Simplicius, *In Phys.* 1213.5–8; Scholium 591 Rashed). Section F of the fragment supports this interpretation. Here, Alexander uses the verb νεύειν (to incline), which has no animistic connotations, to describe the tendency of natural processes towards a form and explains it through the ana-

cify the cause of the one form that the process brings about (*In Phys.* 312.27–28 Diels). Alexander does not explicitly address this issue, but his view that all things long for eternity can explain why the end of the process is not one of the intermediate stages, but the form of the efficient cause that initiates the process.

logy to marionettes. Thus, in view of the above analysis of this analogy, nature as a nonrational power accounts for this inner tendency. Just as the power in the sinews accounts for animals' goal-directed motion, by determining the specific teleological reaction of the sinews to heat and coldness that the objects of desire cause, so nature as a non-rational power, accounts for goal-directed processes by regulating the causal interactions in the specific way that guarantees the production of a form in conformity with the desire to imitate the unmoved mover's eternity. Consequently, through this understanding of longing, Alexander secures a conception of teleology that is neither animistic nor involves design.[44]

5 Conclusion

Understanding Alexander's account of natural production from the viewpoint of his rejection of the Epicureans' non-teleological stance casts doubt on its mechanical interpretation and shows that he models it after his account of animals' motion. This model underlies the analogy to marionettes, his contention that nature is a non-rational power and his description of natural processes as proceeding through a transmission of changes. Through this model, he steers a middle course between the Platonic and the Epicurean understandings of natural production. With respect to the Platonists, he attributes thought and desire to an animate entity that can truly have them, i.e. to the divine body. He holds that by thinking of the unmoved mover and longing to imitate its eternity, this body moves its circular motion that guarantees the preservation of the species in a non-rational manner. In his view, providence is not essential. The divine body does not set as a target the benefit of the object of its concern and act for this sake, as he says in *Quaestio* II 21 (66.21–25 Bruns). Rather, it brings about the end (the preservation of the species) through the natural effects of its motion. Further, Alexander understands specific natural

44 From the viewpoint of contemporary interpretations of Aristotle's conception of teleological causation, Alexander's interpretation is closer to the model that David Charles calls "goal-directed efficient cause", whose main proponent is Allan Gotthelf, than it is to the model that Charles himself endorses, called "sensitivity to goodness"—see Charles (2012), 235–243. While, unlike Gotthelf, Alexander does not regard goodness as derivative, like him he does identify the end with perfection and understands its realisation in terms of an imitation of the unmoved mover. Moreover, Alexander, too, holds that natural processes are directive rather than directed (i.e. bringing about an end without intentionally aiming to do so) and grounds teleology in an irreducible power/potential for form. See Gotthelf (2012), 3–29, 59.

processes along these lines, holding that they arise from the longing of natural things for their proper good. This understanding secures the directive character of natural processes without assuming, as the Platonists do, that nature produces in relation to a paradigm. With respect to the Epicureans, his view that nature is a non-rational power is distinct from their view that material necessity brings about natural processes. According to Alexander, matter and other proximate causes are indeterminate causal factors in the sublunary world that nature governs. Therefore, while it is non-rational, nature does not operate by material necessity, as the Epicureans hold. It is an additional factor, in fact divine, that brings about well-ordered processes by overriding material necessity. Moreover, he holds that efficient causes produce for the sake of an end because they depend on the final cause, thereby securing teleology in nature. He understands natural processes as chains of derivative efficient causes, and thus guarantees that these processes and each of their stages come to be for the sake of the end. Finally, in propounding this account of natural production, he adapts Aristotle's mechanical analogies and his understanding of the contribution of the divine body to the sublunary world to this aim. He presents the analogy to marionettes which illustrates, better than automatic devices, the directive character of teleological processes. In the same vein, he holds that the divine body accounts not only for the continuity of motion, as Aristotle holds, but also for the order and regularity of natural processes. In so doing, he counters not only the views of these rival philosophical schools, but also defends the Peripatetic position against objections that his predecessors failed to tackle.[45]

45 I am grateful to the participants of the workshop "Peripatetic Physics after Theophrastus" for their valuable comments and suggestions. I would especially like to thank Gweltaz Guyomarc'h, whose question about the role of longing in Alexander's philosophy of nature shaped the last section of this article. I am greatly indebted to István Bodnár for helpful discussions of this subject and instructive comments on an earlier draft of this article.

Appendix: Alexander *ap*. Simplicium, *In Phys*. 310.31–311.37

A. For, Alexander says, things that produce by nature do not first have a conception of the things that they produce and thus produce, so that one could say that the concept is a paradigm of the things that come to be according to it, as it is in the case of the arts. Rather, he [scil. Aristotle] calls the form that comes to be in matter a paradigm because nature produces everything that it produces by desiring it. This is clear from the fact that, after this came into being, nature stops the production, since the form is a certain determinate thing and is like a given target toward which nature is directed, and for this reason it is called "paradigm".

οὐδὲ γὰρ τὰ φύσει ποιοῦντα, φησὶν Ἀλέξανδρος, πρῶτον νοήσαντα ἃ ποιεῖ οὕτως ποιεῖ, ἵνα τις εἴπῃ τὸ νόημα παράδειγμα τῶν γινομένων κατ᾽ αὐτόν, ὡς ἐπὶ τῶν τεχνῶν ἔχει, ἀλλὰ τὸ γινόμενον περὶ τῇ ὕλῃ εἶδος τοῦτο καλεῖ παράδειγμα διὰ τὸ τὴν φύσιν τούτου ἐφιεμένην ποιεῖν ἃ ποιεῖ πάντα. δῆλον δὲ ἐκ τοῦ γενομένου τούτου παύεσθαι τῆς ποιήσεως, ὡς ὡρισμένου τινὸς ὄντος τοῦ εἴδους καὶ οἷον σκοποῦ προκειμένου, εἰς ὃν τέταται ἡ φύσις, καὶ παράδειγμα διὰ τοῦτο λεγομένου.

B. Alexander says further that the end and the paradigm are not similarly [present] in all the things that produce for the sake of something. Rather, those that produce by choice, art, and reason, regarding these, the end for the sake of which the other things come to be is necessarily first conceived by the maker and is set forth as a target and paradigm of the things that will come to be, but regarding things that come to be by nature, the case is not similar. For nature acts neither by choice nor by reasoning in it. For nature, Alexander says, is a nonrational power.

οὐχ ὁμοίως δέ, φησὶν Ἀλέξανδρος ἐν τοῖς ἕνεκά του ποιοῦσι πᾶσιν ἔχει τὸ τέλος καὶ τὸ παράδειγμα, ἀλλ᾽ ὅσα μὲν κατὰ προαίρεσιν καὶ τέχνην καὶ λόγον ποιεῖ, ἐπὶ τούτων τὸ τέλος ἀνάγκη οὗ χάριν τὰ ἄλλα γίνεται πρῶτον νοηθὲν ὑπὸ τοῦ ποιοῦντος τοῦτο σκοπὸν ἐκκεῖσθαι καὶ παράδειγμα τῶν ἐσομένων, ἐπὶ δὲ τῶν φύσει γινομένων οὐχ ὁμοίως. οὐ γὰρ κατὰ προαίρεσιν ἡ φύσις ἐργάζεται οὐδὲ κατὰ τὸν ἐν αὐτῇ λόγον. ἄλογος γάρ, φησί, δυνάμις ἡ φύσις.

C. But indeed, when the first principle is sown in the matter that receives both the principle and the things that will

ἀλλ᾽ ἀρχῆς καταβληθείσης τῆς πρώτης ἐν ὕλῃ τῇ δεκτικῇ καὶ τῆς ἀρχῆς καὶ τῶν ὑπ᾽ αὐτῆς τε καὶ ἐξ

(*cont.*)

come to be by it and from it, this first thing having been sown produces the thing of which it is productive, being a determinate thing, and from this something else comes to be. Each of them is the maker and mover of that which comes after it, if nothing prevents; and this is up to a certain end, i.e., the natural form of which was the principle that was first sowed in matter as in the case of marionettes: after the craftsman gives the principle of motion to the first thing, this thing becomes a mover of what comes after it, and that thing becomes the mover of the next, until the motion passes through all the parts, if nothing prevents, since what comes before moves what comes after it not by reason and choice in themselves.	αὐτῆς ἐσομένων, τοῦτο τὸ πρῶτον αὐτὸ καταβληθὲν αὐτὸ μὲν ἐποίησε τοῦτο, οὗ ἐστιν αὐτὸ ποιητικὸν ὄντος ὡρισμένου, τὸ δὲ ἐκ τούτου γενόμενον ἄλλο· ἕκαστον γὰρ αὐτῶν τοῦ μεθ᾿ ἑαυτὸ ποιητικόν τε καὶ κινητικόν ἐστι, εἰ μηδὲν ἐμποδίζοι· καὶ τοῦτο μέχρι τέλους τινὸς καὶ εἴδους τοῦ φυσικοῦ, οὗ ἦν ἀρχὴ τὸ πρῶτον ἐν τῇ ὕλῃ καταβληθέν, ὥσπερ ἐν τοῖς νευροσπαστουμένοις τὴν ἀρχὴν τῆς κινήσεως ἐνδόντος τοῦ τεχνίτου τῷ πρώτῳ τοῦτο τοῦ μετ᾿ αὐτὸ κἀκεῖνο τοῦ ἐφεξῆς κινητικὸν γίνεται, ἕως ἂν διὰ πάντων ἡ κίνησις διέλθῃ, ἂν μή τι ἐμποδίσῃ, οὐ κατὰ λόγον τινὰ καὶ προαίρεσιν τὴν ἐν ἑαυτοῖς τοῦ πρὸ αὐτοῦ τὸ μετ᾿ αὐτὸ κινοῦντος.
D. Similarly, the nature and power that are sown in the semen that comes to be in the proper matter being its mover, move as it is natural, one moves and the other is moved. And the power that comes to be from the first motion makes another motion and can [do so], until it makes something similar to that from which it is sown, and it is identical either in species or in genus, as in the case of animals that are generated from dissimilar [begetters], such as mules, which are identical to their begetters in genus. And this transmission comes to be according to a certain sequence and order until the generated thing is perfected with respect of form. And this is not by reason and	οὕτως δὲ καὶ ἡ συγκαταβληθεῖσα τῷ σπέρματι φύσις τε καὶ δύναμις ἐν τῇ οἰκείᾳ ὕλῃ γενομένη κινητικὴ ταύτης οὖσα κινεῖ καθάπερ πέφυκε τὸ μὲν κινεῖν, ἡ δὲ κινεῖσθαι. ἡ δὲ ἐγγενομένη δύναμις ἐκ τῆς πρώτης κινήσεως ἄλλην πάλιν κίνησιν ἐμποιεῖ καὶ δύναται, ἕως οὗ ὅμοιον ποιήσῃ τῷ ἀφ᾿ οὗ κατεβλήθη, καὶ ταὐτὸν ἢ εἴδει ἢ γένει. ὡς ἐπὶ τῶν ἐξ ἀνομοίων ζῴων γεννωμένων, ὁποῖαι αἱ ἡμίονοι. τῷ γὰρ γένει αἱ αὐταὶ τοῖς ποιήσασι. καὶ αὕτη ἡ διαδοχὴ κατά τινας ἀριθμοὺς καὶ τάξιν γίνεται μέχρι τοῦ τελειωθῆναι κατὰ τὸ εἶδος τὸ γινόμενον, εἰ μή τι γένοιτο ἐμποδών. τοῦτο δὲ οὐ κατὰ λόγον καὶ προαίρεσιν τὴν

(*cont.*)

	choice being present in the movers and makers, as has been said above.	ἐν τοῖς κινοῦσί τε καὶ ποιοῦσιν, ὡς προείρηται.
E.	And yet, it does not immediately follow that since nature does not [produce] this way, it produces by chance and not for the sake of something. For the term "for the sake of something" does not express rational and choice-based generation but anything that comes to be according to a certain order and for a certain purpose comes to be for the sake of something. And this is similarly so, whether it comes to be by reason and choice or without reason, as we say about nature. For even if certain natural things follow material necessity and do not have a reference to something (as it is said to be the case with regard to armpit hairs, if indeed they do not provide any utility), one should not argue, for this reason, that nature does not produce for the sake of something.	ἀλλ᾽ οὐκ ἐπεὶ μὴ τοῦτο, διὰ τοῦτο ἤδη κατὰ τύχην ποιεῖ, καὶ οὐχὶ ἕνεκά του. τὸ γὰρ ἕνεκά του οὐ λογικῆς καὶ προαιρετικῆς γενέσεώς ἐστιν ὄνομα, ἀλλὰ πᾶν τὸ γινόμενον κατὰ τάξιν τινὰ καὶ ἄλλου χάριν ἕνεκά του γίνεται. καὶ τοῦτο ὁμοίως, ἄν τε κατὰ προαίρεσιν καὶ λόγον ἄν τε καὶ χωρὶς λόγου, ὡς ἐπὶ τῆς φύσεως ἡμεῖς λέγομεν. κἂν γὰρ ἐπακολουθῇ τοῖς φυσικοῖς τινα κατὰ τὰς ὑλικὰς ἀνάγκας οὐκ ἔχοντα τὴν ἀναφορὰν ἐπί τι (ὡς ἐπὶ τῶν μασχαλῶν λέγονται τρίχες, εἴπερ μὴ καὶ αὗται χρείαν παρέχονταί τινα), οὐ δεῖ διὰ ταῦτα τὸ μὴ ἕνεκά του ποιεῖντὴν φύσιν κατασκευάζειν.
F.	Thus, the form is a paradigm because nature has inclined towards it not by choice, but rather like marionettes.	παράδειγμα οὖν τὸ εἶδος, διότι ἐπὶ τοῦτο νένευκεν ἡ φύσις οὐ προαιρέσει, ἀλλὰ μᾶλλον ὡς τὰ νευροσπαστούμενα.
G.	The paradigm in things that come to be by nature could also be, he [scil. Alexander] says, the form of the producer, being identical to the [form] of the generated thing in species or in genus. And generally, since those who produce in relation to a paradigm produce in relation to something determinate, and coming to be in relation to	εἴη δὲ ἄν, φησί, παράδειγμα ἐν τοῖς φύσει γινομένοις καὶ τὸ τοῦ ποιοῦντος εἶδος ταὐτὸν ὂν τῷ τοῦ γινομένου εἴδει ἢ γένει. καὶ καθόλου ἐπεὶ οἱ πρὸς παράδειγμα ποιοῦντες πρὸς ὡρισμένον τι ποιοῦσι, καὶ ἔστιν ἴδιον τοῦ πρὸς παράδειγμα γινομένου τὸ πρὸς ὡρισμένον τε γίνεσθαι καὶ

(*cont.*)

something determinate and similar to that thing is the property of what comes to be in relation to a paradigm; and if it comes to be in relation to something determinate and is similar to it, then it comes to be in relation to a paradigm; and things that come to be by nature are like this, hence they come to be in relation to a paradigm.	ὅμοιον ἐκείνῳ, καὶ εἴ τι πρὸς ὡρισμένον γίνεται καὶ ὅμοιον ἐκείνῳ, πρὸς παράδειγμα ἂν εἴη γινόμενον· οὕτως δὲ γίνεται τὰ γινόμενα φύσει· πρὸς παραδείγματα ἄρα

CHAPTER 15

Alexander of Aphrodisias and Peripatetic Meteorology on Optical Phenomena: The Case of the Halo

Inna Kupreeva

In his *Meteorology*, explaining optical phenomena in the sublunary sky such as haloes and rainbows, Aristotle uses the concept of sight-streams emitted from the eye and reflected from bright and smooth surfaces. This view, which is ostensibly in conflict with his theory of vision in the *De anima*,[1] is often seen as overcome or improved upon in the later Peripatetic tradition. In particular, Theophrastus has been taken by scholars to disagree with Aristotle's use of sight-streams in his explanation and offer his own, realist, alternative based on the physical explanation of optical phenomena.[2]

Alexander of Aphrodisias in his commentary on Aristotle's treatise aims to eliminate this tension in Aristotle's theory of sight arguing that Aristotle uses the mathematical idiom of sight-streams for convenience, while 'in truth', and also the way Aristotle means it, it is the sunlight, not our sight-streams, that is reflected from air and water and smooth surfaces to produce those sky phenomena. According to Alexander, this usage is theoretically neutral and need not affect the correct physical account. The physical theory of the halo which Alexander presents as Aristotelian can be traced back to Theophrastus, although Theophrastus is not mentioned by name. Alexander seems to be reconciling Aristotle's and Theophrastus' known approaches to the explanation of the halo. His remarks concluding the discussion of the halo indicate that his criticism of sight-streams may have as its target some recent interpreters of Aristotle's optical theory.

The task of this paper is to reconstruct the logic of Alexander's position by reviewing some evidence we have for different Peripatetic explanations of the halo that Alexander might have used or that might serve to support his discussion. Following the introduction of the concept of reflection (in the first section), I discuss Aristotle's account of the halo in *Meteorology* III 2 and III 3 (sec-

1 This contrast is well appreciated in the literature on Aristotle's *Meteorology*. Gilbert (1907/1967), 587, n. 2, Lindberg (1976), 6–7 and n. 39, Wilson (2013), 240.
2 On Aristotle vs Theophrastus: Steinmetz 197–201, Daiber (1992), Kidd (1992).

tion 2), evidence for Theophrastus's explanation of haloes (section 3) and the testimonia for Aristotelian views on the halo preserved in the later Peripatetic tradition (section 4). Section 5 is devoted to Alexander's discussion of the halo.

1 Aristotle on Reflection

Meteorology is the most 'traditionalist' work in Aristotle's physical corpus: thunder, lightning, earthquakes, comets, rainbows, and other remarkable natural phenomena and puzzles are discussed by the earliest Greek philosophers.[3] Aristotle's exposition of this traditional subject presupposes a rich and rigorous theoretical framework which sets the central concepts of the study of nature and the principles and elements of physical universe.

Aristotle says that the subject of meteorology is a "more disorderly" nature: the comparison is drawn, in the first instance, with the heavenly realm.[4] All the phenomena of the sublunary cosmos are explained on the basis of a new causal factor, the double exhalation (ἀναθυμίασις) made up of dry and moist streams rising from the surface of the earth with warm currents of air. The four 'layers' of the sublunary cosmos distinguished in the *De caelo* (earth, water, air, fire) are presented as pervaded by these two types of exhalation which produce various visible phenomena: fiery (meteors, shooting stars, comets, aurora borealis, and the Milky Way), precipitation (rain, frost, snow, dew), rivers, underground water, and the sea. Book 2 continues with the discussion of the sea (1–3), then goes over winds and earthquakes, and finishes with lightning, but this latter seems to be the first item in a series of 'stormy' effects whose discussion continues and is completed in *Meteor.* III 1.

The third book of Aristotle's *Meteorology* is made up of three heterogeneous parts. In chapter 1 we have the discussion of thunderbolts, hurricanes, firewinds and whirlwinds (started in book 2), in chapters 2–6 the explanation of optical phenomena, and in the end of chapter 6, a general introduction of metals and minerals as two main types of product of vaporous and dry exhalations, respectively (preparing for book 4).

Optical phenomena are clearly the main focus of this short book. Their place in the order of subjects has been a matter of discussion: why are they separated from the sky lights discussed in book 1? It is reasonable that the causal role of exhalations in the production of optical phenomena is not the same as the one

3 See in general Gilbert (1907), recently Taub (2003), Mourelatos, (2005); on Aristotle: Solmsen (1961), 393–439, Wilson (2013).

4 For a discussion of Aristotle's project, see Wilson (2009) and (2013).

they play in the formation of those processes. The optical phenomena are not material processes where exhalations are involved as agents. Their main causal factors, to do with light, are located in the heavenly realm, with exhalations providing the physical medium for reflection. The 'heavenly' connection may also account for a special teleological status of these effects with respect to the rest of the "more disorderly nature".[5] Because of their remarkable regularity of shape these phenomena form a proper object of study for geometrical optics,[6] the "mixed" discipline, which studies the attributes of optical phenomena that are discovered and described by a "physicist" making use of the method and axioms of geometry, a superordinate discipline to geometrical optics. This text thus provides us with an authentic example of the case where "the fact falls under one science (for the underlying kind is different) and the reason falls under the higher science which is concerned with the attributes holding of it in itself" (transl. Barnes).[7] The higher science in our case is geometry and the subordinate science is optics.[8] Moreover, the discussions of the halo and rainbows in chapters 3–5 give us a unique glimpse of Aristotle actually using fully-fledged geometrical proofs of his own.[9]

Defining the common nature of the four classes of phenomena discussed in these chapters (halo, rainbow, mock suns and rods), Aristotle tells us that all of them are produced by reflection (ἀνάκλασις).

> (T1) Aristotle, *Meteor.* III 2, 372a16–21:
>
> (1) Now, the characteristics of each of them are these. The cause of them all is the same: for all of them are a reflection. (2) But they differ by the modes [of reflection], and from what things and how the reflection towards the sun or towards any other of the shining things happens to come about.[10]

---

5 See Wilson (2009) and (2013), 236–239.

6 This also explains why they are treated separately from the other sky lights which are considered in *Meteor.* I.

7 *APo.* I 9, 76a11–13: τὸ μὲν γὰρ ὅτι ἑτέρας ἐπιστήμης (τὸ γὰρ ὑποκείμενον γένος ἕτερον), τὸ δὲ διότι τῆς ἄνω, ἧς καθ' αὑτὰ τὰ πάθη ἐστίν

8 For discussion see Hankinson (2005), Steinkrüger (2018); specifically for the explanation of halo, Johnson (2009).

9 Some of the arguments used by Aristotle have parallels in later mathematical works. See Heath, 181 ff. This has raised questions of authenticity or possible later interpolations in the original Aristotelian text. See Vitrac (2002), Merker (2002), Wilson (2012). The passages in question have to do with the shape of the rainbow, and are thus outside the immediate scope of this paper, but it may be useful to signal this discussion.

10 (1) τὰ μὲν οὖν περὶ ἕκαστον αὐτῶν συμβαίνοντα ταῦτ' ἐστίν· τὸ δ' αἴτιον τούτων ἁπάντων ταὐτό· πάντα γὰρ ἀνάκλασις ταῦτ' ἐστί. (2) διαφέρουσι δὲ τοῖς τρόποις καὶ ἀφ' ὧν, καὶ ὡς συμβαίνει

The concept of reflection, ἀνάκλασις, is used in ancient reports of the views of several earlier thinkers. In these reports, the term ἀνάκλασις refers to an explanatory mechanism for a variety of effects, particularly to do with reflection and propagation of light.[11]

Aristotle himself uses the concept of reflection of light in his discussion of echo and draws a parallel with the reflection of light in *De anima* II 8.[12]

(T1a) Aristotle, *De anima* II 8, 419b27–33:

It seems that there is always echo, but not a clear one, since what happens is the same in the case of sound as in the case of light. Namely, light is always reflected (for otherwise light would not be everywhere, but there would be darkness outside the region that is lit by the sun), but it is reflected not in the same way as from water, bronze or any other of the smooth objects, so as to make a shadow by which we demarcate the light.[13]

Aristotle clearly differentiates this kind of reflection from the one in *Meteorology* 3. The kind of reflection that is excluded from the analogy in the *De anima* II 8 (reflection from smooth surfaces) is exactly the one that that constitutes the common genus of halo, rainbow, mock suns and rods, as is clear from the next passage.

(T2) Aristotle, *Meteor.* III 2, 372a29–b11:

(1) Now, we must assume as established, on the basis of what has been proved in the science of optics, that the sight-stream is reflected from air as well as from water and all [bodies] that have smooth surface (2) and

γίγνεσθαι τὴν ἀνάκλασιν πρὸς τὸν ἥλιον ἢ πρὸς ἄλλο τι τῶν λαμπρῶν. Translations of the texts are mine where not indicated otherwise.

11 E.g. Empedocles A56, 57, 91.2; B42.8, 44, 109; Hippocrates of Cos A6.2.4; Philolaus A19.5; Anaxagoras A1.33, 42.32, 85, 86 DK; Democritus A27.3, 121.3, 135.35; Gorgias B5.2 DK.

12 Aristotle, *De anima* II 8, 419b16–33, comparison of echo with reflection of light: ἔοικε δ' ἀεὶ γίνεσθαι ἠχώ, ἀλλ' οὐ σαφής, ἐπεὶ συμβαίνει γε ἐπὶ τοῦ ψόφου καθάπερ καὶ ἐπὶ τοῦ φωτός· καὶ γὰρ τὸ φῶς ἀεὶ ἀνακλᾶται (οὐδὲ γὰρ ἂν ἐγίνετο πάντῃ φῶς, ἀλλὰ σκότος ἔξω τοῦ ἡλιουμένου), ἀλλ' οὐχ οὕτως ἀνακλᾶται ὥσπερ ἀφ' ὕδατος ἢ χαλκοῦ ἢ καί τινος ἄλλου τῶν λείων, ὥστε σκιὰν ποιεῖν, ᾗ τὸ φῶς ὁρίζομεν.

13 ἔοικε δ' ἀεὶ γίνεσθαι ἠχώ, ἀλλ' οὐ σαφής, ἐπεὶ συμβαίνει γε ἐπὶ τοῦ ψόφου καθάπερ καὶ ἐπὶ τοῦ φωτός· καὶ γὰρ τὸ φῶς ἀεὶ ἀνακλᾶται (οὐδὲ γὰρ ἂν ἐγίνετο πάντῃ φῶς, ἀλλὰ σκότος ἔξω τοῦ ἡλιουμένου), ἀλλ' οὐχ οὕτως ἀνακλᾶται ὥσπερ ἀφ' ὕδατος ἢ χαλκοῦ ἢ καί τινος ἄλλου τῶν λείων, ὥστε σκιὰν ποιεῖν, ᾗ τὸ φῶς ὁρίζομεν.

> that in some of the mirrors shapes too are reflected, while in some only colours. These latter are all the mirrors that are small and have no perceptible division. In these shape cannot be reflected (for it will seem to be divisible: for every shape seems at the same time to be shape and to allow of division). But since something must be reflected, and this is impossible, it remains that colour alone is reflected. (3) The colour of bright objects sometimes appears bright [in reflection], and sometimes either because of the admixture of the colour of the mirror, or because of the weakness of sight, it produces an impression of a different colour. Let us take these matters as established in the study of sense perceptions. Let us therefore state some things and take the rest for granted.[14]

It becomes clear from (T2.1) that the ἀνάκλασις Aristotle refers to here is the concept of geometrical optics. The same is true of the concept of sight-stream (ὄψις) which Aristotle introduces next as that which undergoes this reflection.[15] As Solmsen notes, in the *Meteorology* Aristotle seems to distinguish the concept of reflection used by earlier thinkers from the one that is available at his time and that he himself uses: the doctrines which those earlier philosophers are not familiar with seem to be the concepts of geometrical optics.[16]

It is not entirely clear whether referring to "the science of optics" Aristotle has in mind any specific works by mathematicians. In his discussion of comets in *Meteorology* I 6–7, he criticises a theory which he attributes to Hippocrates of Chios (fl. mid-5th century), according to which our sight-streams reflected towards the sun permit us to see the comet's tail (made of frozen water and thus reflecting the light) when the comet is in the right position with respect

---

14 (1) ὅτι μὲν οὖν ἡ ὄψις ἀνακλᾶται, ὥσπερ καὶ ἀφ᾽ ὕδατος, οὕτω καὶ ἀπὸ ἀέρος καὶ πάντων τῶν ἐχόντων τὴν ἐπιφάνειαν λείαν, ἐκ τῶν περὶ τὴν ὄψιν δεικνυμένων δεῖ λαμβάνειν τὴν πίστιν, (2) καὶ διότι τῶν ἐνόπτρων ἐν ἐνίοις μὲν καὶ τὰ σχήματα ἐμφαίνεται, ἐν ἐνίοις δὲ τὰ χρώματα μόνον· τοιαῦτα δ᾽ ἐστὶν ὅσα μικρὰ τῶν ἐνόπτρων, καὶ μηδεμίαν αἰσθητὴν ἔχει διαίρεσιν· ἐν γὰρ τούτοις τὸ μὲν σχῆμα ἀδύνατον ἐμφαίνεσθαι (δόξει γὰρ εἶναι διαιρετόν· πᾶν γὰρ σχῆμα ἅμα δοκεῖ σχῆμά τ᾽ εἶναι καὶ διαίρεσιν ἔχειν), ἐπεὶ δ᾽ ἐμφαίνεσθαί τι ἀναγκαῖον, τοῦτο δὲ ἀδύνατον, λείπεται τὸ χρῶμα μόνον ἐμφαίνεσθαι. (3) τὸ δὲ χρῶμα ὁτὲ μὲν λαμπρὸν φαίνεται τῶν λαμπρῶν, ὁτὲ δέ, ἢ τῷ μείγνυσθαι τῷ τοῦ ἐνόπτρου ἢ διὰ τὴν ἀσθένειαν τῆς ὄψεως, ἄλλου χρώματος ἐμποιεῖ φαντασίαν. ἔστω δὲ περὶ τούτων ἡμῖν τεθεωρημένον ἐν τοῖς περὶ τὰς αἰσθήσεις δεικνυμένοις· διὸ τὰ μὲν λέγωμεν, τοῖς δ᾽ ὡς ὑπάρχουσι χρησώμεθα αὐτῶν.

15 For a survey of the earlier theories of visual ray and Aristotelian concept, see Wilson (2013), 239–246.

16 See Solmsen (1961), 419 with *Meteor.* II 9 on Cleidemus lacking knowledge of the doctrine of reflection; *Sens.* 2, 437b5–10, on Democritus. The doctrine mentioned here is not the same as the earlier Presocratic concept.

to the sun, to the region on the earth with enough moist to draw, and to the observer on earth. Hippocrates' theory involves a geometrical representation of the position of the comet and the sun relatively to the observer, but the reasoning, at least as reported by Aristotle, seems to be based on the consideration of cases rather than on geometrical propositions.[17] It is not clear what specific works Aristotle could use, but at least the reference to the discipline is clear.[18]

Making the assumption that smooth bodies reflect sight-streams from their surfaces on the basis of what is proved in the science of optics, Aristotle does not say that this position is itself proved or otherwise justified in this science. He means rather that he will be proceeding on the same assumptions as the ones made in those demonstrations. This would involve a view according to which our sight can be represented by geometrical straight lines drawn from the eye towards the object of sight.

In (T2.2) Aristotle describes a specific kind of mirrors which reflect only colour but not shape. These mirrors are small compared to the object they reflect, so they reflect only a part of that object, and they have "no perceptible division" (μηδεμίαν αἰσθητὴν ἔχει διαίρεσιν). This latter expression, occurring twice in this short account, is vague and not very easy to interpret.[19] Presumably, had these small mirrors been perceptibly divisible, the way ordinary mirrors are, the observer would have access to a perceptible feature by which to distinguish between the different parts of the reflected image, but as is, no such feature is available and no such distinction can be drawn.

In (T2.3), Aristotle mentions "the weakness of sight" as one of the factors that may constitute the visible, phenomenal colour. The term ὄψις seems to be referring to a sight-stream as a physical ray emitted from our organ of sight.[20] Its "weakness" presumably would mean its low intensity, which entails its ability of being reflected from the bodies that don't normally reflect sight-streams. Here we seem to have a physical assumption about our sight, not just the possibility of its geometrical representation. This is one of several passages that support a literal, physical, interpretation of Aristotle's theory of sight-streams.[21]

---

17 For discussion of this theory, see Wilson (2008).

18 There is an excellent concise overview of Aristotle's predecessors in Merker (2002), 185–187. Additionally, we may note that Aristotle's catalogue of writings in DL contains the title ὀπτικὸν α′ (DL V 26, 369) and the work *On sight* in four books is attributed to Theophrastus (DL V 49, 266: περὶ ὄψεως α′β′γ′δ′).

19 See Groisard's remarks ad loc. Another occurrence is at *Meteor.* III 3, 373a19–24 (T4.2) below.

20 This term is used also earlier in this text, in the name of the discipline of geometrical optics, τὰ περὶ τὴν ὄψιν (T2.1).

21 *Meteor.* II 9, 370a12–22; *Meteor* III 4, 374b11–15; 373b2–10.

The same point about the weakness of the ὄψις is made before the discussion of the rainbow in *Meteor.* III 4:

> T2.3a Aristotle, *Meteor.* III 4, 373a35–b7 (transl. Lee):
> (1) Our sight-stream, then, is reflected from all smooth surfaces, among them air and water. Air reflects when it is condensed. (2) But even when not condensed it can produce a reflection when the sight is weak. An example of this is what used to happen to a man whose sight was weak and unclear: he always used to see an image going before him as he walked, and facing towards him. (3) And the reason why this used to happen to him was that his sight-stream was reflected back to him; for its enfeebled state made it so weak and slight that even the neighbouring air became a mirror and was unable to thrust it aside.[22]

The reflection of a sight-stream is described as a physical interaction between the two physical bodies: physical smooth surfaces (including air and water) on the one hand and physical sight-streams on the other. In the example which follows in (T2.3a2) the "weakness" of such a sight-stream is understood as an inability to pass through the physical medium of air which is described as too dense for this sight-stream. Thus the ability to see things is directly dependent on the physical condition of the bodily sight-streams emitted by the organ of sight and their "fitness" in attaining the objects of sight. Aristotle here seems to be going beyond the use of the language of geometrical optics as a mere convention in the description of the phenomena of reflection. He shows his commitment to an interpretation of these sight-streams as physical bodies.

Such a physical theory of sight-streams would be in a stark contrast with the account of sight that we find in the *De anima* and *De sensu*, where Aristotle follows his general methodological principle for the study of the powers of the soul, according to which in order to understand the nature of the power we must understand its operation (or activity), and in order to understand the latter, we need to understand its object. The object, in our case, the proper object of sight, colour, acts on our power (first actuality) of sight through the trans-

22 (1) ἀνακλωμένη μὲν οὖν ἡ ὄψις ἀπὸ πάντων φαίνεται τῶν λείων, τούτων δ᾽ ἐστὶν καὶ ἀὴρ καὶ ὕδωρ. γίγνεται δὲ ἀπὸ μὲν ἀέρος, ὅταν τύχῃ συνιστάμενος. (2) διὰ δὲ τὴν τῆς ὄψεως ἀσθένειαν πολλάκις καὶ ἄνευ συστάσεως ποιεῖ ἀνάκλασιν, οἷόν ποτε συνέβαινέ τινι πάθος ἠρέμα καὶ οὐκ ὀξὺ βλέποντι· ἀεὶ γὰρ εἴδωλον ἐδόκει προηγεῖσθαι βαδίζοντι αὐτῷ, ἐξ ἐναντίας βλέπον πρὸς αὐτόν. (3) τοῦτο δ᾽ ἔπασχε διὰ τὸ τὴν ὄψιν ἀνακλᾶσθαι πρὸς αὐτόν· οὕτω γὰρ ἀσθενὴς ἦν καὶ λεπτὴ πάμπαν ὑπὸ τῆς ἀρρωστίας, ὥστ᾽ ἔνοπτρον ἐγίγνετο καὶ ὁ πλησίον ἀήρ, καὶ οὐκ ἐδύνατο ἀπωθεῖν—ὡς ὁ πόρρω καὶ πυκνός·

parent medium with light, and in this way produces the actual perception.[23] There is no room for physical sight-streams in this picture.

In the following passage from his *Generation of Animals*, Aristotle shows that he is acknowledging a theory of sight-streams as a counterpart to the *De anima* theory, but seems to be distancing himself from it.

> (T2.3b) Aristotle, *GA* V 1, 781a3–8 (transl. Peck, lightly modified):
>
> (1) It makes no difference whether we say, as some people do, that seeing is effected 'by the sight-stream issuing forth', (for on this theory, unless there is something projecting in front of the eyes, the sight-stream must get scattered and so less of it strikes the object, with the result that the distant objects are less well seen), (2) or [whether we say] that seeing is effected "by a movement derived from the visible object". For on this theory, the clarity with which the sight sees will of necessity vary directly as the clarity of the movement.[24]

Aristotle here is explaining why animals with sunken eyes (ἔχοντα τὰ ὄμματα ἐν κοίλῳ) see at a distance better than those with prominent eyes (ἐξόφθαλμα). (780b33–781a3). He says that this observation will make good sense whichever theory of sight one adopts: (1) the one whereby we see by the sight-stream emitted from our eyes or (2) the one according to which vision is produced by the motion from the object of sight.[25] The first theory has an attribution ὥσπερ τινές φασι, 'as some say'. The "some" are usually taken to refer to Plato in the *Timaeus*.[26] Aristotle does not subscribe to it here. In (T2.3b2), κίνησις is used without any qualification: it is not κίνησίς τις as in *De anima* 2.5.[27] Nonetheless, it seems to refer to a qualitative change which is responsible for the clarity of sight and on the whole is closer to Aristotle's *De anima* theory of sight than to any other theory. Why does Aristotle acknowledge the other theory here?

---

23 See Gilbert (1907), Lindberg (1976), 6–7 and n. 39, Jones (1994), 61–63; Merker (2002), 191.

24 (1) οὐθὲν γὰρ διαφέρει τὸ λέγειν ὁρᾶν, ὥσπερ τινές φασι, τῷ τὴν ὄψιν ἐξιέναι (ἂν γὰρ μὴ ᾖ τι πρὸ τῶν ὀμμάτων, διασκεδαννυμένην ἀνάγκη ἐλάττω προσπίπτειν τοῖς ὁρωμένοις καὶ ἧττον τὰ πόρρωθεν ὁρᾶν), (2) ἢ τὸ τῇ ἀπὸ τῶν ὁρωμένων κινήσει ὁρᾶν. ὁμοίως γὰρ ἀνάγκη καὶ τὴν ὄψιν τῇ κινήσει ὁρᾶν.

25 This is a very simplified summary of *De anima* II 4 (415a16–22) and II 5, which omits many important and controversial nuances. My purpose here is to contrast this theory with the theory of physical sight-streams which surfaces in the *Meteorology* III.

26 So Peck ad loc. Plato discusses sight-streams in *Tim.* 45B2–46C6.

27 See Burnyeat (2005, 73–74) on the importance of this τις *alienans*.

Could this be a sign of hesitation that might be interpreted as a point in the development of his ideas on sense perception?[28]

It would be useful to see how this implicit physical theory of sight-stream is borne out by the explanation of the optical phenomena. In the following section we'll turn to the case of the halo.

2 Aristotle's Account of the Halo

The meteorological phenomenon of haloes is familiar to us today as a ring with a sharp and bright edge that can be seen around the moon or the sun. It is produced by the refraction and reflection of the sun rays from the mass of tiny crystals of ice in the middle to upper atmosphere. The crystals that produce the most frequently observed kind of halo have a shape of a hexagonal prism. The sun ray, which falls on a face of such a prism, is refracted, i.e. gets inside the prism where it is bent at a constant angle to its original direction; then a part of it is refracted again so that it exits from the other side, again being bent at a new angle. The bent rays can be seen as forming a ring of an angular diameter of 44^0.[29] What we see as a halo is such a fine ring.[30] Needless to say, none of this information on the nature of light and crystals is available to the ancient authors we discuss.

The word ἅλως has the basic meaning of 'threshing floor', which has a circular shape, hence the etymology of the meteorological term.[31] In ancient sources on the subject, the word ἅλως seems to be used as a generic term which may cover a number of different optical phenomena, such as coronae, which have a different appearance from a ring-like halo.[32]

Aristotle's description of the halo as the first item in the list of the four explananda (halo, rainbow, mock sun, rod) that precedes his discussion of optical phenomena in *Meteor.* III is as follows:

---

28 For a developmentalist approach to Aristotle's theory of sight on the basis of the difference between *Meteorology* and *De anima*, see Lindberg (1976), 218 n. 39, Johnson (2009).

29 For the most common type of halo, with a minimal angle of deviation of a light ray in refraction at 22 degrees. On the types of haloes and different mechanisms of their formation, see Greenler (1980).

30 See Greenler (1980) for images, Johnson (2009) for the discussion of geometrical shape.

31 Seneca *NQ* 1.2.3. 'For this reason the Greeks have called these bright phaenomena "threshing-floors", because the places designated for threshing grain are generally round' (Transl. Hine).

32 See below, p. 368.

(T3) Aristotle, *Meteor.* III 2, 371b23–27:

First we must state the characteristics and properties of each of them. The complete circle of a halo is often seen. It comes about around the sun, the moon and the bright stars, furthermore, no less at night than during daytime, i.e. at mid-day or in the afternoon; in the morning and at sunset they occur less frequently.[33]

The two characteristics of the halo that must be explained are its round shape and position around the star (so that both the star and the halo are facing the observer). The expression κύκλος ὅλος in T3 merits attention. The Greek word κύκλος can mean not just a Euclidean circle, but also a circular shape, notably, a ring.[34] If, as his subsequent argument might suggest, Aristotle means a ring, then κύκλος ὅλος will mean a complete circumference of a circle rather than a semi-circle or a smaller arc of the kinds we have in the case of rainbow. If κύκλος means a Euclidean circle, then ὅλος might refer to its appearance as a regular plane figure. At this point Aristotle provides no further description of his halo that might help to draw this distinction. In the geometrical proof he will be using the expressions κύκλος, κύκλου γραμμή, and περιφέρεια as close synonyms.

Aristotle's first question to be answered by the account of the halo has to do with the cause of its circular shape. The second question has to do with its position—why it is seen "around" the star, with both the star and the halo facing the observer:

(T3a) Aristotle, *Meteor.* III 3, 372b15–18:

(1) Let us first speak about the halo, why it is a circle and why it is around the sun or the moon, and in the same way around some of the other stars. For the same reasoning will be suitable for all of them. (2) Now, the reflection of the sight-stream is formed when air and vapour

---

33 πρῶτον δὲ δεῖ λαβεῖν τὰ πάθη καὶ τὰ συμβαίνοντα περὶ ἕκαστον αὐτῶν. τῆς μὲν οὖν ἅλω φαίνεται πολλάκις κύκλος ὅλος, καὶ γίγνεται περὶ ἥλιον καὶ σελήνην καὶ περὶ τὰ λαμπρὰ τῶν ἄστρων, ἔτι δ' οὐδὲν ἧττον νυκτὸς ἢ ἡμέρας καὶ περὶ μεσημβρίαν ἢ δείλην· ἕωθεν δ' ἐλαττονάκις καὶ περὶ δύσιν.

34 This is the first meaning of κύκλος in the LSJ article s.v. along with 'circle'. In Euclid's *Elements* I Def. 15, a distinction is drawn between the circle, κύκλος, i.e. a plane figure circumscribed by one line which is called circumference, such that all the lines drawn from its centre to the circumference (i.e. radii) are lying within the circle and the circumference itself. Heiberg in his edition brackets both expressions containing περιφέρεια as glosses, although the manuscripts have them, pointing out that the secondary tradition does not report these expressions.

are condensed into a cloud, if it happens to be condensed as smooth and made of small parts.[35]

Aristotle's answer to both questions (about the shape of the halo and its position) does include a reference to the physical conditions required for its appearance. Although he can't benefit from our knowledge of crystallography, he seems to be thinking in the right direction: there is *something like* a reflection from tiny little bits of icy mist and this reflection seems to have a distinctive geometrical pattern.

Aristotle does not elaborate further on the physical mechanism of the process of reflection as such, assuming the basic concept from optics, as he promised above. We must understand that when we see a moon in a clear sky, our sight-streams can be represented as going straight to the moon so that we get an undistorted picture of a familiar disk with a face. When there is a mist, the moonlight is reflected in it as in a swarm of tiny mirrors, perhaps vertically orientated, and our sight-streams are therefore also reflected back from that mist: thus we get to see both the moon and its reflections in the mist simultaneously.

The circular shape of the halo is explained in a geometrical demonstration. Aristotle assumes that some of our sight-streams leaving the eye travel straight to the star and some are reflected from the shining mist which becomes visible to us as a halo. He sets out to prove the following geometrical proposition: "When our sight-stream is reflected in the same way on all sides, a circle or a part of a circle must be produced. For equal lines drawn from the same point to the same point will be always broken on a circle" (*Meteor.* III 3, 373a3–5).

The proof goes as follows (see Figure 15.1). Let each of the lines ΑΓΒ, ΑΖΒ, and ΑΔΒ (all drawn from the same point A to the same point B) be forming an angle at the vertices Γ, Ζ, Δ, respectively. The lines ΑΓ, ΑΖ, ΑΔ are equal to each other and the lines ΓΒ, ΖΒ, ΔΒ are equal to each other (this is a given). Let us connect A and B drawing the line ΑΕΒ.[36] The triangles ΑΓΒ, ΑΖΒ, ΑΔΒ are all equal

35 (1) πρῶτον δὲ περὶ τῆς ἅλω τοῦ σχήματος εἴπωμεν, διότι τε κύκλος γίγνεται, καὶ διότι περὶ τὸν ἥλιον ἢ τὴν σελήνην, ὁμοίως δὲ καὶ περί τι τῶν ἄλλων ἄστρων· ὁ γὰρ αὐτὸς ἐπὶ πάντων ἁρμόσει λόγος. (2) γίγνεται μὲν οὖν ἡ ἀνάκλασις τῆς ὄψεως συνισταμένου τοῦ ἀέρος καὶ τῆς ἀτμίδος εἰς νέφος, ἐὰν ὁμαλὴς καὶ μικρομερὴς συνισταμένη τύχῃ.

36 It is remarkable that Aristotle in this passage seems to use triliteral notation for 'composite' lines: 'broken' lines ΑΓΒ, ΑΖΒ, and ΑΔΒ, and straight line ΑΕΒ. "Line ΑΕΒ" at 373a11 seems to have a composite meaning because it designates the straight line connecting A and B (the source the sight-stream and the source of light, respectively) and indicates that this line intersects in point E the plane in which the visible shape of the halo lies. (The task of this proof is to show that this visible shape is indeed a geometrical circle.) Aristotle explains next that the lines ΓΕ, ΖΕ, ΔΕ all intersect in this point. Perhaps the reason for

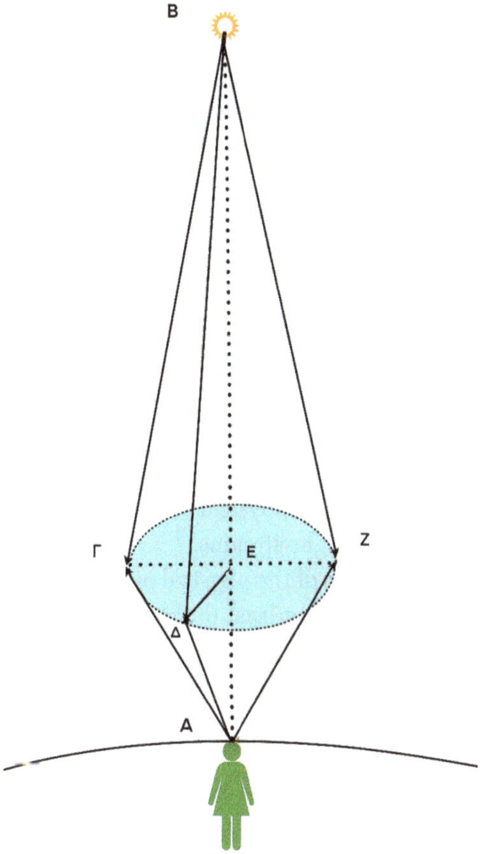

FIGURE 15.1
Circular shape of the halo explained in a geometrical demonstration

(because all their respective sides are equal). Now in each of the triangles AΓB, AZB, AΔB let us draw a perpendicular line from the vertex (Γ, Z, Δ, respectively) to the line AEB in point E. The lines ΓE, ZE, ΔE are of equal length: they are all in equal triangles and in one (and the same) plane (373a14–15). For they are all at a right angle to AEB and they meet at one point, E. Hence a circle will be drawn (through them) with the centre at E. This proof is concluded by an explanation that A is the starting point of a sight-stream, B the sun, and the circumference drawn through ΓZΔ is a halo.

In this generalised form, the proof can explain to us how the observer should be positioned with respect to the star in order for the halo to be seen. Moreover, it can explain how not just a single observer, but many different ones can see

using AEB throughout the proof is to express, in a compact way, the nature of the line. Thanks to Alessio Santoro for raising the point.

the same shining circle at the same time from different positions under the icy mist: the plane in which the halo appears is perpendicular to the straight line between the observer (the starting point of sight-streams) and the star.

We might have some further questions regarding this geometrical argument. What prompts—or licenses—us to choose the lines ΑΓ, ΑΖ, ΑΔ that are equal in length, and similarly with the lines ΓΒ, ΖΒ, ΔΒ?[37] Is it an unstated assumption that the plane ΓΔΖ in which we see the halo is perpendicular to the central sight-stream ΑΒ? The force of the proof would then be to show that from the many sight-streams that may fall onto this plane, it is the ones that connect with the sun in the way described in this proof that form the halo. There are no stated geometrical constraints on the size of the ring or the angles our sight-streams and the sun rays form with the straight line ΑΒ. How is the angular diameter of the halo defined? Aristotle may well think that in each case these parameters are determined by local conditions, the distance of the icy mist from the observer and the sun, perhaps even visibility as the physical property of air on high as making it accessible for the passage of sight-streams.

Aristotle provides us with some further details of the physical picture underlying this mathematical explanation. We do need at least the misty cloud and the external light for the ring to come about, and these are both physical conditions that underwrite the application of the geometrical explanation.

> (T4) Aristotle, *Meteor.* III 3, 373a19–27:
>
> (1) The mirrors must be thought of as continuous. But because of smallness each is invisible, whereas the aggregate of all seems to be one because of being successive. (2) The bright light, i.e. the sun, is manifest as appearing continuously in a circle in each of the mirrors and having no perceptible division, (3) closer to the earth because it is calmer there. For if there is an air current, a clear structure does not obtain. (4) Next to it, there is a dark circumference which seems to be darker because of the brightness of the halo.[38]

---

37 Olympiodorus reports an objection to Aristotle along these lines: his critics show as a counterexample the sight-streams that are emitted simultaneously, but are of different length.

38 (1) δεῖ δὲ νοεῖν συνεχῆ τὰ ἔνοπτρα· ἀλλὰ διὰ μικρότητα ἕκαστον μὲν ἀόρατον, τὸ δ᾽ ἐξ ἁπάντων ἓν εἶναι δοκεῖ διὰ τὸ ἐφεξῆς. (2) φαίνεται δὲ τὸ μὲν λευκόν, ὁ ἥλιος, κύκλῳ συνεχῶς ἐν ἑκάστῳ φαινόμενος τῶν ἐνόπτρων, καὶ μηδεμίαν ἔχων αἰσθητὴν διαίρεσιν, (3) πρὸς δὲ τῇ γῇ μᾶλλον διὰ τὸ νηνεμώτερον εἶναι· πνεύματος γὰρ ὄντος οὐκ εἶναι στάσιν φανερόν. (4) παρὰ δὲ τοῦτο μέλαινα ἡ ἐχομένη περιφέρεια, διὰ τὴν ἐκείνης λευκότητα δοκοῦσα εἶναι μελαντέρα.

In (T4.1–2) Aristotle spells out the role of small mirrors and explains how they are continuous in appearance. In (T4.3) he indicates that the matter of exhalation underlying haloes is vulnerable to weather changes. The role of haloes in weather forecasts is described in many ancient sources. It is likely that in (T4.4), we have a description of the dark circumference next to the bright ring of the halo. Aristotle says that it is dark (black, in fact) and appears darker because of the contrast with the bright ring. The nature of this appearance is not spelled out further. Aristotle's wording here allows for a possibility of explaining the dark ring inside the circle as an optical phenomenon, by reference to the way the inside ring appears to us. This possibility, had it been developed in more detail, could provide a clearer alternative to the physical explanations we shall look at in the next section. But Aristotle's text is tantalisingly vague here: it is not clear what exactly he means by "a dark circumference" (ἡ ἐχομένη περιφέρεια) and how far it extends inside the circle.

On the whole, the geometrical proof seems to be working backwards to solve the problem: assuming we see things with the sight-streams which are straight lines emitted from our eyes, and assuming that these sight-streams are reflected from shining objects and smooth surfaces, what should be our position with respect to the sun and to the reflecting surfaces in order for us to be able to see a halo as a complete regular circle around the sun? In this proof Aristotle needs to make use of some physical properties of the exhalations, but only insofar as these physical properties would help to account for a stable geometrical picture supported by the physical environment.

3 Theophrastus and Later Peripatetic Sources: A Physical Explanation of the Halo

Theophrastus seems to give a purely physical explanation of the halo suggesting a physical mechanism by which a circle is formed in the mist. In the Arabic text of Theophrastus' work translated by Ibn Suwār Ibn al-Khammār[39] he explains that the mechanism by which a halo[40] is produced around the moon is that of a wavelike motion which is produced by the moonlight in the

39 On this Arabic text, see Daiber (1992), 218–227 and Daiber (2021), chap. 12. On Ibn Suwār Ibn al-Khammār see Lettinck (1999), 5–6. The Syriac version of what seems to be the same Greek source, published by Daiber in the same volume does, not have an account of haloes.

40 The Arabic word for halo, الدائرة, literally means 'ring'. I keep 'halo' in the heading and the opening.

mist similarly to the way the ring-shaped motion is produced in the water by a thrown stone (T5.1):

> (T5) Theophrastus, *Meteorology* 14 (part), Arabic version by Ibn Suwār ibn al-Khammār, Daiber (1992) and (2021):
>
> [14] The account of the causes of the halo around the moon.
>
> (1) The halo around the moon is formed when air becomes thick and full of vapour so that in it from the moonlight an undulating movement is produced in the same way as when we throw a stone into the water a ring-shaped motion is produced around the stone. **(2)** And since the moon rays meet the air in a thick condition, the place opposite the moon becomes thin, and the thinner this part of the air, the thicker the parts which are around it. And for this reason the thick air around this part that is already made thin becomes a ring. **(3)** And similar to this is this: when someone blows through a pipe onto a place where there is dust, the place onto which he is blowing becomes clean and the dust that has been swept from it is collected in the place that surrounds the clean place and makes a ring. **(4)** And the halo around the moon grows thin either if a strong wind blows and scatters the thick or if there is a rain. **(5)** And these are the causes of each of these things.[41]

In (T5.2) Theophrastus explains what happens when the moon rays come into contact with the thick air of the mist: the thick air "opposite the moon", i.e. presumably located at the line drawn from the moon to the earth vertically, becomes thin, while the air in the surrounding parts is denser. What causes this thinning? Theophrastus does not tell us directly. In (T5.3) he offers a further analogy of somebody blowing through the pipe onto a place which has

---

41 ذكر اسباب الدائرة الكائنة حول القمر

(١) وأما الدائرة حول القمر متكون إذا غلظ الهواء و امتلأ بخارا حتى تحدث فيه من ضوْ القمر حركة موجية كما انا نحن إذا القينا في الماء حجرا حدثت حول الحجر حركة مستديرة، (٢) وأما لأن شعاع القمر إذا صادف الهواء غليظا لطف الموضع المحاذي للقمر، وكلما لطف هذا الجزء من الهوء غلظت الأجزاء التي حوله أكثر، ومن أجل ذلك يكون الهواء الغليظ الذي حول هذا الجزء الذي قد لطف دائرة، (٣) ونظيرذلك أنه إذا نفخ انسان بأنبوب في موضع فيه تراب نقى الموضع الذي ينفخ فيه واجتمع الغبار المنكنس منه في الموضع المحيط بالموضع النقي وعمل دائرة، (٤) و الدايرة حول القمرتنحل ‹أما› إذا هبّت ريح صعبة وبدّدت الغليظ وأما إذا كان مطر، (٥) فمن هذه الاسباب يكون كلّ واحد من هذه الاشياء

dust in it: we should imagine something like a garden leaf blower of very high precision. This blowing (when undisturbed by other movements of air) would produce a clearing in the dust: presumably in this way we get to see the face of the moon and possibly some area around the moon, surrounded by a shining circle which, for Theophrastus, is made of a real material consisting of icy mist. The icy mist has formed this perfect round shape because it was acted upon by the light of the moon.

Theophrastus' use of the two analogies: the stone thrown into the water producing circles and someone blowing onto the dust, producing wave-like structures, might be meant to illustrate the process that takes place in the mist acted upon by the light of the moon.[42] It may be worth noting that both are close analogies. The direct description of the process in (T5.2) lends itself to different interpretations.

In Aristotelian physics, the action of a moon on the underlying air would be more naturally described as heating as we seem to have a more general description of this process without analogies: the body that lies vertically under the moon becomes thin (maybe being dissolved by the small amount of heat produced by the moonlight when it meets the sublunary sphere?)[43] But in any case, there seems to be an increase of the dense material at the limits of this dissolution.

42 It might also be that each of the two analogies stands for a distinct explanation. I read وأما following Daiber's printed text at (T5.1) and (T5.2) with hamza on the fatḥa. In (T5.2) the meaning of وأما is not easy to understand. At the conference it was suggested to me by Alessio Santoro that the first وأما under (T5.1) could be corrected into وإما, giving a disjunctive construction: (1) 'Either …' (2) 'Or …'. This reading is grammatically possible, and takes care of some unclarities in the interpretation of وأما. Moreover, it might be useful in light of the current discussion of 'multiple explanations' in Theophrastus' meteorology. (For the problem and possible parallels with Epicurus *Letter to Pythocles*, see Bakker (2016), 70–75, further Verde (2022)). I think this option should definitely be kept on the table and discussed further, particularly in comparison with the other sections of this text. It might be that Theophrastus here is explaining not one but two different kinds of phenomena in the same genus of halo: the one where we have several rings (as in the analogy of a thrown stone), for instance, a double halo, and the one where there is a single ring, as in a more frequently observed case. (This case should also stimulate us to work more on the Arabic fortuna of the Greek particles.) Still, since in this section of the text the distinct 'causes' of halo are not as clearly announced as in some previous sections, and for the purposes of this argument, I will provisionally assume that both analogies support the same explanation of a production of a generic annular shape as a physical process rather than optical phenomenon.

43 As we will see shortly, Alexander of Aphrodisias, adopting a version of Theophrastean account, will be taking thinning as a result of the action of heating. Seneca, in the text below (T5b) gives a mechanical interpretation of the action of light on air.

We must note that Theophrastus' argument contains no mention of the reflection of the rays from the halo ring nor any other optical theory. This is sometimes taken as a silent objection to Aristotle's optical theory, but there is no need to suggest that. As we saw, Aristotle's optical theory left the exact parameters of his visual framework unspecified, possibly seeing them as dependent on physical circumstances. Theophrastus' work might be devoted to uncovering such circumstances. It may be telling in this respect that Ibn Suwār in his own meteorological treatise cites almost verbatim Theophrastus' explanation in (T5) as a "physical" explanation of the halo immediately following upon the geometrical explanation without any suggestion of a conflict between the two accounts. He seems to take them as complementary—this is the approach we will shortly find in Alexander.

(T5a) Ibn Suwār, *Treatise on Meteorological Phenomena*, Lettinck 336:
The physical argument, given by Theophrastus, is as follows. (1) He says that the light of the luminary causes in a thin cloud an undulating motion similar to the one that occurs in water when a stone is thrown into it so that a circular motion come about around the stone. (2) And again,[44] the rays from the moon make the air opposite the moon thinner, and thus the cloud opposite the moon, in which **the halo**[45] comes into being, becomes thinner. The thinner the part opposite the luminary becomes, the thicker the parts around it become. These thicker parts are seen as a ring. (3) This is similar to what happens when someone blows with a tube on dust which is under it so as to clean the place on which he is blowing and the dust which is swept away collects in the places surrounding the clearing and forms the ring.[46]

---

44 'And again' (وأيضا) might be an indication of a different explanation from the one we find in (T5a1), as Alessio Santoro points out to me. It could also be signalling an explanation of a different kind of halo. On the other hand, this expression might be referring to the next point in the report of Theophrastus' text which Ibn Suwār is making ('and further'). As before (see n. 42 above) the question needs further discussion.

45 'The halo' (*al-hāla*, الهالة): Ibn Suwār uses a different word in his treatise from the one used in the translation of Theophrastus' text. He explains that 'the term for halo in Greek and Syriac is derived from 'being circular' and 'surrounding' (من الاستدارة والاحاطة.).' (Lettinck 333).

46 فأما الطبيعى فان الذى يقوله ثاوفرسطوس فى ذلك هوهذا. (١) قال انه يحدث من ضوء النير فى الغمام الرقيق اللطيف حركة موجية مثل ما يعرض فى الماء عند قذف الحجر فيه فانه يحدث حول الحجر حركة مستديرة. (٢) وايضا فان الشعاع الخارج من القمر من شأنه ان يلطف الهواء المحاذى له فهو يلطف ما

This text is very close to (T5). The differences are minor: at the beginning the moon is replaced with "luminary" (النير) although القمر is also used in the proof. The analogies are more clearly connected with the parts of the physical account: the thrown stone illustrates the wave-like motion, i.e. the mechanism of generation of the halo, and the analogy of air blows through a tube accounts for the shape of the halo produced by this process of thinning the cloud. There is no explanation of the physical nature of thinning, but the description itself suggests that this is probably not a mechanical impact, but a change in the structure of the cloud.

There is a further striking parallel to Theophrastus' account in Seneca's *Natural Questions*. Theophrastus is not named and the physical account clearly mentions the mechanical impact made by light on air. In this respect it differs from what is attributed to Theophrastus in our Arabic sources, where we have the light "thinning" the air upon coming to contact with it. Still, Seneca does use at least one Theophrastean analogy in support of his account.

> (T5b) Seneca, *Natural Questions* 1.2.1–2, ll. 219–235 Hine (transl. Hine):
> (1) Now let us see how the brightness that encircles stars is formed. It is recorded that on the day when the deified Augustus returned from Apollonia and entered the city, a circle of various colours, such as is usual in the rainbow, was seen around the sun. The Greeks call this a halo; we can most appropriately talk of a garland. I shall explain how it is said to be formed. (2) (a) When a stone is dropped into a pool, we see the water spread out in many circles: first a very small circle is formed, then a wider one, and then other larger ones, until the movement fades away and dissolves into the smoothness of motionless water. (b) Let us think of something similar happening in the air too: when it becomes denser, it can experience a blow; when the light of the sun or the moon or any star encounters it, it forces it to recede in expanding circles. (c) For liquid, air, and everything whose form is altered by an impact is forced into a shape like that of the thing that exerts a force on it. Now, light is all round; therefore the air too will assume that form when struck by light.[47]

يحاذيه من الغمام الذى فيه يكون الهالة وكلما لطف الجزء المحاذى للنير غلظ ما حوله فيرى كالدائرة. (٣) ومثال ذلك ان ينفخ نافخ بانبوب ترابا موضوعا على تحت فنقى الموضع الذى ينفخ فيه ويجتمع الغبار المنكنس منه فى الماضع المحيط بالموضع النقى ويحدث دائرة.

47 (1) *Videamus nunc quemadmodum fiat is fulgor qui sidera circumuenit. memoriae proditum est quo die urbem diuus Augustus Apollonia reuersus intrauit circa solem uisum coloris uarii*

In T5b1 Seneca describes Octavian's arrival in Rome probably in April 44, soon after Caesar's murder: the appearance of a corona is mentioned in several Latin sources.[48] Hine points out in the annotation to his translation of Seneca's treatise that what the ancients called halo does not correspond to our halo; this is something the astronomers call "corona". He is certainly right as far as this passage is concerned. On the other hand, it seems that the term 'halo' could be used generically. Seneca's use of the example with several circles on the water in (T5b2a) and possibly also the one used by Theophrastus in (T5.1) seem to suggest multiple haloes. In Aratus (315–240 BCE), we find a description of simple, double and triple haloes, i.e. haloes consisting of one, two and three concentric circles respectively.[49]

In Seneca's account the Theophrastean analogy of a stone thrown into water (T5b2a) is followed in (T5b2b) by a more detailed description of the physical mechanism underlying the generation of the halo. This generation happens when denser air is forced to recede by the light of the sun or the moon. Seneca describes this affection as a blow (*plaga*), and explains further in (T5b2c) that it is the mechanical blow by which air is struck (*percussus*) that imprints on it the circular form. The imprinted form is circular because all light is round. Seneca here probably refers to the propagation of light on all sides from its source, the sun, the moon, or the bright star. When such a dynamic "sphere" of light strikes the dense air, the imprint left at their intersection is a circle. Seneca's account does not mention perceptible elemental qualities, such as involved in heating or cooling, not does it mention thinning of structure. The Theophrastean analogy is interpreted in the sense of a mechanical interaction between the active factor (the light) and the passive factor (the dense air). The light strikes the air and forces it to move away: the impact produces a circle. This explanation is based on the analogy of two striking bodies, but it is not itself meant to be an analogy. Both the light and the air are physical bodies, and thus must possess physical resistance, ἀντιτυπία. The circle produced by the impact delimits their new boundaries (which will last until dissolved). It is possible

---

circulum, qualis esse in arcu solet. hunc Graeci ἅλω uocant, nos dicere coronam aptissime possumus. quae quemadmodum fieri dicatur exponam. (2) (a) *cum in piscinam lapis missus est, uidemus in multos orbes aquam discedere et fieri primum angustissimum orbem, deinde laxiorem, ac deinde alios maiores, donce euanescat impetus et in planitiem inmotarum aquarum soluatur.* (b) *tale quiddam cogitemus fieri etiam in aëre: cum spissior factus est sentire plagam potest; lux solis aut lunae uel cuiuslibet sideris incurrens recedere illum in circulos cogit.* (c) *nam umor et aër et omne quod ex ictu formam accipit in talem habitum impellitur qualis est eius quod impellit. omne autem lumen rotundum est; ergo et aër in hunc modum lumine percussus exibit.*

48 See Toher (2004), on the halo episode, p. 179 and n. 15.

49 Arat. 811–817: Each kind has a special significance as a weather sign.

that Seneca's interpretation here depends on a Stoic source.[50] Even so, it may be a valuable witness to Theophrastus' account. In line with the Aristotelian tradition, Seneca tells us that a halo is formed not around the moon itself (as in Epicurus), but close to earth, and explains that this s because it is in proximity to earth that such dense air can be found as required for the formation of halo.[51]

One remarkable feature of all the reports about Theophrastus is a complete absence of geometrical optics in all these accounts. The physical accounts explain how the physical circle is created. They seem to leave no room for reflection theory.

4 Arius, Aëtius, Ḥunayn's Anonymus

In this section, I would like to consider three different doxographical reports of Aristotle's theory of the halo. All the three reports are post-Aristotelian and post-Theophrastean. The goal of this brief survey is to use these reports to see some characteristic (different) tendencies in later school developments. All three texts can help us to understand better the context of Alexander's argument against sight-streams. Moreover, some of these developments are indicated in Alexander's own doxographical remark which concludes his discussion of the halo. In some respect, these reports display agreement, but there are also differences and one significant discrepancy.

In the passage from Arius Didymus, we find a summary of the account close to what we find in Aristotle.

> (T6) Arius Didymus, fr. 14 Diels (Stob. *Ecl.* 1.30.1, 240.13–241.3):
>
> Aristotle. (1) Haloes and rainbows and mock sun and rods, and the rest formed in accordance with impressions are produced by the same cause (for all these are the reflection of the sight-stream), but differ in their locations and in respect of those [bodies] from which and how disposed the reflection happens. (2) Now, a halo appears around the sun and around the moon and generally around the bright stars, no less during the daytime than at night nor at midday than in the afternoon, but less frequently

50 Which, for all we know, may be Seneca himself. Williams (2012), chapter 1 for an instructive analysis of Seneca's authorial strategies in *NQ*.

51 *NQ* 2.1, 240 Hine: *non longe a terra fit talis effigies, quam uisus noster solita inbecillitate deceptus circa ipsum sidus putat positum. in uicinia autem stellarum ac solis nihil tale fieri potest, quia illic tenuis aether est. nam formae crasss demum spissique corporibus imprimi possunt, in subtilibus non habent ubi consistant aut haereant.* On Epicurus, see *Pyth.* 110–111 and Verde (2022), 233–235.

> in the morning and at sunset. (3) And it happens when through the surrounding air since it is misty their light shines forth so that the complete circle is seen. For the halo according to the impression is seen from all sides and spherical.[52]

Haloes, rainbows, mock suns and rods are all said to be phenomena of reflection of sight (T6.1): this is in full accordance with Aristotle. Next, in (T6.2) we have the description of the appearance of a halo around the stars and the conditions and circumstances of its appearance. In (T6.3), this description is supplied with some more physical details: a halo happens when the light of the stars is shining through the misty air. The expression ὥσθ' ὅλον ὁρᾶσθαι τὸν κύκλον refers back to τῆς μὲν οὖν ἅλω φαίνεται πολλάκις κύκλος ὅλος (Aristotle, *Meteor.* III 2, T3 above), but here the completeness of the circle seems to be connected with its appearance as a sphere that is seen from all sides, so the geometrical property is placed more clearly into a physical context.

Although there is no distinct physical explanation of the kind we find in Theophrastus, the wording in (T6.3) has some affinities with (T5.2). Perhaps more importantly, the connection between the appearance of the complete circle (geometrical property) and the physical condition (light shining through the misty air) is articulated slightly differently from the way it is in Aristotle and separated from the "optical" account. There is a more articulated connection between the geometrical properties of reflection and the physical mechanism that accounts for their appearance. This might be an indication that according to the school tenet presented here the Aristotle-style "optical" account of the halo was regarded as in some way compatible with the physical account.[53]

52 Ἀριστοτέλους. (1) ἅλω δὲ καὶ ἴριδας καὶ παρήλιον καὶ ῥάβδους καὶ τἄλλα τὰ κατὰ τὰς ἐμφάσεις ὑπὸ μὲν τῆς αὐτῆς αἰτίας γίνεσθαι (πάντα γὰρ εἶναι ταῦτα τῆς ὄψεως ἀνάκλασιν), διαφέρειν δὲ τοῖς τόποις καὶ ἀφ' ὧν καὶ ὡς ἔχει συμβαίνειν τὴν ἀνάκλασιν. (2) ἅλω μὲν οὖν καὶ περὶ ἥλιον φαίνεσθαι καὶ περὶ σελήνην καὶ καθόλου περὶ τὰ λαμπρὰ τῶν ἄστρων καὶ οὐθὲν ἧττον ἡμέρας ἢ νυκτὸς καὶ περὶ μεσημβρίαν ἢ δείλην· ἕωθεν δ' ἐλαττονάκις καὶ περὶ δύσιν. (3) συμβαίνειν δ', ὅταν διὰ τοῦ πέριξ ἀέρος ἀχλυώδους ὄντος ἐπιλάμπηται τὸ φῶς αὐτῶν, ὥσθ' ὅλον ὁρᾶσθαι τὸν κύκλον· περιφανῆ γὰρ εἶναι καὶ σφαιρικὴν κατὰ τὴν ἔμφασιν τὴν ἅλω.

53 Steinmetz 199–200 sees (T6.3) as parallel to the report in Ibn Suwār (while admitting that the main part of Theophrastus' account is missing in Arius). The interpretation of (T6.3) depends on how we take the role of the light shining through the mist. If we take it in a stronger causal sense, then the light may be responsible for the production of a halo, prior to being seen by anyone. This would be close to Theophrastus. But the weaker reading of this sentence is available: the sunlight is a necessary condition for the appearance of halo (ὥσθ' ὅλον ὁρᾶσθαι τὸν κύκλον): in this case, while the physical context is acknow-

In Aëtius, we have a rather different version of the story of the halo.[54]

> (T7) Aëtius, 3.5a Mansfeld/Runia (was 3.18 Diels):
> On the Halo. The halo is formed as follows: between the moon or some other star and the sight-stream thick and misty air is set, next *because the sight-stream is bent and widened in it and then in this way it falls upon the circle of the star at its external circumference*, a circle seems to appear around the star (which circle is called 'halo' because it looks like a [round] threshing floor), the appearance seems to be in the place where the affection of the sight happens.[55]

The account that is reported here treats the halo as an appearance, but the explanation of the way this appearance is produced differs significantly from the reflection theory we've seen in Aristotle and Arius Didymus. Whereas in these latter accounts, the common genus of the explanation was expressed by the term ἀνάκλασις, here it seems this central explanatory role belongs to κατάκλασις.

In Aëtius, as in Arius who follows Aristotle, a halo is said to be produced when the thick and misty air sets in, but the role of this thick mist is different in the two cases (of ἀνάκλασις vs. κατάκλασις). Whereas for Aristotle mist consists of numerous tiny bits of frozen water each of which is a micro-mirror reflecting only the brightness of the source of light and which form a continuum so that these reflections together appear as a regular circle, in this passage the whole body of mist seems to play a role similar to that of a magnifying glass: the sight-stream is said to be "bent" and "widened" in it. This corresponds to the mechanism of refraction rather than reflection.

The concept of refraction is studied systematically and in detail in Ptolemy's *Optics* 5. Ptolemy distinguishes between two ways in which the visual ray is broken:

> One involves rebound and is caused by reflection from bodies that block the [visual ray's] passage and that are included under the single heading

ledged, halo is still an appearance that requires an observer. Moraux (1973) does not see any Theophrastean features in this passage (Moraux (1973), 294–295).

54 I use the new edition of Mansfeld and Runia who have moved this passage from the end of section to the place after the rainbow, thus making it 3.5a, with good reasons.

55 Περὶ ἅλω. ἡ δ' ἅλως οὑτωσὶ ἀποτελεῖται· μεταξὺ τῆς σελήνης ἤ τινος ἄλλου ἄστρου καὶ τῆς ὄψεως ἀὴρ παχὺς καὶ ὁμιχλώδης ἵσταται· εἶτ' ἐν τούτῳ τῆς ὄψεως κατακλωμένης καὶ εὐρυνομένης κᾆθ' οὕτω τῷ κύκλῳ τοῦ ἄστρου προσπιπτούσης κατὰ τὴν ἔξω περιφέρειαν, κύκλος δοκεῖ περὶ τὸ ἄστρον φαίνεσθαι (ὃς κὺκλος φαινόμενος ἅλως καλεῖται, ὅτι ἐστὶν ἅλῳ προσεοικώς), ἐκεῖ δοκοῦντος τοῦ φάσματος γὶνεσθαι, ἔνθα συνέπεσε τὸ πάθος τῆς ὄψεως.

> of "mirrors". The other way, however, involves penetration and is caused by a deflection in media that do not [completely] block the [visual ray's] passage, and those media are included under the single heading "transparent". (Transl. A. Smith)

The explanation of haloes in our passage relies on a physical theory of sight-stream which is said to be bent and widened and in this way fall upon the circle of the star, i.e. at the heavenly orbit with which the star rotates. Because these sight-streams are now refracted at a wider angle the circle which they settle on is also greater than the disc of the star. In this way the circle of the halo is produced by our physical sight-streams.

Ptolemy's *Optics* has reached us in incomplete state: the whole of its first book and a significant part of its last, fifth book (where refraction is discussed) are lost. We do not seem to have any evidence of his treatment of the halo specifically.[56] Some of the effects of refraction which Ptolemy discusses in the extant part of book V are known before him.[57] But the report in Aëtius seems to rely on a more systematic concept of refraction.[58]

It is remarkable that this view is attributed to Aristotle in our source.[59] As we can see, no role of reflection is mentioned, and the role of the sight-streams in the production of the image is different from the one presupposed by the reflection account. This attribution to Aristotle is also supported by the appraisal of the state of the discussion by Alexander of Aphrodisias who seems to think this explanation of the halo is more popular than the one which he himself defends as Aristotelian (T12 below).

One more text that should be mentioned as a further witness to the Peripatetic theory of the halo is from the anonymous Arabic compendium of Aris-

56 Olympiodorus mentions that on Ptolemy's view rainbow has seven colours (*In Meteor.* 242.7), and it is possible (although not necessary, given the shortness of report) that his source is the lost part of the *Optics*. Ptolemy does mention halo in his *Tetrabiblos* 2.14.6–7, but in a way that sheds no light on the way it is produced.

57 Cleomedes *Meteor.* 2.6. 178, see also Bowen &Todd (2004), 101, n. 1.

58 Cf. also Aëtius, 3.5, where prior to the discussion of the rainbow ιν 3.5.3, we have a distinction between the two types of meteorological phenomena—καθ' ἔμφασιν and καθ' ὑπόστασιν, well as the discussion of the types of geometrical lines along which we see: straight, curved or bent, spelling out that these lines are ἀδήλους λόγῳ θεωρητάς καὶ ἀσωμάτους. Lines are thus distinguished from the ὄψεις which are physical.

59 It is remarkable in this case (where there is no evidence of Aristotle subscribing to any such view), but it is not an unusual practice within the school to attribute the dominant school view to its founder. Some discussion of this, in relation to the Aristotelian school after Aristotle, in Sharples (2002).

totelian meteorology which Daiber argues can be attributed to Ḥunayn ibn Isḥāq, a ninth century Baghdādī scholar.[60]

> (T8) Ḥunayn ibn Isḥāq, *Arabic compendium of Aristotelian meteorology* (Daiber (1975)):
>
> The account of the ring around the sun and the moon and around the radiant stars
>
> This ring is seen more at night than during the day. It is seen sometimes during the day, more at midday and in the evening. It is rarely seen at the sunrise and at the sunset. (2) The cause of all these rings is one, namely that when a lot of moist vapour builds up in the air, and the sun and the moon and the shining stars light upon it, their light is spread in the air, then *that light is bent back from the air to this moist vapour and that ring is seen in this way*. (3) And the appearance of this ring is a sign of the occurrence of rain because it comes to be from the moisture of the air. In the same way as its appearance is the sign of the rain, so its disappearance and dissolution is a sign of good weather and the blowing of winds: it (the disappearance of the ring) indicates the arrival of clear weather because it is a sign of dry air, and it indicates the blowing of wind because what is released by this moisture is the hot and dry vapour which is the matter of the wind.[61]

As Daiber remarks, this text (as the whole work) is generally very close to Aristotle's account in *Meteorology*. The halo, the ring around the sun, or the moon, or any of the shining stars, seems to be explained on the basis of reflection. The major difference is that what is reflected is the light, and it is said to be

60 Daiber (1975), 1–17.

61 القول فى الدائرة التى حول الشمس والقمر و حول الكواكب ذوات الشعاع

(١) وهذه الدائرة ترى باليل أكثر مما ترى بالنهار وقد ترى بالنهار أحياناً وأكثر ذلك نصف النهار وبالعشاء. فأما عند طلوع الشمس وعند غروبها فكلما ترى. (٢) وعلة هذه الدائرات كلها واحدة لأن البخار الرطب إذا كثر في الجو فأشرقت الشمس والقمر والكواكب النيرة فيه سطع نورها الهواء ثم عُطِفَ ذلك النور راجعاً من الهواء على ذلك البخار الرطب فترى تلك الدائرة كذلك. ورؤية هذه الدائرة دالّة على كون المطر لأنها انما كانت من رطوبة الهواء وكما ان رؤيتها دالّة على كينونة المطر وكذلك اضمحلا وتحللها دالّ على الصحو وهبوب الريح. (٣) فأما دلالته على حدوث الصحو فلأنه دالّ على يبس الهواء وأما دلالته على هبوب الريح فلأن المحلل لتلك الرطوبة إنما هو البخار الحار اليابس الذي هو مادة الريح.

reflected from the mist (thick air). This account of reflection would fit in with the account of the physical structure underlying the halo that we find in Theophrastus. This text contains no description of the way we get to see this reflection, there is no role for the sight-streams either here or in the account of the rainbow that follows. So, if this text comes from the later Aristotelian school, it seems to have undergone a redaction. (As Daiber points out, it is unlikely that Ḥunayn is responsible for the redaction, the text seems to be a faithful translation of its Greek source.)

Perhaps we can take stock of the three doxographical sources and what they can tell us about the discussion of the halo after Aristotle and Theophrastus. Arius gives us the account that is based on the reflection of the sight-streams which is perhaps regarded as compatible with the physical explanation given by Theophrastus. Aëtius gives us a novel explanation, found neither in Aristotle nor in Theophrastus, based on the theory of refraction (κατάκλασις, bending) of the sight-streams.[62] Daiber's Peripatetic source gives us an account of reflection of light which says nothing about the role of the sight-streams: the light that is reflected "from the air onto the vapour" is the light of the sun, the moon, or the bright stars, as in the title. This account might owe something to Alexander or his circle.

5 Alexander's Account of the Halo

Prior to the discussion of optical phenomena, Alexander explains Aristotle's use of the language of sight-streams as following a scholarly convention which does not affect, and should not be mistaken for, his true theoretical commitment regarding the nature of sight. Alexander has no doubt that Aristotle is committed to the *De anima* theory of sight at the time of writing the *Meteorology*.

This exegetical assumption is in full accordance with his method of interpretation which has little room for a developmental approach.[63] All the texts of Aristotle should be studied as stating his considered views. If there seems to be a conflict between two different positions defended by Aristotle in different works, the case should be examined carefully to see if the conflict is real or merely apparent.

62 It would be tempting to connect it with Ptolemy's influence, but there is not enough material for this.

63 Sharples (2010).

In the following text he states briefly the arguments against the physical theory of sight-streams and formulates Aristotle's true views on these points of controversy.

(T9) Alexander, *In Meteor.* 141.3–20:

(1) He does not favour the view according to which we see by the outflow of the rays, the way the mathematicians teach, the view which the teaching follows which claims that it is by the reflection of these very rays that we see all things seen as reflected images. **(2)(a)** For neither certain rays flowing out from our organ of sight and falling upon the object of sight can be the causes of our sight, **(b)** nor yet can these same rays reflected towards the object of sight from the mirrors and all things seen through them provide a cause for us to see these things insofar as the reflection of the rays is at equal angles. **(3)** For that has been proven to be impossible in *On how we see* which he himself also mentioned in *On the Soul*. **(4)** **(on 2a)** For he thinks that things seen along a straight line are seen not by means of rays, but because the medium between the object seen and the sight being transparent is affected by the colour seen and transmits this affection to the sight, which itself is also transparent. For this is what it is to be transparent.**(5)** **(on 2b)** Those things that are believed to be seen by reflection are seen because (a) first in the mirrors and all the reflecting surfaces the reflection of the object seen comes about by the transparent medium, (b) and then from this it travels back to the sight again through the transparent medium between the mirror and the sight.[64]

64 (1) Οὐκ ἀρέσκεται μὲν τῇ δόξῃ τῇ δι᾽ ἀκτίνων ἐκχύσεως ὁρᾶν ἡμᾶς λεγούσῃ, καθά φασιν οἱ ἀπὸ τῶν μαθημάτων, ᾗ δόξῃ ἀκόλουθός ἐστι καὶ ἡ δι᾽ ἀνακλάσεως τῶν αὐτῶν τούτων ἀκτίνων ὁρᾶν ἡμᾶς πάντα τὰ δι᾽ ἐμφάσεως ὁρώμενα λέγουσα· (2) (a) οὔτε γὰρ ἀπὸ τῆς ὄψεως ἀκτῖνάς τινας ἐκχεομένας καὶ προσπιπτούσας τῷ ὁρατῷ τοῦ ὁρᾶν αἰτίας ἡμῖν οἷόν τε γίνεσθαι (b) οὔτε τὰς αὐτὰς ταύτας ἀνακλωμένας ἀπὸ τῶν κατόπτρων καὶ πάντων τῶν διὰ τοιούτων ὁρωμένων ἐπὶ τὸ ὁρώμενον τὴν αἰτίαν πάλιν ἡμῖν παρέχειν τοῦ ταῦτα ὁρᾶν, τῆς τῶν ἀκτίνων ἀνακλάσεως πρὸς ἴσας γινομένης γωνίας. (3) ὅτι γὰρ ἀδύνατα ταῦτα, ἐδείχθη ἐν τοῖς Περὶ τοῦ πῶς ὁρῶμεν, ὧν καὶ αὐτὸς ἐν τοῖς Περὶ ψυχῆς ἐμνημόνευσε. (4) δοκεῖ γὰρ αὐτῷ τὰ μὲν ἐπ᾽ εὐθείας ὁρώμενα ὁρᾶσθαι οὐ δι᾽ ἀκτίνων, ἀλλὰ τῷ τὸ μεταξὺ τοῦ τε ὁρωμένου καὶ τῆς ὄψεως διαφανὲς ὑπὸ τοῦ ὁρωμένου χρώματος πάσχειν καὶ μεταδιδόναι τούτου τοῦ πάθους τῇ ὄψει, οὔσῃ καὶ αὐτῇ διαφανεῖ· τοῦτο γὰρ εἶναι τὸ εἶναι διαφανεῖ· (5) τὰ δὲ κατὰ ἀνάκλασιν ὁρᾶσθαι δοκοῦντα ὁρᾶσθαι τῷ πρῶτον ἐν τοῖς ἐνόπτροις τε καὶ ἐνοπτρικοῖς πᾶσι τὴν τοῦ ὁρωμένου ἔμφασιν γινομένην διὰ τοῦ μεταξὺ διαφανοῦς εἶτα ἀπὸ τούτου πάλιν ἥκειν ἐπὶ τὴν ὄψιν διὰ τοῦ διαφανοῦς πάλιν τοῦ μεταξὺ τοῦ τε ἐνόπτρου καὶ τῆς ὄψεως.

Aristotle is not committed to the theory of sight-streams: it is wrong to believe that the cause of our ability to see is in some rays that flow out from our eyes to the object (T9.2a), nor yet can the reflection of these rays from a mirror explain how we see the mirror images (T9.2b). On Aristotle's view, we see an object because the colour of this object affects the transparent medium between the object and our sight and transmits the affection to the sight (i.e. the organ of sight) which is itself also transparent (T9.4). This is the view presented in *De anima* and the other work whose title Alexander gives as "On how we see". Thus the sight-streams mentioned in the *Meteorology* do not produce sight. Nor do they produce the mirror images: the images are formed when the object is reflected in the mirror (T9.5a) and is thus also a quasi-standard object of sight ('quasi' because its presence depends on the presence of the thing reflected),[65] and the rest of the process can be described in the standard terms of acting upon the transparent medium which transmits this reflected object to our sight (T9.5b). Thus the theory of sight-streams is a convenient way of presenting what can be visible to an observer who is located at a certain point.

Alexander thinks that the idiom of sight-streams can be translated into the correct terms of senses acted upon by their proper sense objects, without any loss.

> (T9a) Alexander, *In Meteor.* 141.20–30:
>
> But as far as the argument is concerned, it makes no difference whether to say (a) that the sight-stream reflected from a mirror at equal angles, when the visible object happens under such reflection, sees it by falling upon it, or (b) that the visible object itself because of a certain position relatively to the mirror is reflected in it through the affected transparent medium, and this mirror is such that not only can it be affected by the colour so as to transmit a quality from it to another transparent, but is capable of preserving the reflection in virtue of its smoothness and brightness, so that the transparent medium between it and the sight is affected by it again as by a coloured object. So Aristotle uses the view of the rays as it has won a common currency and is favoured by the mathematicians.[66]

65 On Alexander on mirror images, see Ierodiakonou (2023).

66 ἐπεὶ δὲ οὐδὲν ὅσον ἐπὶ τῷ λόγῳ διαφέρει ἢ (α) τὴν ὄψιν λέγειν ἀπὸ τοῦ κατόπτρου ἀνακλωμένην πρὸς ἴσας γωνίας, ὅταν ὑπὸ τὴν τοιαύτην ἀνάκλασιν τύχῃ τὸ ὁρατὸν ὄν, προσπίπτουσαν αὐτῷ ὁρᾶν αὐτό, ἢ (β) αὐτὸ τὸ ὁρατὸν διὰ τὴν ποιὰν σχέσιν πρὸς τὸ κάτοπτρον διὰ τοῦ μεταξὺ διαφανοῦς πάσχοντος ἐμφαινόμενον ἐν ἐκείνῳ, ὄντι τοιούτῳ [διαφανεῖ], ὡς μὴ μόνον πάσχειν ὑπὸ τοῦ χρώματος δύνασθαι οὕτως, ὡς διαδιδόναι τὴν ἀπ᾽ αὐτοῦ ποιότητα ἄλλῳ διαφανεῖ, ἀλλὰ καὶ φυλάσσειν

Here Alexander is using his own text as an illustration of his point showing how much easier it is to express the argument using the sight-streams language than to state the same things in a manner that is absolutely correct from a physical point of view. The second sentence (b) is three times longer than the first (a) and perhaps deliberately tortuous syntactically, so as to make the rhetorical argument particularly convincing.

But as we have seen above, Aristotle sometimes does speak as though the sight-streams had physical reality. In several passages (e.g. see T2.3 and T2.3a above), Aristotle refers to the "weakness" of sight-streams as their inability to pass through a dense medium. In his commentary on *Meteorology* III 4, Alexander gives a detailed explanation of the case of a man whose sight-streams were so weak that they were reflected from ordinary surrounding air and he was seeing his own image any time when he was walking (T2.3a above). Alexander says that "elsewhere" (ἐν ἄλλοις) Aristotle says who this was, a certain Antipheron of Oreus. Antipheron of Oreus is mentioned in *On Memory* 1 in a context which does not have to do with the weakness of sight. Antipheron is cited there as an example of someone contemplating what is not a copy as though it is a copy and is said to be one of the deranged people who talk about their images as though they really occurred and they remembered them.[67] Unless there is any further evidence about this Antipheron which has not come down to us, Alexander's interpretation of (T2.3a) is based on this passage from *On Memory*.[68] On the other hand, Olympiodorus in his commentary on this passage in *Meteor.* III 4 tells this story of Antipheron (whom he says to be not an Oreite, but a Tarentine) as a genuine example of weak sight-streams.[69] Alexander has a medical interpretation of this case:

> (T10) Alexander, *In Meteor.* 148.5–30:
>
> (1) But as we are saying that it is impossible for a sight-stream to be reflected, what was the affection that befell Antipheron? (2) Or as to those who have some scar around the pupil either from a heavy ophthalmia or

δυναμένῳ τὴν ἔμφασιν διὰ λειότητά τε καὶ στιλπνότητα, ὡς ἀπ᾽ αὐτοῦ πάλιν τὸ μεταξὺ αὐτοῦ τε καὶ τῆς ὄψεως διαφανὲς πάσχειν ὡς ἀπὸ κεχρωσμένου, τῇ δόξῃ τῇ τῶν ἀκτίνων καθωμιλημένῃ τε οὔσῃ καὶ τοῖς μαθηματικοῖς ἀρεσκούσῃ προσχρῆται.

67 *Mem.* 1, 451a8–12: γίγνεται δὲ καὶ τοὐναντίον, οἷον συνέβη Ἀντιφέροντι τῷ Ὠρείτῃ καὶ ἄλλοις ἐξισταμένοις· τὰ γὰρ φαντάσματα ἔλεγον ὡς γενόμενα καὶ ὡς μνημονεύοντες. τοῦτο δὲ γίγνεται ὅταν τις τὴν μὴ εἰκόνα ὡς εἰκόνα θεωρῇ.

68 In the commentary on *Parva naturalia* by Michael of Ephesus, the example of Antipheron is further illustrated with the material from *Meteorology* III 4. Michael's commentary owes much to Alexander's lost commentary on these treatises.

69 *In Meteor.* 230.14–18 Stüve.

some injury, and indeed to those who are going to have a cataract, some little mosquito seems to fly before their eyes, since they see what is inside the pupil itself and what is inflicted contrary to nature as what is outside, because of the habit of perception. (3) For it is used to see things that are outside, so it readily sees all the things it sees as being outside. For in a way even these things are external to it because they are imposed upon it against nature. (4) It is a sign that the impression of a little mosquito comes from some such affection that those who see it most of all see it at such a time when some exhalation is brought up from the mouth of the stomach from the undigested remainders in it. For the veins around the pupil as they are filled with greasy and unprocessed residue act against the transparency of the sense organ, and the sight sees that which overshadows it as being before the eyes and outside. (5) The same affection happens when some cloud-like mass is formed around the pupil which although not yet sufficient to block it up and prevent it from seeing, yet already has the density of the kind that usually comes about ahead of the future cataracts. For to those who are in this condition light makes visible as one of the external things, along with the objects of sight that are external, such an affection and that which is imposed in this way. (6) As to those people there something like a little mosquito appears to float in air, sometimes with wings, when the veins are blocked, and sometimes without such an addition, when their fullness is less, in the same way to Antipheron, who sees a little and not sharply, because some greater mass has been spread over his pupils, something greater seemed to go before in front of his sight, which he likened to a face, as those discussed above to a mosquito, experiencing some affection similar to those who liken the cloud masses with their excrescences, on the basis of a small similarity, to Satyrs or some animal forms.[70]

---

70 (1) ἀλλ᾽ ἐπεὶ μὴ οἷόν τε τὴν ὄψιν ἀνακλᾶσθαί φαμεν, ἐπιζητήσειέν τις ἄν, τί ἦν τὸ γενόμενον πάθος περὶ τὸν Ἀντιφέροντα; (2) ἢ ὥσπερ τοῖς οὐλήν τινα περὶ τὴν κόρην ἔχουσιν ἢ διὰ βαρεῖαν ὀφθαλμίαν ἢ διὰ τρῶσίν τινα, καὶ μέντοι τοῖς ὑποχεῖσθαι μέλλουσι κωνώπιόν τι πρὸ τῶν ὀφθαλμῶν ἵπτασθαι δοκεῖ, τὸ ἐν αὐτῇ τῇ κόρῃ ὂν καὶ ἐπικείμενον παρὰ φύσιν ὡς ἔξω ὂν ὁρῶσι διὰ τὴν συνήθειαν τῆς αἰσθήσεως. (3) σύνηθες μὲν γὰρ αὐτῇ τὰ ἔξω ὁρᾶν, ἤδη δὲ καὶ πάντα τὰ ὁρώμενα ὡς ἔξω ὄντα ὁρᾷ. τρόπον γάρ τινα καὶ ταῦτα ἔξω ἐστὶν αὐτῆς διὰ τὸ ἔξωθεν προσκεῖσθαι παρὰ φύσιν. (4) σημεῖον δὲ τοῦ ἀπὸ τοιούτου τινὸς πάθους κωνωπίου γίνεσθαι φαντασίαν τὸ τότε μάλιστα τοὺς ὁρῶντας αὐτὸ ὁρᾶν, ὅταν ἐκ τοῦ στόματος τῆς κοιλίας ἀνενεχθῇ τις ἀναθυμίασις ἀπὸ τῶν ἐν αὐτῇ καταλειφθέντων ἀπέπτων· πληρούμενα γὰρ τὰ περὶ τὰς κόρας φλεβία ῥυπαρᾶς καὶ ἀκατεργάστου περιττώσεως ἀντιπράττει τῇ τοῦ αἰσθητηρίου διαφανείᾳ, καὶ τὸ κατασκιάζον αὐτὴν τοῦτο ἡ ὄψις ὡς πρὸ ὀμμάτων ὂν καὶ ἔξω ὁρᾷ. (5) γίνεται δὲ τὸ αὐτὸ πάθος καὶ νεφελοειδοῦς τινος συστάσεως συνισταμένης περὶ τὴν κόρην οὔπω μὲν ἱκανῆς πρὸς τὸ ἐπιφράττειν καὶ κωλύειν αὐτὴν

In this long passage Alexander deals with the physical interpretation of sight-streams suggested by Aristotle's use of the notion of "weakness of sight-streams". He argues that the experiences that might be interpreted as supporting the literal reading of sight-streams can be explained more naturally as some processes in our body—either in the organ of sight or in some other organs whose malfunctioning can impede the normal operation of these organs. (T10.2) Thus, people whose organ of sight has a scar, or some morbid formation that is in the visual field, will see these structures as external, for such is the "habit" of the sense of sight (T10.3). People who are likely to develop a cataract, prior to that may see a little mosquito floating in front of their eyes. This experience is more likely to occur after meal, it seems, when the exhalation from the mouth of the stomach is brought up and enters the veins around the pupil (T10.4). The expression "mouth of the stomach" is used by Aristotle in *APo.* II 11, in the description of the use of walking after meal (an example of the final cause).[71] By using this expression and by invoking this context Alexander wants to indicate that it is a good Aristotelian explanation that he is offering. In (T10.5–6) Alexander discusses the case when the morbid structure around the organ of sight is greater in size: this will produce the same affection as the one in the case of floating mosquitoes perceived as external objects, but the size of the permanent appearance of this sort will be greater because of the greater affliction of the sight organ. In this way Alexander leads on to the case of a man who saw himself while walking: he must have had some such illness, namely a large morbid formation obscuring sight, and this is what it means to say that his sight-stream is weak. In the end, we seem to have a reference to the context of *On Memory* 1: those who see such things and interpret them as what they are not (cf. contemplating non-copies as copies) are like people who see different creatures in the shapes of the clouds. There are no physical sight-streams. Aristotle beats Aristotle.

ὁρᾶν, ἤδη δὲ πάχος ἐχούσης, ὡς πρὸ τῶν μελλουσῶν ὑποχύσεων γίνεσθαι φιλεῖ. τὸ γὰρ φῶς τοῖς οὕτως ἔχουσι σὺν τοῖς ἔξωθεν ὁρατοῖς καὶ τὸ τοιοῦτον πάθος καὶ τὸ οὕτως προσκείμενον ὁρατὸν ὥς τι τῶν ἔξω ποιεῖ. (6) ὡς οὖν ἐκείνοις ἐοικός τι φαίνεται κωνωπίῳ πρὸ τῆς ὄψεως αἰωρεῖσθαι ποτὲ μὲν ἔχοντι πτερά, ὅταν ᾖ διωγκωμένα τὰ φλεβία, ποτὲ δὲ χωρὶς τοιαύτης προσθήκης, ὅταν ἐλάττων ἡ περὶ αὐτὰ πλήρωσις ᾖ, οὕτως τῷ Ἀντιφέροντι μικρὸν καὶ οὐκ ὀξὺ ὁρῶντι διὰ τὸ πλείω τινὰ σύστασιν ἐπεσκεδάσθαι ταῖς κόραις μεῖζόν τι πρὸ τῆς ὄψεως ἐδόκει προηγεῖσθαι, ὃ εἴκαζε προσώπῳ, ὡς οἱ προειρημένοι κώνωπι, ὅμοιόν τι πάσχοντες τοῖς τὰς τῶν νεφῶν συστάσεις τε καὶ ἐξοχὰς εἰκάζουσιν ἀπὸ μικρᾶς ὁμοιότητος Σατύροις ἢ θηρίων τινῶν μορφαῖς.

71 *APo.* II 11, 94b14–16: διὰ τί δὲ ἀπὸ δείπνου δεῖ περιπατεῖν, καὶ ἕνεκα τίνος δεῖ, οὐδὲν διαφέρει. περίπατος ἀπὸ δείπνου Γ, τὸ μὴ ἐπιπολάζειν τὰ σιτία ἐφ᾽ οὗ Β, τὸ ὑγιαίνειν ἐφ᾽ οὗ Α. ἔστω δὴ τῷ ἀπὸ δείπνου περιπατεῖν ὑπάρχον τὸ ποιεῖν μὴ ἐπιπολάζειν τὰ σιτία πρὸς τῷ στόματι τῆς κοιλίας, καὶ τοῦτο ὑγιεινόν.

It remains for us to see how Alexander explains the halo. In T11, Alexander seems to incorporate the physical explanation of the halo similar (but not identical) to that of Theophrastus into what looks at first glance as a simple presentation of Aristotle's view.[72]

(T11) Alexander, *In Meteor.* 142.21–143.7:

(1) And first he makes an argument about the halo and inquires about its shape, why it is a circle, and why it comes about around the sun itself, or the moon, or some other of the stars at which it is seen and not opposite, say, or sideways. (2) And he says that the reflection of the sight-stream appears in them from the vapour that is condensed into a cloud after being raised up, if the structure happens to be settled as smooth and consisting of small parts. (3) For since [this vapour] is such, if any of the bright stars happens to be carried above it, or the sun, or the moon, that which is perpendicular to the body moving above must be dissolved and become looser and finer in texture, such a motion proceeding in it up to a point, and thereupon it is not affected by anything of the kind, but remains as it is. (4) And since such a process takes place evenly on every side at the perpendicular because of the smoothness of the cloud, the limit of the properly dissolved vapour must be a circle. (5) The part beyond it is more condensed, but being smooth and made of small parts has continuous small mirrors along the circumference of the circle: (a) *according to those who use the explanation by the reflection of the sight-stream*, the sight-stream falls upon these [small mirrors] and being reflected towards the star produces the impression of a halo, (b) while in truth the light of the star falls upon the mirrors mentioned above and then being reflected upon and transmitted to the sight that is seeing produces the impression of a halo. For in the mentioned mirrors because of their smallness only colour is reflected.[73]

---

72 For instance, Boyer sees Alexander as simply rewriting Aristotle's explanation at this point, (Boyer (1959)/(1987), 63–64).

73 (1) Καὶ πρῶτόν γε περὶ τῆς ἅλω ποιεῖται λόγον καὶ ζητεῖ περί τε τοῦ σχήματος αὐτῆς διότι κύκλος, καὶ διὰ τί περὶ αὐτὸν τὸν ἥλιον ἢ τὴν σελήνην ἤ τι ἄλλο τῶν ἄστρων περὶ ὃ ὁρᾶται γίνεται, ἀλλ᾽ οὐχὶ καταντικρὺ φέρε εἰπεῖν ἢ ἐκ πλαγίων. (2) φαίνεσθαι δή φησι τὴν ἀνάκλασιν τῆς ὄψεως ἐν τούτοις ἀπὸ τῆς συνισταμένης ἀτμίδος εἰς νέφος μετὰ τὸ ἀνενεχθῆναι, ἂν ἡ σύστασις αὐτῆς ὁμαλής τε καὶ μικρομερὴς συνισταμένη τύχῃ. (3) τοιαύτης γὰρ οὔσης αὐτῆς, ἂν ὑπερφερόμενον αὐτῆς ἄστρον τι τῶν λαμπρῶν ἢ ὁ ἥλιος ἢ ἡ σελήνη τύχῃ, τὸ μὲν κατὰ τὴν κάθετον τοῦ ὑπερφερομένου σώματος ἀνάγκη διακρινόμενον μανότερόν τε καὶ λεπτότερον γίνεσθαι, μέχρι τινὸς τῆς τοιαύτης προϊούσης ἐν αὐτῇ κινήσεως, τὸ δ᾽ ἐντεῦθεν οὐδὲν τοιοῦτον πάσχει, μένει δὲ οἷόνπερ ἐστί. (4) ἐξ ἴσου δὲ πάντῃ περὶ τὴν κάθετον τοῦ τοιούτου συμβαίνοντος διὰ τὴν ὁμαλότητα τοῦ νέφους, κύκλον

In (T11.1), Alexander restates Aristotle's main questions to be answered by the demonstration. In (T11.2) we find a restatement of the physical conditions necessary for the halo to be seen: a smooth cloud made up of many reflecting surfaces. But next, in sentence (T11.3), we have a physical account of the kind that we do not find in Aristotle. Alexander says that the star running above the cloud dissolves it and makes it thinner in the region opposite the star at the perpendicular. (T11.4) Where the mist is dissolved, there is no reflection (and thus the dark area has a proper physical explanation which can then be strengthened by an optical contrast). After that threshold point, the mist remains thick and reflecting. Because this process is centrally symmetrical, the limit of the dissolved vapour forms a circle. In (T11.5) Alexander goes on to explain the formation of the ring: it is at the rim of the dissolved area, it is more condensed, made up of many tiny mirrors arranged continuously along the circumference of a circle. Alexander gives two versions of the explanation of reflection, (T11.5a) in terms of the sight-streams, where our sight-streams fall upon the tiny mirrors, are reflected towards the sun, and thus the image of the halo is produced, and (T11.5b) the correct one, where the light falls on the mirrors, then is reflected towards our sight and transmitted to it and thus producing an impression of the halo. On Alexander's view, either of these two explanations would work as an interpretation of the geometrical demonstration provided by Aristotle.

In this whole passage (T11), Alexander seems to deal with a problem similar to the one which Theophrastus tried to resolve, and his solution is similar to the Theophrastean solution at least as far as its principle is concerned.

Unlike Theophrastus, Alexander does not draw any analogies: he explains the process more directly using the terminology of dissolution, thinning, being condensed. He does not name the cause of dissolution, although it is natural to assume that it is the heat produced by the stars. Seeing the halo thus produced can be explained on the basis of the theory of sight in the *De anima*. It seems that Alexander came up with a unified account of the halo that includes the description of its physical mechanism (which contains some elements of

ἀναγκαῖον γίνεσθαι τὸ πέρας τῆς ἐπιεικῶς διακεκριμένης ἀτμίδος. (5) ἀφ᾽ οὗ τὸ ἐπέκεινα πάλιν μᾶλλον μὲν συνέστηκεν, ὁμαλὲς δ᾽ ὂν καὶ μικρομερὲς συνεχῆ καὶ μικρὰ κάτοπτρα κατὰ τὴν τοῦ κύκλου περιφέρειαν ἴσχει, πρὸς ἃ (α) κατὰ μὲν τοὺς τὴν τῆς ὄψεως ἀνάκλασιν αἰτιωμένους ἡ ὄψις προσπεσοῦσα ἀνακλασθεῖσα πρὸς τὸ ἄστρον τὴν τῆς ἅλω φαντασίαν ποιεῖ, (β) κατὰ δὲ τὸ ἀληθὲς τὸ φῶς τὸ τοῦ ἄστρου προσπῖπτον τοῖς προειρημένοις κατόπτροις, ἔπειτα ἐπὶ τὴν ὄψιν τὴν ὁρῶσαν ἀνακλασθέν τε καὶ διαδοθὲν ποιεῖ τὴν φαντασίαν τῆς ἅλω. ἐν γὰρ τοῖς εἰρημένοις κατόπτροις διὰ μικρότητα, τοῦ χρώματος ἔμφασις γίνεται μόνου.

geometrical explanation) and the optical account, where Alexander's preferred version is that of the reflection of light from air, water, and smooth surfaces. *De anima* and *Meteorology* are in this way reconciled.

But of course, things are not that simple. While Alexander's theory explains the annular shape of the halo (this explanation is not clearly present in Aristotle), it cannot explain why a halo can be seen by many observers from different places. His explanation still presupposes that the hole in the cloud is made at the intersection of the line drawn between the eye of the observer and the centre of the star, so a halo can be seen as a perfect ring only from this hypothetical point.[74] The observer next to the first one will have to have his own physical hole to see a halo, and a halo-gazing party may destroy the effect altogether. So it seems the reconciliation envisioned by Alexander is in the interests of the theory of the *De anima*, and to some extent at the expense of the project in the *Meteorology*. In fairness, we should probably say that the correct explanation of the halo is not yet known at this time, so the tension between the physical and optical approaches is to some extent inevitable for all natural philosophers who do geometrical optics.

In Alexander's concluding remarks he gives us a glimpse of the context of the contemporary state of discussion on the nature of the halo.

> (T12) Alexander, *In Meteor.* 143.8–14:
>
> (1) And such is Aristotle's view of the halo summarised. (2) Posidonius too followed him, (3) while practically all the rest explained it not with reflection, but with the bendings of the sight-streams, as happens with the things seen through the water. (4) For they assume that the cloud is spherical and hollow, then they say that the star lying above is seen in it as dispersed along the circle. (5) But that such views about the halo are false, our teacher Sosigenes has sufficiently shown in Book 8 of his *On Sight*.[75]

74 Ibn Suwār has a curious report about Alexander which has no parallels in the commentary we have, according to which Alexander, in turn, reported the view of the 'ancients'.

75 (1) καὶ ἡ μὲν Ἀριστοτέλους δόξα περὶ τῆς ἅλω ὡς ἐπὶ κεφαλαίων τοιαύτη. (2) ἐπηκολούθησε δὲ αὐτῷ καὶ Ποσειδώνιος, (3) πάντων σχεδὸν τῶν ἄλλων οὐ κατὰ ἀνάκλασιν, ἀλλὰ κατὰκλάσεις ὄψεων αἰτιωμένων, ὡς ἐπὶ τῶν δι᾽ ὕδατος ὁρωμένων γίνεται· (4) ὑποτίθενται γὰρ σφαιροειδὲς καὶ κοῖλον τὸ νέφος, ἔπειτα τὸ ὑπερκείμενον ἄστρον αὐτοῦ κατὰ κύκλον φασὶ διεσπασμένον ἐν αὐτῷ ὁρᾶσθαι. ἀλλ᾽ ὅτι μὲν αἱ τοιαῦται δόξαι περὶ τῆς ἅλω ψευδεῖς, ἱκανῶς ὁ διδάσκαλος ἡμῶν Σωσιγένης ἐν τῷ ὀγδόῳ Περὶ ὄψεως ἔδειξεν.

3 οὐ κατὰ] οὐκ a fort. recte || 4 read κατὰκλάσεις with BWa κατὰ κλάσεις Hayduck.

From this account, it seems that Alexander, Aristotle and Posidonius[76] are in a minority explaining the halo as a reflection (T12.1–2). The rest of the world has moved on and adopted the theory of refraction (T12.3). In (T12.4) Alexander provides some further curious details about these views, but these details are not enough to identify the source of this theory. Alexander's teacher Sosigenes, who wrote a major work entitled *On Sight* (Περὶ ὄψεως) in at least eight books criticised this explanation of the halo by refraction and influenced Alexander (T12.5).[77]

The doxographical tradition shows that some theory of refraction (not necessarily the same one) was adopted, perhaps by the Peripatetic school, around the time when Aëtius' work was composed. If that were the case, then Sosigenes and Alexander could be seen as resisting some tendencies within the Peripatetic school. We do not know much of Sosigenes' motivation, but he had scientific interests and his interest in ὄψις could be comprehensive, so as to include geometrical optics.[78]

Alexander's project is philosophical and exegetical. He probably sees the theory of refraction as more dependent on the theory of physical sight-streams, in such a way that is not easy to debunk the explanation of the "sight-streams" as a purely geometrical concept. As we have seen, in Aëtius 3.5, the sight-streams are distinguished from the geometrical lines that represent them, and it is the sight-streams that provide a physical explanation of the rainbow and the same pertains to the explanation of the halo. The physical nature of the sight-streams is essential for that explanation. Ptolemy, who had a lot of sympathy for the Aristotelian theory of sight in the *De anima*, was still committed to the physical sight-streams, thus adopting some sort of 'hybrid' theory of vision.[79] It could be considered as a version of Aristotelian theory which had room for physical sight-streams. Perhaps something like this Alexander, in his place and time, perceives as 'mainstream' on this topic, and this is the mainstream he resists in the interests of the integrity of the Aristotelian theory.[80]

---

76 We know from Simplicius that Alexander is at least familiar with Geminus' epitome of Posidonius' *Meteorology*. Simplicius *In Phys*. 291.22–292.31 = F18 EK.

77 On Sosigenes, see Moraux (1984), 335–360; Sharples (2010); Kupreeva (2018).

78 Another of his authored works (as opposed to commentaries) whose title we know is *On Counteracting* [*Spheres*] (Περὶ τῶν ἀνελιττουσῶν).

79 On Ptolemy's *Optics*, Smith (1996); for its sociocultural context, Lehoux (2007).

80 I am grateful to Gweltaz and Alessio for organising the conference, to all the participants for their presentations, especially to István Bodnár, Robert Mayhew, and Alessio Santoro for their discussion at a conference and again to Alessio for careful reading of my draft and for pressing his point in subsequent comments. Thanks also to my physicist friend Irina Kardash for her helpful comments on the physics of halo. Any remaining mistakes are mine.

6 Concluding Remarks

We could see that Aristotle's explanation of the halo relies on the physical mechanism only as much as is necessary to justify the appearances in accordance with the geometrical proof. This has to do with both the physical mechanism of reflection and with psychological mechanism of sight-streams. Both physics and psychology have a supporting role in geometrical proof.

Theophrastus, on the other hand, concentrates on the physical causes of halo, using geometry in a supporting role, be it as a means of stating the explanandum (the circular shape of the halo), or as a heuristic tool in thinking of its physical causes. Theophrastus has no special role for psychology and does not mention sight-streams. He assumes that we see the appearances of halo and that while we may not have direct cognitive access to its physical causes, we may be able to reconstruct it with the help of our empirical concepts—hence the use of analogies. This is a potentially attractive philosophical position, for more than one reason, and it seems to have inspired some followers who took it in different directions.

In Post-Theophrastean Peripatetic doxography we find mainly Aristotle's explanation of the halo in terms of reflection (Arius Didymus, Ḥunayn's Anonymus). On the other hand, 'Aëtius' has a different explanation of halo in terms of refraction rather than reflection—another theory which also involves sight-streams—and Alexander in his commentary confirms that this position is sufficiently widespread in his time.

Alexander aims to reconcile Aristotle's account of halo with his theory of sight in the *De anima*. He sees 'sight-streams' as a useful geometrical idiom which should not be taken literally. Any presumed effects of the sight-streams upon the surrounding air can be explained by a closer look at the internal processes within our organ of sight. The sight-streams are the lines that show how the reflected light from the source reaches our eye. This light is reflected from a relatively stable physical structure up in the clouds below the star. This is where Alexander seems to follow Theophrastus, perhaps to make it absolutely clear that our sight-streams do not produce a halo in the cloud. He does not mention Theophrastus at this point and he probably thinks that Theophrastus' account spells out more clearly what Aristotle says in *Meteorology* III 3. Although Alexander's view has a difficulty as far as explaining the halo is concerned, his position with regard to compatibility between Aristotle's account of reflection and his theory of sight in the *De anima* may very well be correct.

CHAPTER 16

Alexander of Aphrodisias and the Illumination of the *Diaphanes*

Rafaëlle Beauvallet, translated by Chad Jorgenson

1 Introduction

In the section devoted to Alexander of Aphrodisias' commentary on the *De sensu* in the third volume of his series on Aristotelianism among the Greeks, Paul Moraux points out that Alexander introduces only minor changes in the course of his exegesis of the treatise, stating:

> Ebensowenig bemüht er sich, die Lehrmeinungen und die Beobachtungen des Aristoteles zu überprüfen und gegebenenfalls zu ergänzen oder zu korrigieren. Selbständige naturwissenschaftliche Daten und neue Erklärungen des bereits Beobachteten bringt der Exeget praktisch nicht bei. Die Gültigkeit der aristotelischen Lehre steht für ihn außer Frage. Nirgends spricht er die geringste Kritik aus, nirgends meldet er den leisesten Zweifel an: was Aristoteles schreibt, gilt einfach als Wahrheit. Nur ziemlich selten hält er es für angebracht, eine bloße Behauptung des Aristoteles durch eine Reihe von Argumenten zu untermauern. Noch seltener geht er auf Probleme ein, die Aristoteles nicht aufgeworfen hatte: wenn er eine Aporie formuliert, so erfolgt das nicht etwa, weil eine aristotelische These ihm inhaltlich anfechtbar erscheint, sondern vielmehr weil sie anscheinend einer anderen Aussage des Philosophen widerspricht.[1]

My focus here will be on one of these particular *aporiai*, and Alexander's treatment of it. At the beginning of his commentary on the *De sensu* (14.8–20.13), Alexander seems to reject Aristotle's explanation of the view that fire springs forth from the eye when it is moved.[2] This passage invites an unusual exegetical exercise in the context of this commentary, which at first seems to be a *correction* of Aristotle's text. However, the motivation for this rejection is not made

1 Moraux (2001), 401.
2 This phenomenon is known today as phosphene. I am grateful to Julien Devinant for pointing it out to me during the 2024 Paris–Lille–Cambridge doctoral workshop.

explicit, and Alexander vaguely points out that "perhaps it will seem to some hard"[3] to accept Aristotle's solution.

Before considering the possibility that Alexander is targeting adherents of rival philosophical schools, it is worth asking about the principle underlying his reservation. I aim to show that this principle is first and foremost Aristotelian, and that Alexander is justified in discussing Aristotle's solution from the point of view of Aristotelianism itself—as Moraux affirmed. Seen from this perspective, the key question concerns the nature of the apparent correction of the Aristotelian text made by Alexander. From an exegetical point of view, I will thus try to identify what is at stake in this passage, as well as to consider whether Alexander is indeed pointing out what he considers to be an authentic doctrinal contradiction from Aristotle. If this is indeed the case, we can expect Alexander's resolution to take the form of a *correction*, i.e. a *real alternative* that amends the text by breaking with the original hypothesis in one way or another. However, if we take into account the fact that Aristotle himself deliberately keeps the text of the *De sensu* in a rather loose relationship to the reality of the phenomenon it describes, it is more plausible to argue that it is an exercise in *clarification*, i.e. making the passage explicit in more rigorous and less concise terms, while generally keeping with the spirit of it. In this case, it is less a question of *correcting* the text but instead, making an effort to demonstrate its agreement with the text's broader doctrine.

Finally, the hypothesis can be put forward that Alexander is not merely proposing a simple gloss of the Aristotelian text. To the extent that he bases himself on a distinction (that is absent from Aristotle's texts but still can be sensed in and called for by them) between the ways a being can receive a new form, it can be shown that, in this case, the doctrinal unification of the Aristotelian text does not imply its correction, but rather its *creative extension*.[4] The passage in question illustrates two things: first, the significant attention Alexander pays to Aristotle's text and, second, his concern to ensure that Aristotle's doctrine is internally consistent, both at the general, systematic level, and in its most fine-grained details.

3 Alexander of Aphrodisias, *In librum De sensu commentarium* 18.17–18, transl. A. Towey.

4 The passage belongs to what Pierluigi Donini and Paolo Accattino refer to as Alexander's "creative exegesis", an expression that seems particularly appropriate for referring to this exercise, which frequently consists of updating Aristotelianism from within, in order to better establish its truth: e.g. Accattino and Donini (1996), XVIII; Accattino (2009), 58.

2 The Traditional Problem of Fiery Emanation

The Aristotelian text to which Alexander refers is well known: it mentions a phenomenon that was clearly already a traditional topic in classical times. In the second chapter of his *De sensu*, Aristotle announces his intention to discuss the composition of the sense organs, to challenge the Presocratic view that each of the senses corresponds to a particular element.[5] In this context he is led to discuss in more detail the material constitution of the visual organ.[6] In his opinion, his predecessors, with the relative exception of Democritus, always identified the eye with fire. In this passage Aristotle attributes this near consensus to the everyday experience that when the eye is moved in the dark, fire seems to spring forth from it:

> They all make vision consist of fire, because they do not understand the reason of one of the peculiarities of vision. When the eye is pressed and moved, fire seems to flash out. This naturally happens in the dark, or when eyes are closed; for then, too, there is darkness.[7]

To paraphrase the passage, when you press on your eye, you get the impression of seeing flames or sparks. In contrast, in the light, with an open eye, this phenomenon does not seem to occur. The passage has a clear critical dimension: Aristotle points to the ignorance (τὸ ἀγνοεῖν) of his predecessors and grammatically distances the *reality* of the phenomenon of emanation (through the use of φαίνεται with the infinitive). He nonetheless establishes here a first, necessary condition of the phenomenon, which only occurs in darkness (whether due to dark surroundings or the closing of the eyelid). As we shall see, Aristotle's explanation also entails a second condition, which will be addressed later in the text.

Despite the critical dimension of the passage, Aristotle recognises that attributing the fiery element to the eye resolves the traditional difficulty which we will refer to as (1).

5 Aristotle, *Sens.* 2.

6 *Sens.* 2, 437a22 ff.

7 Ποιοῦσι δὲ πάντες τὴν ὄψιν πυρὸς διὰ τὸ πάθους τινὸς ἀγνοεῖν τὴν αἰτίαν· θλιβομένου γὰρ καὶ κινουμένου τοῦ ὀφθαλμοῦ φαίνεται πῦρ ἐκλάμπειν· τοῦτο δ' ἐν τῷ σκότει πέφυκε συμβαίνειν, ἢ τῶν βλεφάρων ἐπικεκαλυμμένων· γίγνεται γὰρ καὶ τότε σκότος (Aristotle, *Sens.*, in *Parva Naturalia*, ed. W.D. Ross (1955), 2, 437a23–27, transl. W.S. Hett, *On Sense and Sensible Objects*, in *On the Soul, Parva Naturalia, On Breath* (1957)).

(1) When the eye is pressed and moved, fire seems to flash out from it.

The traditional solution to (1) is to consider that when the eye is moved, it sees itself—or, more precisely, that it then sees its own composition (i.e., the fire of which it is made). The basic principle underlying this proposed solution is doctrinally unacceptable to Aristotle, for whom the eye is necessarily made of water rather than fire insofar as it is *diaphanes*, "translucent, transparent". This is because water retains the *diaphanes* to a greater extent than the other elements.[8]

But beyond this principled doctrinal opposition, Aristotle mainly argues that the claim that the eye is made of fire, while it resolves difficulty (1) in its application, will lead to a new *aporia* (2), which has two parts:

> But this only raises another difficulty. For unless we suppose that it is possible for a sentient subject to see a visible object without knowing it, (2a) the eye must on this theory see itself. Why then (2b) does this not happen when the eye is at rest?[9]

The first part (2a) concerns the very structure of sensation, and results from an engagement with the fifth chapter of the second book of the *De anima*.[10] How can the eye see itself (i.e., its own composition), when in principle the sense organs do not sense themselves and sensation does not normally occur in the absence of any sensible object?[11] The lesson of chapter II 5 is that, in principle, sensation is for the organ the sensation of something other than itself. The second part (2b) of the *aporia* is rooted in the claim that remains implicit in Aristotle (but that is explicitly endorsed by Alexander in his commentary)[12]

8 *Sens.* 2, 438a12–16.

9 Ἔχει δ' ἀπορίαν τοῦτο καὶ ἑτέραν. εἰ γὰρ μὴ ἔστι λανθάνειν ⟨μὴ⟩ αἰσθανόμενον καὶ ὁρῶντα, ἀνάγκη ἄρ' αὐτὸν ἑαυτὸν ὁρᾶν τὸν ὀφθαλμόν. διὰ τί οὖν ἠρεμοῦντι τοῦτ' οὐ συμβαίνει; (*Sens.* 2, 437a27–31, transl. W.S. Hett).

10 Aristotle, *DA* II 5, 416b32–417a9.

11 Aristotle, *DA* II 5, 417a2–9.

12 In the *De sensu*, the thesis that all sensation is really sensed is proposed by Aristotle only in 437a26–28 (εἰ γὰρ μὴ ἔστι λανθάνειν αἰσθανόμενον καὶ ὁρῶντα ὁρώμενόν τι), then only briefly and hypothetically. Cf. also *DA* III 2, 425b12 ff. However, Alexander takes it for granted, as indicated by his interpretation of this passage of the *De sensu* (*In Sens.* 16.4 ff.): if it is possible for a sentient being to see a visible object *without being aware of it*, then it must be said that this sentient being *seems to see*, but that it does not *in fact* see anything of what is seen. Thus, if we accept that the eye sees all the time, but that the animal is only intermittently aware of it, we must accept the absurd conclusion that it sees without seeing. Alexander sums up the absurdity of this position, claiming that it amounts to saying that

that every sensation implies the awareness that something is sensed—since the very idea of a sensation that is not sensed is absurd: it is contradictory to say both that something is sensed and that it is not sensed. This second aspect of the *aporia* has more to do with the empirical conditions of the phenomenon than with the structure of sensation itself, since it focuses on the fact that the phenomenon in question only occurs when the eye is moved. If the eye can see itself, why should it only see itself when it is moved, and not when it is at rest? If it is really made of visible fire, it should, in principle, always see itself. Thus:

> (2a) If the eye were to see the fire of which it is made, it would see its own composition, which is impossible as the sense organs do not sense themselves (structural dimension of the *aporia*).
> (2b) If the eye were able to see the fire of which it is made, in principle it would have to see it constantly, both at rest and in motion. But the phenomenon of emanation apparently only occurs when it is moved (circumstantial dimension of the *aporia*).

The presentation of this twofold *aporia* results in the traditional identification of the eye and fire being rejected. This leads Aristotle to suggest an alternative explanation for the fiery emanation that occurs when the eye is moved: it occurs because the eyes are smooth rather than due to their elementary composition. This explanation is in accordance with the doctrine laid out in the *De anima*:

> But not everything is visible in the light, but only the proper colour of each individual thing; for some things are not seen in the light, but are only perceptible in the dark, such as those which appear fiery or luminous (there is no single name for these), like fungi, flesh, the heads, scales, and eyes of fishes; but in none of these is the proper colour seen.[13]

This doctrine is confirmed in the passage of the *De sensu* which concerns us here:

---

we see, while we actually do not. See also *In Sens.* 148.9–10, DA 65.4–5, as well as Towey (2000), n. 460, 185, and Ierodiakonou (2022).

13 Οὐ πάντα δὲ ὁρατὰ ἐν φωτὶ ἐστιν, ἀλλὰ μόνον ἑκάστου τὸ οἰκεῖον χρῶμα· ἔνια γὰρ ἐν μὲν τῷ φωτὶ οὐχ ὁρᾶται, ἐν δὲ τῷ σκότει ποιεῖ αἴσθησιν οἷον τὰ πυρώδη φαινόμενα καὶ λάμποντα (ἀνώνυμα δ᾽ ἐστὶ ταῦτα ἑνὶ ὀνόματι), οἷον μύκης, κρέας, κεφαλαὶ ἰχθύων καὶ λεπίδες καὶ ὀφθαλμοί· ἀλλ᾽ οὐδενὸς ὁρᾶται τούτων τὸ οἰκεῖον χρῶμα (Aristotle, DA II 7, 419a1–6, ed. W.D. Ross, transl. W.S. Hett).

> The reason for this, and the solution of our difficulty, and the cause of the theory that vision is fire, must be found in the following considerations. It is always smooth surfaces that shine in the dark, though they do not create light; and the centre of the eye which men call the "black" of the eye is clearly smooth.[14]

It now remains to be determined how Aristotle explains the phenomenon at issue on this basis.

3 The Aristotelian Solution

Difficulty (1) is partly dealt with by the solution that Aristotle sketches here, which makes the smoothness of the eye the cause of the phenomenon. The internal impression within the eye that fire bursts forth from it in the dark or when the eyelid is closed actually derives from the fact that smooth objects glow in the dark—i.e. are capable of actualising the *diaphanes* that surrounds them, even when it is dark and in the absence of an external light source. This then raises the question of whether Aristotle explains the fact that shiny objects are perceived by actualising the *diaphanes* that surrounds them. While Aristotle does not explicitly say this, it is the most plausible hypothesis that can be advanced on the basis of *DA* II 7,[15] which claims that shiny objects are also inherently visible. As such, they must behave in the same way as other things that are visible in themselves, such as colours, and they must therefore be sensed in the same way as colours.

Now, according to the account given in the *De anima*, which clearly prevails in the *De sensu*, vision takes place through a certain medium, air or water, in which there is "a common nature or potency, which is not separable",[16] which Aristotle calls the *diaphanes*. As is well known, the concept of the *diaphanes* is employed to eliminate the apparent difficulty posed by the distance separating the two poles of vision. This particular nature links together and ensures the circulation between the things that are seen and the seer (whose eye is itself *diaphanes* insofar as it is made of water). This *diaphanes* nature can be said to be in potentiality as well as in actuality: in potentiality, it is darkness; in

14 Τὰ δ᾽ αἴτια τούτου, καὶ τῆς ἀπορίας καὶ τοῦ δοκεῖν πῦρ εἶναι τὴν ὄψιν, ἐντεῦθεν ληπτέον. τὰ γὰρ λεῖα πέφυκεν ἐν τῷ σκότει λάμπειν, οὐ μέντοι φῶς γε ποιεῖ, τοῦ δ᾽ ὀφθαλμοῦ τὸ καλούμενον μέλανκαὶ μέσον λεῖον φαίνεται (*Sens*. 2, 437a31–437b2, transl. W.S. Hett).

15 *DA* II 7, 419a1–6.

16 *Sens*. 3, 439a23–24, transl. W.S. Hett, and *DA* II 7.

actuality, light. Light is an actuality, i.e. a certain state of the *diaphanes*, the result of an instantaneous qualitative change in the medium under the influence of a luminous element, which does not, strictly speaking, act on the medium, but whose luminosity occurs in it as a "presence".[17] Vision requires a luminous element (such as fire) capable of actualising the capacity of a potentially luminous medium (e.g. air or water, insofar as they have within them that particular nature that is the *diaphanes*). In the passage from *De sensu* cited earlier, Aristotle argues that shiny objects glow, albeit without producing light (τὰ λεῖα πέφυκεν ἐν τῷ σκότει λάμπειν, οὐ μέντοι φῶς γε ποιεῖ).[18] Thus, in the absence of any indication to the contrary, their visibility must be explained in the same way as colours. Smooth objects do not produce light, i.e. they do not in themselves constitute an igneous source (such as fire or the stars) capable of illuminating the *diaphanes* in an absolute sense. They instead have the capacity to move the external *diaphanes* medium contiguous with them, which consequently moves the *diaphanes* medium of the eye, in line with the ordinary functioning of the visual chain that connects the being that sees to the things that are seen.

To resolve traditional difficulty (1), Aristotle draws on an alternative solution (A):

> (A) Since the "black" of the eye is smooth, it has the distinctive property of being visible in the dark.

At this stage, we are apparently able to understand why fire seems to flash out from the eye and why it only does so in the dark (whether it is actually dark or the eyelids are merely closed). This happens because the eye is smooth, and smooth objects are visible in the dark. However, Aristotle also identifies a second condition for the phenomenon to occur, namely ocular movement, whose theoretical role in resolving *aporia* (1) must now be determined.

Aristotle formulates a second condition to complement his alternative solution (A), as it would lead us back to the *aporia* or difficulty (2) posed by the hypothesis that the eye is made of fire, on both the structural (2a) and the circumstantial (2b) level. This is because the alternative solution (A) can be criticised on the same grounds as the hypothesis that the eye is made of fire. In the first place, how is it possible for the eye to see itself—i.e. in this case to see its own glow (as opposed to its own composition)—when this is, in principle,

17 Παρουσία, *Sens.* 3, 439a20 and *DA* II 7, 418b20, for example.

18 *Sens.* 2, 437a31–32.

impossible? Second, why does this phenomenon *only* occur when the eye is moved—or, to put it positively, why is its glow not always visible, i.e. *including* when it is at rest?

Aristotle's solution to this double *aporia* consists in the establishment of a local duplication of the eye:

> The phenomenon occurs when the eye is moved because then the effect is as though one thing became two. This is due to the rapidity of the movement, which causes the seeing subject to appear different from the object seen. Hence the phenomenon does not occur, unless the movement is rapid and in the dark; for it is in the dark that a smooth surface appears to shine, for instance the heads of certain fishes, and the dark fluid of the cuttlefish; when the movement of the eye is slow, it is impossible that the seeing organ and the object seen should appear to be both one and two at the same moment. But in the other case (when the movement is rapid) the eye sees itself just as in reflection.[19]

The principle underlying this solution can be spelled out as follows: the organs do not sense themselves. Then, if the eye is to be able to see itself, we must conceive that it sees itself *as something other than itself* (which Aristotle understands here in a local sense)—and this only happens when it is in rapid motion. If the eye does not see itself when it is at rest, this is because when it is closed and in its ordinary position, it has nothing *other* than it to see—hence there is no visual sensation. Conversely, when the eye is compressed and moved rapidly, Aristotle sees it as being affected by a pendulum-like movement, in which it oscillates between its initial position A and a position B that is distinct from A.[20] When it is in its initial position, the eye *sees*, but when it is not in this position, it is *visible* and, in fact, *seen*, insofar as it is smooth and glows. This spatial duplication and the rapidity of the eye movement explains why the eye in position A sees its glow in position B as if it had been produced by something else. Thus making it possible to preserve the structure of sensation, which is in principle the sensation of *something else*.

---

19 Φαίνεται δὲ τοῦτο κινουμένου τοῦ ὄμματος διὰ τὸ συμβαίνειν ὥσπερ δύο γίνεσθαι τὸ ἕν. τοῦτο δ᾽ ἡ ταχυτὴς ποιεῖ τῆς κινήσεως, ὥστε δοκεῖν ἕτερον εἶναι τὸ ὁρῶν καὶ τὸ ὁρώμενον. διὸ καὶ οὐ γίνεται, ἂν μὴ ταχέως καὶ ἐν σκότει τοῦτο συμβῇ· τὸ γὰρ λεῖον ἐν τῷ σκότει πέφυκε λάμπειν, οἷον κεφαλαὶ ἰχθύων τινῶν καὶ ὁ τῆς σηπίας θολός· καὶ βραδέως μεταβάλλοντος τοῦ ὄμματος οὐ συμβαίνει ὥστε δοκεῖν ἅμα ἓν καὶ δύο εἶναι τό θ᾽ ὁρῶν καὶ τὸ ὁρώμενον· ἐκείνως δ᾽ αὐτὸς αὑτὸν ὁρᾷ ὁ ὀφθαλμός, ὥσπερ καὶ ἐν τῇ ἀνακλάσει (*Sens.* 2, 437b1–10, transl. W.S. Hett).

20 This is, at least, the interpretation given by Alexander (*In Sens* 17.13–23). It will be dealt with in more detail later.

However, there are reasons to question the status of Aristotle's statements in this text and to wonder whether he himself unreservedly subscribes to this hypothesis for dealing with difficulty (1). In reality, the very syntax of the text suggests at a minimum that its author is distancing himself from the solution sketched here. The constructions involving ὥσπερ and δοκεῖν indeed suggest a certain implicit caution. The discourse here is not free of detours: on the rhetorical level, Aristotle is not directly assertive, and the mobilised ontological register is notably not that of being. Aristotle does not directly say that, in this process, one eye *really* becomes two, but that it becomes *like* two; nor does he say that the seer and the object seen *are* different, but only that they *seem* so.

Should we therefore think that, for Aristotle, the eye really sees itself as something else? Should we not rather take seriously what is suggested here by the use of the two modal qualifiers δοκεῖν and ὥσπερ? Taking this syntactic expression of caution into account requires contextually to ascertain whether rapid movement *actually* causes the eye to split in two or if it only produces the *phenomenal impression* that it splits. It is also possible to see these are mere sketches of a solution or plausible but provisional hypotheses.

The hypothesis that Aristotle distances himself from his solution to difficulty (1) is corroborated by the problematic consequences of the latter, which are identified by Alexander of Aphrodisias in a remarkable passage in his commentary on the *De sensu*, in which he seems to propose an alternative explanation of difficulty (1) as Aristotle raises and resolves it in the second chapter of the *De sensu*. In the following, I will first consider the problematic consequences of the Aristotelian hypothesis as identified by Alexander, before examining how Alexander deals with the same problem. I will then interrogate the status of this solution: should we really consider it as a genuine alternative, i.e. a genuine correction that breaks with the Aristotelian text? Or is it rather an exercise in clarifying and further specifying what Aristotle said—even if this means extending the Aristotelian corpus beyond its own boundaries?

4 Alexander of Aphrodisias' Commentary

In Alexander's literal commentary on the treatise, the short passage from the *De sensu* referred to in this paper is the subject of a major attempt at a critical re-elaboration that spans several pages.[21] The text begins by following the structure of the original work, while glossing it: sparks seem to fly out from the

---

21 *In Sens.* 14.8.20–13.

eye when it is moved, a difficulty that all earlier philosophers, except Democritus, addressed by saying that it was made of fire. This solution, in turn, raises a twofold difficulty (Why does the eye see itself when it is moved and not at rest? How can the eye see itself?). Aristotle resolves this double difficulty by claiming that the eye is smooth and therefore glows in the dark, and that only when it is moved can it see itself—because it sees itself as something else. Alexander thus begins by paraphrasing Aristotle's solution, developing the idea mentioned earlier of a *real* spatial duplication of the eye: when it is pressed and moved, it oscillates between its natural position and a different position, from which it is pushed. In its natural position, the eye sees, but when it is pushed outside of it, it is seen insofar as it glows. As a shiny object, it becomes visible *per se*, and is then perceived as something else.

Yet after elucidating Aristotle's hypothesis, Alexander departs from the order of exposition of the text he is commenting, raising an objection that is grounded in the fact that Aristotle does not provide a complete explanation of the phenomenon and that, in reality, the solution he gives can be interpreted literally in two different ways. Should we think, for example, that the eye that sees and the eye that is seen are in two different places at the same time? This possibility seems to be ruled out in principle by the Aristotelian definition of place as the limit of the enveloping body.[22] Thus, it seems more likely that the glimmer of the eye persists for a time after it has returned to its proper place. The text reads as follows:

> Perhaps it will seem to some hard to accept that [the eye] can come to be both seeing and seen because it comes to be one thing and another because of the rapidity of the pushing aside. [...] But this is impossible. (For no such thing comes to be in the case of the other things that glitter, but the glitter from them ceases at the same time as that which glitters is turned away. For this comes about not only in the case of the things that glitter but also in the case of everything which is seen, since that which is transparent does not admit any colour in itself in a way that involves its being affected, because [it does] not [admit] even light [in this way] at all. For when [light] is removed from the transparent it ceases in conjunction with that which naturally illuminates.) If this [is] so in all cases, how could the pupil, coming to be in its natural [position], if it were moved rapidly, apprehend the glitter which is in it and which travels together with it?

22 Aristotle, *Physics* IV 4. Since the enveloped body (the eye) and the enveloping body (the space occupied by the eye) are coextensive and determined by each other, the same body cannot simultaneously occupy two distinct places.

> For just as (1) it is not able at the same time both to be and not to be in its proper place, even if it is moved rapidly, so too (2) [it is not able] to see its own glitter as being in another place. And so perhaps one should not suppose that the eye comes to be two in the movement, as if the same thing which sees and is seen at the same time in the same respect comes to be two complete things, when [in fact] a part is seen.[23]

In his commentary, Alexander seems less to oppose Aristotle's solution *as such*, and more to oppose this solution to the extent that it is taken *literally*. This is indicated by the first sentence: "perhaps it will seem to some hard to accept that [the eye] can come to be both seeing and seen." In Alexander's text, the statement is not accompanied by any rhetorical caveats comparable to those presented by Aristotle, nor is it modulated in any way. Alexander can therefore be considered to be engaged in a literal interpretation of the passage, which does not reflect the distance that Aristotle himself was careful to establish with regard to his own statements. Moreover, it is noteworthy that it is the second condition set by Aristotle (the speed of the eye motion) that Alexander considers difficult to accept. I will return to this point later.

With regard to the structure of the text, we see that Alexander, after rejecting the two possible literal interpretations of Aristotle's text, proposes his own explanation, which is compatible with the original only on the strict condition that Aristotle's syntactical caution is taken into account. In this text, Alexander asks, in a general way, how we can understand the claim that the eye sees itself in a position other than the one it currently occupies (i.e. that of the *seeing* eye) and thus how the eye can simultaneously see and be seen. Moreover, he shows that Aristotle's solution can be interpreted in two ways, which he makes explicit by rejecting each of these two possibilities in turn. So how, on a literal reading of Aristotle, is the duplication of the eye possible according to Alexander?

---

23 Ἴσως δ' ἐκεῖνο δυσπαράδεκτόν τισι δόξει τὸ δύνασθαι αὐτὸν ⟨καὶ⟩ ὁρῶντα ὁρώμενον γίνεσθαι διὰ τὸ τάχος τῆς παραγωγῆς ὡς ἄλλον γινόμενον καὶ ἄλλον. [...] ἐπεὶ δὲ τοῦτο ἀδύνατον (οὐδὲ γὰρ ἐπὶ τῶν ἄλλων τῶν στιλβόντων τοιοῦτον γίνεται, ἀλλ' ἅμα τῷ παρενεχθῆναι τὸ στίλβον καὶ ἡ ἀπ' αὐτῶν παύεται στιλβηδών· οὐ μόνον γὰρ ἐπὶ στιλβόντων, ἀλλὰ καὶ ἐπὶ πάντων τῶν ὁρωμένων γίνεται, τῷ μηδὲν χρῶμα τὸ διαφανὲς παθητικῶς ἐν αὐτῷ δέχεσθαι, ὅτι μηδὲ τὴν ἀρχὴν τὸ φῶς· καὶ γὰρ ἐκεῖνο ἐκ⟨στὰν⟩ τοῦ διαφανοῦς συμπαύεται τῷ φωτίζειν πεφυκότι), εἰ δὲ ἐπὶ πάντων οὕτω, πῶς ἄν ἡ κόρη ἐν τῷ κατὰ φύσιν γινομένη, εἰ καὶ ταχέως κινηθείη, δύναιτο καταλαμβάνειν τὸ στίλβον ὃ ἐν ἐστι καὶ συμμεταφέρεται αὐτῇ; ὡς γὰρ οὐ δύναται ἅμα καὶ ἐν τῇ οἰκείᾳ χώρᾳ εἶναι μὴ εἶναι, κἀν ταχέως κινῆται, οὕτως τὴν στιλβηδόνα ἀπ' αὐτῆς ὁρᾶνώς οὖσαν κατ' ἄλλο. μήποτ' οὖν οὐχ οὕτω χρὴ δύο γίνεσθαι τὴν ὄψιν ὑπολαμβάνειν ἐν τῇ κινήσει, ὡς ὁλοκλήρων γινομένων τοῦ αὐτοῦ κατὰ τὸ αὐτὸ ὁρῶντός τε ἅμα καὶ ὁρωμένου, μορίου δὲ ὁρωμένου (*In Sens.* 18.17–19.7, transl. A. Towey slightly modified).

Number (1) in the citation above refers to the first interpretation, which is the most literal. It ignores the cautionary clauses ὥσπερ and δοκεῖ, claiming, on this basis, that what is one (the eye) *really* becomes two, leading us to (naively) interpret the duplication as a real duplication—that is, that the eye's movement is so rapid that it simultaneously occupies positions A and B (with the result that the eye sees itself as another, i.e. as an ordinary sensible). The second interpretation is denied at (2); it seems less literal, insofar as the duplication of the eye is no longer considered real. This interpretation involves giving up on the idea that the eye is simultaneously in position A (where it sees) and position B (where it is visible). It therefore requires us to assume that the glow persists in the *diaphanes* once the eye is removed from position B. In reality, both of these interpretations lead to contradictions. If we accept (1), we have to say that the eye *is* and *is not*, at the same time and in the same respect, in its proper position A. The principle of non-contradiction therefore requires us to abandon this interpretation, which cannot ipso facto be the one suggested by Aristotle,[24] leading us to the conclusion that option (2) should be retained.

In brief, the general problem is that the *seeing* eye in position A never coincides temporally with the *visible* eye in position B (indeed, it is one and the same being, alternating successively between two distinct positions). Since it is necessary to explain how the eye sees something that is, in reality, no longer there to be seen, Aristotle's solution would suggest that the eye's gleam persists in the *diaphanes* medium after the eye has returned to its proper place. If the seeing eye has something to see when the visible eye is no longer there, it is necessary to suppose the existence of a persistence, however slight, of the gleam in the *diaphanes*. In other words it must be assumed that the *diaphanes* is affected by the gleaming of the eye that persists in it. Further, once the cause of this affection has disappeared, its effect will persist for some time.

Now Alexander argues in the parenthesis that the *diaphanes* is not affected in this way by the objects that move it: thus, it is not παθητικῶς (i.e. by being affected) that the *diaphanes* receives light. The principle governing his opposition to this idea is, in any case, primarily empirical, as he expresses it in parentheses: taken literally, Aristotle's hypothesis seems to run counter to the ordinary experience we have of glowing, vision and illumination, which are phenomena experienced as instantaneous and which do not give rise to a persisting effect once the cause has been removed. At the same time, beyond these experiential considerations, Alexander's opposition to the idea of a πάθος of the *diaphanes* is supported by a number of statements made by Aristotle himself,

24 *In Sens.* 19.10–12.

which we will discuss next. For now, let me briefly state how Alexander resolves the problem of fiery emanation, before saying a few words about what justifies, on the one hand, this last thesis about the affection of the *diaphanes* and, on the other, Alexander's opposition to the literal interpretation of the original passage.

For Alexander, the phenomenon of fire seeming to flash out from the eyes in the dark is thoroughly explained by the fact that the visual organ is smooth and that smooth things glow in the dark. However, this does not occur insofar as the eye remains one and the same thing ("whole", ὁλόκληρον), i.e. insofar as *all of it* swings from A to B. According to an argument that he also mobilises in *De anima*, within the context of his discussion on the common sense and how we are to understand the possibility of perceiving contrary sensibles with the same organ,[25] Alexander argues in the *Commentary* that "we do not see by means of a body without magnitude".[26] In other words: because it is a magnitude, the eye has parts. While one part is in its natural position and can therefore still see, insofar as it is situated in a straight line with the veins that lead to the heart, it is possible that another part is driven out of this natural position by friction, and that it is therefore visible to the first part. Thus, Alexander does not spatially dissociate the eye, so that it both is and is not in two distinct places at the same time and in the same relation (which would violate the principle of non-contradiction), and he does not need to account for the phenomenon by appealing to the persistence of the effect in the *diaphanes* in the absence of its cause, because the eye does, in fact, see something, i.e. a visible part of itself. Alexander's solution thus avoids the two pitfalls resulting from a literal reading of Aristotle by maintaining the idea of a spatial duplication of the eye, so as to avoid having to say that what the eye sees is its own persistent glow in the *diaphanes* medium, while arguing that the duplication involves *parts* and not the whole eye, so as to not fall foul of the principle of non-contradiction.[27]

The first component of the principle underlying Alexander's opposition to Aristotle's solution can easily be traced back to Aristotle's authority, and, more broadly, to the demands of reason as such, since it is contradictory to think that the same thing can and cannot be found at the same time and in the same respect in the same place. Our focus should therefore rather be on examining what led Alexander to think that the glow of the eye cannot, even briefly, continue to actualise the *diaphanes* once the eye has been removed.

25 Alexander of Aphrodisias, *DA* 64.4–5.

26 Οὐ γὰρ ἀμέγεθες σῶμα δι' οὗ ὁρῶμεν (*In Sens.* 20.1–2, transl. A. Towey).

27 *In Sens* 19.14–20.5.

5 The πάθος of the *Diaphanes*

In the discussion of sight in his own *De Anima*, Alexander opposes two kinds of modification according to the sort of πάθος that causes it.[28] This distinction is given in the general context of a discussion about the perception of colours by seing animals. As we know, it occurs because visible objects first move and illuminate the *diaphanes* medium, and then because the *diaphanes* medium, in turn, moves the *diaphanes* nature contained in the eye that sees. Alexander goes on to argue that the two following phenomena occur in virtue of a unique type of instantaneous change: (a) on one hand, the potential *diaphanes* is illuminated; (b) on the other hand, the illuminated *diaphanes* receives and transmits colours. He then wonders about what happens in the *diaphanes* medium when it receives either colours or light. Here, Alexandre immediately contests and dismisses two possible explanations for one or the other of these phenomena, both of which are *in terms of matter*:

> For the colour comes to be present in what is illuminated and in light in the same way that light comes to be present in what is transparent, though what is transparent does not receive light or light colour (1) in virtue of an effluence or (2) in the way matter [receives something].[29]

These two explanations, marked respectively by (1) and (2), account for the change in the *diaphanes* substance, either by appealing to the reception of effluxes (i.e. the illuminated *diaphanes* substance really becomes coloured or luminous, because it receives films of matter that genuinely alter its material composition) or by saying that the *diaphanes* is affected "as matter" (ὡς ὕλη) by light or colour. What would it mean for the *diaphanes* to be affected as matter in this way? In order to understand this statement, let's examine the other occurrences of this ὡς ὕλη expression in the treatise. We find it on two other occasions in passages where Alexander considers the possibility of the existence of another type of πάθος for the affected object, which is characterised as οὐχ ὡς ὕλη[30] ("not as matter") and that would cause a non-ordinary type of change for the affected thing.

---

28 For a discussion of Theophrastus' views on the types of change involved in sense perception, and particularly on those occurring in the medium and in the sense organs, see Robert Roreitner's paper in this volume.

29 Γίνεται δὲ τὸ χρῶμα ἐν τῷ πεφωτισμένῳ τε καὶ φωτὶ οὕτως ὡς καὶ φῶς διαφανεῖ, οὔτε κατὰ ἀπόρροιάν τινα, οὔτε ὡς ὕλης ἢ τοῦ διαφανοῦς δεχομένου τὸ φῶς ἢ φωτὸς τὸ χρῶμα (Alexander, DA 42.19–22, transl. V. Caston).

30 Alexander, DA 62.2 and 83.22.

The expression occurs again on pages 61–62 of *DA*, when Alexander grapples with the problem of the possibility for an animal to simultaneously perceive two contrary homogeneous qualities, other than by violating the principle of non-contradiction:

> If sight were to be stirred in such a way as to be affected and to receive white and black, it would receive opposites at the same time, which is impossible. But if it is in a different way that sense perception is changed by perceptibles, and the organs do not serve as matter receiving the qualities of the perceptibles, it would no longer be such an impasse. For it is evident that sight does not serve as matter receiving the qualities, for we see that sight does not become black or white when it perceives these. But neither does the illuminated air, in spite of serving sight for the grasping of colours, do this through itself being first changed by the colours and itself becoming black or white. At least nothing prevents one person apprehending black and another white through the same air, when the white and black lie directly in front of the observers, but each looks at the colour lying not by himself but by the other person. Even if a black and a white person were to see each other, the intervening air is still not prevented from serving both of them simultaneously, because it is not changed by the colours qualitatively, and does not come to act as the matter of the colours.[31]

Alexander wonders how it is possible to simultaneously see black and white, if we consider that vision is produced by an ordinary change occurring in the eye, on the one hand, and in the actualised *diaphanes*, which is designated here

31 Ἢ εἰ μὲν οὕτως κινοῖτο, ὡς πάσχειν τε καὶ δέχεσθαι τὸ λευκὸν καὶ τὸ μέλαν, ἅμα ἂν ἀναλαμβάνοι τὰ ἐναντία, ὅ ἐστιν ἀδύνατον, εἰ δ᾽ ἄλλος ὁ τρόπος τῆς ὑπὸ τῶν κινήσεως τῇ αἰσθήσει καὶ οὐχ ὡς ὕλη τῶν αἰσθητῶν τὰ πάθη δέχεται αἰσθητήρια, οὐκέτ᾽ ἂν ὁμοίως ἄπορον εἴη. ὅτι γὰρ οὐχ ὡς ὕλη δέχεται τὰ πάθη ἡ ὄψις ἐναργές. ὁρῶμεν γὰρ ὅτι οὐ γίνεται ἡ ὄψις μέλαινα καὶ λευκή, ὅταν ἐκείνων αἰσθάνηται. ἀλλ᾽ οὐδὲ ὁ πεφωτισμένος ἀήρ, καίτοι διακονούμενος τῇ ὄψει πρὸς τὴν ἀντίληψιν τῶν χρωμάτων διὰ τοῦ αὐτὸς ὑπ᾽ αὐτῶν πρῶτος κινεῖσθαι καὶ μέλας αὐτὸς ἢ λευκὸς γίνεσθαι τοῦτο ποιεῖ. οὐδὲν γοῦν κωλύει διὰ τοῦ αὐτοῦ τὸν μὲν μέλανος ἀντιλαμβάνεσθαι τὸν δὲ λευκοῦ, ὅταν κειμένων τοῦ τε λευκοῦ καὶ τοῦ μέλανος ἐπ᾽ εὐθείας τῶν ὁρώντων αὐτὰ μὴ τὸ κατ᾽ αὐτὸν ἑκάτερος αὐτῶν κείμενον χρῶμα βλέπῃ, ἀλλὰ τὸ κατὰ τὸν ἕτερον. ἀλλ᾽ οὐδ᾽ εἰ μέλας καὶ λευκὸς ἀλλήλους ὁρῷεν, κεκώλυται ὁ μεταξὺ ἀὴρ ἀμφοτέροις αὐτοῖς ἅμα διακονεῖσθαι τῷ μὴ παθητικῶς μηδὲ ὡς ὕλη γινόμενος αὐτῶν ὑπ᾽ αὐτῶν κινεῖσθαι. (Alexander, *DA* 61.30–62.13, transl. R. Sorabji in Sorabji (2004)). The argument evokes that of Aristotle in *DA* II 5 concerning the type of alteration to which sensation corresponds. As we know, sensation is said to constitute an alteration only insofar as there is no more appropriate term to capture precisely the type of change in which it consists. See in particular 417b29–418a6.

as the "illuminated air", on the other hand. Indeed, if vision were produced by an ordinary change, then the formal qualities of black and white would have to be simultaneously transmitted in the same *diaphanes* medium, and simultaneously received by the same organ: since this reception takes the form of an assimilation, we would have to say that the eye really becomes black and white simultaneously. It is in order to find a way out of this *aporia*, that Alexander opens the way to a distinction between two types of change: (i) a change by which the organ or the *diaphanes* would receive or transmit sensible qualities ὡς ὕλη and (ii) another possible type of πάθος called οὐχ ὡς ὕλη.

What we find in this text is the equation of *being acted upon as matter*, on the one hand, and *becoming the sensible qualities* in virtue of which the πάθος takes place, on the other. What Alexander indicates here, is that, to be moved as matter by black and white would mean, both for the *diaphanes* and for the eye, to *really becoming* these qualities, in such a way that the principle of non-contradiction would be violated. The question, then, is obviously what it means to *become the quality* here. How should we understand this expression, if both Aristotle and Alexander understand sensation as an assimilation? Should we not say, then, that assimilating a quality is precisely to actually become that quality?

To answer these questions, we may refer to the third and final occurrence of the expression ὡς ὕλη in *DA* 83–84. There, Alexander seeks to distinguish grasping forms with the sense from doing it with the intellect. He then argues that sensation grasps forms as being in matter (ὡς ὄντων ἐν ὕλῃ), but not as matter (ὡς ὕλη). In this passage, receiving or grasping forms as matter is referred to as "being acted upon without qualification" (ἁπλῶς) and is characterised as follows:

> Now, to receive a form as matter is the same as becoming matter for what is received—as can be seen in the case of affections that do not occur in virtue of a soul. For things that are affected without qualification become matter for these affections. For instance what is heated, when it itself becomes hot, becomes matter for this affection—something that neither sensation nor intellect undergoes.[32]

32 Ἔστι δὲ τὸ μὲν ὡς ὕλην εἶδός τι λαμβάνειν τὸ αὐτὸ τῷ γίνεσθαι τῷ λαμβανομένῳ, ὃ ἐπὶ τῶν παθῶν τῶν οὐ κατὰ ψυχὴν γινομένων ἰδεῖν ἔστι. τὰ γὰρ ἁπλῶς πάσχοντα ὗλαι γίνονται τῶν παθῶν. τὸ γὰρ θερμαινόμενον θερμὸν γινόμενον αὐτὸ ὕλη τῷ πάθει, ὃ οὔτε ἡ αἴσθησις οὔτε ὁ νοῦς ἔχουσιν (Alexander, *DA* 83.23–84.4, translation is mine).

For a sensible substance x, receiving a form as matter is equivalent to a case of ordinary change, i.e. the type of change in virtue of which substance x receives a new form through which it is informed. This is how, for example, a heated object becomes matter for the formal quality of heat, that it materially assimilates, by means of a qualitative change or alteration. Therefore, receiving a form as matter is to be matter for the received quality, i.e. to materially receive the quality in question. Thus, to be acted upon ὡς ὕλη is to undergo the kind of πάθος in virtue of which the affected being materially takes on the received qualities.

By contrast, what would it mean for a being to change without materially taking on the change to which it is submitted? To answer this question, we can look at a passage from the *Mantissa*.[33] There, Alexander seeks to attribute its own modality to a certain type of change, that seems to deviate from the general rules of sublunary change (interpreted as the acquisition by a substance x of a new form, that can only be removed by the acquisition of an opposite form). Here, Alexander thinks of how light and shadow[34] are produced, as well as how images appear on reflective surfaces:[35] these phenomena do not seem to occur as the result of the acquisition of a new form, but in virtue of the mere presence of the change cause. When the cause is removed, the being that has been changed seems to instantly return to its initial state. Here, no change in the object seems to be required: it is not necessary to touch the mirror to make the image it reflects disappear.

Consequently, Alexander distinguishes, in the vocabulary of this passage, between two kinds of τροπαί or modifications, referring to both time and the substances they affect. He calls "alteration" (ἀλλοίωσις) of the object the type of πάθος that causes a qualitative change in it. This type of change is characterised as a movement (κίνησις), i.e. following the *Physics*, as a *process*.[36] Alexander interprets this processual nature of movement as involving that any modification of this kind produces its effects *in time*.[37] Now, the type of modification that matches the illumination of the *diaphanes* stands in contrast to alteration, insofar as Alexander sees it as instantaneous.[38] While a plant gradually heats

33 Alexander of Aphrodisias, *Mantissa* 141.30–147.25.

34 Alexander, *Mant.* 143.26–28.

35 *Mant.* 145.13–16.

36 I am here thinking of the famous Aristotelian definition of movement in *Physics* III 1, 201a10–11, as the entelechy of what is potentially as such (ἡ τοῦ δυνάμει ὄντος ἐντελέχεια, ᾗ τοιοῦτον, κίνησίς ἐστιν), i.e., as the process that enables a given substance to pass from potentiality to actuality.

37 *Mant.* 143.6–7 and 143.24–26.

38 According to Alexander, experience itself belies the fact that the illumination of the

up under the effect of the sun's rays, the *diaphanes* is illuminated and transmits colours without undergoing a qualitative change, without delay and all at once (ἀθρόως),[39] which is why Alexander attributes to Aristotle the idea that "seeing comes about timelessly", "for a person sees [what he sees] in however [small] a part of the time in which he sees [it], and in just any part".[40] Following Alexander, we must therefore consider another type of change that is not processual in nature—in other words, that cannot be read as an actualisa*tion* (understood as a modification involving a passage from one state to another or from one quality to another), but rather as a pure actuality. This type of change does not require the imposition of a new form on the substance that displays it, which it would acquire progressively, but the mere presence of the cause of the change. Thus, Alexander characterises the illumination of the *diaphanes* and its transmission of colours as a merely apparent alteration (δοκοῦσα ἀλλοίωσις).[41] It is only apparent in the sense that it is not a standard case of qualitative change or alteration; nonetheless, it is *real*, since the illumination of the *diaphanes* or even the appearance of images in a mirror are not an illusion or a pure appearance. These phenomena are pure effects of presence (or absence, in the case of shadows): therefore, they are not the effect of the actualisation of the beings that manifest them, but of the efficacy of their simple co-presence. It is remarkable that Alexander does not use the same vocabulary to distinguish these two kinds of change in the *De anima* and in the *Mantissa* (speaking alternatively of affection ὡς ὕλη / οὐχ ὡς ὕλη and of ἀλλοίωσις / δοκοῦσα ἀλλοίωσις), but it does seem possible to find in these two texts analogous characterisations that, to this extent, give us grounds to identify them.

For Alexander, therefore, it is not in virtue of a real πάθος, i.e. an ordinary affection, that the *diaphanes* medium is modified by what illuminates it.[42] Instead, Alexander thinks the actuality of the *diaphanes* as a relation (σχέ-

diaphanes occurs over time (*Mant.* 143.23–30). Furthermore, and in virtue of the thesis already mentioned concerning the fact that every sensation implies awareness, it is not possible to claim that this illumination occurs in a time that is too short to be identified and sensed.

39 *Mant.* 143.21.

40 Διὰ τοῦτο δὴ καὶ ἀχρόνως δείκνυσιν ὁ Ἀριστοτέλης γινόμενον τὸ ὁρᾶν. τοῦ γὰρ χρόνου, ἐν ᾧ τις ὁρᾷ, τούτου ἐν ᾡτινιοῦν μορίῳ καὶ τῷ τυχόντι τοῦτο ὁρᾷ (*Mant.* 143.30–31, transl. R.W. Sharples, *Alexander of Aphrodisias: Supplement to* On the Soul (2004)).

41 *Mant.* 143.17–18. Once again, we think of Aristotle's *DA* II 5: specifically, the formula seems to parallel the characterisation of sensation as an ἀλλοίωσίς τις at the beginning of the chapter, in 416b34. Alexander thus seems to be attempting to attribute its own modality to this particular type of change that Aristotle himself only hinted at.

42 As is well known. For example, Schroeder (1981), Accattino (1992), Moraux (2001), Towey (2019), Cordonier (2019), Robert (2020), Ierodiakonou (2020, 2021).

σις) and, following Aristotle, in terms of presence:[43] the actuality comes about from a relation between what is illuminated and what is illuminating, which is either realised or not, at a given moment, without there being any intermediate state between the two. As a result, the *diaphanes* does not undergo any *process* of actualisation, by which it becomes illuminated.[44] Thus, for Alexander, the πάθος of the *diaphanes* is apparently not reducible to a πάθος, in the ordinary sense in which the phenomenon is described by Aristotle in the *Physics*, namely as the imposition of a new form on a given matter.

Throughout Alexander's writings, we repeatedly find the assertion that the *diaphanes* neither is illuminated nor receives colours in such a way as to be affected by them (παθητικῶς),[45] an assertion that, as such, has no direct equivalent in Aristotle's works. However, while Aristotle is never as explicit as Alexander about this point, we can reconstruct his position on the illumination of the *diaphanes* from a number of passages that addresses the question: in *DA* II 7[46] and *Sens*. 6,[47] Aristotle argues that the illumination of the *diaphanes* medium is instantaneous, in some cases in opposition to Empedocles and in others to draw a contrast between visual sensation, which is perceived as immediate, and the propagation of odours and sounds, which seems to take place over time. On this basis, it seems possible to affirm that Alexander's theoretical account of this non-ordinary type of affection represents an extension of the Aristotelian doctrine of change, much more than a real breakthrough, especially given that the Aristotelian corpus at its minimum opens up this theoretical possibility.[48] Therefore, we can argue that, here, we are dealing less with a correction of Aristotle's doctrine, than with its creative extension, i.e. with the development of the orthodox theory of the *diaphanes* by Alexander.

43 This has led various commentators to consider this type of change as a mere Cambridge change. See Caston (2012), 138, 156–157.

44 As Moraux (2001), 413, states: "Sie besteht in einer Relation, die entweder verwirklicht ist oder nicht; das Licht ist im Durchsichtigen vorhanden oder nicht, etwa wie ein Gegenstand als Ganzes rechts von einem anderen steht oder nicht."

45 Among others and quoting only these occurrences, *DA* 62.12–13, *In Sens* 42.27, 50.16, 131.22–132.5, 134.11–19, *Mant.* 141.30–147.25, *Quaestio* I 2, 6.25.

46 Aristotle, *DA* II 7, 418b20–26.

47 *Sens*. 6, 446a25–447a11.

48 See also Sharples (1998), 77–79.

6 Conclusion

The foregoing analysis suggests that it is possible to reconstruct, on the basis of various passages in the corpus, a clear-cut distinction, maintained on several occasions by Alexander, between two types of sublunary modification. He proposes to draw a line between two types of change, the first, an ordinary type of change in which a given being is qualitatively affected in the long term, by the acquisition of a new form (at the end of a temporal process). The second type of change corresponds to cases in which a real modification can be observed without requiring anything other than the co-presence of beings that lend themselves to it, instantaneously and without any necessity whatsoever that the being that receives it acquire a new form. This distinction is, as such, absent from Aristotle's corpus, but it nonetheless constitutes a possible interpretation of his writings, which minimally seem to allow for this possibility.

But what, then, is the status of Alexander's specific reorganisation of Aristotle's thoughts on the subject of the affection of the *diaphanes* at the beginning of the *De sensu*? It is not clear that Alexander sees any inconsistency or even a doctrinal error on Aristotle's part. The very nature of his discourse and the fact that Aristotle clearly establishes a form of distance between what he is saying and the phenomenon described, invites the reader to be cautious to a certain extent with regard to this point. Instead, it shows Alexander's more general concern as a commentator to consolidate that which, in part, was established through him as Aristotelianism—i.e. that particular set of characteristic theses attributed by posterity to Aristotle—above all by identifying and reducing its ambiguities. Here, this is clearly the case: confronted with an Aristotelian discourse that can, at a minimum, be described as imprecise, we see how it is first clarified. Then, its theoretical consequences are identified; finally, what Aristotle must have meant is restituted and reproduced in a discourse that is seen as more rigorous and more closely corresponding to the reality of the phenomenon described.

On the whole, it is clear that, with this apparent modification, Alexander is simply following Aristotle and giving an explanation of the phenomenon of fiery emanation that conforms to what is already found in Aristotle. The cost of this procedure is, however, remarkable: in order to give an explanation that fit with the phenomenon in question, it was necessary, as Ross notes in his commentary,[49] to abandon one of the two conditions presented by Aristotle as necessary, namely the speed of ocular movement. Alexander's solution to the

49 Ross (1906), 135.

problem essentially consists in saying that one part of the eye sees the other: this is explained by the fact that the eye is pressed and spatially divided by such pressure. The speed of the eye's movement then no longer plays a role in explaining the phenomenon. Clearly, the exegesis and explanation of Aristotle does not require us to follow the text literally as it comes down to us. Beyond the agreement of principles with Aristotle's doctrines, which are truer than all others,[50] their truth implies an exercise in critical accommodation on the part of the commentator. In this case, exegesis therefore takes the form of a creative rearrangement of the Aristotelian doctrine.[51]

50 Alexander, *DA* 2.4–6.

51 I would like to thank the organisers for their invitation, members of the audience for their comments, and Chad Jorgenson for his translation. I am particularly grateful to Valérie Cordonier, Gweltaz Guyomarc'h and Pierre-Marie Morel for their insightful suggestions on earlier drafts of this paper.

Bibliography

Sources

Alexander of Aphrodisias

De anima

Accattino, Paolo and Pierluigi Donini. 1996. *Alessandro di Afrodisia, L'anima*, Rome–Bari: Biblioteca universale Laterza.

Caston, Victor. 2012. *Alexander of Aphrodisias, On the Soul. Part 1, Soul as Form of the Body, Parts of the Soul, Nourishment and Perception*, London: Bloomsbury.

De mundo

Genequand, Charles. 2001. *Alexander of Aphrodisias, On the Cosmos*. Arabic text with English Translation, Introduction and Commentary, Leiden–Boston: Brill.

De providentia

Fazzo, Silvia and Mauro Zonta. 1999. *Alessandro di Afrodisia, La Provvidenza. Questioni Sulla Provvidenza*, Milan: BUR.

Thillet, Pierre. 2003. *Alexandre d'Aphrodise, Traité De la providence*. Introduction, édition et traduction, Paris: Verdier.

In De caelo

Rescigno, Andrea. 2004. *Alessandro di Afrodisia, Commentario al De caelo di Aristotele, Frammenti del primo libro*, Amsterdam: Hakkert.

In De sensu

Towey, Alan. 2000. *Alexander of Aphrodisias, On Aristotle On Sense Perception*, London–New York: Duckworth–Cornell University Press.

In Physica

Rashed, Marwan. 2011. *Alexandre d'Aphrodise, Commentaire perdu à la Physique d'Aristote (Livres IV–VIII). Les Scholies Byzantines*. Édition, traduction et commentaire, Berlin–Boston: De Gruyter.

Mantissa

Sharples, Robert W. 2004. *Alexander of Aphrodisias, Supplement to 'On the Soul'*, London: Bloomsbury.

Sharples, Robert W. 2008a. *Alexander Aphrodisiensis, De anima libri mantissa*, Berlin–New York: De Gruyter.

Quaestiones

Sharples, Robert W. 1992–1994. *Alexander of Aphrodisias, Quaestiones 1.1–2.15* and *2.16–3.15*, London: Duckworth.

Scripta minora

Bruns, Ivo. 1892. *Alexandri Aphrodisiensis praeter commentaria scripta minora.* CAG, *Supplementum Aristotelicum* II.2, Berlin: Reimer.

Anaxagoras of Clazomenae

Curd, Patricia. 2007. *Anaxagoras of Clazomenae, Fragments and Testimonia*, Toronto: University of Toronto Press.

Aristotle

De anima

Miller, Fred. 2018. *Aristotle on the Soul and Other Psychological Works*, Oxford: Oxford University Press.

De caelo

Federspiel, Michel. 2017. *Aristote, Du ciel*, Paris: Les Belles Lettres.

De generatione et corruptione

Rashed, Marwan. 2005. *Aristote, De la génération et la corruption.* Texte établi, traduit et commenté, Paris: Les Belles Lettres.

De motu animalium

Corcilius, Klaus and Oliver Primavesi. 2018. *Aristoteles, De motu animalium.* Historisch-kritische Edition des griechischen Textes und philologische Einleitung von O. Primavesi; deutsche Übersetzung, philosophische Einleitung und Kommentar von K. Corcilius, Hamburg: Meiner.

De sensu; De memoria

Ross, W. David. 1906. *Aristotle de Sensu and de Memoria.* Text and Translation with Introduction and Commentary, Cambridge: Cambridge University Press.

Ethica

Broadie, Sarah and Christopher Rowe. 2002. Aristotle, *Nicomachean Ethics.* Translation, Introduction and Commentary, Oxford: Oxford University Press.

Burnet, John. 1900. *The Ethics of Aristotle.* Edited with and Introduction and Notes, London: Methuen & Co.

Bywater, Ingram. 1894. *Aristotelis Ethica Nicomachea.* Recognovit brevique adnotatione critica instruxit, Oxford: Clarendon Press.

Frede, Dorothea. 2020. *Aristoteles, Nikomachische Ethik*. Übersetzung mit Einleitung und Kommentar, Berlin: De Gruyter.

Gauthier, René-Antoine and Jean-Yves Jolif. 1970. *L'Éthique à Nicomaque*. Introduction, traduction et commentaire, Louvain–Paris: Publications universitaires et Béatrice-Nauwelaerts.

Rowe, Christopher. 2023. *Aristotelis Ethica Eudemia*. Recognovit brevique adnotatione critica instruxit, Oxford: Clarendon Press.

Tricot, Jules. 1959. Revised ed. 1987. *Aristote, Éthique à Nicomaque*. Nouvelle traduction, avec introduction, notes et index, Paris: Vrin.

Metaphysica

Baghdassarian, Fabienne. 2019. Aristote. *Métaphysique, Livre Lambda*. Présenté et traduit, Paris: Vrin.

Judson, Lindsay. 2019. Aristotle, *Metaphysics, Book Λ*. Translated with an Introduction and Commentary, Oxford: Clarendon Press.

Meteorologica

Groisard, Jocelyn. 2008. *Aristote: Météorologiques*, Paris: Flammarion.

Physica

Charlton, William. 1970. *Aristotle's Physics Books I and II*, Oxford: Clarendon Press.

Pellegrin, Pierre. 2000. Revised ed. 2021. *Aristote, Physique*, Paris: Flammarion.

Ross, William D. 1936. *Aristotle's Physics*. A revised Text with Introduction and Commentary, Oxford: Clarendon Press.

Wagner, Hans. 1967. *Aristoteles, Physikvorlesung*, Berlin: Akademie Verlag.

The Complete Works of Aristotle

Barnes, Jonathan (ed.). 1984. The Revised Oxford Translation, Princeton: Princeton University Press.

Aspasius

Konstan, David. 2006. *Aspasius, On Aristotle Nicomachean Ethics 1–4, 7–8*, London: Bloomsbury.

Longpré, Hélène. 1997. *Aspasius, Commentateur du livre III de l'Éthique à Nicomaque* (58–80.3 Heylbut). Traduction, notes et commentaire. Thèse de doctorat de l'Université d'Ottawa (dir. Y. Lafrance).

Atticus

Des Places, Edouard. 1977. *Atticus, Fragments*. Texte établi et traduit, Paris: Les Belles Lettres.

Boethus of Sidon

Chiaradonna, Riccardo and Marwan Rashed. 2020. *Boéthos de Sidon—Exégète d'Aristote et philosophe*, Berlin–Boston: De Gruyter.

Critolaus of Phaselis

Wehrli, Fritz. 1959. Revised ed. 1969. *Die Schule des Aristoteles, Vol. x: Hieronymos von Rhodos, Kritolaos und seine Schüler*, Basel: Schwabe.

Diogenes Laertius

Dorandi, Tiziano. 2013. *Diogenes Laertius. A New Text edition*, Cambridge: Cambridge University Press.

Doxographi Graeci

Diels, Hermann. 1879. *Doxographi Graeci: Collegit Recensuit Prolegomenis Indicibusque Instruxit*, Berlin: Reimer.

Eudemus of Rhodes

Wehrli, Fritz. 1955. Revised ed. 1969. *Die Schule des Aristoteles, Vol. VIII: Eudemos von Rhodos*, Basel: Schwabe.

Stork, Peter (draft), *Eudemus of Rhodes: The Sources, Text and Translation.*

Galen

Bazou, Athena. 2011. *Γαληνοῦ Ὅτι ταῖς τοῦ σώματος κράσεσι αἱ τῆς ψυχῆς δυνάμεις ἕπονται*, Athens: Academy of Athens Press.

Singer, Peter (ed.). 2013. *Galen, Psychological Writings*, Cambridge: Cambridge University Press.

Ocellus Lucanus

Harder, Richard. 1926. *Ocellus Lucanus. Text und Kommentar*, Berlin: Weidmann.

Philoponus

Broadie, Sarah. 2011. *On Aristotle Physics 4.10–14*, London: Bloomsbury.

Vitelli, Johannes G. 1887. *Ioannis Philoponi in Aristotelis Physicorum Libros Tres Priores Commentaria*. CAG, Vol. XVI, Berlin: Reimer.

Priscian of Lydia

Bywater, Ingram. 1886. *Prisciani Lydi quae extant: Metaphrasis in Theophrastum et Solutionum ad Chosroem liber*, Berlin: Reimer.

Huby, Pamela. 1997. *Priscian: On Theophrastus on Sense-Perception*, London: Bloomsbury.

Pseudo-Aristotle

De coloribus

Ferrini, Maria Fernanda. 1999. *Pseudo Aristotele, I colori*. Edizione critica, traduzione e commento, Pisa: ETS.

Hett, Walter S. 1936. *Aristotle, On Colours*, Loeb Classical Library, 307, Cambridge, MA: Harvard University Press.

Loveday, Thomas and Edward S. Forster. 1984. "Aristotle, *On Colours*". In: Jonathan Barnes (ed.), *The Complete Works of Aristotle. The Revised Oxford Translation*, Princeton: Princeton University Press, 2659–2680.

Wöhrle, Georg. 1999. *Aristoteles, Opuscula* V (*De Coloribus*), Berlin: Akademie Verlag.

De mundo

Federspiel, Michel. 2018. *Pseudo-Aristote, Du monde. Positions et dénominations des vents. Des plantes*, Paris: Les Belles Lettres.

Forster, Edward S. 1914. *De Mundo*. The Works of Aristotle in English, Vol. 3, Oxford: Clarendon Press.

Furley, David. 1955. *Aristotle, On Sophistical Refutations. On Coming-to-Be and Passing-Away. On the Cosmos*, Cambridge MA–London: Harvard University Press.

Reale, Giovanni and Abraham P. Bos. 1995. *Il trattato Sul cosmo per Alessandro attribuito ad Aristotele*, Milan: Vita e Pensiero.

Strohm, Hans. 1970. *Aristoteles, Meteorologie. Über die Welt*, Berlin: Akademie Verlag.

Pseudo-Simplicius

Steel, Carlos, in collaboration with James O. Urmson, notes by Peter Lautner. 1997. *"Simplicius", On Aristotle On the Soul 2.5–12*, London: Bloomsbury.

Ptolemy

Lejeune, Albert. 1956. *L'Optique de Claude Ptolémée dans la version latine d'après l'arabe de l'émir Eugène de Sicile*, Leiden: Brill.

Smith, A. Mark. 1996. *Ptolemy's Theory of Visual Perception. An English Translation of the Optics. With Introduction and Commentary*, Philadelphia: The American Philosophical Society.

Ptolemy al-Gharīb

Rashed, Marwan. 2021. *Ptolémée «al-Gharīb»: Épître à Gallus sur la vie, le testament et les écrits d'Aristote*, Paris: Les Belles Lettres.

Simplicius

In Categorias

Hadot, Ilsetraut (dir.), in collaboration with Philippe Hoffmann and Concetta Luna. 1990. *Simplicius, Commentaire sur les Catégories*. Fascicule III: Préambule aux *Cat-*

égories / Commentaire au premier chapitre des *Catégories* (pp. 21–40, 13 Kalbfleisch), Leiden–Boston: Brill.

In Physica

Baltussen, Han, Michael Share, Michael Atkinson and Ian Mueller. 2013. *Simplicius, On Aristotle Physics 1.5–9*, London–Ithaca: Duckworth–Cornell University Press.

Bodnár, István, Michael Chase and Michael Share. 2012. *Simplicius, On Aristotle Physics 8.1–5*. Bristol: Bristol Classical Press.

Diels, Hermann. 1882a. *Simplicii in Aristotelis Physicorum Libros Quattuor Priores Commentaria*. CAG, Vol. IX, Berlin: Reimer.

Fleet, Barrie. 1997. *Simplicius, On Aristotle Physics 2*, Ithaca: Cornell University Press.

Konstan, David. 1989. *Simplicius, On Aristotle Physics 6*. London: Duckworth.

McKirahan, Richard. 2001. *Simplicius, On Aristotle Physics 8.6–10*. London: Duckworth.

Urmson, James O. 2002. *Simplicius, On Aristotle Physics 3*. London: Duckworth.

Urmson, James O. and Péter Lautner. 1992. *Simplicius, On Aristotle Physics 4.1–5, 10–14*. London: Duckworth.

Corollaries on Place and Time. Urmson, James O. 1992. *Simplicius, Corollaries on Place and Time*. London–Ithaca: Duckworth.

Stoics

von Arnim, Hans F.A. 1903–1905. *Stoicorum veterum fragmenta*, Leipzig: Teubner.

Strato of Lampsacus

Sharples, Robert W. 2017. "Strato of Lampsacus: The Sources, Texts and Translations". In: Marie-Laurence Desclos and William W. Fortenbaugh. (eds.). 2011. *Strato of Lampsacus*. Text, Translation, and Discussion, New Brunswick: Transaction Publishers. Revised ed.: London–New York: Routledge, 5–229.

Wehrli, Fritz. 1950. Revised ed. 1969. *Die Schule des Aristoteles, Vol. V: Straton von Lampsakos*, Basel: Schwabe.

Themistius

In De anima

Transl. Todd, Robert B. 1996. *Themistius, On Aristotle On the Soul*, London: Duckworth.

In Physica

Transl. Todd, Robert B. 2003. *Themistius. On Aristotle Physics 4*, London: Duckworth.

Theon of Smyrna

Martin, Thomas H. 1849. *Theonis Smyrnaei Platonici Liber de astronomia cum Sereni fragmento*. Textum primus edidit, latine vertit Th.H. Martin, Paris: Reipublicae typographeo.

Petrucci, Federico M. 2012b. *Teone di Smirne. Expositio rerum mathematicarum ad legendum Platonem utilium*. Introduzione, traduzione, commento, Sankt Augustin: Academia Verlag.

Theophrastus

Coutant, Victor. 1971. *Theophrastus: De igne*, Assen: Royal Van Gorcum.

Eigler, Ulrich and Georg Wöhrle. 1993. *Theophrast, De odoribus*. Edition, Übersetzung, Kommentar, Stuttgart: Teubner.

Einarson, Benedict and George K.K. Link. 1990. *Theophrastus, De causis plantarum. Books V–VI*, Cambridge, Mass.: Harvard University Press.

Fortenbaugh William, Pamela M. Huby, Robert W. Sharples, and Dimitri Gutas. 1992. *Theophrastus of Eresus, Sources for his Life, Writings, Thought, and Influence*, Leiden–Boston: Brill.

Fortenbaugh, William W., Robert W. Sharples, and Michael G. Sollenberger. 2003. *Theophrastus of Eresus. On Sweat, On Dizziness and On Fatigue*, Leiden–Boston: Brill.

Gutas, Dimitri. 2010. *Theophrastus, On First Principles (Known as His Metaphysics). Greek Text and Medieval Arabic Translation, Leiden–Boston: Brill.*

Hort, Arthur. 1916. *Theophrastus, Enquiry Into Plants and Minor Works on Odours and Weather Signs*, London: Heinemann.

Ierodiakonou, Katerina, and Börje Bydén. Forthcoming. *Theophrastus, De sensibus*. Critical Edition, English Translation and Commentary.

Repici, Luciana. 2013. *Teofrasto, Metafisica*. Introduzione, traduzione e commento, Rome: Carocci.

Schneider, Johann G. 1818–1821. *Theophrastou Eresii ta sooizomena. Theophrasti Eresii quae supersunt opera et excerpta librorum quatuor tomis comprehensa*. Ad fidem librorum editorum et scriptorum emendavit coniuncta opera D.H.F. Linkii excerpta solus explicare conatus est Io. Gottlob Schneider, Saxo. Leipzig: Vogelli.

Van Raalte, Marlein. 1993. *Theophrastus, Metaphysics*. Introduction, Translation and Commentary, Leiden–Boston: Brill.

Thomas Aquinas

Commissio Leonina. 1969. *Sententia libri Ethicorum, Opera Omnia*. T. 47/1. Rome–Paris.

Secondary Literature

Accattino, Paolo. 1988. "Alessandro di Afrodisia e la trasmissione della forma nella riproduzione animale", *Atti dell'Accademia delle Scienze di Torino* 122, 79–94.

Accattino, Paolo. 1992. "Alessandro di Afrodisia e la luce", *Atti dell'Accademia delle Scienze di Torino* 126, 39–62.

Accattino, Paolo. 2009. "Alejandro de Afrodisia intérprete del *De Anima* de Aristóteles", *Estudios de Filosofía* 40, 53–78.

Alberti, Antonina and Robert W. Sharples (eds.). 1999. *Aspasius. The Earliest Extant Commentary on Aristotle's Ethics*, Berlin–New York: De Gruyter.

Algra, Kempe. 1995. *Concepts of Space in Greek Thought*, Leiden–Boston: Brill.

Algra, Kempe. 2014. "Plutarch and the Stoic Theory of Providence". In: Pieter d'Hoine and Gerd Van Riel (eds.), *Fate, Providence and Moral Responsibility in Ancient, Medieval and Early Modern Thought*. Studies in Honour of Carlos Steel, Leuven: Leuven University Press, 117–135.

Angioni, Lucas. 2010. "Aristóteles e o progresso da investigação científica: o caso do *De caelo*". *Scientiae Studia* 8 (3), 319–338.

Annas, Julia. 1990. "The Hellenistic version of Aristotle's Ethics", *The Monist* 73, 80–96.

Auffret, Thomas. 2015. "Un témoignage négligé de Théophraste sur la théorie platonicienne des lignes". In: Annick Jaulin and David Lefebvre (eds.), *La Métaphysique de Théophraste. Principes et apories*, Louvain: Peeters, 17–36.

Auffret, Thomas. 2020. "La doctrine de l'âme". In: Riccardo Chiaradonna and Marwan Rashed (eds.) *Boéthos de Sidon—Exégète d'Aristote et philosophe*, Berlin–New York: De Gruyter, 361–402.

Babut, Daniel. 1974. 2019. *La Religion des philosophes grecs. De Thalès aux Stoïciens*, Paris: Les Belles Lettres.

Bakhouche, Béatrice. 2011. *Calcidius: Commentaire au Timée de Platon*, Paris: Vrin.

Bakker, Frederik A. 2016. *Epicurean Meteorology*, Leiden–Boston: Brill.

Baltussen, Han. 1998. "The Purpose of Theophrastus' *De sensibus* Reconsidered", *Apeiron* 31 (2), 167–200.

Baltussen, Han. 2000. *Theophrastus Against the Presocratics and Plato. Peripatetic Dialectic in the De sensibus*, Leiden–Boston: Brill.

Baltussen, Han. 2002a. "Philology or Philosophy? Simplicius on the Use of Quotations". In: Ian Worthington (ed.), *Epea and Grammata: Oral and Written Communication in Ancient Greece*, Vol. IV, Leiden–Boston: Brill, 173–189.

Baltussen, Han. 2002b. "Wehrli's Edition of Eudemus of Rhodes. The physical fragments in Simplicius' *Commentary on Aristotle's Physics*". In: William W. Fortenbaugh and István Bodnár (eds.), *Eudemus of Rhodes*, New Jersey: Transaction Publishers, 127–156.

Baltussen, Han. 2002c. "Theophrastean Echoes? Theophrastus' *De sensibus* in the Platonic and Aristotelian Tradition (Alcinous, Alexander, Plotinus)". In: William W. Fortenbaugh and Georg Wöhrle (eds.), *On the Opuscula of Theophrastus*. Stuttgart: Steiner, 39–58.

Baltussen, Han. 2003. "Early Reactions to Plato's *Timaeus*. Polemic and Exegesis in Theophrastus and Epicurus". In: Robert W. Sharples and Ann Sheppard (eds), *Ancient Approaches to the "Timaeus"* (BICS Suppl. 78), London, 49–71.

Baltussen, Han. 2005. "The Presocratics in the Doxographical Tradition. Sources, Controversies, and Current Research", *Studia Humaniora Tartuensia* 6, URL = https://ojs.utlib.ee/index.php/sht/article/view/6.A.6.

Baltussen, Han. 2006. "Addenda Eudemea", *Leeds International Classical Studies*, Vol. V.1, URL = https://web.archive.org/web/20141129235438/http://lics.leeds.ac.uk/2006/200601.pdf.

Baltussen, Han. 2008. *Philosophy and Exegesis in Simplicius. The Methodology of a Commentator*, London: Duckworth.

Baltussen, Han. 2009. "Simplicius and the Subversion of Authority", *Antiquorum Philosophia. An International Journal* 3, 121–136.

Baltussen, Han. 2010. "Simplicius". In: Lloyd Gerson (ed.), *Cambridge History of Philosophy in Late Antiquity*, Vol. II, Cambridge: Cambridge University Press, 711–732.

Baltussen, Han. 2015a. "Ancient Philosophers on the Sense of Smell". In: Mark Bradley (ed.), *Smell and the Ancient Senses*, New York: Routledge, 30–45.

Baltussen, Han. 2015b. "Simplicius on Elements and Causes in Greek Philosophy: Critical Appraisal or Philosophical Synthesis?". In: Anna Marmodoro and Brian D. Prince (eds.), *Causation and Creation in Late Antiquity*, Cambridge: Cambridge University Press, 213–244.

Baltussen, Han. 2016a. "Philosophers, Scholars, Exegetes. The Ancient Philosophical Commentary from Plato to Simplicius". In: Christina Kraus and Christopher Stray (eds.), *Classical Commentaries*, Oxford: Oxford University Press, 173–194.

Baltussen, Han. 2016b. *The Peripatetics. Aristotle's Heirs 322 BCE–200 CE*, London: Routledge.

Baltussen, Han. 2019. "Theophrastus' *De sensibus* in A-fragments of Diels-Kranz: Revisiting the Testimonia and their Value", *Rhizomata. Journal for Ancient Philosophy and Science* 7(2), 120–145.

Baltussen, Han. 2022. "'Received Opinions' in Aristotle, Theophrastus and Simplicius: Doxography or Endoxography?". In: Andreas Lammer and Mareike Jas (eds), *Received Opinions: Doxography in Antiquity and the Islamic World*, Leiden–Boston: Brill, 151–174.

Barker, Andrew. 1989. *Greek Musical Writings II: Harmonic and Acoustic Theory*, Cambridge: Cambridge University Press.

Barker, Andrew. 2007. *The Science of Harmonics in Classical Greece*, Cambridge: Cambridge University Press.

Barnes, Jonathan. 1982. *The Presocratic Philosophers*, Revised ed., New York: Routledge.

Barnes, Jonathan. 1992. "Metacommentary", *Oxford Studies in Ancient Philosophy* 10, 267–281.

Barnes, Jonathan. 1999. "An Introduction to Aspasius". In: Antonina Alberti and Robert W. Sharples (eds.), *Aspasius. The Earliest Extant Commentary on Aristotle's Ethics*, Berlin–New York: De Gruyter, 1–50.

Barnes, Jonathan. 1997. "Roman Aristotle". In: Jonathan Barnes and Miriam Griffin (eds.) *Philosophia Togata II: Plato and Aristotle at Rome*, Oxford: Clarendon Press, 1–69.

Battegazzore, Antonio M. 1984. "Aristotelismo e anti-aristotelismo del *De igne* teofrasteo", *Elenchos* 5, 45–102.

Becchi, Francesco. 1994. "Aspasio, commentatore di Aristotele." In: Wolfgang Haase and Hildegard Temporini (eds.), *Aufstieg und Niedergang der römischen Welt*, Band II.36.2, Berlin–New York: De Gruyter, 5365–5396.

Bergk, Theodor. 1850. "Rezension von T. Martin, *Theonis Platonici Liber de astronomia*", *Zeitschrift für die Altertumswissenschaft* 8, 174–181.

Berryman, Silvia. 1996. *Rethinking Aristotelian Teleology: The Natural Philosophy of Strato of Lampsacus*, Texas.

Berryman, Sylvia. 2007. "Teleology without Tears: Aristotle and the Role of Mechanistic Conceptions of Organisms", *Canadian Journal of Philosophy* 37, 351–370.

Betegh, Gábor and Pavel Gregorić. 2014. "Multiple Analogy in Ps.-Aristotle, *De Mundo* Ch. 6", *Classical Quarterly* 64, 574–591.

Betegh, Gábor and Pavel Gregorić. 2020. "God's Relation to the Cosmos (Chapter 6)". In: Pavel Gregorić and George Karamanolis (eds.), *Pseudo-Aristotle, De Mundo (On the Cosmos). A Commentary*, Cambridge: Cambridge University Press, 176–212.

Betegh, Gábor. 2004. *The Derveni Papyrus. Cosmology, Theology and Interpretation*, Cambridge: Cambridge University Press.

Betegh, Gábor. 2021. "Thinking With Empedocles: Aristotle on the Soul as *Harmonia*", *Oxford Studies in Ancient Philosophy* 59, 1–44.

Bodéüs, Richard. 1992. *Aristote et la théologie des vivants immortels*. Québec–Paris: Bellarmin–Les Belles Lettres.

Bodnár, István and William W. Fortenbaugh (eds.). 2002. *Eudemus of Rhodes*, New Brunswick: Transaction Publishers.

Bodnár, István. 1997. "Movers and Elemental Motions in Aristotle", *Oxford Studies in Ancient Philosophy* 15, 81–117.

Bodnár, István. 2002a. "Eudemus' Unmoved Movers: Frs. 121–123b Wehrli". In: István Bodnár and William W. Fortenbaugh (eds.), *Eudemus of Rhodes*, New Brunswick: Transaction Publishers, 171–189.

Bodnár, István. 2002b. "Theophrastus' *De igne*". In: W.W. Fortenbaugh and G. Wöhrle (eds.), *The Opuscula of Theophrastus*, Stuttgart: Steiner, 75–90.

Bodnár, István. 2014. "Alexander's Unmoved Mover," in Cristina Cerami (ed.), *Nature et sagesse: Les rapports entre physique et métaphysique dans la tradition aristotélicienne—Recueil de textes en hommage à Pierre Pellegrin*, Louvain-la-Neuve: Peeters, 387–416.

Bodnár, István. 2015. "The *Problemata physica*: An Introduction". In: Robert Mayhew (ed.), *The Aristotelian Problemata Physica: Philosophical and Scientific Investigations*, Leiden–Boston: Brill, 1–9.

Bonazzi, Mauro. 2007. "Eudorus' psychology and Stoic ethics". In: Mauro Bonazzi and Christopher Helmig (eds.), *Platonic Stoicism—Stoic Platonism. The Dialogue between Platonism and Stoicism in Antiquity*, Leuven: Leuven University Press, 109–132.

Bonazzi, Mauro. 2020. "The End of the Academy". In: Paul Kalligas, Chloe Balla, Effie Baziotopoulou-Valavani and Vassilis Karasmanis (eds.), *Plato's Academy: Its Workings and Its History*, Cambridge: Cambridge University Press, 242–255.

Bonelli, Maddalena. 2014. "Alexandre d'Aphrodise et la cause ποιητική". In: Carlo Natali and Cristina Viano (eds.), *Aitia II. Avec ou sans Aristotle: Le Débat sur les causes à l'âge hellénistique et imperial*, Louvain-la-Neuve: Peeters, 119–135.

Bos, Abraham. 2021. "Review of Gregorić and Karamanolis (2020)", BMCR 2021.06.24, URL = https://bmcr.brynmawr.edu/2021/2021.06.24.

Boureau, Mai-Lan. 2022. "Une occasion manquée? La critique implicite de la *chôra* dans le livre III du *De caelo* d'Aristote", *Chôra* 20, 71–91.

Boureau, Mai-Lan. Forthcoming. "Un ou plusieurs mouvements contre nature? Sur une contradiction apparente dans le traité *Du ciel* d'Aristote". In: Ulysse Chaintreuil and Alice de Fornel (eds.), *La nature et ses autres. Penser le «contre-nature» chez Aristote*, Paris: Éditions de la Sorbonne.

Bowen, Alan C. and Robert B. Todd. 2004. *Cleomedes' Lectures on Astronomy: a Translation of The Heavens*, Berkeley: University of California Press.

Boyer, Carl B. 1959. *The Rainbow. From Myth to Mathematics*, The Sagamore Press, Inc. Revised ed.: Princeton, 1987.

Boys-Stones, George. 2018. *Platonist Philosophy 80 BC to AD 250. An Introduction and Collection of Sources in Translation*, Cambridge: Cambridge University Press.

Brandis, Christian A. 1860. *Handbuch der Geschichte der griechisch-römischen Philosophie, III.1. Übersicht über das Aristotelische Lehrgebäude und Erörterung der Lehren seiner nächsten Nachfolger*, Berlin: Reimer.

Broadie, Sarah. 1993. "Aristotle's Perceptual Realism", *The Southern Journal of Philosophy* 31, 137–159.

Broadie, Sarah. 1993. "Que fait le premier moteur d'Aristote? (Sur la théologie du livre Lambda de la *Métaphysique*)", *Revue philosophique de la France et de l'étranger* 2, 375–411.

Brumana, Selene. 2023. *La scienza del divino nel Peri kosmou pseudo-aristotelico*, Milan: Vita e Pensiero.

Burnikel, Walter. 1974. *Textgeschichtliche Untersuchungen zu neun Opuscula Theophrasts*. Wiesbaden: Steiner.

Burns, Dylan M. 2020. *Did God Care? Providence, Dualism, and Will in Later Greek and Early Christian Philosophy*, Leiden–Boston: Brill.

Burnyeat, Myles F. 1992. "Is an Aristotelian Philosophy of Mind Still Credible? (A Draft)". In: Martha C. Nussbaum and Amelie Oksenberg Rorty (eds.), *Essays on Aristotle's De Anima*, Oxford: Clarendon Press, 15–26.

Burnyeat, Myles F. 1995. "How Much Happens When Aristotle Sees Red and Hears Middle C? Remarks on *De Anima* 2. 7–8". In: Martha C. Nussbaum and Amelie Oksenberg Rorty (eds.), *Essays on Aristotle's De Anima*, Oxford: Clarendon Press, 421–434.

Burnyeat, Myles F. 2001. "Aquinas on 'Spiritual Change' in Perception". In: Dominik Perler (ed.), *Ancient and Medieval Theories of Intentionality*, Leiden–Boston: Brill, 129–153.

Burnyeat, Myles F. 2002. "*De Anima* II 5", *Phronesis* 47 (1), 28–90.

Burnyeat, Myles F. 2004. "Aristotle and the Foundation of Sublunary Physics". In: Jaap Mansfeld and Frans A.J. de Haas (eds.), *Aristotle, On Generation and Corruption, Book I*. Symposium Aristotelicum, Oxford: Oxford University Press, 7–24.

Carter, Jason W. 2018. "Does the Soul Weave? Reconsidering *De Anima* 1.4, 408a29–b18", *Phronesis* 63, 25–63.

Caston, Victor. 1997. "Epiphenomenalisms, ancient and modern", *The Philosophical Review* 106, 309–363.

Caston, Victor. 2005. "The Spirit and the Letter: Aristotle on Perception". In: Ricardo Salles (ed.), *Metaphysics, Soul and Ethics: Themes from the Work of Richard Sorabji*, Oxford: Oxford University Press, 245–320.

Caston, Victor. 2019. "Theophrastus on Perceiving", *Rhizomata* 7, 188–225.

Caston, Victor. 2020. "Aristotle on the Transmission of Information: Receiving Form without the Matter". In: David Bennett and Juhana Toivanen (eds.), *Philosophical Problems in Sense Perception: Testing the Limits of Aristotelianism* (Studies in the History of Philosophy of Mind), Cham: Springer, 15–55.

Charles, David. 1988. "Aristotle on Hypothetical Necessity and Irreducibility", *Pacific Philosophical Quarterly* 69, 1–53.

Charles, David. 2000. *Aristotle on Meaning and Essence*, Oxford: Clarendon Press.

Charles, David. 2012. "Teleological Causation". In: Christopher Shields (ed.), *The Oxford Handbook of Aristotle*, Oxford: Oxford University Press, 227–266.

Chiaradonna, Riccardo. 2009. "Le traité de Galien *Sur la démonstration* et sa postérité tardo-antique". In: Riccardo Chiaradonna and Franco Trabattoni (eds.), *Physics and Philosophy of Nature in Greek Neoplatonism. Proceedings of the European Science Foundation Exploratory Workshop*, Leiden–Boston: Brill, 43–77.

Chiaradonna, Riccardo. 2020. "La substance et la forme". In: Riccardo Chiaradonna and Marwan Rashed (eds.), *Boéthos de Sidon—Exégète d'Aristote et philosophe*, Berlin–New York: De Gruyter, 143–178.

Chiaradonna, Riccardo. 2023. "Plotinus on Demiurgic Causation". In: Riccardo Chiaradonna, *Ontology in Early Neoplatonism. Plotinus, Porphyry, Iamblichus*. Berlin–Boston: De Gruyter. 29–50.

Chiaradonna, Riccardo and Rashed, Marwan. 2020. "Introduction". In: Riccardo Chiaradonna and Marwan Rashed (eds.), *Boéthos de Sidon—Exégète d'Aristote et philosophe*, Berlin–New York: De Gruyter, 1–16.

Connell, Sophia. 2016. *Aristotle on Female Animals. A Study of the Generation of Animals*, Cambridge: Cambridge University Press.

Coope, Ursula. 2004. "Aristotle's Account of Agency in *Physics* III.3", *Proceedings of the Boston Area Colloquium in Ancient Philosophy* 20, 201–227.

Coope, Ursula. 2005. *Time for Aristotle*. Oxford: Oxford University Press.

Coope, Ursula. 2012. "Aristotle on the Infinite". In: Christopher Shields (ed.), *The Oxford Handbook of Aristotle*, Oxford: Oxford University Press, 267–286.

Cooper, John M. 2016. "Aristotelian Infinites", *Oxford Studies in Ancient Philosophy* 51, 161–206.

Corcilius, Klaus. 2019. "Teleology of the Practical in Aristotle: The Meaning of 'πρᾶξις'", *Manuscrito* 42 (4), 352–386.

Cordonier, Valérie. 2019. "Problématique passion: les catégories du changement selon Alexandre d'Aphrodise (annexe: traduction de la 'Question I, 21')". In: Véronique Brière and Juliette Lemaire (eds.), *Qu'est-ce qu'une catégorie? Interprétations d'Aristote*, Leuven: Peeters, 261–314.

Cornford, Francis M. 1937. *Plato's Cosmology: The Timaeus of Plato*. London: Routledge.

D'Ancona, Cristina. 2022. "Greek Sources in Arabic and Islamic Philosophy". In: Edward N. Zalta (ed.), *The Stanford Encyclopedia of Philosophy*, URL = https://plato.stanford.edu/archives/spr2022/entries/arabic-islamic-greek/.

Daiber, Hans. 1975. *Ein Kompendium der aristotelischen Meteorologie in der Fassung des Ḥunain ibn Isḥâq*, Amsterdam, Oxford: North-Holland Publishing Company.

Daiber, Hans. 1992. "The Meteorology of Theophrastus in Syriac and Arabic translation". In: William W. Fortenbaugh and Dimitri Gutas (eds), *Theophrastus: His Psychological, Doxographical, and Scientific Writings*, New Brunswick: Transaction Publishers, 166–293.

De Haas, Frans A.J. 2010. "Priscian of Lydia and Pseudo-Simplicius on the Soul". In: Lloyd P. Gerson (ed.), *The Cambridge History of Philosophy in Late Antiquity*, Cambridge: Cambridge University Press, 756–764.

De Haas, Frans A.J. 2021. "Deduction and Common Notions in Alexander's Commentary on Aristotle's *Metaphysics* A 1–2", *History of Philosophy & Logical Analysis* 24, 71–102.

De Haas, Frans A.J. Forthcoming. "The Notion of Praxis and the Position of *De sensu* in Aristotle's Project (436a1–b1)". In Klaus Corcilius (ed.), *Text and Ideas of Aristotle's De sensu*.

Desclos, Marie-Laurence and William W. Fortenbaugh. (eds.). 2011. *Strato of Lampsacus. Text, Translation, and Discussion*, New Brunswick: Transaction Publishers.

Dicks, D.R. 1970. *Early Greek Astronomy to Aristotle*, Ithaca–New York: Cornell University Press.

Diels, Hermann. 1882b. *Zur Textgeschichte der Aristotelischen Physik*, Berlin: Verlag der Königlichen Akademie der Wissenschaften.

Diels, Hermann. 1893. "Über das physikalische System des Straton", *Sitzungsberichte der Königlichen Preußischen Akademie der Wissenschaften zu Berlin*, 101–127.

Donini, Pierluigi. 1988. "The History of the Concept of Eclecticism," in John M. Dillon and Anthony A. Long (eds.), *The Question of Eclecticism: Studies in later Greek Philosophy*, Berkeley–Los Angeles: University of California Press, 15–33. Reprinted in: Pierluigi Donini, *Commentary and Tradition: Aristotelianism, Platonism, and Post-Hellenistic Philosophy*, edited by Mauro Bonazzi, Berlin: De Gruyter, 2010, 197–210.

Donini, Pierluigi. 1974. *Tre studi sull'aristotelismo nel II secolo d.C.*, Turin: Paravia.

Donini, Pierluigi. 1986. "Il *De anima* di Alessandro d'Afrodisia e Michele Efesio", *Rivista di Filologia e di Istruzione Classica* 96, 316–323.

Donini, Pierluigi. 1994. "Testi e commenti, manuali e insegnamento: la forma sistematica e i metodi filosofici nella filosofia postellenistica". In: Wolfgang Haase and Hildegard Temporini (eds.), *Aufstieg und Niedergang der römischen Welt*, Band II.36.7, Berlin–New York: De Gruyter, 5027–5100.

Donini, Pierluigi. 1996. "*Theia dynamis* in Alessandro di Afrodisia". In: Francesco Romano and R. Loredana Cardullo (eds.), *Dunamis nel neoplatonismo*. Florence: La Nuova Italia, 11–30. Reprinted in Pierluigi Donini, *Commentary and Tradition: Aristotelianism, Platonism, and Post-Hellenistic Philosophy*, edited by M. Bonazzi, Berlin: De Gruyter, 2010, 125–138.

Dragona-Monachou, Myrto. 1994. "Divine Providence in the Philosophy of the Empire". In: Wolfgang Haase and Hildegard Temporini (eds.), *Aufstieg und Niedergang der römischen Welt*, Band II.36.7, Berlin–New York: De Gruyter, 4418–4490.

Duhem, Pierre. 1913–1915. *Le Système du Monde. Histoire des Doctrines cosmologiques de Platon à Copernic*, Paris: Hermann.

Düring, Ingemar. 1966. *Aristoteles. Darstellung und Interpretation seines Denkens*. Heidelberg: Carl Winter.

Eastwood, Brad S. 1981. "Galen on the Elements of Olfactory Sensation", *Rheinisches Museum für Philologie* 124, 268–290.

Ebert, Theodor. 1983. "Aristotle on What Is Done in Perceiving", *Zeitschrift Für Philosophische Forschung* 37, 181–198.

Effe, Bernd. 1970. *Studien zur Kosmologie und Theologie der Aristotelischen Schrift "Über die Philosophie"*, München: C.H. Beck.

Eliasson, Erik. 2013. "The Account of the Voluntariness of Virtue in the Anonymous Peripatetic Commentary on *Nicomachean Ethics* 2–5", *Oxford Studies in Ancient Philosophy* 44, 195–231.

Ellis, John. 1990. "The Trouble with Fragrance", *Phronesis* 35, 290–302.

Engberg-Pedersen, Troels. 1987. "New Light on Theophrastus?", *The Classical Review* 37(1), 53–57.

Everson, Stephen. 1997. *Aristotle on Perception*, Oxford: Clarendon Press.

Falcon, Andrea (ed.). 2016. *Brill's Companion to the Reception of Aristotle in Antiquity*, Leiden–Boston: Brill.

Falcon, Andrea. 2005. *Aristotle and The Science of Nature: Unity without Uniformity*, Cambridge University Press, Cambridge.

Falcon, Andrea. 2011. *Aristotelianism in the First Century BCE: Xenarchus of Seleucia*, Cambridge: Cambridge University Press.

Falcon, Andrea. 2014. "The Reception of Aristotle in Antiquity", *The Oxford Handbook of Topics in Philosophy*. Online edition: Oxford Academic.

Falcon, Andrea. 2015. "The Argument of *Physics* 8". In: Mariska Leunissen (ed.), *Aristotle's Physics. A Critical Guide*, Cambridge: Cambridge University Press, 265–283.

Falcon, Andrea. 2016. "The Reception of Aristotle's *Physics* in Antiquity: Ps.-Ocellus and the Doctrine of the Eternity of the World in the Late Hellenistic and Early Post-Hellenistic World (2nd and 1st Century BC)", *Philosophia* 46, 154–169.

Falcon, Andrea. 2018. "Aristotle: An Overview". In: Alexander Jones and Liba Taub (eds.), *The Cambridge History of Science*, Cambridge: Cambridge University Press, 181–195.

Falcon, Andrea. 2020. "The Lost *On Philosophy* and Aristotle's Natural Philosophy". In: António Pedro Mequita, Simon Noriega-Olmos, Christopher Shields (eds.), *Revisiting Aristotle's Fragments. New Essays on the Fragments of Aristotle's Lost Works*, Berlin–New York: De Gruyter, 93–108.

Falcon, Andrea. 2022. "Eternalism in Aristotle and After," in Joseph Verheyden, Geert Roskam and Gerd van Riel (eds.), *From Protology to Eschatology: Competing Views on the Origin and the End of the World in Platonism and Christian Thought*, Stuttgart: Mohr Siebeck, 7–22.

Falcon, Andrea. 2024. *The Architecture of the Science of Living Beings: Aristotle and Theophrastus on Animals and Plants*, Cambridge: Cambridge University Press.

Fazzo, Silvia. 2002. *Aporia e sistema. La materia, la forma, il divino nelle Quaestiones di Alessandro di Afrodisia*. Pisa: ETS.

Fazzo, Silvia. 2008a. "L'exégèse du livre Lambda de la *Métaphysique* d'Aristote dans le *De principiis* et dans la *Quaestio* I.1 d'Alexandre d'Aphrodise", *Laval théologique et philosophique* 6 (3), 607–626.

Fazzo, Silvia. 2008b. "Nicolas l'auteur du *Sommaire de la philosophie d'Aristote*. Doutes sur son identité, sa datation, son origine", *Revue d'études grecques*, 99–126.

Fernandez, Patricio A. and Jorge Mittelmann. 2020. "When Life Imitates Art: Vital Locomotion and Aristotle's Crafts Analogy" In Hynek Bartoš and Colin. G. King (eds.), *Heat, Pneuma, and the Soul in Ancient Philosophy and Science*, Cambridge: Cambridge University Press, 260–287.

Ferrari, Franco. 2000. "I commentari specialistici alle sezioni matematiche del *Timeo*". In: Aldo Brancacci (ed.), *La filosofia in età imperiale*, Naples: Bibliopolis, 169–224.

Ferrini, Maria Fernanda. 2020. "Περὶ χρωμάτων (*Peri chrōmatōn*): Colour Formation and Investigation Method", *Cultura e Scienza del Colore—Color Culture and Science Journal* 12 (2), 16–20.

Festugière, André Jean. 1949. *La révélation d'Hermès Trismégiste. II Le Dieu cosmique*, Paris: Lecoffre.

Festugière, André Jean. 1981. *L'Idéal religieux des Grecs et l'Évangile*. Paris: Gabalda & Co.

Freudenthal, Gad. 1995. *Aristotle's Theory of Material Substance: Heat and Pneuma, Form and Soul*, Oxford: Oxford University Press.

Frey, Christopher. 2015. "Capacities and the Eternal in *Metaphysics* Θ.8 and *De Caelo*." *Phronesis* 60. 88–126.

Furley, David. 1969. "Aristotle and the Atomists on Infinity". In: Ingemar Düring (ed.), *Naturphilosophie bei Aristoteles und Theophrast*, Proceedings of the Fourth Symposium Aristotelicum, Heidelberg: Lothar Stiehm, 85–96.

Furley, David. 1985. "Strato's Theory of the Void". In: Jürgen Wiesner (ed.), *Aristoteles und seine Schule*, Vol. 1. *Aristoteles. Werk und Wirkung*, Berlin–New York: De Gruyter.

Gatzemeier, Matthias. 1970. *Die Naturphilosophie des Straton von Lampsakos. Zur Geschichte des Problems der Bewegung im Bereich des frühen Peripatos*. Monographien zur Naturphilosophie, Vol. 10, Meisenheim am Glan: Anton Hain.

Gendler, Tamar Szabo and John Hawthorne (eds.). 2002. *Conceivability and Possibility*. Oxford: Oxford University Press.

Genequand, Charles. 1984. "Quelques aspects de l'idée de nature, d'Aristote à Al-Ghazâlî", *Revue de théologie et de philosophie* 116, 105–129.

Gilbert, Otto. 1907. *Die meteorologischen Theorien des griechischen Altertums*. Leipzig: Teubner. Reprinted by G. Olms, Hildesheim, 1967.

Golitsis, Pantelis. 2008. *Les commentaires de Simplicius et de Jean Philopon à la Physique d'Aristote: tradition et innovation*. Berlin: De Gruyter.

Gotthelf, Allan. 2012. *Teleology, First Principles, and Scientific Method in Aristotle's Biology*, Oxford: Oxford University Press.

Gottschalk, Hans B. 1964. "The *De Coloribus* and Its Author", *Hermes* 92 (1), 59–85.

Gottschalk, Hans B. 1967. "Review of Steinmetz, *Theophrastos of Eresos* (1964)", *Gnomon* 39, 17–26.

Gottschalk, Hans B. 1987. "Aristotelian Philosophy in the Roman World from the Time of Cicero to the End of the Second Century AD". In: Wolfgang Haase and Hildegard Temporini (eds.), *Aufstieg und Niedergang der römischen Welt*, Band II.36.2, Berlin–New York: De Gruyter, 1079–1174.

Gourinat, Jean-Baptiste. 2015. "Les limites du finalisme chez Aristote et Théophraste". In: Annick Jaulin and David Lefebvre (eds.), *La Métaphysique de Théophraste. Principes et apories*, Leuven: Peeters, 149–178.

Greenler, Robert. 1980. *Rainbows, Halos, and Glories*, Cambridge: Cambridge University Press.

Gregorić, Pavel and George Karamanolis (eds.). 2020. *Pseudo-Aristotle: De Mundo (On the Cosmos). A Commentary*, Cambridge: Cambridge University Press.

Gregorić, Pavel and Martin Kuhar. 2014. "Aristotle's Physiology of Animal Motion: On *Neura* and Muscles", *Apeiron* 47 (1), 94–115.

Gregorić, Pavel. 2020. "The Eternity of the Cosmos (Chapter 5)". In: Pavel Gregorić and George Karamanolis (eds.), *Pseudo-Aristotle: De Mundo (On the Cosmos). A Commentary*, Cambridge: Cambridge University Press, 149–175.

Gregorić, Pavel. 2025. "Two Types of Criticism in the Pseudo-Aristotelian *De Mundo*", *Classical Quarterly* 75 (1).

Griffin, Michael. 2015. *Aristotle's Categories in the Early Roman Empire*. Oxford: Oxford University Press.

Guthrie, Kenneth S. 1987. *The Pythagorean Sourcebook. An Anthology of Ancient Writings Which Relate to Pythagoras and Pythagorean Philosophy*, Grand Rapids: Phanes Press.

Guyomarc'h, Gweltaz. 2015. *L'Unité de la Métaphysique selon Alexandre d'Aphrodise*. Paris: Vrin.

Hadot, Ilsetraut. 2002. "Simplicius or Priscianus? On the Author of the Commentary on Aristotle's *De anima*", *Mnemosyne* 55, 159–199.

Hahm, David E. 1978. "Early Hellenistic Theories of Vision and of the Perception of Color". In: Peter K. Machamer and Robert G. Turnbull (eds.), *Studies in Perception. Interrelations in the History of Philosophy and Science*, Columbus: Ohio State University Press, 60–95.

Hahm, David E. 1985. "The Stoic Theory of Change", *The Southern Journal of Philosophy* 23, 39–56.

Hahm, David E. 2007. "Critolaus and Late Hellenistic Peripatetic Philosophy". In: Anna Maria Ioppolo and David Sedley (eds.) *Pyrrhonists, Patricians, Platonizers. Hellenistic Philosophy in the Period 155–86 BC.*, Naples: Bibliopolis, 47–101.

Hankinson, Robert J. 1998. *Cause and Explanation in Ancient Greek Thought*, Oxford: Oxford University Press.

Hankinson, Robert J. 2005. "Aristotle on Kind-Crossing". In: Robert W. Sharples (ed.), *Philosophy and the Sciences in Antiquity*, Ashgate, 2005, 23–25. Reprinted: Routledge, 2019, 8–22.

Hankinson, Robert J. 2017. "Teleology and Necessity in Greek Embryology". In: Julius Rocca (ed.), *Teleology in the Ancient World: Philosophical and Medical Approaches*, Cambridge: Cambridge University Press, 242–271.

Harari, Orna. 2016. "Bodies and Explanation: Alexander of Aphrodisias on Elemental Motion", *Archiv für Begriffsgeschichte* Sonderheft 13, 508–523.

Harari, Orna. 2020. "Mathematics and the Universe: Alexander of Aphrodisias on Euclid's *Elements* I.1", *Iyyun: The Jerusalem Philosophical Quarterly* 68, 1–17.

Harriman, Benjamin. 2019. *Melissus and Eleatic Monism*. Cambridge: Cambridge University Press.

Hatzimichali, Myrto. 2013. "The texts of Plato and Aristotle in the First Century BC". In: Malcolm Schofield (ed.), *Plato, Aristotle and Pythagoras in the First Century BC*, Cambridge: Cambridge University Press, 1–27.

Hatzimichali, Myrto. 2016. "Andronicus of Rhodes and the Construction of the Aristotelian Corpus". In: A. Falcon (ed.), *Brill's Companion to the Reception of Aristotle in Antiquity*, Leiden–Boston: Brill, 81–100.

Heath, Thomas. 1949. *Mathematics in Aristotle*, Oxford: Oxford University Press.

Heidegger, Martin. 1953. *Einführung in die Metaphysik*. Tübingen: Max Niemeyer Verlag, 1953. (Gregory Fried and Richard Polt, transl. 2014. *An Introduction to Metaphysics*, 2nd ed. New Haven: Yale University Press.)

Henry, Devin. 2005. "Embryological Models in Ancient Philosophy", *Phronesis* 50 (1), 1–42.

Hensley, Ian. 2020. "The Physics of *Pneuma* in Early Stoicism". In: Sean Coughlin, David Leith and Orly Lewis (eds.), *The Concept of Pneuma after Aristotle*, Berlin: Edition Topoi, 171–201.

Hiller, Eduard. 1871. "De Adrasti peripatetici *in Platonis Timaeum commentario*", *Rheinisches Museum Für Philologie* 26, 582–589.

Horn, Christoph. 2016. "The Unity of the World-Order According to *Metaphysics* Λ 10". In: Christoph Horn (ed.), *Aristotle's Metaphysics Lambda—New Essays*. Berlin–Boston: De Gruyter, 269–293.

Huby, Pamela M. 1985. "Medieval Evidence for Theophrastus' Discussion of the Intellect". In: William W. Fortenbaugh, Pamela M. Huby and Anthony A. Long (eds.), *Theophrastus of Eresus. On his Life and Work*, New Brunswick: Transaction Publishers, 165–182.

Huby, Pamela M. 1999. *Theophrastus of Eresus: Sources for His Life, Writings, Thought and Influence. Commentary Volume 4. Psychology (Texts 265–327)*, Leiden–Boston: Brill.

Huffman, Carl. 2005. *Archytas of Tarentum: Pythagorean, Philosopher and Mathematician King*. Cambridge: Cambridge University Press.

Huffman, Carl. 2025. "Archytas". In: Edward N. Zalta & Uri Nodelman (eds.), *The Stanford Encyclopedia of Philosophy*, URL = https://plato.stanford.edu/entries/archytas/.

Hussey, Edward. 1983. *Aristotle, Physics Books III and IV*. Oxford: Clarendon Press.

Ierodiakonou, Katerina. 1993. "The Stoic Division of Philosophy", *Phronesis* 38, 57–74.

Ierodiakonou, Katerina. 2011. "Remarks on the History of an Ancient Thought Experiment". In: Katerina Ierodiakonou and Sophie Roux (eds.), *Thought Experiments in Methodological and Historical Contexts*, Leiden–Boston: Brill, 37–49.

Ierodiakonou, Katerina. 2020. "Alexander of Aphrodisias on the Reality of Mirror Images". In: Maria Gerolemou and Lilia Diamantopoulou (eds.), *Mirrors and Mirroring from Antiquity to the Early Modern Period*, London–New York: Bloomsbury, 19–28.

Ierodiakonou, Katerina. 2021. "Aristotle and Alexander on Sight as a Relative". In: Pavel

Gregorić and Jacob Leth Fink (eds.), *Encounters with Aristotelian Philosophy of Mind*, New York: Routledge, 99–118.

Ierodiakonou, Katerina. 2022. "Perceptual Attention and Reflective Awareness in the Aristotelian Tradition". In: Caleb Cohoe (ed.), *Aristotle's On the Soul: A Critical Guide*, Cambridge: Cambridge University Press, 174–194.

Ierodiakonou, Katerina. Forthcoming (a). "Plato on Taste and Smell, and the Aristotelian Criticism". In: Riccardo Chiaradonna, Alexander Nehamas, and Voula Tsouna (eds.), *New Essays on the Platonic Tradition in Honour of Paul Kalligas*, Leiden–Boston: Brill.

Ierodiakonou, Katerina. Forthcoming (b). "Flavours and the Sense of Taste". In: Klaus Corcilius (ed.), *Text and Ideas of Aristotle's De Sensu*.

Ioppolo, Anna Maria and David Sedley. (eds.). 2007. *Pyrrhonists, Patricians, Platonizers. Hellenistic Philosophy in the Period 155–86 BC*, Naples: Bibliopolis.

Jaeger, Werner. 1934. Revised ed. 1948. *Aristotle: Fundamentals of the History of His Development*, transl. Richard Robinson, Oxford: Oxford University Press.

Jaulin, Annick. 2011. "Straton et la question du temps comme nombre du mouvement". In: Marie-Laurence Desclos and William W. Fortenbaugh (eds.), *Strato of Lampsacus. Text, translation, and discussion*, London–New York: Routledge.

Jaulin, Annick and David Lefebvre (eds.). 2015. *La Métaphysique de Théophraste. Principes et apories*, Louvain: Peeters.

Johansen, Thomas K. 1996. "Aristotle on the Sense of Smell", *Phronesis* 41, 1–19.

Johansen, Thomas K. 1997. *Aristotle on the Sense-Organs*, Cambridge: Cambridge University Press.

Johansen, Thomas K. 2012. *The Powers of Aristotle's Soul*, Oxford: Oxford University Press.

Johansen, Thomas K. 2014. "Why the Cosmos Needs a Craftsman", *Phronesis* 59, 297–320.

Johansen, Thomas K. 2019. "The Principle That 'Like Perceives Like' in Theophrastus' *De sensibus*", *Rhizomata* 7, 226–248.

Johnson, Monte R. 2009. "The Aristotelian explanation of the Halo", *Apeiron* 42 (4), 325–358.

Johnstone, Mark A. 2012. "Aristotle on Odour and Smell", *Oxford Studies in Ancient Philosophy* 43, 143–183.

Jones, Alexander. 1994. "Peripatetic and Euclidean Theories of Visual Ray", *Physis* XXXI, Fasc. 1, 47–76.

Judson, Lindsay. 1993. "Eternity and Necessity in *De caelo* I 12", *Oxford Studies in Ancient Philosophy* 1, 217–255.

Judson, Lindsay. 2015. "Aristotle's Astrophysics", *Oxford Studies in Ancient Philosophy* 49, 151–192.

Kahn, Charles H. 1960. *Anaximander and the Origins of Greek Cosmology*, New York.

Kalderon, Mark E. 2017. "Priscian on Perception", *Phronesis* 62, 443–467.

Kechagia, Eleni. 2011. *Plutarch Against Colotes: A Lesson in History of Philosophy*, Oxford: Oxford University Press.

Kidd, Ian G. 1992. "Theophrastus' Meteorology, Aristotle and Posidonius". In: William W. Fortenbaugh and Dimitri Gutas (eds), *Theophrastus: His Psychological, Doxographical, and Scientific Writings*, New Brunswick: Transaction Publishers, 294–306.

Koyré, Alexandre. 1957. *From the Closed World to the Infinite Universe*, New York: Harper.

Kuhn, Thomas S. 1977. *The Essential Tension: Selected Studies in Scientific Tradition and Change*, Chicago: University of Chicago Press.

Kuhn, Thomas S. 2000. *The Road since Structure: Philosophical Essays, 1970–1993, with an Autobiographical Interview, edited by J. Conant and J. Haugeland*, Chicago: University of Chicago Press.

Kupreeva, Inna. 2003. "Qualities and Bodies: Alexander against the Stoics", *Oxford Studies in Ancient Philosophy* 25, 297–343.

Kupreeva, Inna. 2009. "Stoic Themes in Peripatetic Sources". In: Ricardo Salles (ed.), *God and Cosmos in Stoicism*, Oxford: Oxford University Press, 135–170.

Kupreeva, Inna. 2016. "Aristotelianism in the Second Century AD: Before Alexander of Aphrodisias." In Andrea Falcon (ed.), *Brill's Companion to the Reception of Aristotle in Antiquity*, Leiden–Boston: Brill, 138–159.

Kupreeva, Inna. 2018. "Kaiserzeitlicher Aristotelismus." In Christoph Horn, Christoph Riedweg and Dietmar Wyrwa (eds), *Grundriss der Geschichte der Philosophie ("Neuer Überweg"): Die Philosophie der Kaiserzeit und der Spätantike*, Vol. 5, Basel: Schwabe, 255–428.

Laks, André. 2002. "La physique de la sensation aristotélicienne selon Théophraste (*Physique*, livre 5)". In: Monique Canto-Sperber and Pierre Pellegrin (eds.), *Le style de la pensée: recueil de textes en hommage à Jacques Brunschwig*, Paris: Les Belles Lettres, 353–374.

Laks, André. 2019. "The Lever, or How to Act at a Distance: A Backdrop to Theophrastus' *De sensibus*", *Rhizomata* 7, 168–187.

Lautner, Péter. 2002. "Andronicus of Rhodes on Aristotle's *Physics* 202a13–21, b23–30", *Acta Antiqua Academiae Scientiarum Hungaricae* 42, 125–132.

Lautner, Péter. Forthcoming. "Problems with Smell. Aristotle, *De sensu* 5, 442b27–445b2". In: Klaus Corcilius (ed.), *Text and Ideas of Aristotle's De sensu.*

Lefebvre, David. 2015. "Recherches péripatéticiennes sur le moteur immobile: Aristote et Théophraste". In: Annick Jaulin and David Lefebvre (*eds.*), *La Métaphysique de Théophraste. Principes et apories*, Louvain: Peeters, 37–69.

Lefebvre, David. 2016. "Aristotle and the Hellenistic Peripatos: From Theophrastus to Critolaus." In A. Falcon (ed.), *Brill's Companion to the Reception of Aristotle in Antiquity*, Leiden–Boston: Brill, 13–34.

Lefebvre, David. 2017. "Aristote, Théophraste, Straton et la philosophie des êtres divins".

In: Fabienne Baghdassarian and Gweltaz Guyomarc'h (ed.), *Réceptions de la théologie aristotélicienne. D'Aristote à Michel d'Éphèse*, Louvain: Peeters, 59–88.

Lefebvre, David. 2026. "Aristote sur les monstres dans La *Génération des animaux*. Du contre-nature à l'autre nature". In: Ulysse Chaintreuil and Alice de Fornel (eds.), *La nature et ses autres. Penser le «contre-nature» chez Aristote*, Paris: Classiques Garnier.

Lehoux, Daryn. 2007. "Observers, Objects, and the Embedded Eye; or, Seeing and Knowing in Ptolemy and Galen", *Isis* 98 (3), 447–467.

Lennox, James. 1985. "Theophrastus on the Limits of Teleology". In: William W. Fortenbaugh, Pamela M. Huby and Anthony A. Long (eds.), *Theophrastus of Eresus: On His Life and Work*, New Brunswick: Transaction Publishers, 143–163.

Lettinck, Paul. 1999. *Aristotle's Meteorology and its Reception in the Arab World. With an Edition and Translation of Ibn Suwār Treatise on Meteorological Phenomena and Ibn Bâjja's Commentary on the Meteorology*, Leiden–Boston: Brill.

Lewis, Frank A. 1988. "Teleology and Material/Efficient Cause in Aristotle", *Pacific Philosophical Quarterly* 69, 54–98.

Lindberg, David C. 1976. *Theories of Vision from Al-Kindi to Kepler*, Chicago: The University of Chicago Press.

Lloyd, Geoffrey E.R. 1973. *Greek Science after Aristotle*. London: Chatto & Windus.

Lorenz, Hendrik. 2007. "The Assimilation of Sense to Sense-Object in Aristotle", *Oxford Studies in Ancient Philosophy* 33, 179–220.

Lynch, John P. 1972. *Aristotle's School, A Study of a Greek Educational Institution*, Berkeley/Los Angeles: University of California Press.

Maguire, Joseph P. 1939. "Sources of Pseudo-Aristotle *De mundo*", *Yale Classical Studies* 6, 111–167.

Mansfeld, Jaap and David T. Runia. 1997. *Aëtiana. The Method and Intellectual Context of a Doxographer. Vol. I, The Sources*, Leiden–Boston: Brill.

Mansfeld, Jaap and David T. Runia. 2020. *Aëtiana V. An Edition of the Reconstructed Text of the* Placita *with a Commentary and a Collection of Related Texts*, Leiden–Boston: Brill.

Mansfeld, Jaap. 1989. "Gibt es Spuren von Theophrasts *Phys. Op.* bei Cicero?". In: William W. Fortenbaugh and Peter Steinmetz (eds.), *Cicero's Knowledge of the Peripatos*, New Brunswick: Transaction Publishers, 133–158.

Mansfeld, Jaap. 1990. "Doxography and Dialectic: The *Sitz im Leben* of the *Placita*". In: Wolfgang Haase and Hildegard Temporini (eds.), *Aufstieg und Niedergang der römischen Welt*, Band II.36.4, Berlin–New York: De Gruyter, 3056–3230.

Mansfeld, Jaap. 1992. "*Physikai doxai* and *Problêmata physika* from Aristotle to Aëtius (and Beyond)". In: William W. Fortenbaugh and Dimitri Gutas (eds.), *Theophrastus. His Psychological, Doxographical and Scientific Writings*, New Brunswick: Transaction Publishers, 63–111.

Mansfeld, Jaap. 1994. *Prolegomena: Questions to be Settled before the Study of an Author, or a Text*, Leiden–Boston: Brill.

Mansfeld, Jaap. 1996. "Aristote et la structure du *De sensibus* de Théophraste", *Phronesis* 41, 158–188.

Mansfeld, Jaap. 2005. "From Milky Way to Halo". In: Aldo Brancacci (ed), *Philosophy and Doxography in the Imperial Age*, Florence: Leo S. Olschki, 2005, 23–58.

Mayhew, Robert. Forthcoming. "Critolaus on God and Divine Providence". In: Robert Mayhew and Stefan Schorn (eds.), *Critolaus of Phaselis: Essays on Second Century BC Peripatetic Philosophy*, London–New York: Routledge.

McDiarmid, John B. 1953. "Theophrastus on the Presocratic Causes", *Harvard Studies in Classical Philology* 61, 85–156.

Menn, Stephen. 2012. "Aristotle's Theology". In: Christopher Shields (ed.), *The Oxford Handbook of Aristotle*, Oxford: Oxford University Press, 412–464.

Menn, Stephen. 2022. *Simplicius. On Aristotle's Physics 1–8. General Introduction to the 12 Volumes of Translations*. London–New York: Bloomsbury.

Merker, Anne. 2002. "Aristote et l'Arc-en-Ciel: enjeux philosophiques et étude scientifique", *Archive for the History of Exact Sciences* 56, 183–238.

Michalewski, Alexandra. 2017. "Faut-il préférer Épicure à Aristote? Quelques réflexions sur la providence". In: Fabienne Baghdassarian and Gweltaz Guyomarc'h (eds), *Réceptions de la théologie aristotélicienne. D'Aristote à Michel d'Éphèse*, Louvain: Peeters, 123–142.

Michalewski, Alexandra. 2024. *Le dieu, le mouvement, la matière. Atticus et ses critiques dans l'Antiquité tardive*. Paris: Les Belles Lettres.

Mignucci, Mario. 1990. "Aristotle's *De caelo* I 12 and his Notion of Possibility". In: Daniel Devereux and Pierre Pellegrin, *Biologie, logique et métaphysique chez Aristote*, Paris: Éditions du CNRS, 321–334.

Minio-Paluello, Lorenzo. 1972. "La tradition aristotélicienne dans l'histoire des idées". In: *Opuscula. The Latin Aristotle*, Amsterdam: Hakkert, 405–424.

Modrak, Deborah. 2002. "*Phantasia*, Thought and Science in Eudemus". In: István Bodnár and William W. Fortenbaugh (eds.), *Eudemus of Rhodes*, New Brunswick: Transaction Publishers, 191–206.

Moraux, Paul. 1949. "L'exposé de la philosophie d'Aristote chez Diogène Laërce (V, 28–34)", *Revue Philosophique de Louvain* 47 (1), 5–43.

Moraux, Paul. 1967. "Alexander von Aphrodisias *Quaest.* 2,3", *Hermes* 95, 159–169.

Moraux, Paul. 1970. *D'Aristote à Bessarion. Trois exposés sur l'histoire et la transmission de l'Aristotélisme grec*, Québec: Les Presses de l'Université Laval.

Moraux, Paul. 1973. *Der Aristotelismus bei den Griechen. Von Andronikos bis Alexander von Aphrodisias, I. Band: Die Renaissance des Aristotelismus im I. Jh. v.Chr.*, Berlin–New York: De Gruyter.

Moraux, Paul. 1984. *Der Aristotelismus bei den Griechen. Von Andronikos bis Alexander von Aphrodisias, II. Band: Der Aristotelismus im I. und II. Jh. n. Chr.*, Berlin–New York: De Gruyter.

Moraux, Paul. 2001. *Der Aristotelismus bei den Griechen. Von Andronikos bis Alexander von Aphrodisias, III. Band: Alexander von Aphrodisias*, edited by J. Wiesner, Berlin–Boston: De Gruyter.

Morison, Benjamin. 2020. "Completing the Argument that Locomotion Requires an External and Unmoved Mover: *De motu animalium* 4–5". In: Christof Rapp and Oliver Primavesi (eds.), *Aristotle's De motu animalium*, Oxford: Oxford University Press.

Natali, Carlo. 2023. "Un commento di età imperiale al libro secondo dell'*Etica Nicomachea*. Traduzione con introduzione e note", *Elenchos*, 44 (1), 1–44.

Netz, Reviel. 2001. "On the Aristotelian Paragraph," *Proceedings of the Cambridge Philological Society* 47, 211–232.

Paparazzo, Emiliano. 2023. "Aristotle's Threefold Conception of Fire in *Meteor.* I–III", *Physis* 63 (1), 1–26.

Perkams, Matthias. 2005. "Priscian of Lydia, Commentator on the *De anima* in the Tradition of Iamblichus", *Mnemosyne* 58, 510–530.

Perkams, Matthias. 2019. "The Date and Place of Andronicus' Edition of Aristotle's Works According to a Neglected Arabic Source", *Archiv für Geschichte der Philosophie* 101, 445–468.

Perry, Bruce M. 1983. *Simplicius as a Source for and Interpreter of Parmenides*, PhD thesis, University of Washington.

Petrucci, Federico M. 2012a. "Il *Commento al Timeo* di Adrasto di Afrodisia", *Documenti e studi sulla tradizione filosofica medievale* 23, 1–33.

Petrucci, Federico M. 2012b. *Teone di Smirne. Expositio rerum mathematicarum ad legendum Platonem utilium. Introduzione, traduzione, commento*, Sankt Augustin: Academia Verlag.

Petrucci, Federico M. 2015. "Adrastus on Aristotle's Cosmology. The Peripatetic Exegesis of *De caelo* and *Metaphysics* Λ", *Rhizomata* 3 (2), 159–199.

Petrucci, Federico M. 2016. "L'esegeta e il cielo del *Timeo*: riargomentazione ed esegesi astronomica κατὰ ζητήματα nel Medioplatonismo", *Athenaeum* 104, 158–186.

Petrucci, Federico M. 2019. "Making Sense of the Soul's Numbers. Middle Platonist Readings of Plato's *Divisio Animae*", *Apeiron* 52 (1), 65–91.

Petrucci, Federico M. 2020. "Ascoltare l'anima cosmica: riargomentazione ed esegesi tecnica κατὰ ζητήματα della *divisio animae* platonica". In: Christoph Helmig (ed.), with the collaboration of Laura Marongiu, *World Soul—Anima Mundi. On the Origin and Fortunes of a Fundamental Idea*, Berlin–New York: De Gruyter, 91–134.

Philippson, Ludwig. 1831. *Ὕλη ἀνθρωπίνη*, Berlin: Sumtibus J.A. List.

Prantl, Karl. 1849. *Aristoteles. Über die Farben. Erläutert durch eine Übersicht der Farbenlehre der Alten*, München: Christian Kaiser.

Rand, Ayn. 1990. *Introduction to Objectivist Epistemology*, Revised ed., Harry Binswanger and Leonard Peikoff (eds.), New York: New American Library.

Rapp, Christof. 2019. "The Explanatory Value of Developmental Hypotheses as Exemplified by the Interpretation of Aristotle". In: Pantelis Golitsis and Katerina Ierodiakonou (eds.), *Aristotle and His Commentators*, Berlin–Boston: De Gruyter, 3–18.

Rashed, Marwan. 2004. "Priorité de l'*eidos* ou du *genos* entre Andronicos et Alexandre: vestiges arabes et grecs inédits", *Arabic Sciences and Philosophy* 14, 9–63.

Rashed, Marwan. 2007. *Essentialisme. Alexandre d'Aphrodise entre logique, physique et cosmologie*, Berlin–Boston: De Gruyter.

Rashed, Marwan. 2011. *Alexandre d'Aphrodise, Commentaire perdu à la Physique d'Aristote (Livres IV–VIII). Les Scholies Byzantines*. Édition, traduction et commentaire, Berlin–Boston: De Gruyter.

Rashed, Marwan. 2021. *Ptolémée «al-Gharīb»: Épître à Gallus sur la vie, le testament et les écrits d'Aristote*, Paris: Les Belles Lettres.

Regenbogen, Otto. 1940. "Theophrastos of Eresos", *Paulys Realencyclopädie der Altertums-wissenschaft*, Supplement Band 7, cols. 1354–1562.

Reger, Gary. 2005. "The Manufacture and Distribution of Perfume". In: Zofia H. Archibald, John K. Davies, and Vincent Gabrielsen (eds.), *Making, Moving and Managing: The New World of Ancient Economies, 323–31 B.C.*, Oxford: Oxbow Books, 253–297.

Repici, Luciana. 1988. *La natura e l'anima: saggi su Stratone di Lampsaco*. Turin: Tirrenia.

Repici, Luciana. 1990. "Limits of Teleology in Theophrastus' *Metaphysics*", *Archiv für Geschichte der Philosophie* 72, 182–213.

Robert, Aurélien. 2020. "John of Jandun on Sense Perception and Instantaneous Change". In: David Bennett and Juhana Toivanen (eds.), *Philosophical Problems in Sense Perception: Testing the Limits of Aristotelianism*, Cham: Springer, 193–212.

Roreitner, Robert. 2023. "The Perceptive Soul's Impassivity in Late Ancient Reception of Aristotle's *De Anima*", *Ancient Philosophy* 43, 219–249.

Roreitner, Robert. 2024. "Human Ontogeny in Aristotle and Theophrastus", *Apeiron* 57, 427–477.

Roreitner, Robert. 2025. *Aristotle on the Nature and Causes of Perception*. Cambridge: Cambridge University Press.

Sambursky, Shmuel. 1962. *The Physical World of Late Antiquity*. Princeton: Princeton University Press.

Sandbach, Francis H. 1985. *Aristotle and the Stoics*, Cambridge: Cambridge Philological Society.

Schorlemmer, Richard. 2022. *Transmission und Transformation: Überlieferungsanalysen und Rekonstruktionen frühperipatetischer Seelenlehren*, Basel: Schwabe.

Schorlemmer, Richard. Forthcoming. "Matter and Perception of Objects: The Rise of

Matter-Bound Quality Transmission in the Early Peripatos and the Quality-Mixture-Problem".

Schroeder, Frederic M. 1981. "The Analogy of the Active Intellect to Light in the '*De Anima*' of Alexander of Aphrodisias", *Hermes* 109, 215–225.

Sedley, David N. 1985. "Three Notes on Theophrastus' Treatment of Tastes and Smells". In: William W. Fortenbaugh, Pamela M. Huby and Anthony A. Long (eds.), *Theophrastus of Eresus: On His Life and Work*, New Brunswick: Transaction Publishers, 205–207.

Sedley, David N. 1997. "Plato's Auctoritas and the Rebirth of the Commentary Tradition". In: Jonathan Barnes and Miriam Griffin (eds), *Philosophia Togata II. Plato and Aristotle at Rome*, Oxford: Oxford University Press, 110–129.

Sedley, David N. 1998a. *Lucretius and the Transformation of Greek Wisdom*, Cambridge: Cambridge University Press.

Sedley, David N. 1998b. "Theophrastus and Epicurean Physics". In: Johannes van Ophuijsen and Marlein van Raalte (eds.), *Theophrastus: Reappraising the Sources*, New Brunswick: Transaction Publishers, 131–142.

Sedley, David N. 2003. "Philodemus and the Decentralisation of Philosophy", *Cronache Ercolanesi* 33, 31–41.

Sedley, David N. 2007. *Creationism and Its Critics in Antiquity*, Berkeley–Los Angeles–London: University of California Press.

Sharples, Robert W. and Ann Sheppard (eds.). 2003. *Ancient Approaches to Plato's Timaeus. Bulletin of the Institute of Classical Studies. Supplement* 78, London: Institute of Classical Studies.

Sharples, Robert W. 1982. "Alexander of Aphrodisias on Divine Providence: Two Problems", *The Classical Quarterly* 32 (1), 198–211.

Sharples, Robert W. 1983. "Nemesius of Emesa and Some Theories of Divine Providence", *Vigiliae Christianae* 37, 141–156.

Sharples, Robert W. 1985. "Theophrastus on Tastes and Smells". In: William W. Fortenbaugh, Pamela M. Huby and Antony A. Long (eds.), *Theophrastus of Eresus: On His Life and Work*, New Brunswick: Transaction Publishers, 183–204.

Sharples, Robert W. 1987. "Alexander of Aphrodisias: Scholasticism and Innovation". In: Wolfgang Haase and Hildegard Temporini (eds.), *Aufstieg und Niedergang der römischen Welt*, Band II.36.2, Berlin–New York: De Gruyter, 1176–1243.

Sharples, Robert W. 1994. "On Body, Soul and Generation in Alexander of Aphrodisias", *Apeiron* 27 (2), 163–170.

Sharples, Robert W. 1995. *Theophrastus of Eresus. Sources for his Life, Writings, Thought and Influence. Commentary Volume 5. Sources on Biology (Human Physiology, Living Creatures, Botany: Texts 328–435)*, Leiden–Boston: Brill.

Sharples, Robert W. 1998. *Theophrastus of Eresus, Commentary Volume 3.1: Sources on Physics (Texts 137–223)*, Leiden–Boston: Brill.

Sharples, Robert W. 1999. "The Peripatetic School". In: David Furley (ed.), *Routledge History of Philosophy. Volume II: From Aristotle to Augustine*, London: Routledge, 147–187.

Sharples, Robert W. 2002a. "Aristotelian Theology after Aristotle". In: Dorothea Frede and André Laks (eds.), *Traditions of Theology. Studies in Hellenistic Theology, its Background and Aftermath*, Leiden–Boston: Brill, 1–40.

Sharples, Robert W. 2002b. "Eudemus' *Physics*: Change, Place and Time". In: István Bodnár and William W. Fortenbaugh (eds.), *Eudemus of Rhodes*, New Brunswick: Transaction Publishers, 107–126.

Sharples, Robert W. 2003. "Threefold Providence: The History and Background of a Doctrine". In: Robert W. Sharples and Ann Sheppard (eds)., *Ancient Approaches to Plato's Timaeus. Bulletin of the Institute of Classical Studies. Supplement* 78, London: Institute of Classical Studies, 107–127.

Sharples, Robert W. 2006. "Common to Body and Soul: Peripatetic Approaches after Aristotle". In: Richard A.H. King (ed.), *Common to Body and Soul*, Berlin: De Gruyter, 165–186.

Sharples, Robert W. 2007a. "Peripatetics on Fate and Providence". In: Robert W. Sharples and Richard Sorabji (eds), *Greek & Roman Philosophy 100 BC–200 AD, Bulletin of the Institute of Classical Studies. Supplement* 94, Vol. II, London: Institute of Classical Studies, 595–605.

Sharples, Robert W. 2007b. "Peripatetics on Soul and Intellect". In: Robert W. Sharples and Richard Sorabji (eds), *Greek & Roman Philosophy 100 BC–200 AD, Bulletin of the Institute of Classical Studies. Supplement* 94, Vol. II, London: Institute of Classical Studies, 607–620.

Sharples, Robert W. 2008b. "*Habent sua fata libelli*: Aristotle's Categories in the First Century BC," *Acta Antiqua Academiae Scientiarum Hungaricae* 48, 273–287.

Sharples, Robert W. 2008c. "Philo and Post-Aristotelian Peripatetics". In: Francesca Alesse (ed.), *Philo of Alexandria and Post-Aristotelian Philosophy*, Leiden–Boston: Brill, 55–73.

Sharples, Robert W. 2010. *Peripatetic Philosophy 200 BC to AD 200: An Introduction and Collection of Sources in Translation*, Cambridge: Cambridge University Press.

Sharples, Robert W. 2012. "Alexander on *Physics* 2.9", *Bulletin of the Institute of Classical Studies* 55, 19–30.

Siegel, Rudolph. 1970. *Galen on Sense Perception, His Doctrines, Observations and Experiments on Vision, Hearing, Smell, Taste, Touch and Pain, and Their Historical Sources*, Basel: S. Karger.

Slakey, Thomas J. 1961. "Aristotle on Sense Perception", *Philosophical Review* 70, 470–484.

Smith, E.S. 1971. "Galen's Account of the Cranial Nerves and the Autonomic Nervous System", *Clio Medica* 6, 77–98.

Sollenberger, M.G. 1985. "Diogenes Laertius 5.36–57: The *Vita Theophrasti*". In: William W. Fortenbaugh, Pamela M. Huby and Anthony A. Long (eds.), *Theophrastus of Eresus: On His Life and Work*, New Brunswick: Transaction Publishers, 1–62.

Solmsen, Friedrich. 1961. *Aristotle's System of the Physical World: A Comparison with his Predecessors*, Ithaca: Cornell University Press.

Sorabji, Richard. 1974. "Body and Soul in Aristotle", *Philosophy* 49, 63–89.

Sorabji, Richard. 1988. *Matter, Space and Motion: Theories in Antiquity and Their Sequel*. Ithaca, NY: Cornell University Press.

Sorabji, Richard. 1992a. "Intentionality and Physiological Processes: Aristotle's Theory of Sense-Perception". In: Martha C. Nussbaum and Amelie Oksenberg Rorty (eds.), *Essays on Aristotle's de Anima*, Oxford: Clarendon Press, 195–225.

Sorabji, Richard. 1992b. "Introduction". In: James O. Urmson (transl.), *Simplicius: Corollaries on Place and Time*, London: Duckworth, 1–10.

Sorabji, Richard. 2007. "Adrastus: Modifications to Aristotle's Physics of the Heavens by Peripatetics and Others, 100 BC to 200 AD". In: Robert W. Sharples and Richard Sorabji (eds), *Greek & Roman Philosophy 100 BC–200 AD, Bulletin of the Institute of Classical Studies. Supplement* 94, Vol. II, London: Institute of Classical Studies, 575–594.

Squillace, Giuseppe. 2020. *Il profumo nel mondo antico. Con la traduzione italiana del "Sugli odori" di Teofrasto*. Florence: Leo S. Olschki.

Steel, Carlos. 1978. *The Changing Self. A Study on the Soul in Later Neoplatonism: Iamblichus, Damascius and Priscianus*, Brussels: Paleis der Academiën.

Steinkrüger, Philipp. 2018. "Aristotle on Kind-Crossing", *Oxford Studies in Ancient Philosophy* 54, 107–158.

Steinmetz, P. 1964. *Die Physik des Theophrastos von Eresos* (*Palingenesia I*), Bad Homburg: Gehlen.

Stratton, George M. 1917. *Theophrastus and the Greek Physiological Psychology before Aristotle*, London: Allen & Unwin.

Switalski, Bronislaus W. 1902. *Des Chalcidius Kommentar zu Plato's Timaeus*, Münster: Aschendorff.

Syme, Ronald. 1972. "Fraud and Imposture". In: Kurt von Fritz (ed.), *Pseudepigrapha I*, Genève: Fondation Hardt, 1–21.

Taub, Liba. 2004. *Ancient Meteorology*, London: Routledge.

Taub, Liba. 2024. "Paradigms in the Eye of the Beholder: Aristotle, Ptolemy and Copernicus", *Journal of Physics: Conference Series* 2877 (October). IOP Publishing. DOI: 10.1088/1742–6596/2877/1/012069

Theiler, Willy. 1926. "Ocellus Lucanus. Text und Kommentar von Richard Harder", *Gnomon* 2, 585–597.

Théry, Gilbert. 1926. *Autour du décret de 1210: II. Alexandre d'Aphrodise. Aperçu sur l'influence de sa noétique*, Bibliothèque Thomiste Vol. 7, Kain: Le Saulchoir.

Thom, Johan C. (ed.). 2014. *Cosmic Order and Divine Power: Pseudo-Aristotle, On the Cosmos*, Tübingen: Mohr Siebeck.

Thompson, George R. 1941. *Theophrastus on Plant Flavors and Odors. Studies on the Scientific Significance of De Causis Plantarum VI Accompanied by Translation and Notes*, Doctoral Dissertation, Princeton University.

Todd, Robert B. 1982. "Alexander of Aphrodisias, *De anima* 76.16: Michael of Ephesus' Text Defended", *Liverpool Classical Monthly* 7, 48–49.

Todd, Robert B. 1984. "Alexander of Aphrodisias and the Case for the Infinite Universe (*Quaestiones* III.12)", *Eranos* 82, 185–193.

Toher, Mark. 2004. "Octavian's Arrival in Rome, 44 B.C.", *Classical Quarterly* 54 (1), 174–184.

Towey, Alan. 2019. "The Physiology of Vision in Alexander's *Commentary on the De sensu*", *Ancient Philosophy* 39 (1), 211–223.

Usener, Hermann. 1858. *Analecta Theophrastea*, Diss. Bonn, 1858. Reprinted in: *Kleine Schriften I: Arbeiten zur griechischen Philosophie und Rhetorik*, Berlin, 1912; reprinted Osnabrück 1965.

Van Raalte, Marlein. 2010. "The Nature of Fire and Its Complications: Theophrastus' *De Igne* 1–10", *Bulletin of the Institute of Classical Studies* 53 (1), 47–97.

Verbeke, Gerard. 1978. "Doctrine du *pneuma* et entéléchisme chez Aristote". In: Geoffrey E.R. Lloyd and Gwilym E.L. Owen (eds.), *Aristotle on Mind and the Senses. Proceedings of the VIIth Symposium Aristotelicum*, Cambridge: Cambridge University Press, 191–214.

Verde, Francesco. 2022. *Peripatetic Philosophy in Context: Knowledge, Time, and Soul from Theophrastus to Cratippus*, Berlin–Boston: De Gruyter.

Vigo, Alejandro G. 1996. *Zeit und Praxis bei Aristoteles. Die Nikomachische Ethik und die zeit-ontologische Voraussetzungen des vernunftgesteuerten Handelns. Symposion.* Philosophische Schriftenreihe, Vol. 105, Freiburg–München: Alber.

Vitrac, Bernard. 2002. "Note textuelle sur un (problème de) lieu géométrique dans les *Météorologiques* d'Aristote (III.5 375b16–376)", *Archive for the History of Exact Sciences* 56, 239–283.

Waszink, Jan H. 1964. *Studien zum Timaioskommentar des Calcidius: die erste Hälfte des Kommentars*, Leiden–Boston: Brill.

Waterlow, Sarah. 1982a. *Nature, Change, and Agency in Aristotle's* Physics: *A Philosophical Study*, Oxford: Clarendon.

Waterlow, Sarah. 1982b. *Passage and Possibility: A Study of Aristotle's Modal Concepts*, Oxford: Clarendon.

Wehrli, Fritz. 1983. "Theophrastus". In: Hellmut Flashar (ed.), *Die Philosophie der Antike* (III. Band: *Ältere Akademie, Aristoteles–Peripatos*), Basel–Stuttgart, 474–522.

White, Stephen. Forthcoming. "Eudemian Arguments in Athens and Rhodes: Logic, Physics, and Ethics". In: Oliver Hellmann, Robert Mayhew and Arnaud Zucker (eds.), *Peripatetic Ethics in Theophrastus and after*, London: Routledge.

Whittaker, James. 1989. "The Value of Indirect Tradition in the Establishment of Greek

Philosophical Texts or the Art of Misquotation". In: John N. Grant (ed.), *Editing Greek and Latin Texts*, Papers given at the Twenty-Third Annual Conference on Editorial Problems, University of Toronto, 6–7 November 1987, New York: AMS Press, 63–95.

Wiesner, Jürgen. 1987. "Theophrast und der Beginn des Archereferats von Simplikios' Physikkommentar", *Hermes* 117 (3), 288–303.

Williams, Gareth D. 2012. *The Cosmic Viewpoint: A Study of Seneca's Natural Questions*, Oxford: Oxford University Press.

Wilson, Malcolm. 2008. "Hippocrates of Chios' Theory of Comets", *Journal for the History of Astronomy* 39, 141–160.

Wilson, Malcolm. 2009. "A Somewhat Disorderly Nature: Unity in Aristotle's *Meteorologica* I–III", *Apeiron* 42, 63–88.

Wilson, Malcolm. 2013. *Structure and Method in Aristotle's Meteorologica*, Cambridge: Cambridge University Press.

Wöhrle, Georg. 1988. "The Structure and Function of Theophrastus' Treatise *De odoribus*". In: William W. Fortenbaugh and Robert W. Sharples (eds.), *Theophrastean Studies on Natural Science, Physics and Metaphysics, Ethics, Religion, and Rhetoric*, New Brunswick: Transaction Publishers, 3–13.

Wöhrle, Georg. 2002. "Ps.-Aristoteles *De Coloribus*—A Theophrastean Opusculum?". In: William W. Fortenbaugh and Georg Wöhrle (eds.), *On the Opuscula of Theophrastus. Akten der 3. Tagung der Karl-und-Gertrud-Abel-Stiftung vom 19.–23. Juli 1999 in Trier*, Stuttgart: Steiner, 91–97.

Wray, K. Brad. 2021. *Kuhn's Intellectual Path: Charting The Structure of Scientific Revolutions*. Cambridge: Cambridge University Press.

Zeller, Eduard. 1921. *Die Philosophie der Griechen in ihrer geschichtlichen Entwicklung II.2. Aristoteles und die Alten Peripatetiker*, Leipzig: Reisland. Reprinted Hildesheim: Olms, 1963.

Zeller, Eduard. 1923. *Die Philosophie der Griechen in ihrer geschichtlichen Entwicklung III.1. Die nacharistotelische Philosophie, erste Hälfte*, Leipzig: Reisland. Reprinted Hildesheim: Olms, 1963.

Zingano, Marco. 2008. "O tratado do impulso e da faculdade impulsiva de Alexandre de Afrodísia e sua versão em Miguel de Éfeso", *Journal of Ancient philosophy* 2, 1–22.

Index of Ancient Authors

Index of Main Passages

Index of Topics

www.ingramcontent.com/pod-product-compliance
Lightning Source LLC
Chambersburg PA
CBHW071336230926
51849CB00002B/407
9789004739307